Studies in Latin American Ethnohistory & Archaeology

Joyce Marcus
General Editor

Volume I *A Fuego y Sangre: Early Zapotec Imperialism in the Cuicatlán Cañada, Oaxaca*, by Elsa Redmond, Memoirs of the Museum of Anthropology, University of Michigan, No. 16. 1983.

Volume II *Irrigation and the Cuicatec Ecosystem: A Study of Agriculture and Civilization in North Central Oaxaca*, by Joseph W. Hopkins, Memoirs of the Museum of Anthropology, University of Michigan, No. 17. 1984.

Volume III *Aztec City-States*, by Mary G. Hodge, Memoirs of the Museum of Anthropology, University of Michigan, No. 18. 1984.

Volume IV *Conflicts over Coca Fields in Sixteenth-Century Peru*, by María Rostworowski de Diez Canseco, Memoirs of the Museum of Anthropology, University of Michigan, No. 21. 1988.

Volume V *Tribal and Chiefly Warfare in South America*, by Elsa Redmond, Memoirs of the Museum of Anthropology, University of Michigan, No. 28. 1994.

Volume VI *Imperial Transformations in Sixteenth-Century Yucay, Peru*, transcribed and edited by R. Alan Covey and Donato Amado González, Memoirs of the Museum of Anthropology, University of Michigan, No. 44. 2008.

Volume VII *Domestic Life in Prehispanic Capitals: A Study of Specialization, Hierarchy, and Ethnicity*, edited by Linda R. Manzanilla and Claude Chapdelaine, Memoirs of the Museum of Anthropology, University of Michigan, No. 46. 2009.

Volume VIII *Yuthu: Community and Ritual in an Early Andean Village*, by Allison R. Davis, Memoirs of the Museum of Anthropology, University of Michigan, No. 50. 2011.

Volume IX *Advances in Titicaca Basin Archaeology–III*, edited by Alexei Vranich, Elizabeth A. Klarich, and Charles Stanish, Memoirs of the Museum of Anthropology, University of Michigan, No. 51. 2012.

Volume X *Regional Archaeology in the Inca Heartland: The Hanan Cuzco Surveys*, edited by R. Alan Covey, Memoirs of the Museum of Anthropology, University of Michigan, No. 55. 2014.

Volume XI *The Northern Titicaca Basin Survey: Huancané-Putina*, by Charles Stanish, Cecilia Chávez Justo, Karl LaFavre, and Aimée Plourde, Memoirs of the Museum of Anthropology, University of Michigan, No. 56. 2014.

Volume XII *Coastal Ecosystems and Economic Strategies at Cerro Azul, Peru: The Study of a Late Intermediate Kingdom*, edited by Joyce Marcus, Memoirs of the Museum of Anthropology, University of Michigan, No. 59. 2016.

Memoirs of the Museum of Anthropology
University of Michigan
Number 59

STUDIES IN LATIN AMERICAN ETHNOHISTORY & ARCHAEOLOGY
Joyce Marcus, General Editor
Volume XII

Coastal Ecosystems and Economic Strategies at Cerro Azul, Peru

The Study of a Late Intermediate Kingdom

edited by
Joyce Marcus

Ann Arbor, Michigan
2016

©2016 by the Regents of the University of Michigan
The Museum of Anthropology
All rights reserved

Printed in the United States of America
ISBN 978-0-915703-88-3

Cover design by John Klausmeyer and Katherine Clahassey; sky image: Mykola Mazuryk/Shutterstock.com; cliff image: photo eye/Shutterstock.com; ocean image: Seaphotoart/Shutterstock.com.

Original paintings by John Klausmeyer.

The Museum currently publishes two monograph series: Anthropological Papers and Memoirs. For permissions, questions, or catalogs, contact Museum publications at 1109 Geddes Avenue, Ann Arbor, Michigan 48109-1079; umma-pubs@umich.edu; www.lsa.umich.edu/ummaa/publications

Library of Congress Cataloging-in-Publication Data

Names: Marcus, Joyce, editor of compilation.
Title: Coastal ecosystems and economic strategies at Cerro Azul, Peru : the study of a Late Intermediate kingdom / edited by Joyce Marcus.
Description: Ann Arbor, Michigan : University of Michigan, Museum of Anthropology, 2016. | Series: Memoirs of the Museum of Anthropology, University of Michigan ; number 59 | Series: Studies in Latin American ethnohistory & archaeology ; volume 12 | Includes bibliographical references and index.
Identifiers: LCCN 2016051173 | ISBN 9780915703883 (alkaline paper)
Subjects: LCSH: Cerro Azul Site (Peru) | Coastal archaeology--Peru--Cerro Azul Region. | Environmental archaeology--Peru--Cerro Azul Region. | Biotic communities--Peru--Cerro Azul Region--History--To 1500. | Coastal ecology--Peru--Cerro Azul Region--History--To 1500. | Plant remains (Archaeology)--Peru--Cerro Azul Region. | Animal remains (Archaeology)--Peru--Cerro Azul Region. | Indians of South America--Peru--Cerro Azul Region--Antiquities. | Indians of South America--Peru--Cerro Azul Region--Economic conditions. | Ethnoarchaeology--Peru--Cerro Azul Region.
Classification: LCC F3429.1.C44 C63 2016 | DDC 985/.25--dc23
LC record available at https://lccn.loc.gov/2016051173

The paper used in this publication meets the requirements of the ANSI Standard Z39.48-1984 (Permanence of Paper).

Dedication

I dedicate this book to María Rostworowski, a generous friend and *co-directora* of the Cerro Azul project. She was a brilliant and relentless advocate for the study of coastal señoríos and *curacazgos*. I miss her exceptional contributions, which shed so much light on coastal societies. Her much-needed coastal focus complemented the highland focus of John Murra and many others.

I was introduced to María in 1980 by Ramiro Matos Mendieta, who thought, unbeknownst to me, that she and I would make a good research team. I met her at the Instituto de Estudios Peruanos in Lima. She was at her desk and she said that we should talk at length about the 16th-century documents. She discussed her most recent publication, a study of the señoríos of Huarco and Lunaguaná in the Cañete Valley (Rostworowski de Diez Canseco 1978–1980), stressing that she was still interested in that document and wanted to find an archaeologist who would excavate Cerro Azul, a key site in the señorío of Huarco. With real enthusiasm in her voice she urged me to excavate Cerro Azul. She suggested that I visit the site the very next day, and I did. When I saw the site and its spectacular setting overlooking the Pacific Ocean, I was sold. I recognized that Cerro Azul's excellent preservation would allow me to study prehispanic fishing and evaluate the extent to which ethnohistoric data could be projected into the past. María Rostworowski (1978–1980) had already done much of the ethnohistoric work; what remained to do was excavate Cerro Azul.

Table of Contents

List of Illustrations — viii
List of Tables — xvi
Acknowledgments — xix

Part I: An Introduction to the Kingdom of Huarco and its Ecosystems

1 The Ecosystems of the Kingdom of Huarco — 3
  Joyce Marcus

2 Provenience and Context of the Plant and Animal Remains at Cerro Azul — 20
  Joyce Marcus

Part II: How Cerro Azul Made Use of Trophic Levels 2, 3, and 4

3 The Collection of Shellfish — 35
  Kent V. Flannery and Joyce Marcus

4 The Collection of Crustaceans — 54
  Kent V. Flannery and Jeffrey D. Sommer

5 Crayfish Trapping — 65
  Joyce Marcus and Ramiro Matos

6 The Fish Resources of Cerro Azul in the 1980s — 72
  Kent V. Flannery and Joyce Marcus

7 Fishing Strategies and Fishing Gear — 98
  Joyce Marcus

8 The Drying of Fish for Export — 116
  Joyce Marcus

9 The Archaeological Fish Remains from Cerro Azul — 120
  Jeffrey D. Sommer and Kent V. Flannery

10 The Hunting of Birds and Mammals — 158
  Joyce Marcus

11 The Bird Life of Cañete and the Avifauna of Cerro Azul — 172
  Joyce Marcus and Christopher P. Glew

12 The Hunting of Marine Mammals 186
    Christopher P. Glew and Kent V. Flannery

Part III: The Use of Plants at Cerro Azul

13 Edible, Ritual, and Medicinal Plants 201
    C. Earle Smith, Jr. and Joyce Marcus

14 Comments on the Late Intermediate Maize 232
    C. Earle Smith, Jr. and Joyce Marcus

15 *Phaseolus* and *Erythrina* from Cerro Azul 254
    Lawrence Kaplan

16 Industrial Plants 262
    C. Earle Smith, Jr. and Joyce Marcus

Part IV: The Domestic Animals, Their Skeletal Remains, and Their Byproducts

17 Camelids and *Ch'arki* at Cerro Azul 287
    Christopher P. Glew and Kent V. Flannery

18 Domestic Dogs 318
    Kent V. Flannery and Christopher P. Glew

19 The Raising of Guinea Pigs 324
    Christopher P. Glew and Kent V. Flannery

20 Macrofossil and Palynological Analysis of the Coprolites from Cerro Azul 334
    John G. Jones

Part V: The Interface of Ecology and Economy

21 The Economy of the Kingdom of Huarco 340
    Joyce Marcus

Appendix A: Artisanal Fishing at Cerro Azul, 1984–1986 352
    Khalid Kattan, Robert G. Reynolds, and Joyce Marcus

References Cited 359

Index 371

List of Illustrations

front cover   Designed by John Klausmeyer and Kay Clahassey

1.1. Map of the lower valley of the Río Cañete, *4*
1.2. An 1893 map of the Kingdom of Huarco, *5*
1.3. The *costa* or cobble beach environment lying between Cerro Azul and the Río Cañete, *7*
1.4. The *playa* or sandy beach environment of the Kingdom of Huarco, *8*
1.5. Remnants of Cerro Azul's defensive wall, *9*
1.6. The *peña*, or rocky coast environment, of the Kingdom of Huarco, *13*
1.7. The 15–20 year cycles during which anchovetas and sardines alternate as the dominant small fish, *15*
1.8. Monthly ocean temperatures at Callao, Peru, during the years 1981–1987, *17*
1.9. Cross-section of the Kingdom of Huarco, *18*

2.1. Contour map of the ruins of Cerro Azul, *21*
2.2. An artist's conception of the southwest quadrant of Structure D, *22*
2.3. Human fingernail and toenail clippings from Feature 6, *23*
2.4. An artist's conception of the southeast quadrant of Structure D, *24*
2.5. An artist's conception of the northeast quadrant of Structure D, *25*
2.6. An artist's conception of the northwest quadrant of Structure D, *26*
2.7. An artist's conception of Structure 9, *27*
2.8. Profile of Squares N6 and N7 in Terrace 16, Quebrada 5, Cerro Azul, *29*
2.9. Cross-section of Terrace 9, Quebrada 5a, Cerro Azul, *31*

3.1. Desiccated chiton (*Enoplochiton niger*) from Feature 6, Structure D, *36*
3.2. Two views of the keyhole limpet *Fissurella crassa*, *37*
3.3. Two views of the keyhole limpet *Fissurella limbata*, *37*
3.4. Two views of the keyhole limpet *Fissurella limbata*, *38*
3.5. Two chanques (*Concholepas concholepas*), *38*
3.6. A sample of the 26 chanques recovered from Feature 22, Terrace 16, Quebrada 5, *39*
3.7. Two views of the sea snail *Tegula atra*, *40*
3.8. Two views of the sea snail *Calyptraea trochiformis*, *40*
3.9. Two views of the sea snail *Thais chocolata* (from archaeological debris), *40*
3.10. The mussel *Aulacomya ater* used as pigment dish, *40*
3.11. The giant Chilean mussel *Choromytilus chorus* used as pigment dish, *41*
3.12. Four examples of the mussel *Semimytilus algosus*, *42*
3.13. The mussel *Semimytilus algosus*, *42*
3.14. The mussel *Perumytilus purpuratus*, *42*
3.15. *Perumytilus* mussels below the pier, Cerro Azul Bay, *42*
3.16. Two examples of the coquina clam *Donax obesulus* (= *peruvianus*), *43*
3.17. The coquina clam *Donax obesulus* (from archaeological debris), *43*
3.18. The clam *Mesodesma donacium* (from archaeological debris), *44*
3.19. Two views of the clam *Mulinia edulis* from Late Intermediate specimens, *44*
3.20. The clam *Mulinia edulis* (from Late Intermediate archaeological debris), *45*
3.21. Two marisqueros from the upper fill of Room 8, Structure D, *52*

4.1. Two species of crab caught at Cerro Azul in 1984, *55*
4.2. Fragments of stone crab from Structure D, *56*
4.3. Men searching for stone crabs with their bare feet, *57*
4.4. Edgar's harvest: a dozen stone crabs, the product of only 15 minutes of barefoot searching, *57*
4.5. Two desiccated specimens of *muy muy* or mole crab (*Emerita analoga*) from Room 7, Structure D, *58*
4.6. A pair of maruchas or ghost shrimp (*Callianassa islagrande*) from a sandy beach habitat at Cerro Azul, *59*
4.7. A fisherman uses a metal piston to extract ghost shrimp from a sandy beach at Cerro Azul, *60*

5.1. *Cryphiops caementarius*, the crayfish most frequently eaten at ancient Cerro Azul, *66*
5.2. Fragments of large chelate hands (pincers) of *Cryphiops caementarius* from Feature 6, Structure D, *66*
5.3. Portion of damaged *chauchu* (crayfish trap) from Collca 1, Structure D (Late Intermediate period), *67*
5.4. The remains of an ancient canal near Cerro Azul, *68*
5.5. *Chauchu* crayfish trap, 1986, *69*
5.6. Simplified diagram of a *chauchu*, a traditional crayfish trap, *69*
5.7. The embudo or funnel from a *chauchu*, *69*
5.8. Heavy cobblestones alongside a *chauchu*, ensuring that it will not be moved by the current, *70*
5.9. The *chauchu* is now secure, *70*
5.10. Donning scuba goggles, Carlos searches for crayfish, *70*
5.11. Two live crayfish, caught by Carlos while wearing scuba goggles, *70*

6.1. The Capitanía del Puerto, Cerro Azul Bay, *74*
6.2. The tollo or smoothhound shark, *75*
6.3. The eagle ray, *75*
6.4. Artisanal fisherman drying more than 100 eagle rays, *76*
6.5. The pejegallo or plow-nosed chimaera, *77*
6.6. Pacific sardine, one of the small fish dried for export at ancient Cerro Azul, *78*
6.7. The anchoveta negra (*Engraulis ringens*), small fish dried for export at ancient Cerro Azul, *78*
6.8. The anchoveta blanca (*Anchoa nasus*) visited Cerro Azul during the 1980s, *79*
6.9. The sea catfish (*Galeichthys peruvianus*), *80*
6.10. Two mullets, *81*
6.11. The barbudo or threadfin (*Polynemus approximans*), *81*
6.12. The pejerrey or silversides (*Odontesthes regia regia*), *82*
6.13. The ojo de uva (*Hemilutjanus macrophthalmos*), *83*
6.14. The cabrilla or Peruvian rock bass (*Paralabrax humeralis*), *83*
6.15. The jurel or Chilean jack mackerel, *83*
6.16. The paloma pompano, *84*
6.17. The cojinova or blackruff (*Seriolella violacea*), *84*
6.18. The dorado or dolphinfish, *85*
6.19. The chita or grunt, *86*
6.20. The mismis, a member of the drum family, *87*
6.21. The zorro (*Menticirrhus rostratus*), a relative of the mismis, *87*
6.22. The coco (*Paralonchurus peruanus*), a medium-sized member of the drum family, *88*
6.23. The ayanque (*Cynoscion analis*), another medium-sized member of the drum family, *88*
6.24. The burro (*Sciaena fasciata*), another medium-sized member of the drum family, *88*
6.25. The lorna (*Sciaena deliciosa*), *88*
6.26. The róbalo (*Sciaena starksi*), *90*
6.27. The corvina (*Cilus gilberti*), *90*
6.28. The mojarrilla (*Stellifer minor*), *91*
6.29. The pintadilla (*Cheilodactylus variegatus*), *91*
6.30. Two species of blenny caught at Cerro Azul, *92*
6.31. The caballa or Pacific mackerel (*Scomber japonicus peruanus*), *93*
6.32. The Pacific bonito (*Sarda sarda chiliensis*), *93*
6.33. The sierra, a member of the mackerel family, *94*

6.34. The pejesapo or clingfish, *94*
6.35. The lenguado or left-eye flounder, *96*
6.36. The merluza or Pacific hake, *96*

7.1. A fisherman throws his circular net into the Pacific, *99*
7.2. A fisherman of the *peña* zone retrieves a grunt from his atarraya, *99*
7.3. Four types of fishing gear: the atarraya, the espinel, the red de cortina, and the chinchorro, *100*
7.4. A caballito de totora, kept in the home of a Cerro Azul fisherman, *101*
7.5. Close-up of the comienzo of an atarraya from Structure 5, Quebrada 5a, *103*
7.6. Two sketches of the comienzo de atarraya shown in Fig. 7.5, *104*
7.7. Lower border of a cotton trammel net from Structure 4, Quebrada 5a, *105*
7.8. Stone weight from a trammel net, with cords still attached to both ends, *105*
7.9. Eighteenth-century fishermen using a chinchorro, *106*
7.10. Lower border of a trammel net from Terrace 9 of Quebrada 5a, *107*
7.11. Fragment of trammel net with sherd attached as weight, Structure 12, Quebrada 5-south, *108*
7.12. Pot sherd used as a weight for a trammel net, Structure 12, Quebrada 5-south, *108*
7.13. Carrier net from Collca 1, Structure D, Cerro Azul, *109*
7.14. Pottery vessel from a burial at Qurna, Egypt, inside its carrier net, *109*
7.15. Close-up of a net found in Room 7 of Structure D, *110*
7.16. Two views of Net 1 from a looted burial on Terrace 9 of Quebrada 5a, *111*
7.17. Net 1 from a looted burial near Structure G of Cerro Azul, *112*
7.18. Close-up photograph of the sheet bend knots used in the net seen in Fig. 7.17, *112*
7.19. The border of the net seen in Fig. 7.17. Such a border is typical of an atarraya, *113*
7.20. Close-up of square knots used in Net 2 from a looted burial near Structure G, *113*
7.21. Close-up of Net 2 from another looted burial on Terrace 9 of Quebrada 5a, *114*
7.22. Two malleros, templates for standardizing the spacing of knots in netmaking, *115*

8.1. Remains of dried fish from Room 1, Structure 9, *117*
8.2. Clay floor of Room 11, Structure 9, showing remains of fish, *118*
8.3. Remains of dried fish sticking to the floor of Room 11, Structure 9, *118*
8.4. Vertebrae from a corvina resting on the clay floor of Room 11, Structure 9, *119*

9.1. Bones of Pacific sardine (*Sardinops sagax*) from Feature 6, Structure D, *122*
9.2. Bones of Pacific sardine from Feature 6, Structure D, *123*
9.3. Cleithrum and pectoral fin rays of Pacific sardine from Feature 6, Structure D, *124*
9.4. Remains of anchoveta from Feature 6, Structure D, *124*
9.5. Remains of dried anchovetas from Feature 6, Structure D, *124*
9.6. Lenses from the eyes of anchovetas from Feature 6, Structure D, *124*
9.7. Bones of grunt (*Anisotremus scapularis*) from Feature 6, Structure D, *125*
9.8. Bones of grunt from Feature 6, Structure D, *126*
9.9. Fish of the drum family display occasional exostoses or abnormal swellings, *127*
9.10. Articular of ayanque (*Cynoscion analis*) from Feature 6, Structure D, *128*
9.11. Articular of burro (*Sciaena fasciata*) from Feature 6, Structure D, *128*
9.12. Bones of the lorna (*Sciaena deliciosa*) from Feature 6, Structure D, *128*
9.13. Still-articulated preopercular, hyomandibular, quadrate, symplectic, and metapterygoid of lorna from Feature 6, Structure D, *129*
9.14. Articular and dentary of large róbalo still articulated, from Room 3, Structure D, *133*
9.15. Maxilla of large róbalo from Late Intermediate refuse, Structure D, *134*
9.16. Premaxilla of corvina (*Cilus gilberti*) from Feature 6, Structure D, *135*
9.17. Premaxilla of mojarrilla (*Stellifer minor*) from Feature 6, Structure D, *135*
9.18. Drum otoliths from Feature 6, Structure D, *135*
9.19. Bones of the pintadilla (*Cheilodactylus variegatus*) from Feature 6, Structure D, *136*
9.20. Bones of the scaleless blenny (*Scartichthys gigas*), *136*

9.21. Bones of the scaled blenny (*Labrisomus philippii*) from Feature 6, Structure D, *137*
9.22. Bones of the Pacific bonito (*Sarda sarda chiliensis*) from Feature 6, Structure D, *138*
9.23. Bones of the pejesapo or clingfish (*Sicyases sanguineus*) from Feature 6, Structure D, *139*
9.24. Premaxillae of left-eye flounders showing size range of specimens caught at Late Intermediate Cerro Azul, *139*
9.25. Bones of the left-eye flounder from Feature 6, Structure D, *140*
9.26. Fish bones displaying cut marks, all from Feature 6 of Structure D, *141*

10.1. Complete sling with webbed cradle from Structure 5 on Terrace 9, Quebrada 5a, *160*
10.2. Complete sling with slit cradle from Zone A of Terrace 9, Quebrada 5a, *161*
10.3. Agave fiber sling from Tomb A-16 at Cerro del Oro, Cañete, *162*
10.4. Webbed cradles from Late Intermediate slings, *162*
10.5. Slit cradles from Late Intermediate slings, *163*
10.6. Components of Late Intermediate slings, *164*
10.7. Tassel of vegetal fiber, possibly part of a ritual sling, from Structure 5, Terrace 9, Quebrada 5a, *165*
10.8. Two-stone bolas from Structure 6, Terrace 9, Quebrada 5a, *167*
10.9. Close-up of the bolas from Figure 10.8, showing repair to one of the stone-carrying loops, *168*
10.10. Two-stone bolas from Structure 6, Terrace 9, Quebrada 5a, *169*
10.11. Close-up of bolas from Structure 6, *170*
10.12. Close-up of a bolas from Collca 1, Structure D, *171*

11.1. This Humboldt penguin (*Spheniscus humboldti*) died during the El Niño of 1982–83, *173*
11.2. Patch of skin from a Humboldt penguin. Found in Feature 6, Structure D, *174*
11.3. A paloma del cabo (*Daption capensis*) on Structure D, Cerro Azul, *174*
11.4. A starving guanay cormorant (*Phalacrocorax bougainvillii*) rests on a cobble beach at Cerro Azul, *175*
11.5. A chuita or red-legged cormorant (*Phalacrocorax gaimardi*) sits on the rocky ledge at Cerro Azul, *175*
11.6. Cranium of guanay cormorant from Terrace 11, Quebrada 6, *175*
11.7. Bones of guanay cormorant from Feature 6, Structure D, *176*
11.8. Sternum of red-legged cormorant from Feature 6, Structure D, *177*
11.9. A group of piqueros or Peruvian boobies (*Sula variegata*) resting on the cliffs of Cerro del Fraile, *178*
11.10. Peruvian pelicans (*Pelecanus thagus*) flying low over Cerro Azul Bay, *178*
11.11. Bones of the Peruvian pelican from archaeological refuse, *179*
11.12. The zarcillo or Inca tern (*Larosterna inca*) frequently rests on the sea cliffs at Cerro Azul between dives, *180*
11.13. The cranium of an Inca tern from the Northeast Canchón, Structure D, *180*
11.14. Inca tern that died in the ruins of Structure 1 and became completely mummified, *180*
11.15. Gulls on the beach at Cerro Azul Bay, *180*

12.1. This rocky island near the Cerro Azul lighthouse is a favorite "haul out" locality for sea lions, *187*
12.2. A young sea lion (*Otaria byronia*) resting on a beach boulder at Cerro Azul, *187*
12.3. Two humeri from sea lions (*Otaria byronia*), illustrating sexual dimorphism, *188*
12.4. Abandoned juvenile sea lion that died of starvation, *189*
12.5. Right mandible of sea lion (*Otaria byronia*) from Feature 22, Terrace 16, Quebrada 5, *190*
12.6. Left femur of sea lion from Stratigraphic Zone A, Terrace 16, Quebrada 5, *190*
12.7. Phalanx and innominate of female sea lion, from the surface of Cerro Azul, *191*
12.8. Carcass of butchered dolphin, discarded on the beach at Cerro Azul during the 1982–83 El Niño, *192*
12.9. The bottlenosed dolphin (*Tursiops* sp.) and the Pacific white-sided dolphin (*Lagenorhynchus* sp.), *192*
12.10. Heads of butchered dolphins (*Tursiops* sp. and *Lagenorhynchus* sp.) discarded on beach at Cerro Azul, *193*
12.11. Vertebrae from two genera of dolphins eaten at Late Intermediate Cerro Azul, *194*

13.1. A seed from the cherimoya fruit (*Annona cherimolia*) found in Feature 20, Structure 9, *202*
13.2. Achira (*Canna edulis*) is still grown in the Cañete Valley, *202*
13.3. Leaf, probably achira, from Room 4, Structure D, *203*

13.4. The contents of a "medicine bundle" from Structure 12, a looted burial cist in Quebrada 5-south, *203*
13.5. Mass of coca leaves from the interior of amphora from Stratigraphic Zone A of Terrace 9, Quebrada 5a, *205*
13.6. Two seeds of ciruela del fraile (*Bunchosia armeniaca*), found in Terrace 11, Quebrada 6, *205*
13.7. A sample of *Phaseolus* pods and valves from the floor of the South Corridor, Structure D, *205*
13.8. Pod and seed of common bean (*Phaseolus vulgaris*) from Structure 12, Quebrada 5-south, *206*
13.9. Seed of lima bean (*Phaseolus lunatus*) from Feature 6, Structure D, *206*
13.10. Three seeds of *Canavalia* beans, found on the west side of "Room 2," Structure D, *206*
13.11. *Erythrina* pod from Structure 5, Quebrada 5a, *206*
13.12. Both elite families and commoners at Cerro Azul enjoyed peanuts (*Arachis hypogaea*), *207*
13.13. A pacay leaf (*Inga feuillei*), found in Terrace 11, Quebrada 6, *208*
13.14. Three stems of butternut squash (*Cucurbita moschata*), *208*
13.15. Two squash seeds from the Southwest Canchón, Structure D, *208*
13.16. Desiccated fruit of yuca de monte (*Apodanthera* sp.) from Collca 1, a storage cell in Structure D, *209*
13.17. Ruptured pod of achiote (*Bixa orellana*) from Room 4, Structure D, *209*
13.18. Lúcuma (*Pouteria lucuma*) from Room 4, Structure D, *210*
13.19. A sweet potato (*Ipomoea batatas*) from the northeast corner of the Southwest Canchón, Structure D, *211*
13.20. A white potato (*Solanum tuberosum*) found near Structure 5, Quebrada 5a, *211*
13.21. Stems and seeds of chile pepper (*Capsicum* sp.) from Feature 6 of Structure D, *212*
13.22. Four stems of an unidentified plant, each bearing a wasp gall, found in Room 4, Structure D, *212*
13.23. Unidentified plants from Feature 6, Structure D, *213*
13.24. Unidentified nuts from the fill of Room 4, Structure D, *213*
13.25. Squash stems from Feature 6, Structure D, *214*
13.26. Two incomplete squash seeds from Feature 6, Structure D, *214*
13.27. Two views of a dried chile pepper from Feature 6, Structure D, *214*
13.28. A sample of bean pods from Collca 1, Structure D, *216*
13.29. Five of the 13 lúcuma seeds from Collca 1, Structure D, *217*
13.30. Eleven seeds of *Canavalia* sp. from Room 7, Structure D, *218*
13.31. Sample of *Phaseolus* pods from Feature 4 in "Room 2" of Structure D, *218*
13.32. Four pods of common bean (*Phaseolus vulgaris*) from Feature 4, "Room 2," Structure D, *218*
13.33. Eight seeds of *Canavalia* sp. from Feature 4 in "Room 2" of Structure D, *218*
13.34. Lúcuma from Feature 4, "Room 2," Structure D, *219*
13.35. Beans from Feature 20 of Structure 9, the midden left by a commoner household, *220*
13.36. Seeds of what appears to be butternut squash, *220*
13.37. Dried fruit and seed of lúcuma from Feature 20, Structure 9, *220*
13.38. Fifty-one *Canavalia* beans from Room 2, Structure 11, Quebrada 5-south, *222*
13.39. Food plants from Room 2, Structure 11, a storage facility in Quebrada 5-south, *223*
13.40. Plants associated with a kincha house, Room 3, Structure 11, a storage facility in Quebrada 5-south, *223*
13.41. Three lúcuma seeds from Structure 12, a burial cist in Quebrada 5-south, *223*
13.42. *Canavalia* seeds from a refuse deposit in Terrace 11, Quebrada 6, *223*
13.43. Leaves of pacay (*Inga feuillei*) used to line the interior of burial cist, Structure 6, Quebrada 5a, *225*
13.44. Structure 1 of Cerro Azul, *227*
13.45. Two seeds of cherimoya (*Annona cherimolia*) from 16th-century A.D. refuse, Room 9, Structure 1, *228*
13.46. Pods, valves, and seeds of vetch (*Vicia* sp.) from 16th-century A.D. refuse, Room 9, Structure 1, *228*
13.47. Herbs used as seasonings by 16th-century squatters in Room 9, Structure 1, *229*
13.48. Nine seeds of common beans (*Phaseolus vulgaris*) from 16th-century refuse in Room 9, Structure 1, *229*
13.49. Crop plants used by 16th-century squatters in Structure 1, *229*
13.50. Barley (*Hordeum* sp.) from 16th-century refuse in Room 10, Structure 1, *230*

14.1. Remains of *hatun maccma*, a chicha storage vessel with a capacity of nearly 2000 liters, North Central Canchón, Structure D, *233*
14.2. Maize ears still bearing kernels from Terrace 9 of Quebrada 5a, *234*
14.3. An immature maize ear still in its husk, from Structure 12, a burial cist in Quebrada 5-south, *235*
14.4. An empty maize husk found near Burial 4, Quebrada 5a, *235*
14.5. Maize cobs found near burials in Quebrada 5a, *236*

14.6. A badly frayed maize stalk and husk from a midden deposit on Terrace 11, Quebrada 6, *237*
14.7. A maize tassel found near Burial 9, Quebrada 5a, *237*
14.8. Four cobs each of Maize Types 1 and 2 from Feature 6, Structure D, *238*
14.9. Four cobs each of Maize Types 1 and 2 from the South Corridor, Structure D, *239*
14.10. Four cobs each of Maize Types 4 and 5 from Feature 6, Structure D, *241*
14.11. Four cobs of Maize Type 6 and one cob of Maize Type 7 from Feature 6, Structure D, *242*
14.12. Two cobs of Maize Type 2a, two cobs of Maize Type 3, and four cobs of Maize Type 4 from the South Corridor, Structure D, *243*
14.13. Four cobs each of Maize Types A and B from Collca 1, Structure D, *244*
14.14. Four cobs each of Maize Types C and D from Collca 1, Structure D, *245*
14.15. Four cobs of Maize Type E and one cob of Maize Type F from Collca 1, Structure D, *246*
14.16. Six cobs from Feature 20, Structure 9, *247*
14.17. Examples of Maize Types 1, 2, 4, 5, 6, and 7 from Room 4, Structure 9, *248*
14.18. One maize tassel and six cobs from a midden on Terrace 11, Quebrada 6, *250*
14.19. Six cobs from Structure 10, a storage cist on Terrace 12 of Quebrada 6, *251*
14.20. Three cobs from Structure 5, a looted burial cist in Quebrada 5a, *252*
14.21. Six cobs from Structure 6, a burial cist in Quebrada 5a, *253*

15.1. Sample 001, a pod of *Phaseolus lunatus* carrying two seeds, found in Feature 6 of Structure D, *255*
15.2. Sample 002, five of six dark red-brown *Phaseolus vulgaris* seeds, from Feature 6 of Structure D, *255*
15.3. Sample 003, nine dark red-brown *Phaseolus vulgaris* seeds found in Feature 6 of Structure D, *256*
15.4. Sample 004, eight dark red-brown *Phaseolus vulgaris* seeds found in Feature 6 of Structure D, *256*
15.5. Sample 005, two curved and elongate *Phaseolus lunatus* seeds with yellow seed coats and spotted dark areas, from Feature 6, Structure D, *257*
15.6. Sample 006, *Phaseolus vulgaris* seeds found in Feature 6 of Structure D, *257*
15.7. Sample 007, a pod of *Phaseolus lunatus* containing four seeds, from Feature 6 of Structure D, *258*
15.8. Sample 008, three *Phaseolus lunatus* seeds found in Feature 6 of Structure D, *258*
15.9. Sample 009, seven *Erythrina* seeds found in Feature 6 of Structure D, *259*
15.10. Sample 010, one *Phaseolus vulgaris* pod and one dark red-brown seed, found in Structure 12, Quebrada 5-south, *260*
15.11. Sample 011, seven large *Phaseolus vulgaris* seeds, Room 9, Structure 1, *261*

16.1. Stem of sacuara or cattail from a refuse deposit, Terrace 11, Quebrada 6, *263*
16.2. Totora or bulrushes (*Scirpus californicus*), *264*
16.3. Bulrush and cattail growing intermixed, *264*
16.4. Bundle of totora from Structure 4 in Quebrada 5a, *265*
16.5. Braided rope from Feature 20, Structure 9, *266*
16.6. Twilled mat from Zone A of Terrace 9, Quebrada 5a, *267*
16.7. Junco or sedge (*Cyperus* sp.), *268*
16.8. Inflorescence of sedge (*Cyperus* sp.) from Terrace 11, Quebrada 6, *268*
16.9. Carrizo or caña hueca (*Phragmites australis*), *268*
16.10. Carrizo or caña hueca (*Phragmites australis*), *268*
16.11. A sample of monocotyledonous plants from Feature 6, Structure D, *269*
16.12. A sample of reeds and canes from Feature 20, Structure 9, *269*
16.13. Three specimens of caña hueca from the remains of a kincha house, *270*
16.14. Two views of a fragment of kincha wall, found on the central platform of Structure 9, *270*
16.15. Three fragments of kincha wall from the Feature 4 squatters' house in "Room 2," Structure D, *271*
16.16. Caña brava (*Gynerium sagittatum*), *272*
16.17. Two litter poles made from caña brava, Structure 6, a burial cist in Quebrada 5a, *273*
16.18. Deliberately cut section of caña brava found in Room 4, Structure D, *273*
16.19. A bundle of willow twigs from Structure 6, Quebrada 5a, *274*
16.20. Three broken huarango or algarrobo spines from Feature 20, Structure 9, *274*
16.21. The bottle gourd (*Lagenaria siceraria*) at Cerro Azul, *275*
16.22. A shallow gourd bowl found with Burial 2, Quebrada 5a, *276*

16.23. A shallow gourd bowl found with Individual 1 of Burial 6, Quebrada 5a, *277*
16.24. Two large spines of the prickly pear (*Opuntia* sp.) found with Individual 2, Burial 4, Quebrada 5a, *278*
16.25. Large spine of prickly pear found in "Room 2," Structure D, *278*
16.26. Unmodified cactus spines and needles made from cactus spines from workbasket in Burial 4, Quebrada 5a, *278*
16.27. Boll of cotton (*Gossypium barbadense*) from Terrace 11 of Quebrada 6, *279*
16.28. Segment of cotton boll (*Gossypium barbadense*) from the South Corridor, Structure D, *279*
16.29. Seeds of cotton (*Gossypium barbadense*) from Feature 6, Structure D, *280*
16.30. Cotton seeds from Feature 20, Structure 9, *280*
16.31. Two locks of kidney cotton found with Burial 8, Quebrada 5a, *280*
16.32. A seed of *suirucu* or soapberry from Room 2, Structure 11, Quebrada 5-south, *280*

Color plates following page 284

Plate I. Cañete's rocky coast supports hundreds of species of molluscs, crustaceans, fish, sea birds, and marine mammals.
Plate II. Cerro Azul's nobles imported *Choromytilus chorus* to use as a cosmetic pigment palette.
Plate III. The embudo, or funnel, from a crayfish trap.
Plate IV. The burrowing of *Callianassa islagrande* has a profound effect on sandy beach environments at Cerro Azul.
Plate V. The arched blue crab expands its range to Cerro Azul during El Niño years.
Plate VI. A fisherman on Cerro Centinela launches his atarraya, or cast net, into the Pacific.
Plate VII. A cast net found in a burial cist at Cerro Azul.
Plate VIII. The most common member of the grunt family captured at Cerro Azul is *Anisotremus scapularis*.
Plate IX. The largest member of the drum family captured at Cerro Azul is *Sciaena starksi*.
Plate X. The Peruvian booby lives in colonies on the sea cliffs at Cerro Azul.
Plate XI. During the El Niño of 1982–83, many Humboldt penguins died of starvation.
Plate XII. Mummified Inca tern that died in the ruins of Cerro Azul during the 1850s.
Plate XIII. Sacuara or cattails are among the "industrial" plants harvested near Cerro Azul.
Plate XIV. Some elite burials at Cerro Azul were provided with litters combining caña brava and totora.
Plate XV. One ear of dent corn and two ears of imbricated purple corn from burials at Cerro Azul.
Plate XVI. Many of the lima beans at Cerro Azul featured brown seeds.
Plate XVII. A white potato from a looted burial at Cerro Azul.
Plate XVIII. A sweet potato found in Structure D at Cerro Azul.
Plate XIX. Guinea pig with its mouth stuffed with coca leaves, buried in the ruins of Structure 1, Cerro Azul.

17.1. Pellets of camelid dung from Feature 6, Structure D, *289*
17.2. Two views of the right innominate of adult camelid, found near Features 15 and 16, North Central Canchón, Structure D, *291*
17.3. Drawing of the same camelid innominate shown in Fig. 17.2, *292*
17.4. Right innominate of adult camelid, found near Features 15 and 16, North Central Canchón, Structure D, *293*
17.5. Complete right scapula of camelid, found in the fill of Structure D's Northwest Canchón, *294*
17.6. Right and left mandibles from the same camelid, found in Room 4, Structure D, *295*
17.7. Left metacarpal of a camelid, found in fill of Room 4, Structure D, below the level of Feature 5, *296*
17.8. Two views of the right innominate, adult camelid, Room 4, Structure D, below the level of Feature 5, *297*
17.9. Complete right radius and ulna of camelid, found in Stratigraphic Zone A1 of Terrace 16, Quebrada 5, *301*
17.10. Right metacarpal of llama, found in midden between Structures 11 and 12 in Quebrada 5-south, *302*
17.11. Distal metapodial of llama, found on the surface of Cerro Azul, *302*
17.12. Measurements of scapula, humerus, radius, ulna, cervical vertebra, and innominate used to establish Wing's "decision rule," *304*
17.13. Measurements of the calcaneum, astragalus, first phalanx, femur, metapodials, and tibia used to establish Wing's "decision rule," *305*
17.14. Anterior end of llama cervical vertebra, showing signs of arthritis, found in Room 7, Structure 9, *307*
17.15. First phalanx of a llama from Feature 6, Structure D, showing arthritic calcification, *307*
17.16. The relative abundance of bones from seven anatomical regions in Structure D, Structure 9, and a complete llama skeleton, *308*

17.17. Two proximal tibiae of adult llamas, butchered in much the same way, from Structure D, *309*
17.18. Fused proximal radius and ulna of a llama, showing cut marks, from Structure D's Northeast Canchón, *309*
17.19. Distal end of llama metacarpal displaying a cut mark, from Room 4, Structure D, *310*
17.20. Left calcaneum of a llama with cut marks from Feature 6 of Structure D, *310*
17.21. Llama vertebra, displaying cut marks, from Southwest Canchón, Structure D, *311*
17.22. Llama atlas vertebra displaying numerous cut marks, *311*
17.23. Sacrum of llama from Feature 6 of Structure D, showing cut marks, *312*
17.24. Vertebra of a llama from Feature 6 of Structure D, displaying cut marks, *313*
17.25. Cervical vertebra of llama, displaying cut marks, found in Burial 2 of Quebrada 5a, *313*
17.26. Cervical vertebra of llama, displaying cut marks, found in Room 7 of Structure 9, *314*
17.27. Pigment pouch made from what appears to be a camelid bladder, *314*
17.28. On this drawing of a llama skeleton, the bones present in the Feature 6 midden are indicated, *315*
17.29. On this drawing of a llama skeleton, the bones present in the Feature 20 midden are indicated, *316*

18.1. The cranium from a dog associated with Burial 1, Quebrada 5a, *320*
18.2. Two views of dog cranium, Zone A, Terrace 9, Quebrada 5a, *321*
18.3. Two views of dog cranium, Zone A, Terrace 9, Quebrada 5a, *322*

19.1. Guinea pigs were kept in Room 9, while Room 10 was used as a place to store bedding and food, *325*
19.2. Pellets of guinea pig dung from the floor of Room 9, Structure D, *326*
19.3. This gourd vessel, found with Individual 2 of Burial 7, contained the remains of a guinea pig, *330*
19.4. Cranium and mandible of guinea pig from a burial in Zone A, Terrace 9, Quebrada 5a, *331*
19.5. This vessel, Gourd 2 from the floor of Structure 12, contained a guinea pig, *332*
19.6. Guinea pig wrapped in bandana as an offering in the ruins of Structure 1, *333*

21.1. Wooden agricultural implements found in the Chincha Valley, *341*
21.2. Pyroengraved gourd, found with Individual 1 of Burial 7 (Terrace 9, Quebrada 5a), *346*
21.3. Pyroengraved gourd bowl found near Structure 4, a burial cist on Terrace 9 of Quebrada 5a, *347*
21.4. Working model of the sociopolitical hierarchy in the Kingdom of Huarco, *349*

A.1. A comparison of the monthly catches of shrimp and silversides at Cerro Azul, 1984–86, *354*
A.2. Monthly catches of Chilean jack mackerel at Cerro Azul, 1984–86, *355*
A.3. Monthly catches of bonito at Cerro Azul, 1984–86, *356*
A.4. A comparison of the monthly catches of lorna and mismis at Cerro Azul, 1984–86, *357*
A.5. A comparison of the monthly catches of grunt and morwong at Cerro Azul, 1984–86, *358*

List of Tables

3.1. Marine molluscs from archaeological contexts at Cerro Azul, *35*
3.2. Chitons from archaeological contexts at Cerro Azul, *36*
3.3. Shellfish from the Feature 6 midden (Structure D), *46*
3.4. Shellfish from the Feature 20 midden (Structure 9), *48*
3.5. Shellfish from Stratigraphic Zone B, Terrace 9, Quebrada 5a, *49*
3.6. Shellfish from the mortar used in the building of Structure 5, *50*
3.7. Shellfish from Stratigraphic Zone A of Terrace 9, Quebrada 5a, *50*
3.8. Estimated weight of meat present in each species of shellfish at Cerro Azul, *51*
3.9. Comparison of Features 6 and 20, based on estimated amount of meat provided by each genus of mollusc, *52*
3.10. Estimates of the weight of meat provided by each genus of mollusc in Stratigraphic Zone B, Terrace 9, Quebrada 5a, *53*

4.1. Crustaceans from Feature 6 of Structure D, *60*
4.2. Crustaceans from Feature 20 of Structure 9, *63*

5.1. Archaeological occurrences of crayfish at Cerro Azul, *71*

6.1. Fish species observed by the University of Michigan Project, *73*

9.1. Taxa of identified fish from archaeological contexts at Cerro Azul, *121*
9.2. Fish bones from Feature 6, Structure D, *130*
9.3. Fish bones from Room 3, Structure D, *142*
9.4. Fish bones from Room 8, Structure D, *143*
9.5. Fish bones from Feature 20, Structure 9, *144*
9.6. Comparison of Features 6 and 20 at Cerro Azul, *149*
9.7. Fish bones from the layer between Floors 1 and 2 of Room 3, Structure 11, *151*
9.8. Fish bones from Terrace 11 of Quebrada 6, *155*

11.1. Birds identified from archaeological contexts at Cerro Azul, *173*
11.2. Bird bones from Feature 6, Structure D, *181*
11.3. Bird bones from Feature 20, Structure 9, *183*
11.4. Bird bones from Upper Zone B of Terrace 9, Quebrada 5a, *184*
11.5. Bird bones from Zone B below Structure 5 on Terrace 9, Quebrada 5a, *185*

12.1. Marine mammal bones from Feature 22, Terrace 16, *196*
12.2. Marine mammal bones from Stratigraphic Zone A2, Terrace 16, *196*

13.1. Edible, ritual, and medicinal plants from Late Intermediate contexts at Cerro Azul, *201*
13.2. Edible, ritual, and medicinal plants from Feature 6 of Structure D, *213*
13.3. Edible, ritual, and medicinal plants from Feature 20 of Structure 9, *220*
13.4. Food plants from Terrace 11 of Quebrada 6, *226*
13.5. Food plants from 16th-century Colonial refuse at Cerro Azul, *226*

16.1. Industrial plants from Late Intermediate contexts at Cerro Azul, *263*
16.2. Industrial plants from Feature 6 of Structure D, *281*

17.1. Camelid bones from Feature 6, *288*
17.2. Camelid bones from Feature 20, *298*
17.3. Measurements of camelid bones from Cerro Azul, *306*

19.1. Guinea pig bones from Feature 6 of Structure D, *326*
19.2. Guinea pig bones from Feature 20 of Structure 9, *328*

20.1. Proveniences of the Cerro Azul coprolites, *335*
20.2. Macrofossil component analysis of the coprolites from Cerro Azul, *336*
20.3. Fish bone from the Cerro Azul coprolites, *337*
20.4. Results of the Cerro Azul coprolite sample pollen analysis, *338*

Acknowledgments

From 1982 through 1986 the University of Michigan conducted five seasons of work at Cerro Azul. In addition to my co-director María Rostworowski, the project included Ramiro Matos Mendieta (archaeologist), Kent V. Flannery (archaeologist and zooarchaeologist), Sonia Guillén (physical anthropologist), Charles Hastings (cartographer), Lawrence Kaplan (bean specialist), C. Earle Smith, Jr. (ethnobotanist), Jim Stoltman (petrographic analyses), Dwight Wallace (textiles), John G. Jones (analysis of coprolites), and Linda Perry (starch grains and phytoliths). We also received outstanding contributions from two zooarchaeologists, Jeffrey Sommer (crustaceans and fish) and Christopher Glew (birds and mammals). All these specialists made Cerro Azul's ecology and economy come alive.

I used a University of Michigan Faculty Fund Grant to complete the preliminary mapping of Cerro Azul. This 1982 season was vital because it supplied us with the kinds of data we needed to write a detailed grant proposal to secure external funding from the National Science Foundation.

Following the 1982 mapping season came the seasons dedicated to excavation, more detailed mapping, and artifact analyses. The four seasons from 1983 to 1986 were generously supported by the National Science Foundation (BNS-8301542). I appreciate not only the funding but the excellent advice offered by Charles Redman, Mary Greene, John Yellen, and Craig Morris throughout the project.

Permission to excavate Cerro Azul was granted by Peru's Instituto Nacional de Cultura (Credencial No. 102-82-DCIRBM, Credencial No. 041-83-DCIRBM, Credencial No. 018-84-DPCM, and Resolución Suprema No. 357-85-ED).

Several colleagues provided unflagging enthusiasm and encouragement. Among them were Michael E. Moseley, Craig Morris, Christopher B. Donnan, Charles Stanish, Robert L. Carneiro, Guillermo Cock, Charles S. Spencer, Elsa M. Redmond, Ramiro Matos, Rogger Ravines, Duccio Bonavia, Jorge Silva, Sonia Guillén, John O'Shea, John Hyslop, R. Alan Covey, Christina Elson, Bruce Mannheim, Luis Jaime Castillo Butters, Marc Bermann, Jordan Dalton, Robert D. Drennan, Jeremy A. Sabloff, Geoffrey Braswell, E. Wyllys Andrews, Geoffrey W. Conrad, Jo Osborn, Allison Davis, Howard Tsai, Conrad Kottak, Betty Kottak, Evon Z. Vogt, Brian Bauer, Véronique Bélisle, Patrick Ryan Williams, Kenny Sims, Loa Traxler, Robert J. Sharer, and Don and Prudence Rice.

In the town of Cerro Azul we found informants, friends, and industrious people who wanted to work for us, either in town or at the archaeological site. I want to thank all the residents of Cerro Azul, but especially Diomides Aguidos, Edalio Aguidos, Marcelina Aguidos, Urbano Aguidos, Zenobio Aguidos, Pedro Álvarez, Alberto Barraza, Adolfo Casella, don José Chumpitaz y su familia, Emilio Cordero, Rosa Cordero, Ruperto Corral, Cirilo Cruz, Victor de la Cruz Álvarez, Pablo Cubillas, Victor Cubillas, Ramón Espinosa, César (Chinaco) Francia, Iván Francia, Roberto García, Luis Gómez, José Huaratapaira, Ramón Landa, Carlos Manco Flores, José Antonio Manco Flores, Rufino Manco Flores, Francisco Padilla, Camilo Quispe, Edgar Zavala, and Pedro Manuel (Pato Loco) Zavala.

I want to end this acknowledgment section by thanking three very talented people—Elizabeth Noll, our Museum editor, as well as John Klausmeyer and Kay Clahassey, two wonderful artists who make all my archaeological reports come to life.

Part I

An Introduction to the Kingdom of Huarco and its Ecosystems

1

The Ecosystems of the Kingdom of Huarco

Joyce Marcus

From 1982 through 1986 the University of Michigan carried out five seasons of research at Cerro Azul, a Late Intermediate and Late Horizon archaeological site in Peru's Cañete Valley (Marcus 1987a, 1987b, 2008, 2009, 2015, 2016, Marcus et al. 1999). It was my good fortune to co-direct that project with the late ethnohistorian María Rostworowski, who had brought the site to my attention (Rostworowski de Diez Canseco 1978–1980).

A key member of the University of Michigan team was Professor Ramiro Matos Mendieta; recruited as an archaeologist, he actually became a jack-of-all-trades and our guardian angel (Marcus and Flannery 2010). In charge of plant remains was the late Dr. C. Earle Smith, Jr., who turned to Dr. Lawrence Kaplan for help with the races of beans. In charge of faunal remains was Kent V. Flannery, who turned to Jeffrey Sommer and Chris Glew for help with the crustaceans, fish, birds, and mammals. Our human burials were analyzed by Dr. Sonia Guillén, our human coprolites by Dr. John G. Jones, and our llama and guinea pig feces by Dr. Linda Perry. I mention these collaborators prominently because their work forms the core of this volume. My first book on Cerro Azul (Marcus 2008) focused on the site's architecture and pottery. The focus of this volume is on the site's ecology and economy, based on the analysis of our plant and animal remains.

The Kingdom of Huarco

Rostworowski (1978–80) suggests that during the Late Intermediate period (A.D. 1000–1470) the lower valley of the Río Cañete was divided between two neighboring señoríos or petty kingdoms (Fig. 1.1). The *yunga* or coast proper was controlled by the Kingdom of Huarco, while the *chaupi yunga* or piedmont was controlled by the Kingdom of Lunahuaná (Chu 2015, Díaz Carranza 2015, Harth-Terré 1933, Hyslop 1985, Marcone Flores and Areche Espinola 2015, Urton and Chu 2015). Both kingdoms were conquered by the Inca around A.D. 1470.

At the time the Inca arrived, the Kingdom of Huarco occupied roughly 140 km^2. In Figure 1.2 we see the best available map of the ancient señorío, an ethnohistoric reconstruction prepared by Eugenio Larrabure y Unanue in 1893 and published in 1935. The inland limits of Huarco were set by its border with Lunahuaná; from there it extended to the sea. The Río Cañete (labeled on the map as the "Río de Lunahuaná") was close to being the kingdom's limit on its southeast side. The bay of Cerro Azul was close to its limit on the northwest side.

Key landmarks shown in Figure 1.2 are the ruins of Canchari, where the residence of the Huarco ruler was built; the hilltop fortress of Ungará, which guarded the takeoff point

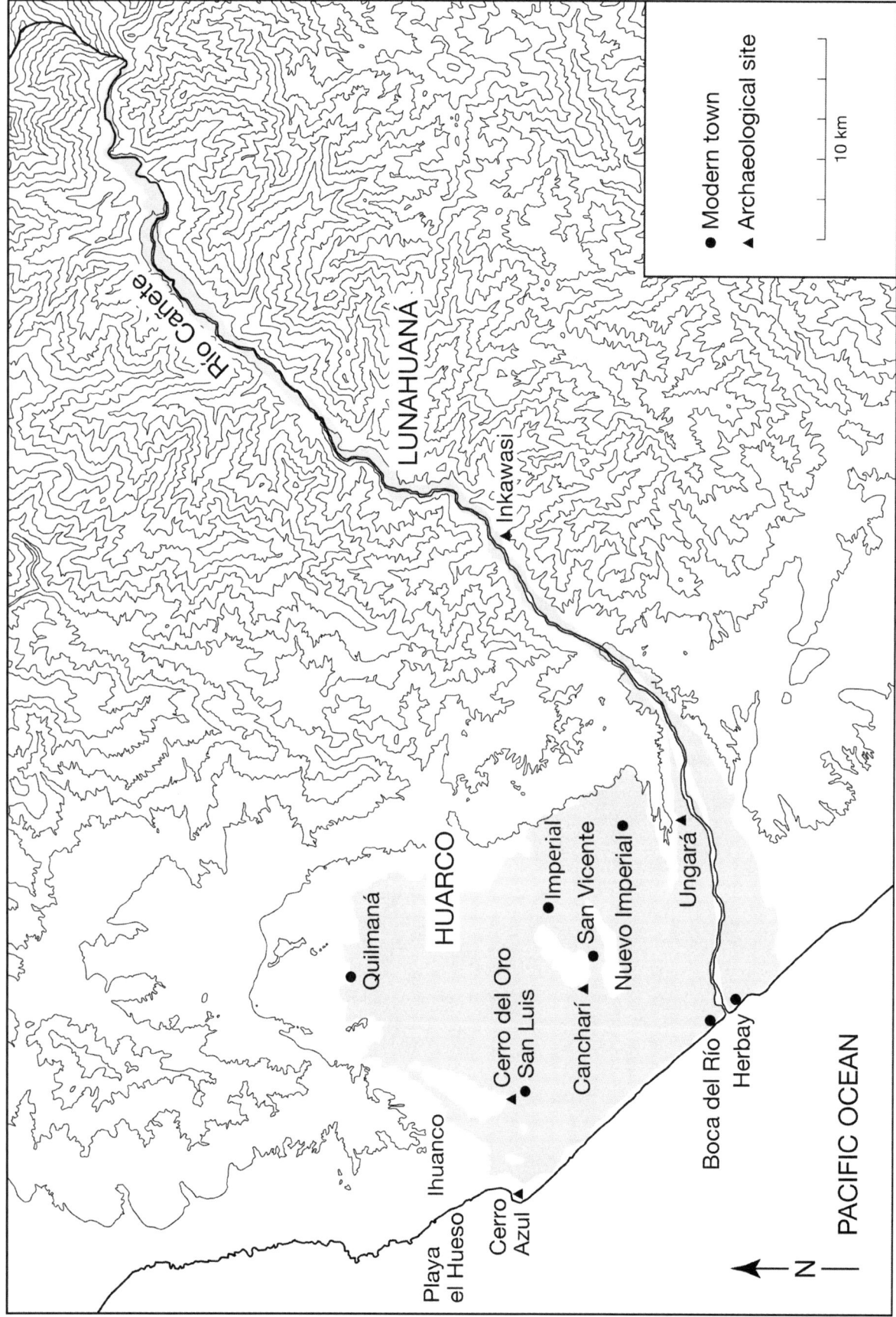

Figure 1.1. The lower valley of the Río Cañete, showing the relationship of the Kingdoms of Huarco and Lunahuaná. (Artwork by John Klausmeyer, based on work by Marcone Flores and Areche Espinola [2015]).

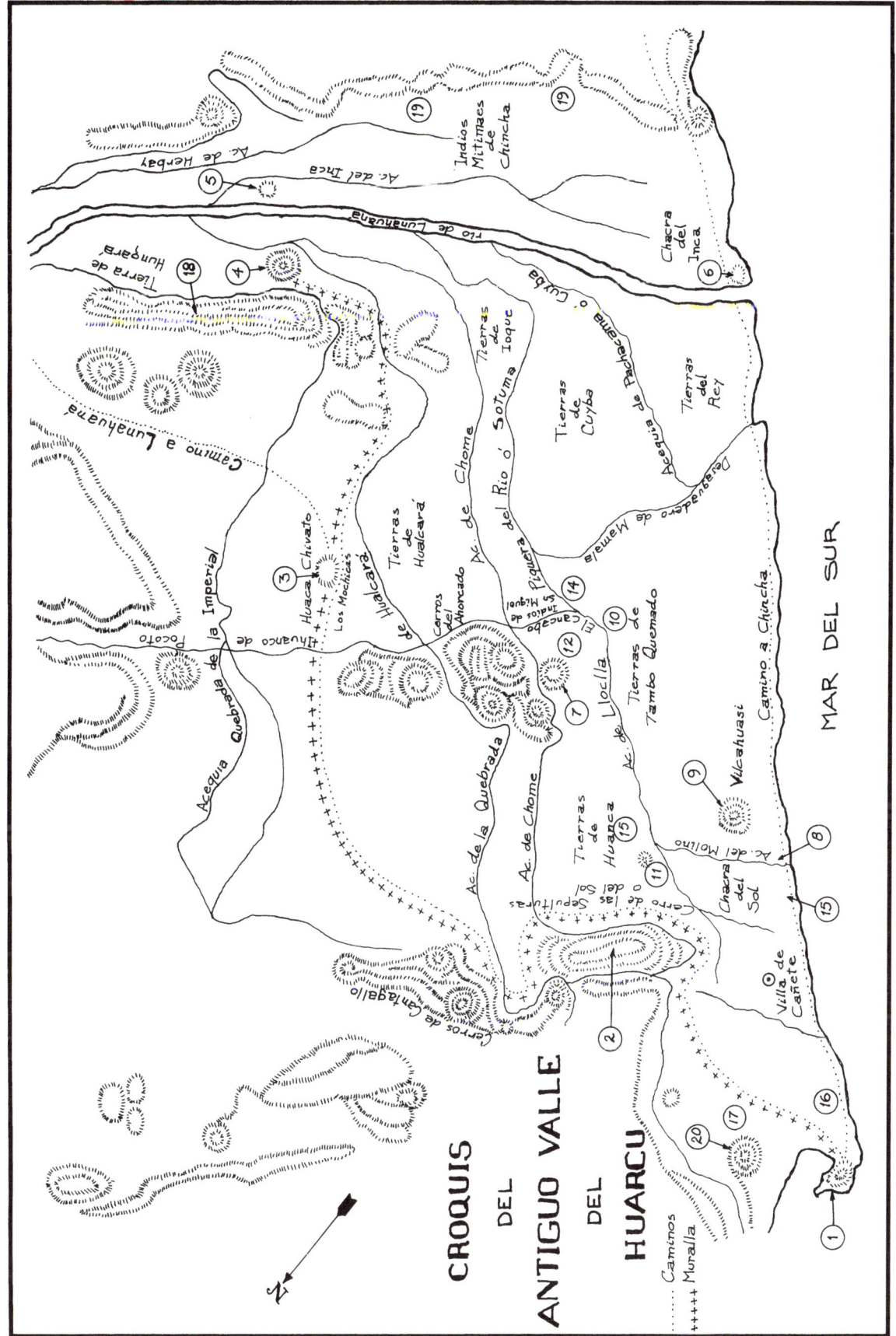

Figure 1.2. This 1893 map of the Kingdom of Huarco shows its defensive wall (muralla), half a dozen irrigation canals (acequias) and a series of numbered archaeological sites: Cerro Azul (1), Cerro del Oro (2), Canchari (7), Huaca Chivato (3), and Ungará (4). (Redrawn by Kay Clahassey from Larrabure y Unanue [1935:270]).

of the kingdom's key irrigation canals; the hilltop fortress of Huaca Chivato, where the shortest road to Lunahuaná left the Kingdom of Huarco; Cerro del Oro, a major hilltop center whose occupation goes back to the Early Intermediate period; and the Great Wall of Huarco, a defensive feature running along the entire inland margin of the kingdom.

Also featured in Figure 1.2 are a number of irrigation canals that drew water from the Río Cañete and carried it northwest across the Kingdom of Huarco. Some of these canals date back at least to the Late Intermediate period, while others are Colonial versions of prehispanic canals.

One of the most important canals took off from the Río Cañete near the Fortress of Ungará (indicated on the map by the number 4). Immediately south of the fortress, this canal bifurcated into northern and southern branches. The northern branch was labeled "Acequia de la Quebrada de Hualcará" by Larrabure y Unanue; however, according to documents in the Archivo General de Indias in Seville, its more ancient name was Chumbe (AGI Escribanía de Cámara 498-B, folios 792 and 797v [A.D. 1575]). In Colonial times this canal was renamed María Angola; it is said to have been 24 km long and featured 34 water takeoff points (ONERN 1970). Fully defended by the Great Wall of Huarco, the María Angola canal passed to the east of the ruler's residence at Canchari and continued on to Cerro del Oro, where it once again bifurcated; one of those branches continued on to Cerro Azul Bay.

The southern branch of the Quebrada de Hualcará canal was labeled "Acequia de Chome" by Larrabure y Unanue, but came to be renamed San Miguel during Colonial times. This canal passed even closer to Canchari, with one branch encircling the ruler's palace and providing it with water (Harth-Terré 1933, Marcus 2008: Fig. 1.3). The Acequia de San Miguel continued on to Cerro del Oro, wrapping around the site before joining up with the Acequia María Angola. The total length of the San Miguel canal is said to have been 35 km.

Two more of the canals shown in Figure 1.2 are likely to have been ancient: the "Acequia de Lloclla" and the "Acequia de Pachacama o Cuyba." These canals also took their water from the Río Cañete, but at points downstream from the Acequia María Angola/San Miguel, and neither was as extensive as the latter.

These four ancient canals were crucial to the ecology of the Kingdom of Huarco, for they brought water from the Río Cañete to lands that would otherwise have been uncultivated desert. The canals also carried live crayfish (*Cryphiops caementarius*) from the river to distant parts of the valley (Chapter 5), and it is said that their waters supported reeds, bulrushes, and even bottom feeders like catfish and mullet.

The Inland Ecosystem

The Río Cañete originates at Lake Ticllacocha in the Andes, some 4600 m above sea level. From there its course to the Pacific is roughly 220 km long, traversing the provinces of Yauyos, Lunahuaná, and Huarco, all the while laying down nearly 8000 ha of alluvium (ONERN 1970).

The entire valley of the Río Cañete is estimated to cover 6192 km^2. Only the stretch from 2000 m elevation to the Pacific Ocean is relevant here, and all of that is classified as subtropical desert (Tosi 1960). The average temperature of the lower Cañete Valley is 19.7° C and the annual rainfall 27.9 mm—just a little over an inch.

During the Andean summer—December to March—the Río Cañete reaches a maximum flow of 850 m^3 per second and discharges 69 percent of its volume. During the Andean winter—from June to November—the flow can dwindle to 5.80 m^3 and the river discharges only 12 percent of its volume. The river, of course, constitutes only a part of Cañete's water supply, since there is an aquifer of 43 million m^3 below the valley floor.

Despite these resources, the Kingdom of Huarco was likely to have been an area of water deficit. In 1970 the Cañete Valley's four main irrigation canals, which at that time totaled 213.3 km, could not satisfy the water demands of the region's farmers. They fell short by an average of 29.7 million m^3 per year (ONERN 1970).

Geological Stratigraphy

To understand the geology of the Cañete Valley, we need to go back to the Late Tertiary era. The Pliocene Cañete Formation consists of conglomerates of rounded cobbles and pebbles, interdigitated with friable sand lenses. These conglomerates are overlain by Plio-Pleistocene volcanics, including andesites, rhyolites, lavas, tuffs, and volcanic ash. Later came Quaternary alluvial deposits, including sands, gravels, and clays (ONERN 1970).

These formations are responsible for much of the geological diversity seen on the coast of Huarco. The sea cliffs and offshore islands are made up largely of Plio-Pleistocene volcanics. The Pacific shore is a narrow band of some 480 ha, 230 ha of which are the cobble and pebble beaches known locally as *costa* (Fig. 1.3). The origins of the *costa* can be found in the eroded Pliocene conglomerates, as well as the cobbles and pebbles washed to the sea by the Río Cañete, all moved into place by the currents of the Pacific Ocean.

The remaining 250 ha of the Pacific shore are sandy beaches of marine origin, stretching north from Cerro Azul Bay and known locally as *playa* (Fig. 1.4). Just inland from Cerro Azul are areas of poor drainage, a number of which were once covered by the sea; after the sea receded they were cut off, and some became bodies of standing water.

One of these water-filled depressions, known as Huaca Chola, was considered a sacred place by the people of Cerro Azul. Old timers considered its mineral waters to be health-giving; it was reportedly bordered by useful stands of cattails (*Typha* cf. *angustifolia*), bulrushes (*Scirpus californicus*), and carrizo (*Phragmites australis*) (Heiser 1974, 1978; Macía and Balslev 2000). It is said that ducks, mullet, and crayfish were available

Figure 1.3. The *costa*, or cobble beach environment, characterizes much of the coast between Cerro Azul and the Río Cañete. The men in the background are using fishing gear called an espinel (see Chapter 7) to catch the róbalo, a large drum that grazes on the crustaceans and bristle worms that seek refuge beneath the cobbles.

at Huaca Chola. Unfortunately, during the cotton boom of the 1930s this depression was deliberately filled in so that a nearby hacienda could expand its plantings. The workmen filling in the depression used adobes from Cerro Azul's defensive wall, leaving gaps in the ancient structure (Fig. 1.5).

Just inland from Huaca Chola was a major alluvial fan known locally as Ihuanco (Fig. 1.1). This fan, which is considered Class 1 soil (ONERN 1970), is believed to be the clay source for the tapia walls of the residential compounds at Cerro Azul.

The Native Vegetation

Native vegetation is extremely sparse in the immediate vicinity of Cerro Azul. Without irrigation, the most one can hope for is occasional bunches of saltgrass (*Distichlis spicata* and *Salicornia fruticosa*).

In many coastal valleys of Peru, humid fog from the ocean can produce what the local population calls lomas vegetation, a green carpet covering the foothills of the mountains at certain seasons of the year. As Weberbauer (1945:249) pointed out in his classic early twentieth-century study, however, genuine lomas vegetation does not exist near Cerro Azul. Only by taking the route to Lunahuaná via the Quebrada de Pocotó (Fig. 1.2) does one have a chance of seeing anything approaching lomas vegetation. Occasionally the fog will support the growth of bromeliads (*Tillandsia* sp.) on inland sand dunes. *Tillandsia* was one of ancient Huarco's "industrial" plants, used (among other things) to make decorative tassels for Andean slings (Chapter 10).

As one travels inland to Lunahuaná, annual precipitation rises to 82.3 mm (a little over 3 inches). On the slopes of the piedmont here one can find not only *Tillandsia* but also columnar cacti (*Cereus macrostibas*) and prickly pear (*Opuntia* spp.). The spines of these cacti were used by the occupants of Cerro Azul to make sewing needles (Chapter 16). The fact that many grow on land belonging to the Kingdom of Lunahuaná would not have limited access to these cacti, since the rulers of Huarco and Lunahuaná were on friendly terms (Rostworowski de Diez Canseco 1978–1980).

If, instead of taking the Pocotó road, one took the inland route along the Cañete River, he would have encountered additional useful plants at elevations of 1000 m (ONERN 1970). Here the vegetation of the riverine alluvium includes willow (*Salix humboldtiana*); the Peruvian pepper tree (*Schinus molle*); *Croton* sp.; *Capparis* sp.; the leguminous trees *Caesalpina tinctoria*, *Acacia* spp. (huarango), and *Prosopis juliflora* (algarrobo); and a cane called caña brava (*Gynerium* cf. *sagittatum*).

Many of these plants were used by the ancient people of Cerro Azul (Chapter 16). Bundles of willow twigs were used as pillows; the fruits of the pepper tree were used to flavor chicha, or maize beer; the woody spines of huarango and algarrobo were made into spindles for textile fibers; and caña brava served as construction material.

Figure 1.4. The *playa*, or sandy beach environment (A), is home to countless coquina clams, mole crabs, ghost shrimp, and other invertebrates. A special habitat within the *playa* zone is Cerro Azul Bay (B), whose shallow waters attract stone crabs, anchovetas, rays, and dogfish sharks.

Figure 1.5. Remnants of Cerro Azul's adobe defensive wall can still be found on the slopes of Cerro Camacho.

Soils and Irrigation

While the native vegetation of the subtropical desert was understandably meager, the complex of plants introduced through human intervention was large indeed. Canals like the Acequia de la Quebrada de Hualcará (María Angola) and the Acequia de Chome (San Miguel) crosscut soils described by ONERN (1970) as Class 1, Class 2, and Class 3. These canals made possible the growing of staples such as common beans, lima beans, *Canavalia* beans, squash, peanuts, chile peppers, potatoes, and sweet potatoes; "industrial" plants such as gourds and cotton; maize for brewing chicha; and fruit crops such as guava, lúcuma, pacay, and ciruela del fraile (Chapters 13–16).

The presence of these crops, in turn, increased the bird life of the Kingdom of Huarco by attracting white-winged doves (*Zenaida meloda*), eared doves (*Zenaida auriculata*), and several species of pigeons (*Columba* spp.) to the irrigated fields. Also drawn to the croplands of Huarco was the rice rat *Oryzomys xanthaeolus*, which then became food for the burrowing owl *Speotyto cunicularia* (Chapter 11).

We cannot specify, of course, exactly how many hectares of the Kingdom of Huarco were irrigated during the Late Intermediate. In the 1960s the María Angola canal was 24 km long, had a flow of 4.2 m^3 per second, and irrigated 2254 hectares. The San Miguel canal was 35 km long, had a flow of 7 m^3 per second, and irrigated 3868 hectares (ONERN 1970:271).

Class 1 soils could be found on the alluvial fan of Ihuanco, mentioned earlier, and a strip of land on the right bank of the Río Cañete north of Ungará. A second alluvial fan called San Isidro—just inland from the modern town of San Luis—was considered Class 2 soil, and 1153 ha of it were planted in potatoes and cotton in the 1960s. This land, not far from the site of Cerro del Oro, lay within reach of both major canals.

Class 2 soils could also be found on the hacienda of Hualcará (watered by the canal of the same name) and at Cantagallo near San Isidro. One of today's most productive canals passes through the town of Nuevo Imperial, but since it lies outside the defensive wall of the Kingdom of Huarco, it is likely too recent a construction to be relevant to the Late Intermediate period.

The irrigated crop schedule for the Cañete Valley is as follows. Cotton is planted in August or September and harvested from March to May. Potatoes are planted in April or May and harvested in August or September. Squash and maize can be planted throughout the year.

Llama Herding

We know from the camelid dung on the floor of the Southwest Canchón in Structure D that llama pack trains reached Cerro Azul (Marcus 2008:84). The key question is, where were those llamas raised and what were they fed? None of their preferred foods grow near Cerro Azul, so it seems likely that some kind of fodder was kept there for them. According to Weberbauer (1945:249–250), suitable forage can be found in the lomas of the Cerros de Quilmaná, the mountain range separating the valleys of Cañete and Asia. There, at elevations between 200 and 550 m above sea level, one can find grasses of the genera *Poa* and *Festuca*, which llamas enjoy eating.

Suitable forage for llamas could also be found in the Kingdom of Lunahuaná, upstream from Huarco on the Río Cañete. Roadcuts through middens in the Lunahuaná region reveal hundreds of llama bones, while those on the coast near Cerro Azul yield predominantly fish bone, shellfish, and plant remains (Marcus 2008:1). I suspect that llamas could be raised in the *chaupi yunga* or piedmont, but not on the *yunga* or coast proper. Almost certainly, the Kingdom of Huarco sent dried fish to the Kingdom of Lunahuaná in return for *ch'arki*, or dried camelid meat.

The Coastal Ecosystem

The southwestern limit of the Kingdom of Huarco was the rim of the Pacific Ocean. The ruler's access to the sea ran for 15 to 20 km, from the beaches north of Cerro Azul Bay to the hills south of the Río Cañete. This is one of those stretches of coast where the outlying ranges of the Andes sometimes come right to the ocean, forming sea cliffs and small offshore islands. In addition to the Río Cañete, at least three ancient irrigation canals emptied their waters into the sea.

The Late Intermediate site of Cerro Azul lies in a protected hollow among three rocky hills (Chapter 2). To the west are the sea cliffs called Cerro del Fraile and Cerro Centinela; inland lies Cerro Camacho. The bedrock of these hills is described by Stoltman (2008:63) as "extrusive igneous rock, originally of andesitic composition, that had been subjected to metamorphic alteration." This rock appears blue to the ships at sea, which is how Cerro Azul got its name.

The fishing environment of Cerro Azul's vertical cliffs is known locally as *peña*, and both the cliffs and the rocky ocean floor below them constitute a distinct marine habitat (Fig. 1.6). The *peña* is the third major fishing environment for the Kingdom of Huarco, providing an alternative habitat to the cobble beach and sandy beach environments.

While the fishermen of Cerro Azul give each of these landscapes a different name, oceanographers insist that neither rocky shores nor sandy beaches are complete ecosystems; one is often the breakdown product of the former (Sumich 1976:102). Together they form a larger ecosystem, with two complementary pathways for energy transfer: a grazing food chain and a detritus food chain.

Within rocky shore communities one sees well-developed grazing food chains, but little detritus; the very erosional nature of rocky shores prevents its accumulation. Sandy beaches and mudflats, on the other hand, are depositional features and show well-developed detritus food chains. Distinctive plant and animal communities develop in *peña*, *playa*, and *costa* habitats, but even those communities should not be considered the mature end-products of nature; instead, they continue to change with the currents. During the 1980s, many owners of beachfront houses at Cerro Azul complained that their once-sandy property was being converted to cobbles.

Even the ocean off Cerro Azul should not be thought of as predictable, for it is made up of several different kinds of water masses. The most famous of these is the Humboldt Current—a highway of cool water paralleling the coast—which rises from the ocean depths loaded with nutrients. This coastal upwelling, however, is only one of "a labile mix of several water types of different temperature and salinity ranges" (Swartzman et al. 2008:230). The cool water of the Humboldt Current (usually 14°–18° C) competes with masses of subtropical surface water (18°–27° C), and still warmer equatorial water that periodically displaces the cool. As Ayón et al. (2008a:250) have put it, upwelling systems like the Humboldt Current are stochastic, and our current oceanographic models have difficulty predicting their behavior for more than a few months at a time.

Trophic Levels and Marine Habitats

The fact that the Huarco coast had rocky cliffs, sandy bays, and pebble beaches unquestionably enriched its fish resources. All those fish, however, were ultimately dependent on the energy provided by the waters of Peru's continental shelf, and much of that energy comes in the form of creatures so small as to be invisible to the casual observer. According to Peru's Instituto del Mar, the seas off Cañete feature a food chain of at least five trophic, or energy-transfer, levels. At their most efficient, the creatures at each level capture only about 10 percent of the energy stored in the level below them, and it is not unusual for some predators to draw on two or more trophic levels (Medina et al. 2007, Tam et al. 2008, Taylor et al. 2008).

My 2008 volume on Cerro Azul concluded that its Late Intermediate occupants were drying fish on an "industrial" scale for shipment to inland communities. In this volume my collaborators and I discuss both the fish resources available in the 1980s (Chapter 6) and the totality of the fish remains from our Late Intermediate refuse (Chapter 9). One cannot fully understand what brought these fish within reach of the Cerro Azul fishermen, however, without considering the ecology of the Huarco coast. I hope to make my coverage of this topic as uncomplicated as possible, since I am addressing archaeologists and not marine biologists. Unfortunately, even a simplified account requires me to define a whole series of specialized terms.

Thankfully—given the unseaworthy nature of prehistoric watercraft and the simplicity of Late Intermediate fishing technology—I need not discuss the ecology of the open sea. It is clear from the archaeological remains that Cerro Azul got its fish, crustaceans, molluscs, and sea mammals either from the intertidal zone (the area between the lines of highest and lowest tides) or the neritic zone (the comparatively shallow waters of the continental shelf).

Level One: The Producers

At the lowest level of Huarco's marine food chain came the plants whose photosynthesis converted solar energy and nutrients into food for the creatures of higher trophic levels. The most visible plants of this level are the seaweeds that cling to sea cliffs like El Fraile and La Centinela. At the highest intertidal level are the blue-green algae; lower down comes a belt of brown algae, sometimes referred to as the Fucoid zone. It is worth noting that two genera of seaweed, one brown (*Durvillaea* sp.) and one red (*Chondracanthus* sp.), were considered edible by the Inca, who called them *cochayuyu* (Masuda 1981, 1985). Some herbivorous fish like the borracho or scaleless blenny (*Scartichthys gigas*) dine on brown algae; even carnivorous fish who do not eat algae may seek protection from predators in the kelp beds of the Fucoid zone. Local fishermen say that grunts come to the cliffs of Cerro Azul because they are "yuyu-eaters."

Still lower in the intertidal zone come beds of red algae, which continue to depths where even the lowest tides leave them submerged. These kelp beds harbor no end of invertebrate life and provide shelter for the smaller fish.

It is not only such visible algae that serve as food, of course. Many rocks of the intertidal zone are covered with a viscous algal film that is rich in nutrients, allowing thousands of small crustaceans and molluscs to graze upon it.

And then there is phytoplankton, the hundreds of plant species so tiny that they simply drift by the billions wherever the currents take them. Their incredible abundance and rapid life cycle make them the cornerstone of the Pacific's food supply. Restricted by their need for sunlight to the upper 100 m of the ocean, they use photosynthesis to produce food for hundreds of herbivores and omnivores in the trophic levels above them.

Three of the leading role players among these microscopic plants are diatoms (unicellular algae surrounded by a silica capsule, 10 microns to 1 mm in diameter), dinoflagellates (unicellular algae surrounded by a cellulose membrane, sporting a whiplike flagellum for catching nutrients), and unicellular versions of the blue-green algae already mentioned. Crustaceans, molluscs, and small fish of the Huarco coast consume tons of these plants.

Levels Two and Three: Lower-level Consumers

Dining enthusiastically upon the plants listed above are hundreds of tiny animal species: the zooplankton. The smallest of these are one-celled organisms called *foraminifera* (shell-bearing protozoans) and *radiolaria* (protozoans 50 microns to several millimeters in diameter). There are also *rotifers*, tiny multicellular creatures encased in a transparent cuticle.

Only slightly larger are the crustacean species of the plankton world. When we hear the word "crustacean" we think of crabs and shrimp, but marine biologists will tell you that perhaps 90 percent of all crustaceans are smaller than a cigarette filter.

Major players in the zooplankton of the Huarco coast are the copepods, crustaceans measuring only 0.2–2.0 mm. Most of them are filter feeders, meaning that their appendages allow them to fine-screen diatoms from the water. In turn, copepods are among the most common prey of sardines (*Sardinops sagax*) and anchovetas (*Engraulis ringens* and *Anchoa nasus*). Note that when a sardine bites a copepod—which is an animal—it places itself in Level 3 of the trophic food chain; when it filter-feeds directly on diatoms—which are plants—it is acting as a Level 2 consumer. Sardines are therefore regarded as *heterotrophic*, feeding on organisms of multiple trophic levels.

Somewhat larger than copepods are the *euphausiids*, shrimplike crustaceans 1–6 cm in length. Known in the Arctic as "krill," euphausiids often swarm by the millions and dine on phytoplankton. While mature krill is one of the favorite foods of fish like the lorna (*Sciaena deliciosa*) and the corvina (*Cilus gilberti*), a euphausiid begins life as a tiny free-swimming larva called a *nauplius*, then molts its way through 10 pre-adult stages before reaching maturity. For every stage in the life of krill, therefore, there is a fish whose mouth is just the right size.

For still bigger fish there are *mysids*, or opossum shrimp, which lie buried during the day for protection and become active at night. Opossum shrimp range from 2–3 cm to 15 cm long, meaning that they, too, can serve as prey for a wide variety of fish.

Finally, the sea off Cerro Azul features *ostracods* (crustaceans in a hinged carapace, up to 2.3 cm in length) and *amphipods* (small crustaceans, some of which live in protective tubes and others of which walk freely on the ocean floor).

Adding to the variety of zooplankton is the fact that many sea creatures—including species that we think of as relatively immobile—have larval forms that swim or drift with the current. Each has its own name. The free-swimming larva of a clam is called a *veliger*. The larva of a jellyfish is called a *planula*. The larva of a bristle worm is called a *trochophore*, while that of a ribbon worm is called a *pilidium*. The larva of a starfish is called a *pluteus*. Add the fact that many fish species release thousands of tiny eggs into the ocean, and you have a veritable smorgasbord of food for even the smallest fish.

Included in Trophic Level 2 of the Huarco coast are many creatures that are not planktonic. Among these are the polychaetes, or bristle worms, which swim, crawl, burrow, or live in tubes on the ocean floor. Some polychaetes can filter-feed on the nutrients in ocean-floor sediments, while others are mobile carnivores. Spend time plucking submerged cobbles from the *costa* floor at Cerro Azul and you will find it impossible to count all the bristle worms you uncover. Small wonder that the róbalo (*Sciaena starksi*) grazes the ocean floor south of Cerro Centinela.

The sandy beach, or *playa*, has its attractive invertebrates as well. The *muy muy* or mole crab (*Emerita analoga*) lies buried just below the surface of the sand on wave-swept beaches (Chapter 4). As the waves recede, the mole crab extends its long, feathery antennae to catch suspended food particles. The grey gull (*Larus modestus*) has learned how to detect those antennae, and can be seen running along the tide line at Cerro Azul Bay, eating mole crabs as fast as it can (Chapter 11).

When the water above them is calm, mole crabs may emerge part way from the sand (Brusca 1980:286). At this point they become vulnerable to members of the drum family (Sciaenidae) and to several cartilaginous fish who scour the bay for crustaceans and molluscs.

The marucha or ghost shrimp (*Callianassa islagrande*) burrows more deeply into the submerged beaches of the *playa* zone. It feeds both on particles suspended in the water and those deposited in the sand, and can turn over prodigious quantities of sediment, making it a key player in the detritus ecosystem. Predators from Trophic Level 3 have learned to detect the burrows of marucha, watching for air bubbles or protruding appendages.

In addition to crustaceans, molluscs play a role in both Levels 2 and 3 of the food chain (Chapter 3). Chitons and mussels are Level 2 herbivores that graze on algal film or filter plant material from the water; the false abalone and rock snail are Level 3 carnivores that prey on small mussels.

Molluscs figure prominently in both the grazing food chain and the detritus food chain. Let us begin with the zonation of grazers on the sea cliff of La Centinela (Fig. 1.6). At the highest point splashed by the waves the cliff is covered with periwinkles (*Littorina* spp.), barnacles (*Balanus* spp.), chitons (*Enoplochiton niger*), and the agile little Sally Lightfoot crab (*Grapsus grapsus*), which eats algae and small invertebrates. Somewhat lower on the cliff comes a zone with brown algae, chitons, small mussels (*Perumytilus purpuratus*), and the carnivorous rock snail *Thais chocolata* (Cantillánez et al. 2011; Paredes 1974a, 1974b). Even lower, but still within the intertidal zone, we find the small mussel *Semimytilus algosus*, the black turban shell *Tegula atra*, two varieties of limpets (*Fissurella* spp.), and more chitons. As we reach the lowest level of the intertidal zone we begin to see red algae as well as larger mussels of the species *Aulacomya ater*.

The detritus food chain leads to the sandy beaches of the *playa*. Today that environment still features millions of small coquina clams (*Donax obesulus*). In the Late Intermediate it also had a larger clam, *Mulinia edulis*, which filtered its food from the water like so many Level 2 consumers.

Before turning to the Level 4 predators in this ecosystem we should mention two prominent scavengers or "reducers," one of whose roles it is to clean up dead or decaying matter. The purple stone crab (*Platyxanthus orbigyni*) patrols the floor of Cerro Azul Bay in large numbers. The arched blue swimming crab (*Callinectes arcuatus*) visits Cañete in El Niño years. It likes the shallow waters of the bay, but may also bury itself in the sand and filter feed (Chapter 4).

Levels Four and Five: Top Predators

Finally, we come to the predators that eat the creatures of Trophic Level 3. As so often happens in nature, many of these top predators are heterotrophic, consuming species of Level 2 as well. Marine biologists sometimes deal with heterotrophy by giving each predator a numerical score, one that reflects the percentage of Level 2 and Level 3 prey in its diet.

I refer the reader to two trophic simulation models, one for the Peruvian coast (Tam et al. 2008, Taylor et al. 2008) and one for the northern Chilean coast (Medina et al. 2007). Both sets of researchers fed quantitative dietary data into a computer model called ECOPATH, which then evaluated each species' position in the food chain.

In the Peruvian model, the top predators (Trophic Levels 4–5) were sharks, jumbo squid, large hake (*Merluccius gayi*), flounder (*Paralichthys adspersus*), and whales. Medium predators (Trophic Levels 3–4) were octopus; squid; jack mackerel (*Trachurus murphyi*); mackerel (*Scomber japonicus peruanus*); small hake; medium-sized drums (Sciaenidae); sea catfish (*Galeichthys peruvianus*); seabirds like the guanay cormorant, booby, and pelican; and sea lions (*Otaria byronia*). Lower level predators (Trophic Levels 2–3) included sardines (*Sardinops sagax*), anchovetas (*Engraulis ringens*, *Anchoa nasus*), and zooplankton. At the bottom of the food chain (Trophic Level 1) was the phytoplankton.

In the Chilean model, sharks received a rating of 4.4 (since they take prey from both Trophic Levels 4 and 5), while sea lions received a 4.0. The same sea birds featured in the Peruvian study were given a score of 3.8 (closer to Level 4 than to 3), as were two fish, the bonito (*Sarda sarda*) and the mackerel; for its part, the jack mackerel received a 3.7. Another fish, the blackruff (*Seriolella violacea*) received a score of 3.1, making it very close to a prototypic Level 3 predator. Anchovetas and sardines, which prey on creatures of both Levels 2 and 3, received scores of 2.9 and 2.3, respectively. Zooplankton scored 2.1 and phytoplankton 1.0.

Human beings were not included in either simulation. It is safe to say that the Late Intermediate inhabitants of Cerro Azul would have been found heterotrophic, since they ate everything from sea lions and cormorants to bonito, drums, sardines, anchovetas, crabs, mussels, and possibly seaweed.

Figure 1.6. Cerro Centinela and its adjacent offshore islands epitomize the *peña*, or rocky coast environment, of the Kingdom of Huarco. Dozens of species of molluscs and crustaceans make this habitat a year-round buffet for fish and marine mammals.

The Place of Fish in the Huarco Ecosystem

Such is the importance of fish in the economies of Peru and Chile that their habitat preferences, behavior, and stomach contents have been extensively studied (e.g., Berrios C. and Vargas 2004, Vargas F. et al. 1999). From these studies we learn that the fish available to the Late Intermediate fishermen of Huarco included omnivores, carnivores, herbivores, and detritus feeders. Some fish preferred the rocky intertidal zone and some the shallow waters of the continental shelf; still others were open-sea species that made infrequent visits to the coast. Some fish formed part of the grazing food chain; others found their niche in the detritus food chain.

Let us begin with the intertidal fish of the sea cliffs and rocky islands. The most herbivorous was the borracho or scaleless blenny (*Scartichthys gigas*), whose stomach contents were approximately 94 percent seaweed by weight. Apparently borrachos like their salad topped off with seafood, however, for the remaining 6 percent of their stomach contents consisted of small molluscs, amphipods, and porcelain crabs.

In the omnivore category was the chita or grunt (*Anisotremus scapularis*). While grunts like seaweed, algae were part of a broader spectrum diet including zooplankton (copepods, amphipods, and ostracods), bristle worms, crustacean eggs and larvae, and small fish.

Among the intertidal carnivores were the free-swimming pintadilla or morwong (*Cheilodactylus variegatus*); the trambollo or scaled blenny (*Labrisomus philippii*), which frequents kelp beds; and the pejesapo or clingfish (*Sicyases sanguineus*), which attaches itself to rocks by its sucker disc and eats every small mollusc and crustacean within reach. Both the pintadilla and the trambollo eat zooplankton, bristle worms, small crustaceans, and an occasional porcelain crab.

Beyond the intertidal zone one finds a whole series of free-swimming fish, divided by Vargas F. et al. (1999) into "plankton lovers," "crustacean lovers," and "small fish lovers." Among the plankton lovers are small, schooling fish like the sardine, the anchoveta, and the pejerrey or silversides (*Odontesthes regia regia*). All these fish can be present in huge numbers, so long as the cool Humboldt Current is the dominant water mass. The pejerrey eats foraminifera, copepods, amphipods, fish eggs, and the nauplius larvae of mussels. Sardines and anchovetas eat different proportions of many of the same planktonic species, and undergo multi-decadal cycles of dominance (to be described later).

Chief among the crustacean lovers are members of the Sciaenidae, or drum family. Three drums whose stomach contents were studied by Vargas F. et al. (1999) were the mismis (*Menticirrhus ophicephalus*), the lorna (*Sciaena deliciosa*), and the corvina (*Cilus gilberti*). Their love of crustaceans notwithstanding, all three species eat a varied diet that includes mussels and bristle worms along with the mole crabs, opossum shrimp, and amphipods. Mismis and lorna were judged to eat the widest variety of items, which may account for their abundance at Cerro Azul both in the Late Intermediate and the 1980s (Chapters 6 and 9).

The small fish lovers studied by Vargas F. et al. (1999) were the cabrilla or rock bass (*Paralabrax humeralis*), the cabinza or grunt (*Isacia conceptionis*), and the caballa or mackerel (*Scomber japonicus peruanus*). Among the small fish found in their stomachs were anchoveta and pejerrey, identifying them as Level 3 consumers.

Finally, let us turn to the fish that form part of the detritus food chain. The most obvious are the lisas or mullets (*Mugil cephalus* and *Mugil curema*). These fish literally filter organic detritus from the ocean floor, including phytoplankton (diatoms), zooplankton (foraminifera, copepods, and amphipods), and the eggs and larvae of larger creatures.

Sea catfish (*Galeichthys peruvianus*) also feed on the soft sandy floor of Cerro Azul Bay, eating not only marine worms but the larval and subadult forms of the anchoveta.

The left-eye flounder (*Paralichthys adspersus*) does not survive by filtering detritus; it preys on mole crabs, pejerrey, and anchoveta. However, it is adapted to lying half-buried in the sand of Cerro Azul Bay, so it is physically dependent on the soft detritus of the ocean floor.

Cyclic Change in the Ocean: The Longer Term

Like most of the world's upwelling systems, the Humboldt Current cannot be considered a "mature" ecosystem (Medina et al. 2007:36). It is more of a work in progress, "low in the efficiency of energy transfer" and featuring "short trophic chains." One can observe in the seas off Peru at least two cycles of change—both unpredictable—for which very different time scales apply.

It is the cycle of short-term change that is the best known, since its environmental signals are strong and its effects can be drastic. At unpredictable intervals, the cool waters of the Humboldt Current are overrun by warm water masses from the southern Pacific. This temporary warming of the ocean, which usually lasts no more than a year, is called the El Niño-Southern Oscillation (ENSO). Sometimes a warmer-than-average El Niño year will be preceded or followed by a year of colder-than-average upwelling, called La Niña in order to contrast it with El Niño. During our excavations at Cerro Azul, we experienced one strong El Niño year (1982–83) and two mild La Niña years (1983–84 and 1984–85).

More recently, however, oceanographers and marine biologists have identified cycles of longer-term change in the ocean off Peru. These cycles, which can last 15 to 20 years, are characterized by the numerical dominance of either anchovetas or sardines (Alheit and Ñiquen 2004, Ayón et al. 2011, Espinoza and Bertrand 2008; Espinoza et al. 2009, Gutiérrez et al. 2009, Swartzman et al. 2008). Specifically, anchovetas seem to have dominated from 1950 to 1970 and again from 1985 to 2004; sardines dominated from 1970 to 1985, a period that included the University of Michigan's excavations at Cerro Azul. Some oceanographers now propose to call the periods of anchoveta dominance "La Vieja" and the periods of sardine dominance "El Viejo," providing an analogy with El Niño and La Niña.

The Sardine-Anchoveta Cycles

Despite the fact that the multi-decadal cycles of sardines and anchovetas have less-easily-detected environmental signals than ENSO events, oceanographers consider them more ecologically significant. Anchovetas generally rebound from El Niño events in one or two years. The longer-term cycles, in contrast, "restructure the entire ecosystem from phytoplankton to the top predators" (Alheit and Ñiquen 2004:201). With the discovery of these longer cycles, Alheit and Ñiquen conclude, the highly publicized population crash of anchovetas during the 1970s can no longer be attributed to a strong 1972–73 El Niño; rather, the ultimate cause of that crash was the 1970–1985 cycle of sardine dominance (Fig. 1.7).

What lies at the heart of these cycles of sardine or anchoveta dominance? According to marine biologists, it is explained not only by changes in water temperature and plankton, but also by something called the "size-selective feeding hypothesis" (Ayón et al. 2011:220).

At the heart of that hypothesis is the fact that sardines are more efficient filter-feeders than anchovetas. The morphology of their gill rakers allows them to filter a wide size range of zooplankton, e.g. from 10 microns to 1.23 mm in diameter. Anchovetas, on the other hand, cannot filter zooplankton larger than 0.7 mm. As a result, a lot of anchoveta feeding is by direct biting, and they like their zooplankton to be less than 1.0 mm in diameter. That fact means that a lot of the competition between sardines and anchovetas is mediated by changes in the size distribution of zooplankton (Ayón et al. 2011:212).

Sardines are more ubiquitous than anchovetas, and have a predilection for subtropical offshore water masses (Swartzman et al. 2008). Their filter-feeding of phytoplankton is focused on dinoflagellates, with diatoms second in frequency. They eat smaller copepods and less krill than anchovetas, and get lots of their zooplankton by filter-feeding (Espinoza et al. 2009). It is the sardine's affinity for smaller zooplankton that attracts it to warmer water.

Anchovetas, on the other hand, tend to occur in the cooler waters of the Humboldt upwelling, which feature larger zooplankton. When anchovetas filter-feed on phytoplankton, they consume mostly diatoms; when they use their mouths for zooplankton, they take in larger copepods and more krill than sardines. Anchovetas tend to consume more copepods during the day and more krill at night, when euphausiids leave the protection of deeper and darker water (Espinoza and Bertrand 2014).

Now the reasons for the cycles of sardine and anchoveta

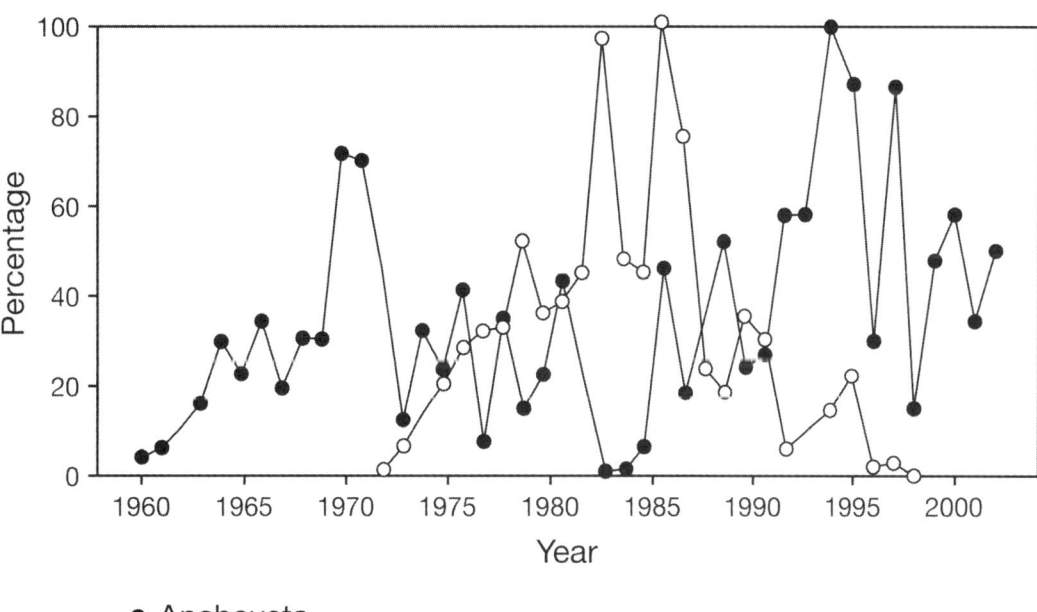

Figure 1.7. The waters of Peru's continental shelf undergo 15–20 year cycles, during which anchovetas and sardines alternate as the dominant small fish. The graph above records catches of the southern Peruvian/northern Chilean varieties of these two species. (Redrawn from Alheit and Ñiquen [2004: Fig. 3]).

dominance become clear. Cooler water (such as occurred from 1950 to 1970 and 1985 to 2004) means larger zooplankton and favors anchoveta; warmer water (such as occurred from 1970 to 1985) means smaller zooplankton and favors sardines.

These two fish also react differently when conditions are not right for them. Sardines are more willing to migrate long distances in search of small plankton; anchovetas are more likely to retreat to refuge areas for larger plankton (Swartzman et al. 2008). On the other hand, anchovetas have shown flexibility in changing their mix of prey species—for example, during 1996–2003, when copepods increased relative to krill (Espinoza and Bertrand 2014).

In the 1960s, Rojas de Mendiola et al. (1969) examined the stomach contents of anchovetas off Tambo de Mora in the Chincha Valley, to the south of Cerro Azul. They found that anchovetas were eating no less than sixteen species of diatoms; eight species of dinoflagellates; five types of crustaceans, including copepods, krill, and nauplius larvae; three types of protozoa; one tunicate; and the eggs of miscellaneous fish.

Obviously, we would like to know whether the Late Intermediate period at Cerro Azul was dominated by sardines or anchovetas. A paleoenvironmental study by Gutiérrez et al. (2009) has concluded that during the four centuries of the so-called "Little Ice Age" (A.D. 1400 to 1800), neither sardines nor anchovetas were especially abundant. No midden dating to the fifteenth century A.D., in other words, should reflect dominance by either fish. The importance of this fact will become apparent in our discussion of Features 6 and 20 of Cerro Azul (Chapter 9).

Cyclic Change in the Ocean: The Short Term

We now know that the short-term perturbations known as El Niño and La Niña are caused by stochastic shifts in the Pacific trade winds (Philander 1990). In a typical year these winds, created by our planet's west-to-east rotation, collect heat and water vapor off the Peruvian coast and blow it westward toward Indonesia, allowing the cool upwelling of the Humboldt Current to dominate the continental shelf.

Periodically, however, the trade winds relax, leaving the heat and water vapor in place (Huyer et al. 1987). Warming by advection soon competes with cool upwelling, and eventually the Humboldt Current is overtaken by tropical or semitropical water masses. The effects are widespread. Not only do the waters off Peru become warmer, but the western Pacific does not receive its usual input of heat and humidity from the trade winds. As atmospheric pressure rises in the eastern Pacific, it declines in the Indian Ocean from Africa to Australia.

Sir Gilbert Walker (1924) was the first to notice this seesaw relationship between atmospheric pressure in the Indian Ocean and the eastern South Pacific, which he called the Southern Oscillation. Forty years later, Bjerknes (1969) discovered a link between Walker's Southern Oscillation and the periodic warming of the eastern Pacific known to Peruvians as "El Niño."

Now that the causal links between these two phenomena have been established, we refer to this short-term warming of the Pacific as the El Niño-Southern Oscillation (ENSO). There are countervalent years when the trade winds are unusually intense, and the eastern Pacific becomes cooler than expected; these are known as La Niña years.

ENSO events, which usually last no longer than a year, have been recorded ever since the Spaniards arrived in Peru. There have been at least 20 such events since 1900 (Glynn 1988, Parsons 1970, Quinn et al. 1990). El Niño suppresses both the upwelling of the Humboldt Current and the removal of water vapor by the trade winds. The result is less nutrition for plankton and devastating rains on normally arid coasts.

While both Peruvian fishermen and farmers regard ENSO as a natural disaster, biologists consider that view an oversimplification. The 1982–83 ENSO (which was one of the twentieth century's strongest) did devastate the sardine and anchoveta harvest; on the other hand, it produced record numbers of tuna and bonito, which loved the warmer water. And while the rains associated with this ENSO did flood many agricultural fields, they also awoke a whole series of plants that have evolved a long dormancy between El Niño years (Weisburd 1984).

Ayón et al. (2008a, 2008b) studied the effects of the 1982–83 El Niño on the plankton of Peru's continental shelf. As ocean temperatures rose from their usual range (14°–18° C) to a subtropical 18°–27° C, copepods declined to one-sixth of their previous abundance and remained low until 1986–87. Soon, cold water phytoplankton was restricted to a few patches of coast (Caviedes 1984:274). Many anchovetas dove to depths of 100–150 m, where the water was cooler. Sardines (which are willing to migrate longer distances) swam southward rapidly until they reached the cooler waters off Chile. Some Pacific jack mackerel (which typically eat sardines and anchovetas) followed the sardines south, while others followed the anchovetas shoreward to refuge areas where they could still find large plankton. Silversides virtually disappeared from the Cerro Azul area and did not return until 1986. On the other hand, as Ayón et al. (2008a, 2008b) point out, the Peruvian scallop *Argopecten purpuratus* flourished during the 1982–83 ENSO because its larvae have higher survival rates in warmer waters.

A second powerful El Niño struck the Peruvian coast in 1997–98; it was followed by a La Niña in 1998–99. Aronés et al. (2009) have documented the effects of these short-term perturbations. The minima for zooplankton occurred when ocean temperatures rose to 25° C. Such was the therapeutic effect of La Niña, however, that afterwards the zooplankton rose to five times the density seen in the years preceding the ENSO of 1997–98.

As an indication of how rapidly sardine and anchoveta populations can rebound from ENSO events, Aronés et al. (2009:596) present the following Peruvian harvests. In 1988, following the powerful 1982–83 El Niño and two mild La Niña years (Fig. 1.8), Peruvian fishermen caught 3×10^6 metric tons of sardines. In 1994, they caught 9.7×10^6 metric tons of anchovetas. Both were records for Peru.

STATION: CHUCUITO — CALLAO

LATITUDE:
12° 03' 30" S.

LONGITUDE:
77° 09' 00" W.

PERIOD:
1981 — 1987

YEAR	JAN	FEB	MAR	APR	MAY	JUN	JUL	AUG	SEP	OCT	NOV	DEC
1981	14.8	16.6	16.4	17.3	17.3	16.5	15.2	15.2	14.8	14.4	14.8	14.9
1982	15.4	16.5	17.1	16.4	16.6	16.6	16.6	16.3	15.9	16.3	19.2	21.6
1983	23.2	23.3	23.2	23.9	24.0	24.2	19.8	17.6	16.0	15.9	15.8	15.5
1984	15.8	16.1	16.5	17.9	16.2	15.8	15.7	15.5	14.8	14.5	15.2	14.6
1985	15.1	15.4	15.5	15.1	14.7	14.8	15.0	14.7	14.3	14.3	14.2	14.3
1986	14.7	16.6	15.8	15.1	15.2	15.4	15.6	16.1	15.3	15.0	15.5	16.4
1987	17.5	28.9	20.8	19.4	17.9	17.6	17.2	16.3	15.8	15.8	15.8	16.2

Figure 1.8. Monthly ocean temperatures at Callao, Peru, during the years 1981–87. This period included the strong El Niño of 1982–83 and the two mild La Niña years of 1983–84 and 1984–85. (Source: Instituto del Mar del Perú)

Marine Birds

It is not only the fish living on sardines and anchovetas that suffer in El Niño years. It is also the Humboldt penguin (*Spheniscus humboldti*), the guanay cormorant (*Phalacrocorax bougainvillii*), the Peruvian booby (*Sula variegata*), and the Peruvian pelican (*Pelecanus thagus*).

Penguins and cormorants pursue fish by swimming after them, while boobies are plunge-divers who drop vertically from the sky (Chapter 11). The problem is that when anchovetas pursue cooler water to depths of 100–150 m, even the most determined plunge-divers cannot reach them; and when the sardines migrate south to Chile, the marine birds that prey on them are left behind. My research team was saddened by the number of birds that died on the beaches of Cerro Azul in 1983.

Ayón and Swartzman (2008) estimate that in the 1960s there were some 15 million cormorants, boobies, and pelicans living on the Peruvian coast. Since 1970, heavy commercial fishing has reduced their numbers by undercutting their food supply.

These birds, as we saw earlier, are consumers of Trophic Levels 3–4. Their role in the transfer of nutrients and energy, however, is more complex than this. All three genera roost on sea cliffs or rocky offshore islands for protection from predators, and with the passage of time their roosting places become covered with excreta. In any well-watered region this bird excreta, called *wanu* (Quechua) or guano (Hispanicized), would be washed away. The hyperaridity of the Peruvian coast, however, "inhibits both the microbiological breakdown of uric acid ($C_5H_4N_3O_3$) and the leaching of water-soluble ammonia (NH_3) and ammonium salts (NH_4) from guano deposits and makes them uniquely rich in nitrogen" (Cushman 2005:482).

Chemical analysis of a typical guano deposit reveals it to be 15 percent nitrogen, 9 percent potassium, and 3 percent phosphorus. Largely the product of digested anchovetas, guano is an excellent agricultural fertilizer, and would once have been available in enormous quantities. According to Ayón and Swartzman (2008:428), one cormorant can produce an average of 17.4 kg dry weight of guano annually, making the potential yield of 15 million cormorants, boobies, and pelicans a staggering 31,330 metric tons of nitrogen per year. This guano could then be spread on the fields of a kingdom like Huarco, increasing agricultural productivity.

Guano was present on the Peruvian coast long before agriculture began; one deep deposit on an island off Chincha has been radiocarbon dated to >19,000 years ago (Murphy 1951). Both Bernabé Cobo (1990:211) and Garcilaso de la Vega (1966[1609]:246) describe the Inca using coastal guano to fertilize highland fields. Garcilaso, in fact, says that the Inca ruler protected the guano birds by prohibiting his subjects from visiting their island homes during the breeding season. To be sure, not everyone was dependent on guano; the valleys of Chilca and Mala fertilized their crops with sardine heads (Garcilaso de la Vega 1966:247).

In his classic article on the history of guano, Kubler (1948:39) argued that use of the Chincha Island resources went back at least to the ninth century A.D. He assigned to the Moche period a red-on-white effigy vessel found 62 feet deep in a guano deposit on one of the islands. Kubler also illustrated a Moche vessel that shows boats visiting a guano island (Kubler 1948: Fig. 29).

For the Cañete Valley specifically, Kubler reports that Cerro Centinela was considered a guano source. He also cites José de Acosta's sixteenth-century report that guano from the central coast was used to fertilize fields in the "Lunaguaná Valley" (Acosta 1940, 1954; Kubler 1948:40). In short, it seems virtually guaranteed that the Kingdom of Huarco used guano as fertilizer.

Human agents, in other words, linked land and sea in the following unexpected feedback loop of calories, nitrogen,

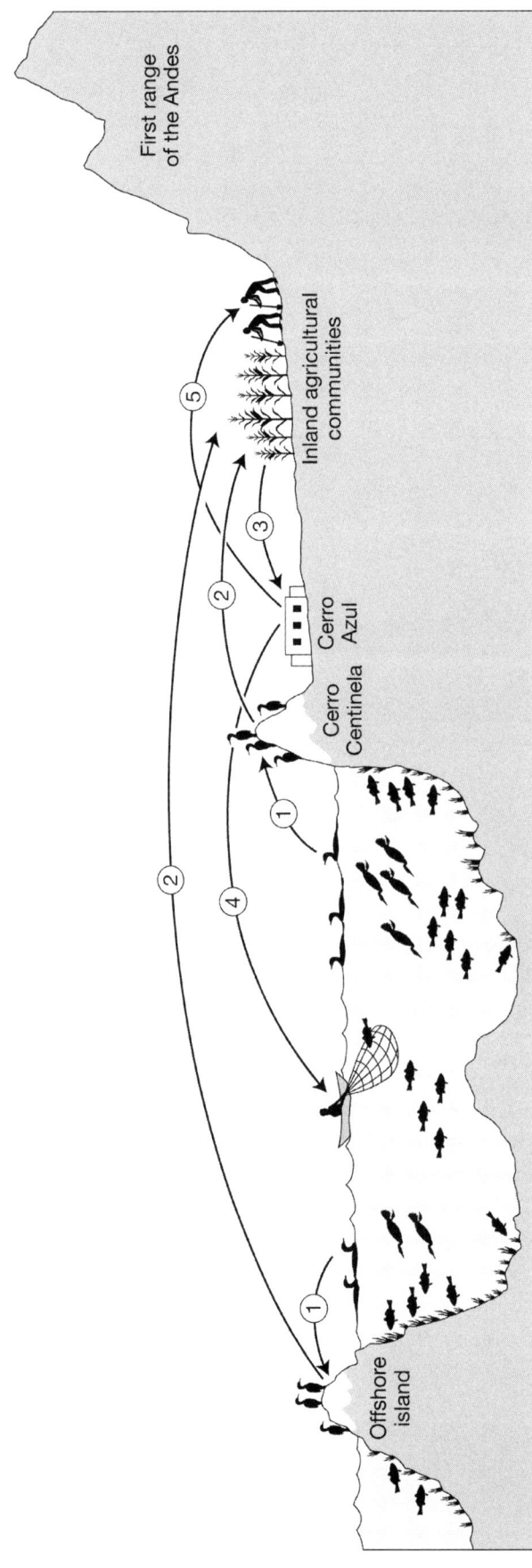

Figure 1.9. Diagrammatic cross-section of the Kingdom of Huarco, showing the way human intervention altered the movement of resources. (1) Sea birds eat anchovetas and deposit nitrogen-rich guano on sea cliffs and offshore islands. (2) Humans retrieve guano and use it to fertilize inland maize fields. (3) Much of the maize is converted to chicha in Cerro Azul's breweries. (4) Cerro Azul's nobles lavish chicha on the fishermen who harvest anchovetas for them. (5) Anchovetas harvested at Cerro Azul are exported to inland communities and paid as tribute to hereditary leaders.

potassium, and phosphorus. The phytoplankton of the Humboldt Current was eaten by zooplankton; both phyto- and zooplankton were eaten by anchovetas and sardines; the anchovetas and sardines were eaten by sea birds; the birds deposited their guano on offshore islands; humans dug up the guano and spread it on their maize and potato fields; their potatoes blossomed; and the nitrogen removed from the soil by the maize crop was restored. In ecological terms, Trophic Level 1 producers came to receive a boost from the excreta of Level 3–4 consumers (Fig. 1.9).

The Species that Visit in ENSO Years

As mentioned earlier, it is simplistic to consider only those species negatively affected by ENSO. Many other species, which thrive in warmer water, take advantage of El Niño to move into areas once dominated by the Humboldt Current.

These invasive species include invertebrates. Tarazona et al. (1988) monitored small animal life on the floor of Ancón Bay from 1981 to 1984. The 1982–83 El Niño increased the oxygen saturation of this bay's deep-water zone from <8.5 percent to >20 percent. As a result, the small mollusc species increased from 0 to 5; the small crustacean species rose from 0 to 10; and the species of bristle worms rose from 4 to 29.

As water temperatures rose on the Peruvian coast, many tropical species expanded south into Peru from their homes off Ecuador. For example, 5 species of tropical swimming crabs showed up in Peruvian waters (Glynn 1988). Shrimp (*Xiphopenaeus* sp.) showed up off the coast of Cerro Azul, an unheard-of occurrence in "normal" years. And while 12 species of cool-water fish declined, 19 warm-water fish both increased in numbers and invaded Peruvian waters (Weisburd 1984). The dorado or dolphinfish (*Coryphaena hippurus*) and the barrilete or skipjack (*Katsuwonus pelamis*), normally restricted to tropical seas, showed up on the Peruvian coast. Tuna, marlin, and swordfish, normally caught 50 km offshore, followed warm currents to the coast as well. The fishermen of Cerro Azul were used to seeing an occasional bonito (*Sarda sarda chiliensis*), but not the hundreds they saw during the El Niño of 1982–83.

As late as March of 1984 the fishermen of Cerro Azul were still catching the paloma pompano (*Trachinotus paitensis*), the pez dama or whale shark (*Rhincodon typus*), the dolphinfish, and plenty of shrimp. Then came two mild La Niña years, and the invasive tropical species were gone. Their place was taken by two drums, the lorna (*Sciaena deliciosa*) and the mismis (*Menticirrhus ophicephalus*), both of which flourished under the cooler conditions of La Niña (Appendix A). Little by little, all the usual species of the temperate Humboldt Current returned to Cerro Azul. Among the last to return was the pejerrey or silversides (*Odontesthes regia*), which had migrated to Chile during the peak of the 1982–83 El Niño. By February of 1986 it was being caught by the dozens (Marcus et al. in press), a good sign that conditions had returned to normal.

Summary

One of the most important conclusions to be drawn from an ecological study of ancient Huarco is that the sea cannot be treated as a mature ecosystem with predictable resources. The more we learn about the sea, the more evident it is that its history is one of stochastic changes at different time scales. The ocean masses off the coast of Peru are an unstable mix of temperate, semitropical, and tropical waters, each with its distinct assemblage of plankton. At intervals of 15–20 years the waters of the continental shelf may be dominated by anchovetas or sardines; at even less predictable intervals, they may suffer episodes of short-term warming (El Niño) or short-term cooling (La Niña). The episodes of short-term warming drive away both sardines and anchovetas; the episodes of short-term cooling restore the plankton and lure back the fish.

The cool Humboldt Current and the trade winds' constant removal of water vapor created the inland desert of the Kingdom of Huarco. Both the inland and marine ecosystems were also linked by human activity. When Late Intermediate farmers tapped the waters of the Río Cañete and brought them 24–35 km across the valley, they altered the distribution of mullet and crayfish, increased the number of places where fresh water entered the ocean, and produced a windfall of food for a whole series of bird species. When those same farmers dug up guano from Cerro Centinela and spread it on their fields, they were making their crops part of a nitrogen cycle that began with plankton and passed through sardines, anchovetas, cormorants, and boobies (Murphy 1920, 1923). The fishermen of Huarco took from the ocean thousands of times as many anchovetas as they could eat, and by drying them for export made them available to people for whom the sea was a foreign province.

2

Provenience and Context of the Plant and Animal Remains at Cerro Azul

Joyce Marcus

One goal of this volume is to reconstruct the subsistence economy of Late Intermediate Cerro Azul. To accomplish this goal, the authors of Chapters 3–20 rely on archaeological plant and animal remains.

Inevitably, the provenience of those remains—and their taphonomic context—affects the archaeologist's reconstruction. Like most sites, Cerro Azul had (1) *primary* contexts, in which objects appeared to be lying roughly where they had been used; (2) *secondary* contexts, in which objects had been moved only a short distance from their original provenience; and (3) *tertiary* contexts, where objects had been moved several times, and might even be mixed with objects from other proveniences (Marcus 2008:62).

My 2008 volume describes all the proveniences at Cerro Azul in detail, and the reader who seeks that level of detail should consult it. What this chapter provides is an abbreviated catalogue of proveniences and contexts at Cerro Azul, aimed at the reader who needs fewer details and prefers not to switch back and forth between two books.

Site Layout

The site of Cerro Azul occupies a defensible location not far from the bay of the same name. It lies in a protected saddle between the sea cliffs of La Centinela and El Fraile and the inland peak called Cerro Camacho (Fig. 2.1). Within that protected saddle lies a 200 x 400 m complex of tapia (poured clay) structures, contoured to rocky spurs and low rises in the natural terrain. The area enclosed by Cerro Azul's ancient defensive wall is somewhat larger than that, since that wall takes advantage of the eastern slopes of Cerro Camacho (Fig. 1.5).

In 1925 A. L. Kroeber (1937) located 10 large tapia-walled compounds at Cerro Azul; he assigned them letters A through J. Kroeber also identified a series of "small ruins," and put several of them on his map under the designation "SM." The 10 large compounds and the abundant small ruins surround an irregular central plaza, which is currently being excavated by Perú's Proyecto Qhapaq Ñan (Marcone Flores and Areche Espinola 2015).

My University of Michigan team excavated one of Kroeber's large compounds (Structure D) and one of his "small ruins" (Structure 9). Both buildings dated to the Late Intermediate period. We also excavated three Late Horizon buildings on Cerros La Centinela and El Fraile; these were designated Structures 1, 2, and 3 (Marcus et al. 1985).

Kroeber discovered that the quebradas on the lower slopes of Cerro Camacho also contained archaeological material, including Late Intermediate burials. He numbered these quebradas from 1 to 8a. My project conducted excavations in Quebradas 5, 5-south, 5a, and 6. We were impressed by the number of artificial terraces created in these quebradas by the Late Intermediate occupants of Cerro Azul.

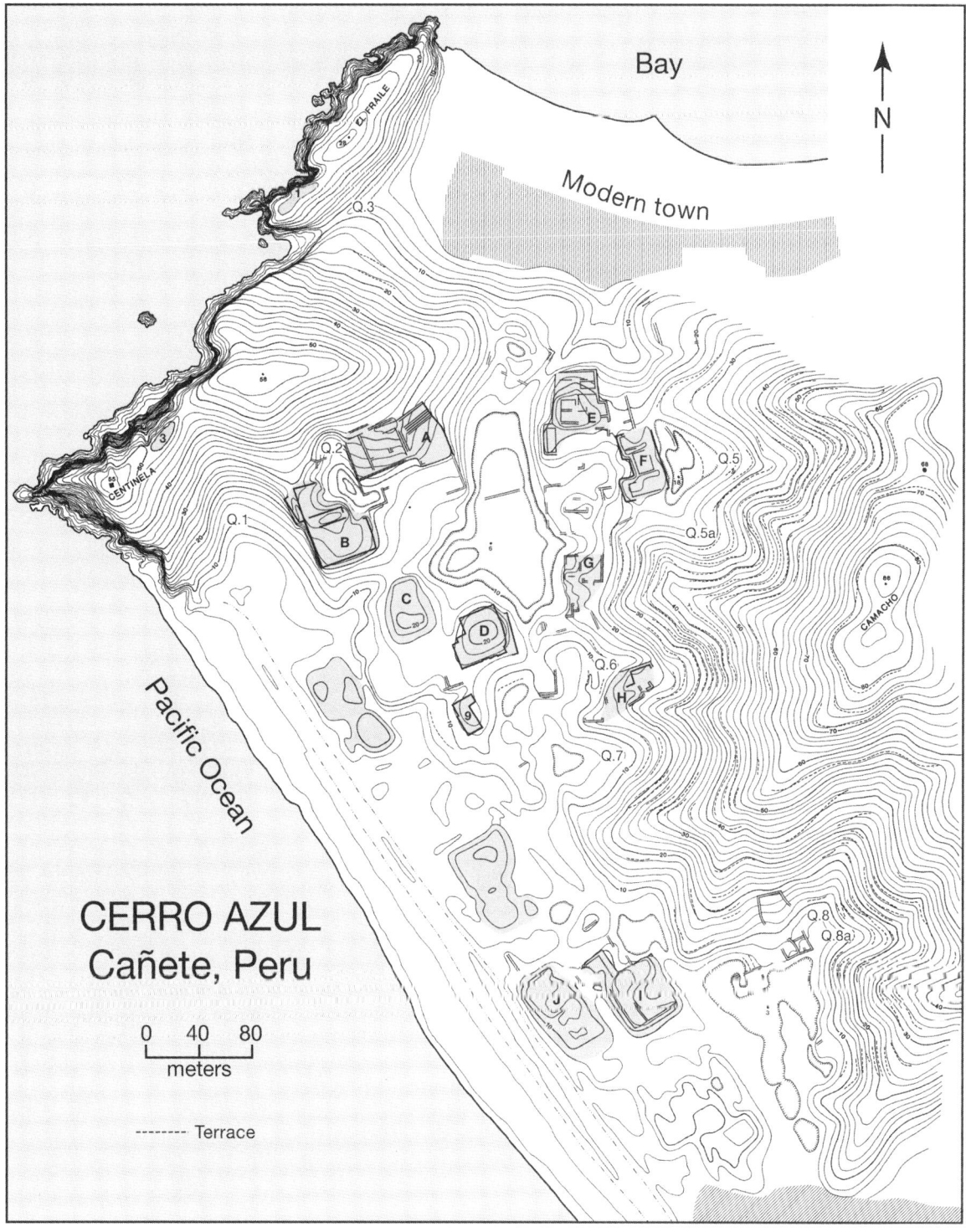

Figure 2.1. Contour map of the ruins of Cerro Azul, showing Structures A-J of Kroeber; Structures 1, 3, and 9 of the University of Michigan Project; and Quebradas 5–8a of Cerro Camacho. Elevations in meters above sea level (cartography by Charles M. Hastings; drafted by Kay Clahassey).

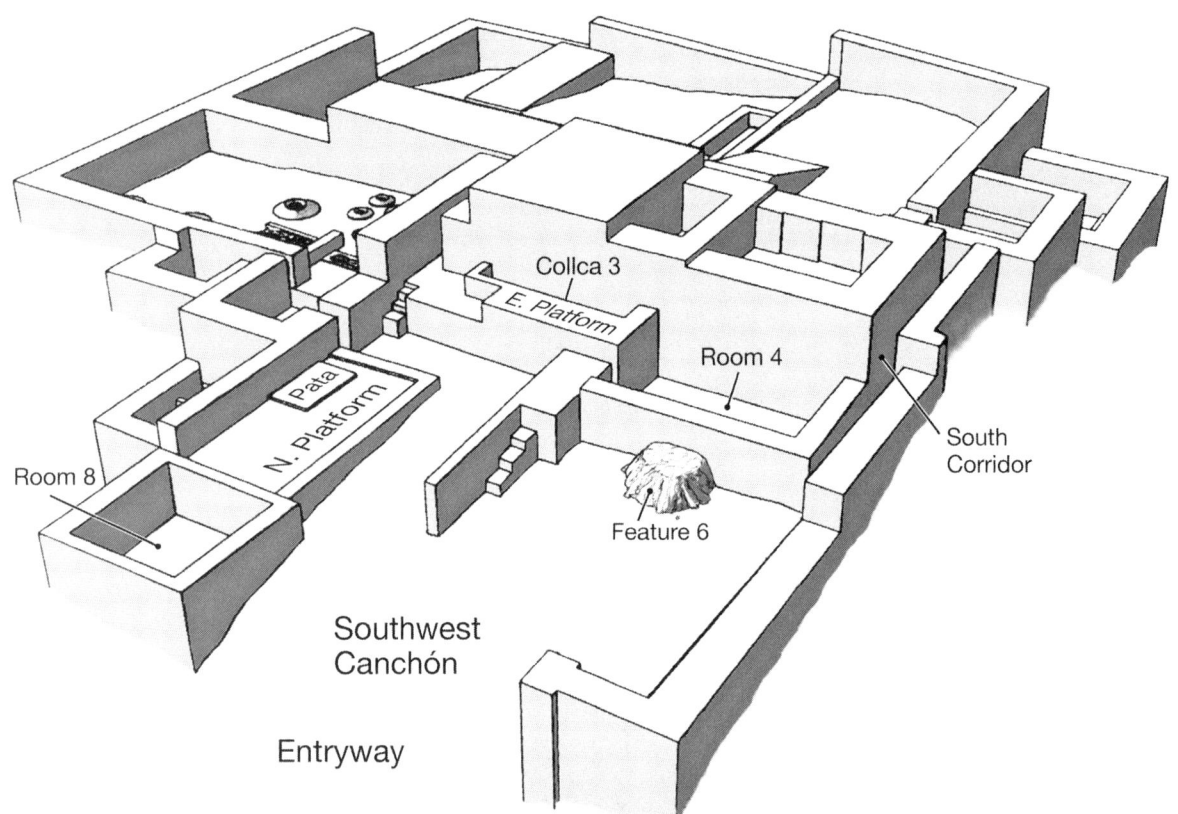

Figure 2.2. An artist's conception of the southwest quadrant of Structure D.

Structure D

Structure D turned out to be a 1640 m² tapia residential compound, divided into at least a dozen rooms and four canchones or open-air work spaces. We interpret this building as the residence of an elite family and its commoner-class support staff. We found clear living quarters with built-in sleeping benches; multiple *collcas* or storage bins; a large chicha brewery and/or kitchen; a guinea pig nursery; and rooms where thousands of anchovetas had been stored in layers of sand.

Entry into Structure D was through the **Southwest Canchón**, a large open-air work area (Fig. 2.2). Not long before the Inca conquest of Huarco and the abandonment of Structure D, the floor of the Southwest Canchón was swept and the resulting refuse piled up against the east wall of the canchón. We labeled this midden **Feature 6**, and consider its contents to be in secondary context. Had the building not been abandoned, the refuse comprising Feature 6 would undoubtedly have been removed and dumped elsewhere.

Feature 6 was one of our richest sources of plant and animal remains—maize, beans, fruits, shellfish, crabs, fish, guinea pigs, llama bones, and even human fingernail clippings (Fig. 2.3). This midden also contained hundreds of pellets of llama dung. That discovery let us know that llama caravans had periodically entered the Southwest Canchón, presumably to deliver agricultural products and carry away dried fish (Marcus 2008:84). We believe that such activities were overseen by an administrator whose "office" was the **North Platform** of the canchón. There he had been provided with a *llamkana pata*, or raised work platform (Marcus 2008: Figs. 4.4, 4.5).

While llama caravans could enter the Southwest Canchón, they could not continue on into the interior of Structure D. The **South Corridor**, a principal route to the interior, was too narrow for anything but single-file human traffic (Marcus 2008:130). This L-shaped corridor led to the small interior patio we called "Room 2."

Late in the occupation of Structure D, someone left a pile of refuse on the floor of the South Corridor. This refuse included hundreds of maize cobs, probably ones whose kernels had recently been used in the brewing of chicha. We consider these cobs to be in secondary context.

"**Room 2**"—which, upon excavation, turned out to be a small interior patio rather than a room—provided access to Rooms 1 and 3 (Fig. 2.4). The latter two rooms were originally part of an elite residential apartment, but had suffered earthquake damage (Marcus 2008:147–161). Because of this damage Rooms 1 and 3 were vacated, and we suspect that Rooms 5 and 6 were repurposed as residences. In order to provide better access to these rooms, an earthen ramp was added to the "Room 2" patio (Marcus 2008: Fig. 5.7).

"Room 2" provided little in the way of Late Intermediate flora and fauna. After the Inca conquest and the abandonment of Structure D, however, Late Horizon squatters built a kincha (wattle-and-daub) house in "Room 2" (Marcus 2008: Fig. 5.9). A midden associated with this house was designated **Feature 4** (Marcus 2008:140–141). This midden provided a sample of Late Horizon plant and animal remains, which we consider to be in secondary context.

Most of the northeast quadrant of Structure D was taken up by the **Northeast Canchón**, a large open-air work space (Fig. 2.5). Artifacts left in this canchón suggested that three of the activities carried out there were spinning, weaving, and embroidery (Marcus 2008:174). The floor in the northwest corner of the canchón had been badly worn, and we found a number of ash-filled hollows there. These ash deposits contained plant and animal remains from the final stages of the Late Intermediate, presumably in secondary context.

Set against the south wall of the Northeast Canchón was a storage bin we called **Collca 1** (Marcus 2008:169). We found clues that this bin's original purpose was to store loom parts, raw materials and artifacts for spinning and weaving, and other items for craft activity. Such materials could be considered virtually in primary context. By the time my excavation team found it, however, Collca 1 also contained debris from general housecleaning, including plant and animal remains. The latter items were treated as being in secondary conquest.

The **North Central Canchón** (Fig. 2.6) was the kitchen/chicha brewery for Structure D. Its features include two hearth trenches and the cavities left by at least nine beer storage vessels that had been set in the floor (Marcus 2008:180–191). The North Central Canchón produced only a small number of plant and animal remains, all probably in secondary context.

The **Northwest Canchón** had been so badly damaged by wind erosion that its original dimensions could not be determined (Marcus 2008:179). Only a few square meters of its original floor remained, making it difficult to determine whether any organic remains were in primary or secondary context.

The Individual Rooms in Structure D

Room 1 was originally designed to be part of the elite residential quarters in Structure D (Marcus 2008:147–153). At some point, however, it suffered severe earthquake damage and was vacated. Room 1 was then filled with sand and used for fish storage. Floral and faunal remains left on the original floor of the room may tell us something about foods eaten by the elite. Fish remains found in the later sand deposits, on the other hand, tell us something about drying fish for export. Still later remains in the fill above the sand were presumably in secondary or tertiary context.

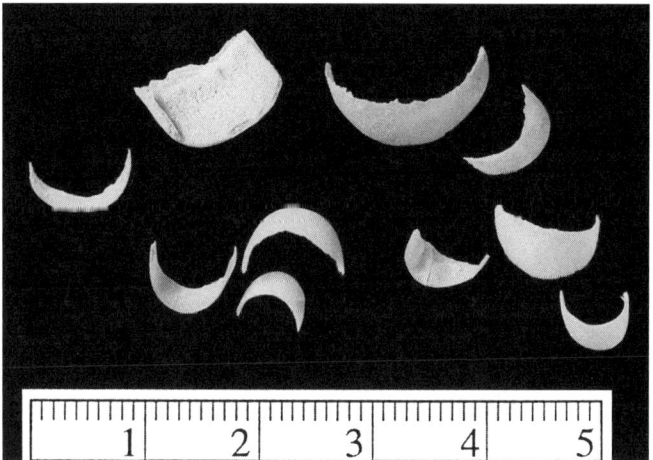

Figure 2.3. Human fingernail and toenail clippings from Feature 6.

Room 3 lay adjacent to Room 1 and had a similar history. Originally part of the elite residential quarters, it had a sleeping bench along one wall (Marcus 2008:153–161). The sherds associated with this stage of the room's history suggest that nobles ate here, making any plant and animal remains of relevance to reconstructing their diet. After this room suffered earthquake damage, it was filled with sand and converted to fish storage. The fish bones in the sand were considered to be in primary context. Very late debris, found stratigraphically above the sand layer, was treated as being in secondary or tertiary context.

Room 4 had been created by adding an L-shaped wall to one corner of the Southwest Canchón (Marcus 2008:110–125). It seems likely that the original purpose of Room 4 was for storage, since its walls were more than 1.5 m high and it had no door. It was not, however, devoted to fish storage; we suspect that it may have been a holding area for commodities coming into (or leaving) the Southwest Canchón.

The original floor of Room 4 lay 3.73 m below the arbitrary datum point established for the excavation of Structure D. At some point, however, Room 4 was partially filled in and a new floor was created at a depth of 2.23 m. The laying of this floor was accompanied by the sacrifice of a guinea pig. The fill between the old and new floors is considered to be in secondary or tertiary context.

Resting on the new floor of Room 4 was **Feature 5**, a mass of partially restorable pottery vessels that had collapsed and fragmented under the weight of the overburden. These vessels are assumed to have been in primary context when originally set on the floor. Sometime after their collapse, the room was filled with debris and Feature 5 remained hidden from view. Any flora and fauna in this later debris is considered to be in secondary or tertiary context.

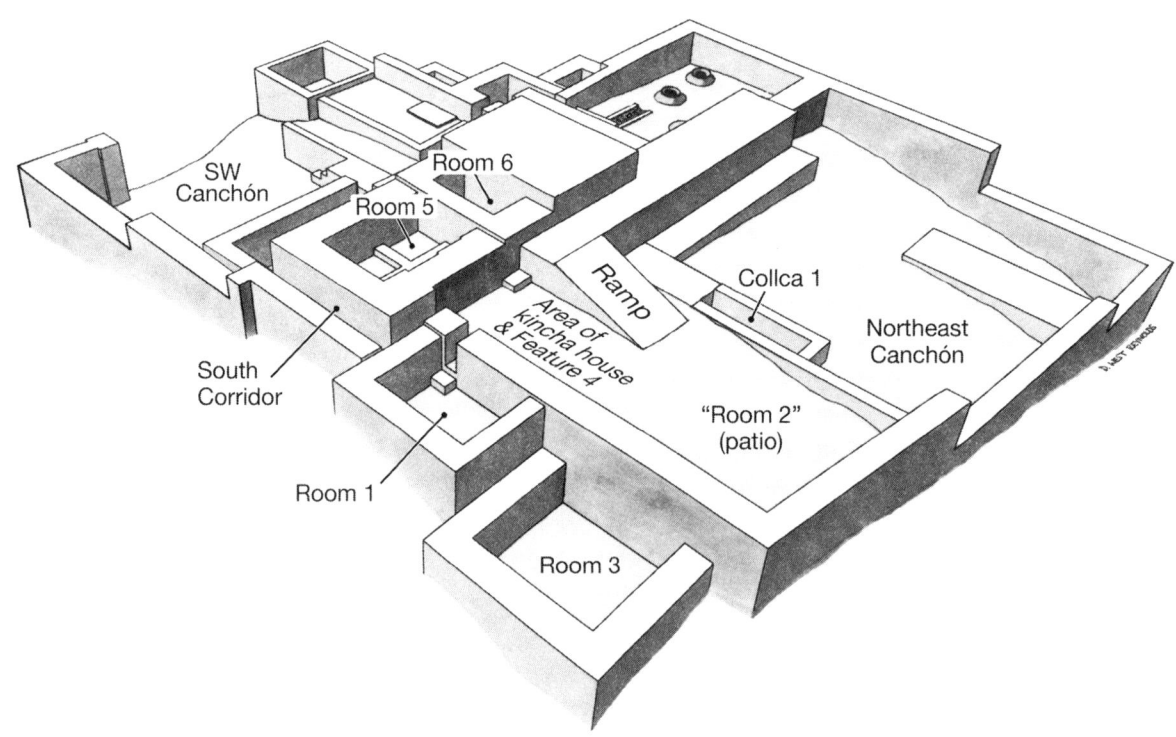

Figure 2.4. An artist's conception of the southeast quadrant of Structure D.

Room 5 had a history that was the reverse of that documented for Rooms 1 and 3. It had originally been designed for storage, and owing to its location near the center of Structure D it had suffered no earthquake damage. At some point in its history, however, it was converted to a residential room—perhaps after Rooms 1 and 3 had been damaged (Marcus 2008:161–167).

On more than one occasion after its transformation into a residential unit, Room 5 was provided with sleeping benches or platforms. Figures 5.32 and 5.33 of Marcus (2008) show the sequence of benches and the order in which they were added. Some organic remains were found in every architectural stage of the room. In fact, it appears that a guinea pig was sacrificed and buried in Room 5 around the time that the room was repurposed as a residence (see Chapter 19).

Room 6 lay adjacent to Room 5 and had a similar history (Marcus 2008:167–168). Originally designed as a doorless domestic storage room, it was later converted to a residential unit. Its original floor lay at a depth of 1.15 m below datum; that floor was later covered with fill, and a new floor was created at a depth of 85 cm below datum. Any flora or fauna in the fill between the floors was considered to be in secondary or tertiary context; remains resting on the new (upper) floor were considered residential debris, although not necessarily in primary context.

Room 7 had originally been connected to the Southwest Canchón by a narrow corridor (Marcus 2008:201–204). After this room suffered earthquake damage, however, the corridor was deliberately blocked. Late in its history, Room 7 became a convenient place to dump refuse. We suspect that most plant and animal remains included in this refuse came from activities in the Southwest Canchón, and should be regarded as in secondary context.

Room 8 was a doorless storage unit that seems to have been devoted to fish storage from the very beginning (Marcus 2008:128–130). This room had more than a meter of fine, clean, greenish-colored sand in it when discovered, and we screened its contents through both 1.5 mm and 0.6 mm mesh screen.

It became clear that thousands of anchovetas had been stored in layers in this sand so that the drying process would be

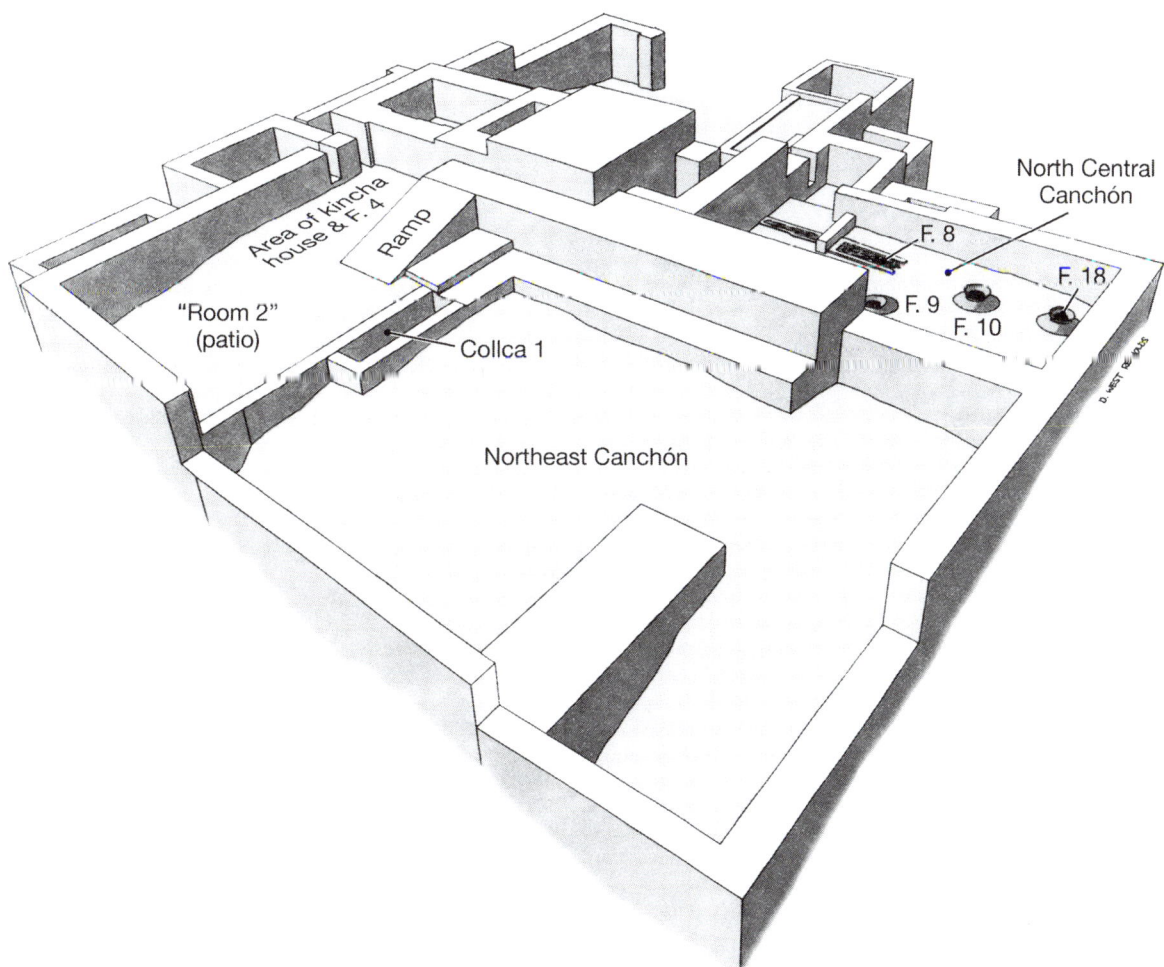

Figure 2.5. An artist's conception of the northeast quadrant of Structure D.

carried to completion (see Chapter 8). Where any small fish had accidentally touched the tapia walls and floor of the room, the residual moisture left by marine fog caused patches of skin and scales to stick to the clay surfaces.

While a number of earthquake-damaged rooms in Structure D had been converted to anchoveta storage over the years, Room 8 had apparently always had this role. We therefore consider the fish remains in this room to be in primary context. Any extraneous or unrelated organic remains, to be sure, might have been accidentally introduced into Room 8 while it was being filled or emptied.

Rooms 9 and 10 amounted to a guinea pig nursery (Marcus 2008:207–210). This area had originally been one room, but was later divided into two by the addition of a crude provisional wall of tapia blocks. The western room of the two new units became Room 9; the eastern unit became Room 10. Room 9, whose floor was covered by a mixture of matted grass and rodent dung pellets, was clearly the room where the guinea pigs had been kept. Room 10, whose floor was covered by a mixture of grass, small leafy herbs, and maize stalks, leaves, and loose kernels, was evidently where the food for the guinea pigs had been stored (see Chapter 19).

We consider any guinea pig remains in Room 9 to be in primary context. After Structure D was abandoned, however, wind erosion caused debris from the Southwest Canchón to drift downslope into Rooms 9 and 10. Any plant and animal remains included in this debris should be regarded as in secondary or tertiary context.

Room 11 appears to have served originally as a pantry or storage room for the North Central Canchón kitchen/brewery (Marcus 2008:204–206). A doorway between Room 11 and the Southwest Canchón would have allowed supplies for the kitchen/brewery to be carried there directly from the canchón after they arrived. Late in the occupation of Structure D, however, that doorway was deliberately blocked with tapia, and Room 11 then became a convenient place to dump refuse. Any organic remains in this refuse should be considered as in secondary context.

Room 12 was a small unit opening into the Northwest

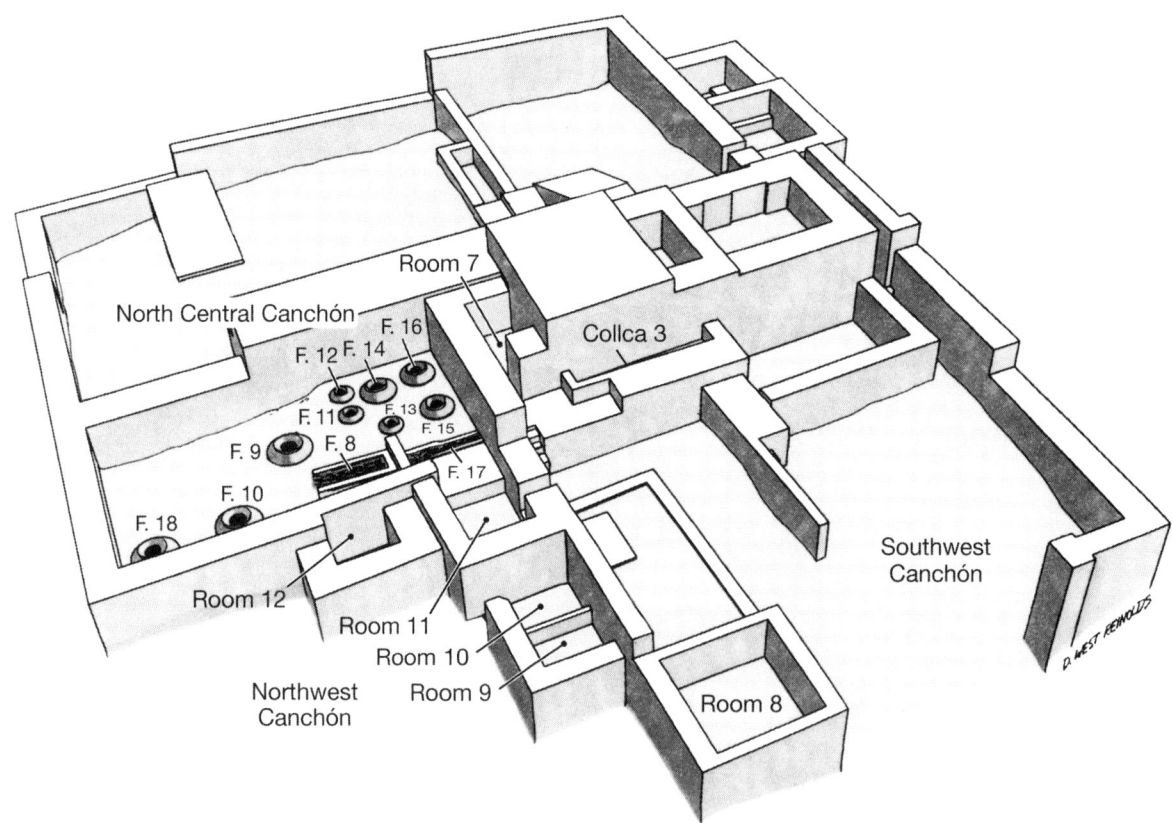

Figure 2.6. An artist's conception of the northwest quadrant of Structure D.

Canchón (Marcus 2008:210). It was very badly wind-eroded, and any organic debris found in it should be regarded as in secondary or tertiary context.

Feature 3

Feature 3 was an isolated feature, found while sweeping the surface of Structure D prior to excavation. It was found near the exact center of the building, and appeared to be the debris from an offering made after Structure D had been abandoned (Marcus 2008:212). The offering included maize cobs and a large number of burned camelid bones, probably in secondary context (see Chapter 17).

Structure 9

Structure 9 was an example of the type of building Kroeber (1937) called a "small ruin." It lay only 22 m southeast of Structure D, and its ceramics indicate that it was contemporaneous with the latter (Marcus 2008: Chapter 8).

Structure 9 turned out to be an L-shaped, multiroom tapia building, too eroded to measure with precision but covering roughly 290 m^2. Access to its interior was through the **South Entryway**, which was flanked by storage rooms (Fig. 2.7). This entryway led to a small patio, on the north side of which stood a low platform bearing a kincha house. We believe that this house was the residence of the commoner-class overseer who supervised activities in the building. The most important of those activities was the drying and storing of fish, primarily anchovetas. To this purpose, roughly half the rooms in the building had been filled with sand like that found in Room 8 of Structure D.

I interpret Structure 9 as a small storage facility whose administrator answered to one of the noble families involved in exporting anchovetas. Like Structure D, it had seen more and more of its rooms converted to fish storage over time, sometimes following earthquake damage.

Just outside Structure 9, and banked up against its east wall, was a midden containing thousands of plant and animal remains. My assumption is that this midden—**Feature 20**—contained household refuse produced by the family of the overseer in the kincha house (Marcus 2008:249–250). Feature 20 therefore provides us with an opportunity to compare the plant and animal remains from a commoner family with those from an elite

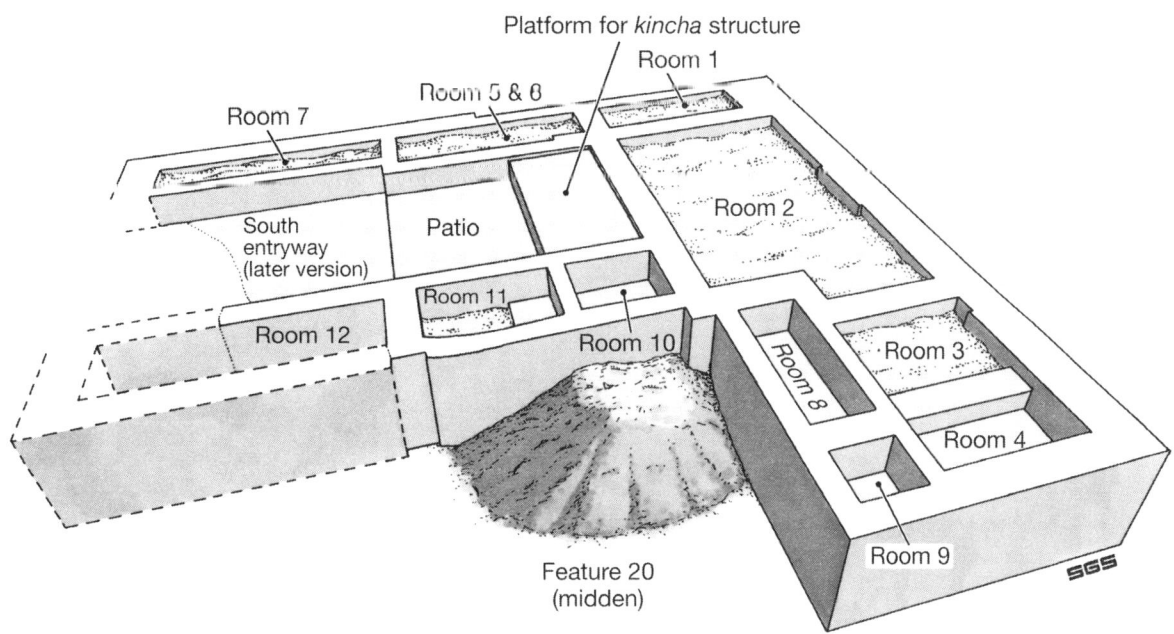

Figure 2.7. An artist's conception of Structure 9 during a late stage of its occupation, when many of its rooms had been modified or converted to fish storage units.

residential compound (Feature 6). The organic remains from Feature 20 should be regarded as in secondary context. Had the building not been abandoned, the refuse comprising Feature 20 would undoubtedly have been carried away and dumped elsewhere.

The Individual Rooms in Structure 9

The South Entryway was a long, narrow space between storage rooms. Its original floor lay at a depth of 1.77 m below the arbitrary datum point established for Structure 9. At a later date, the entryway was widened and a new floor established at a depth of 99 cm below datum (Marcus 2008:220). The fill dirt between the two floors covered up not only the old entryway but also **Room 13**, a narrow storage unit on its west side (Marcus 2008: Fig. 8.2). Plant and animal remains associated with both stages of the South Entryway were considered to be in secondary context.

Room 1 was a small unit whose original function is uncertain. It had suffered considerable earthquake damage, after which it was filled with sand and used for fish storage (Marcus 2008:223–224). This room was a great source of anchoveta bones, scales, and skin, all in primary context.

Room 2 was the largest in Structure 9, and may have been the original living quarters of the building's overseer (Marcus 2008:225–226). Among its features were a sleeping bench and a possible chicha storage vessel. At some point, however, the room's function had changed. Some 1.29 m of sand were poured into the room and it was filled with dried anchovetas, whose bones we considered to be in primary context. It may have been at this point that the former occupant of Room 2 moved to a kincha house on a nearby platform.

We found no earthquake damage in Room 2. Its conversion to fish storage, therefore, may simply represent an effort to increase the building's storage capacity by more than 50 m³.

Rooms 3 and 4 had formerly been parts of one large room and were created by the addition of a later wall (Marcus 2008:227–228). Room 3 had been filled with sand and used for anchoveta storage. Although the anchovetas were presumably in primary context, Room 3 had later become a convenient place to dump refuse; this later refuse was regarded as in secondary context. Room 4—badly damaged by seismic activity—also contained a modest amount of sand, but did not show as serious a commitment to fish storage as Room 3.

Rooms 5 and 6 had at one point been a single long, narrow room (Marcus 2008:233–241). At some point, this room was divided into two units by the addition of two loose tapia blocks (Marcus 2008: Fig. 8.25). Both rooms had then been filled with sand and used for fish storage.

After their use as fish storage units was over, both rooms became convenient places to throw trash. Thus, while the anchoveta remains in the sand were in primary context, any debris in the uppermost strata of Rooms 5 and 6 can be regarded as in secondary context.

Room 7, immediately to the south of Room 6, was a long narrow room that had not been divided into smaller units. This room had apparently been a fish storage room from the very beginning. The small fish bones in its green sand were in primary context. Like so many rooms, however, toward the end of its history Room 7 had been used as a convenient place to dump refuse, which we treated as a secondary deposit.

Room 8 had never been used for fish storage. Its contents included numerous sherds from large, partially restorable storage and cooking vessels. My assessment of this collection was that "even though it may be a secondary deposit, it had not been moved far from its original place of use" (Marcus 2008:229). The same would be true of any plant and animal remains found among the sherds. It is possible that this room contained material discarded by the occupants of the kincha house on the platform nearby.

Room 9 was a small storage unit that produced nothing of note (Marcus 2008:233).

Room 10 was a small unit that produced nothing but windblown dust and disintegrated tapia (Marcus 2008:246).

Room 11 had originally been a "walk-in" storage unit, with a tapia bench occupying its north half (Marcus 2008:246). At a later stage of its history, this room had been filled with sand and converted to fish storage. We fine-screened the sand from this room, recovering considerable amounts of fish bone. The anchovetas and sardines in the sand were presumably in primary context. There were, however, traces of larger fish such as corvina and bonito; whether the occasional bones of larger fish were also in primary context, or had been introduced accidentally when the room was being filled or emptied, is uncertain.

Room 12 had never been used for fish storage (Marcus 2008:247–248). It contained two broken but partially restorable pottery vessels that may actually have been stored in the room.

Room 13 was so thoroughly covered up during the process of reflooring the South Entryway that it proved impossible to separate its contents from those of the entryway.

The Quebradas of Cerro Camacho

The quebradas of Cerro Camacho were terraced nearly to the summit of the hill (Marcus 2008: Chapter 9). Many of these terraces (but not all) consisted of midden debris. They appear to have formed when basketload after basketload of refuse was carried away from Cerro Azul's residences and dumped in the quebradas (Marcus 2008:260–263). Needless to say, plant and animal remains in such middens were in tertiary context, having been moved far from their original place of origin (and perhaps having been moved multiple times). We believe that Feature 6 (in Structure D) and Feature 20 (in Structure 9) both represent refuse that would eventually have been carried off to the quebradas, had their respective buildings not been abandoned after the Inca conquest of A.D. 1470.

Not every terrace, however, was pure midden. Some terraces (such as Terrace 9 of Quebrada 5a) witnessed the accumulation of so many meters of refuse that they became convenient places to install burial cists. Others (such as Terrace 16 of Quebrada 5) became places where large hearths or earth ovens for outdoor feasting could be excavated. Still others (such as Quebrada 5-south) were seen as appropriate places for building tapia storage facilities. We suspect, in fact, that we have only scratched the surface of the uses to which the quebradas were put.

Quebrada 5, Terrace 16

Our excavations on Terrace 16 of Quebrada 5 reached a depth of 2.2 m (Marcus 2008:264–272). We defined nine stratigraphic levels, numbered from G (oldest) to A1 (youngest). These strata revealed two phases of activity. Early in the Late Intermediate period, the terrace was used for the dumping of refuse from one or more of the tapia compounds. Then came a period of disuse, during which sterile deposits formed on the terrace. At a later stage of the Late Intermediate period, Terrace 16 was used as a place to roast sea lions (Chapter 12).

As Figure 2.8 shows, Stratigraphic Zones F to D1 were made up predominantly of ashy midden deposits. Plant and animal remains from these levels were in tertiary context. Zones D1 and C were separated by a layer of mussels, also presumably in tertiary context. Zones C and B were largely sterile.

Late in the history of the terrace, it became the venue for a sea lion "cookout." The cooking was done in Feature 22, a stone-lined hearth. Sea lion bones and false abalones were found not only in the hearth, but also in Zone A2, a layer of refuse associated with it. These remains were considered to be in primary context, or close to it.

Quebrada 5-south

Our excavations in Quebrada 5-south (Marcus 2008:272–283) produced a number of surprises. There was midden material there, to be sure, and a Late Intermediate burial cist (**Structure 12**) had been sunk into the midden until its floor touched bedrock. Although Structure 12 had been looted, the grave robbers had left behind anything they did not consider commercially valuable. Since it appeared that most of the plant and animal remains in the looted cist had been left there as food for the afterlife, we considered them to have been shifted from primary to secondary

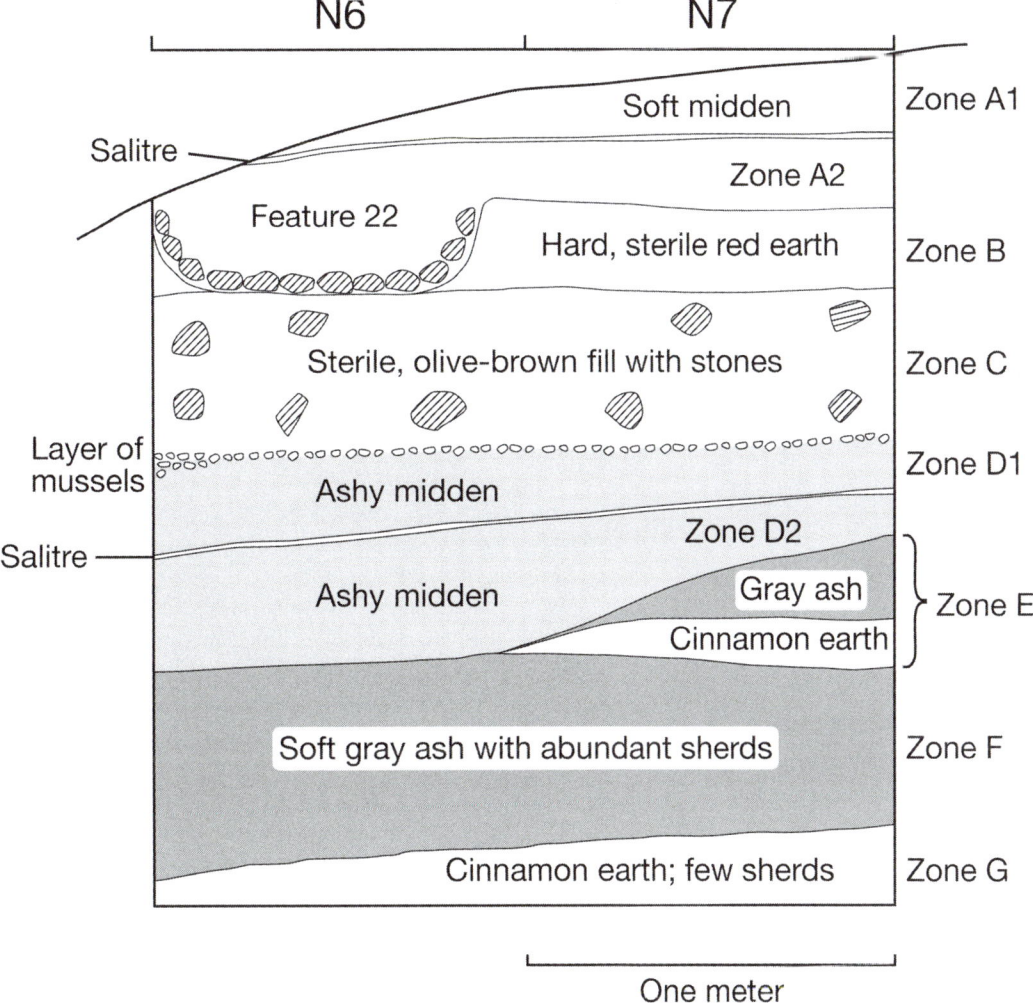

Figure 2.8. A profile of the east wall of Squares N6 and N7 in Terrace 16, Quebrada 5, Cerro Azul (after Marcus 2008: Fig. 9.7).

context by the looters. Organic materials in the adjacent midden, on the other hand, were considered to be in tertiary context.

The most unexpected discovery in Quebrada 5-south was **Structure 11**, a multiroomed tapia building contoured to the slope of the quebrada. It appeared that this structure had been built to store vegetal foods, and one of its rooms still contained both *Canavalia* and *Phaseolus* beans, many still in their pods. My workmen were not surprised by this. They informed me that this area, located in the broad mouth of Quebrada 5, was far enough inland so that any crops stored in it would be less damaged by salt fog than those stored in Structures A through J.

We did not have the time or money to excavate Structure 11 completely; we contented ourselves with exposing three rooms. Room 3 consisted of one long continuous unit, while Rooms 1 and 2 were part of an east-west row of small storage rooms (Marcus 2008: Fig. 9.16). We considered the plants left on the floor of these rooms to be traces of the crops stored there, and hence in primary context.

At some points in the building's history, someone built a kincha house in Room 3. Whether the occupants were squatters, or simply the family of the storage facility's overseer, is uncertain. We treated any plant or animal refuse associated with the kincha house as being in secondary context.

Quebrada 5a, Terrace 9

Terrace 9 of Quebrada 5a turned out to be 2.6 m deep and stratigraphically complex (Fig. 2.9). Its oldest and deepest layer, Stratigraphic Zone C, was a deposit of brown domestic refuse that antedated the occupation of Structures D and 9 (Marcus 2008:286–290). Zone C produced acacia posts and cane wall fragments from a destroyed kincha house. While these house remains may have been in secondary context, the midden debris below them was considered to be in tertiary context.

Stratigraphic Zone B was a largely undisturbed midden, over a meter thick, whose ceramics overlapped with those from the earliest occupation of Structures D and 9 (Marcus 2008:290–293). This midden's plant and animal remains were in tertiary context and contrasted with those of Feature 6 (in Structure D) and Feature 20 (in Structure 9). Our overwhelming impression was that Zone B represented the refuse from commoner households, and that those households were not involved in drying large numbers of anchovetas for export. The dominant mollusc in Zone B was a small coquina clam that is still readily available in the sandy beaches around Cerro Azul Bay (Chapter 3). The midden's range of fish species was limited, and consisted mainly of medium-sized drums, grunts, mullets, and mackerels (Chapter 9). Its plant remains were lacking in most of the tropical fruits desired by the Cerro Azul elite (Chapter 13). Overall, Zone B convinced us that we had yet to sample the full range of commoner households at Cerro Azul.

Stratigraphic Zone A, the youngest and uppermost layer on the terrace, was a midden whose ceramics dated to the final stage of the Late Intermediate. This stratum had been used as the matrix for at least four Late Intermediate burial cists, Structures 4–7. In some cases (for example, Structure 5), it appeared that the mortar holding the stones of the burial cist together was composed of ashy Zone B midden mixed with water. All the plant and animal remains in this midden—and in the mortar—were treated as being in tertiary context.

All of the burial cists we found were totally or partially looted, and we found other looted burials throughout Zone A. Fortunately, however, we also discovered a number of unlooted burials, some of them hidden by the looters' own backdirt. Obviously, any plant or animal offerings found with these unlooted burials could be treated as in primary context. Included were gourd bowls, complete guinea pigs, complete sardines, whole ears of maize, and portions of *ch'arki* or dried camelid meat. In addition, the abdominal cavities of some mummified burials produced coprolites with identifiable plant and animal remains (Chapter 20). We did our best to salvage whatever data we could, even from the partially looted burials.

Quebrada 6, Terrace 11

I chose to excavate Terrace 11 of Quebrada 6 because of its distinctive coal-black color. From a distance it reminded me of some of Junius Bird's exposures at Huaca Prieta (Bird 1985). There, Bird and his colleagues surmised that the black color was caused by a high quantity of rancid fish oil. What we found was that Terrace 11 appeared to include debris from one of the large tapia compounds—almost certainly, one of the elite compounds drying anchovetas for export. This debris featured maize cobs, mussels, clams, beans, canes, bulrushes, fish bones, and pellets of llama dung. In many ways, Terrace 11's contents resembled those from Feature 6 in Structure D (Marcus 2008:295–299).

Where dry, the Terrace 11 midden was dark brown; where dampened and hardened by salt fog at the southern limits of the terrace, it had turned the coal-black color that attracted my attention. I concluded, therefore, that the black color had resulted from the action of salt fog on rich organic remains, including fish oil. The fish bones in the midden, as well as all other plant and animal remains, were considered to be in tertiary context.

Quebrada 6, Terrace 12

Terrace 12 lay just downslope from Terrace 11 (Marcus 2008:308–312). Its uppermost stratigraphic layer, in fact, appeared to be brown-to-beige midden material that had slid down from Terrace 11 (Marcus 2008: Fig. 9.51).

The earlier deposits on Terrace 12, however, were different from those of any other terrace we excavated. Our excavation exposed a looted burial cist, a virtually empty storage cist, and an ash-filled earth oven. These three features had not been excavated into a preexisting midden. Rather, they were held in place by

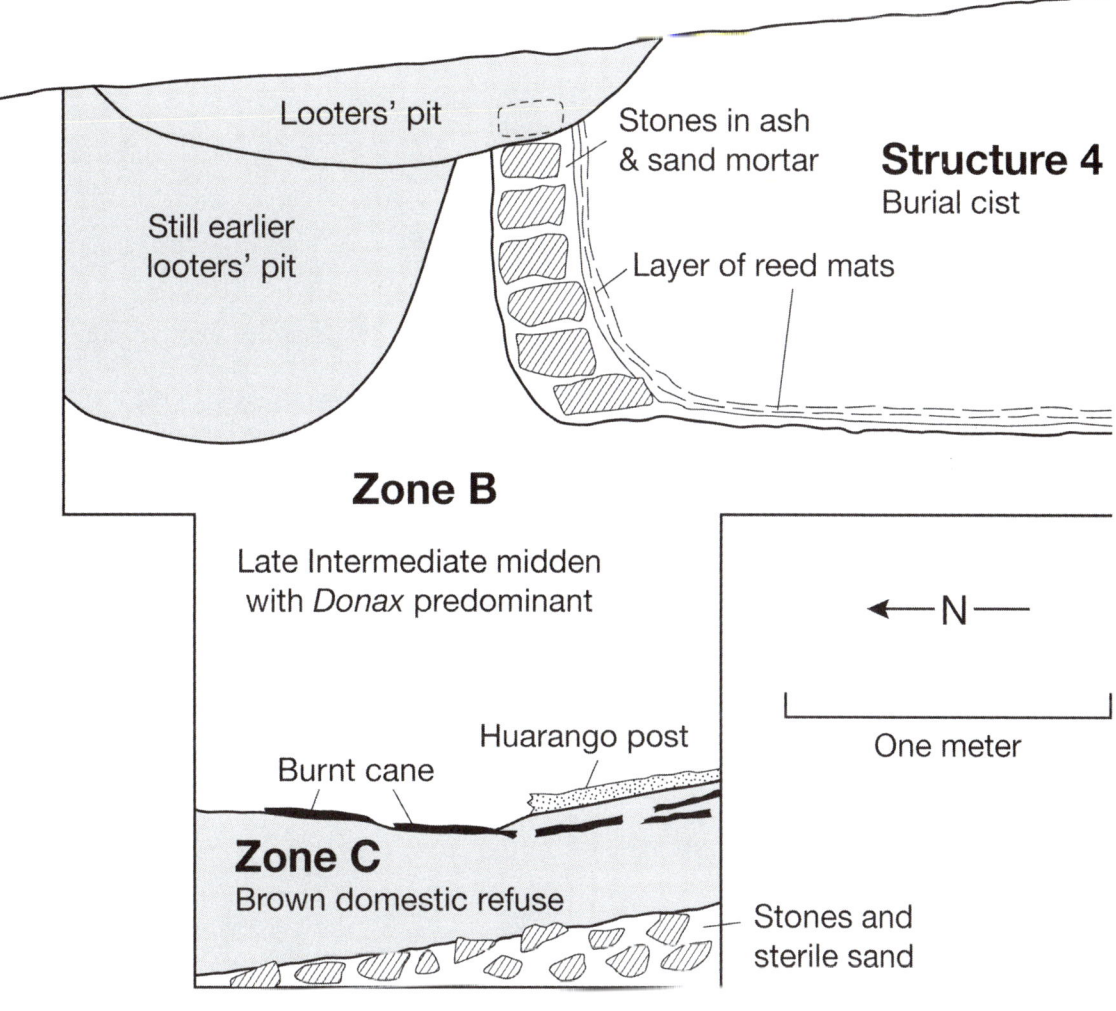

Figure 2.9. The north-south cross-section of a deep sounding into Terrace 9, Quebrada 5a, Cerro Azul. Structure 4 began in Stratigraphic Zone A and intruded slightly into Zone B (after Marcus 2008: Fig. 9.28).

a 40-cm to 60-cm-thick matrix of beach gravel that had been brought to Terrace 12 for that very purpose.

Terrace 12 was not a particularly productive venue for plant and animal remains. **Structure 10**, the storage cist, produced bits of decomposed netting, cordage, and coarse fibrous plants, all of which could have been the remains of containers or packing material for whatever had once been stored in the cist. The cist itself had been re-sealed with a puddled clay cap after being nearly emptied.

While some of the packing material from Structure 10 may have been in primary or secondary context, most of the meager organic remains from Terrace 12 were considered to be in tertiary context.

Summary

Plant and animal remains left in their original primary contexts were relatively rare at Cerro Azul. Among those rare discoveries were complete guinea pigs, buried where they had been sacrificed; anchoveta bones, left in the sand-filled rooms where fish had been dried; sea lion bones, left in the hearth where the animals had been cooked; and food for the afterlife, left with undisturbed burials.

More common at Cerro Azul were plant and animal remains in secondary context, moved a relatively short distance from their original place of use. Included were the leftovers from a meal, tossed into an abandoned or earthquake-damaged room nearby; piles of organic debris that had never been removed from a building, owing to its abandonment; bones from sea lion flippers, trimmed off in the process of using the animal's skin to create an inflatable watercraft; and empty maize cobs, tossed aside after their kernels had been used to make chicha.

Finally, we found thousands of plant and animal remains in tertiary contexts at Cerro Azul. These remains had been moved so far (or so often) that their original proveniences could not be reconstructed. The clearest examples of such tertiary contexts were the middens on the quebrada terraces of Cerro Camacho. They contained refuse swept from the floor of a residential compound, storage facility, or kincha house, transferred to a basket, carried to a quebrada, and dumped.

To be sure, we detected that certain terrace middens featured commoner household debris, while others held refuse from the elite compounds engaged in drying fish for export. We saw no way, however, that we could confidently identify the building (or buildings) from which any terrace midden's contents had come. Such taphonomic considerations played a role in our interpretation of the plant and animal remains from Cerro Azul.

Part II

How Cerro Azul Made Use of Trophic Levels 2, 3, and 4

3

The Collection of Shellfish

Kent V. Flannery and Joyce Marcus

The University of Michigan's excavations at Cerro Azul produced nearly 10,000 specimens of marine molluscs. It does not appear, however, that the Late Intermediate inhabitants of the site concentrated as heavily on shellfish as they could have, nor did they utilize all the locally available species (see Alamo Vásquez 1973; Alamo Vásquez and Valdivieso 1987, 1997; Paredes 1974a, 1974b; Peña 1970; Valdivieso Milla 1987; Valdivieso Milla and Alarcón 1983; Vegas Vélez 1968; Vélez Diéguez 1980). Among the 14 species of molluscs they collected were chitons, keyhole limpets, false abalones, sea snails, large and small mussels, and a variety of clams (Table 3.1).

In our identification of the living and prehistoric molluscs of Cerro Azul we benefitted from an excellent type collection, assembled over the years by our late colleague Duccio Bonavia. We were also assisted by Mario Peña of the Universidad Nacional Agraria "La Molina," and the malacology staff of Peru's Instituto del Mar. Special thanks go to Dr. James McLean of the Los Angeles Natural History Museum for identifying our specimens of *Mulinia edulis*, a clam that seems no longer to be at home in the Cerro Azul region.

Table 3.1. Marine Molluscs Recovered from Archaeological Contexts at Cerro Azul.

Chitons	*Enoplochiton niger*
Keyhole limpets	*Fissurella crassa*
	Fissurella limbata
False abalones	*Concholepas concholepas*
Sea snails	*Tegula atra*
	Calyptraea trochiformis
	Thais chocolata
Mussels	*Aulacomya ater*
	Choromytilus chorus
	Semimytilus algosus
	Perumytilus purpuratus
Coquina clams	*Donax obesulus*
Wedge clams	*Mesodesma donacium*
White clams	*Mulinia edulis*

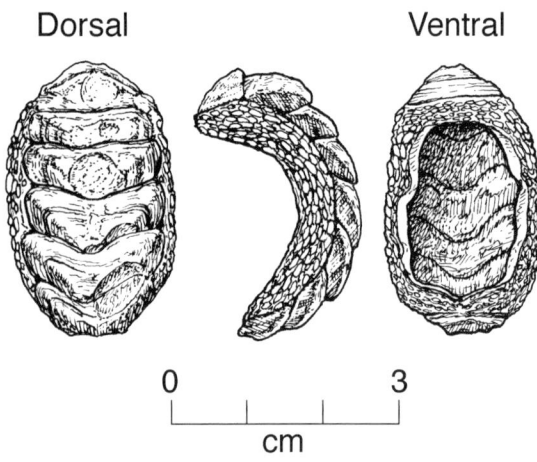

Figure 3.1. Desiccated chiton (*Enoplochiton niger*) from Feature 6, Structure D.

Chitons

Chitons (Class Polyplacophora) were present in the Cerro Azul environment, but not heavily exploited in the Late Intermediate. The species most common in our archaeological refuse was *Enoplochiton niger* (Fig. 3.1). This creature, known locally as the barquillo, averages 5.6 cm long and has eight overlapping armor plates on its dorsal surface. (The fact that each individual has eight plates provided a way of calculating minimum numbers of individuals from isolated plates.) The ventral surface of the chiton consists of a large flat foot to facilitate creeping; this foot is the edible part.

Enoplochiton is an herbivore that lives on rocky intertidal surfaces, clinging to the rocks when the waves are rough and creeping slowly when they are calm. Using its radula—a horny, ribbonlike structure found in the mouths of all molluscs except for bivalves—this chiton grazes on the sporophytes of brown algae. *Enoplochiton* also consumes the film of diatoms, loose algal cells, and bacteria covering the rocks of the Fucoid or brown algae zone. Chitons live on the sea cliffs of La Centinela and El Fraile at Cerro Azul today, but since they do not provide a lot of meat our workmen do not often collect them.

We recovered 229 specimens of chiton at Cerro Azul (Table 3.2). Since many of these specimens were isolated dorsal plates, however, only 47 minimum individuals could be calculated. Our largest sample came from Feature 6, a midden in Structure D; included were 8 whole (or nearly whole) chitons and 125 isolated plates. Smaller samples were found in Rooms 3 and 7 of the same building.

Our second largest sample, 84 dorsal plates, came from Feature 20. This was a midden associated with Structure 9, a fish storage facility housing a commoner-class overseer. Smaller samples were found in Rooms 5 and 8 of the same building.

Obviously chitons were eaten by the occupants of both buildings, including the occupants of Room 3 in Structure D, who were members of the nobility. Unfortunately, the terrace middens on Cerro Camacho did not add substantially to our data on chitons.

Virtually all the specimens we recovered were *Enoplochiton niger*. However, two dorsal plates—one from Terrace 11 of Quebrada 6 and one from Room 7 of Structure D—appeared to be from a larger species of chiton. They remain unidentified.

Table 3.2. Chitons Recovered from Archaeological Contexts at Cerro Azul (all *Enoplochiton* unless stated otherwise).

Structure D	
Feature 6 midden	6 whole specimens (see Fig. 3.1) (MNI = 6)
	2 nearly complete specimens (MNI = 2)
	125 dorsal plates (MNI = 23)
Room 3	8 dorsal plates (MNI = 1)
Room 7	1 dorsal plate of a chiton too large to be *Enoplochiton* (MNI = 1)
Structure 9	
Feature 20 midden	84 dorsal plates (MNI = 11)
Room 5	1 dorsal plate (MNI = 1)
Room 8	1 dorsal plate (MNI = 1)
Quebrada 6, Terrace 11	
Square N8, depth 30 cm	1 dorsal plate of a chiton too large to be *Enoplochiton* (MNI = 1)
Totals	**NISP= 229**
	MNI= 47

Gastropods

At least six species of gastropods were collected at Cerro Azul. Let us begin, in phylogenetic order, with the keyhole limpets. According to Keen (1958), the genus *Fissurella* reaches its most prolific development on the coasts of Peru and Chile. Two species were present in the archaeological refuse at Cerro Azul: *Fissurella crassa* (Fig. 3.2) and *Fissurella limbata* (Figs. 3.3, 3.4).

Both these limpets are herbivores of the rocky intertidal zone, where they cling to rocks that are so heavily wave-battered as to be inhospitable for most molluscs. Both species are sensitive not only to predation from humans but also from sea otters and gulls.

Where both limpets are present (as was evidently the case at Cerro Azul), *F. crassa* and *F. limbata* compete for the same foods. Limpets lower in the water tend to eat the cortical cells of

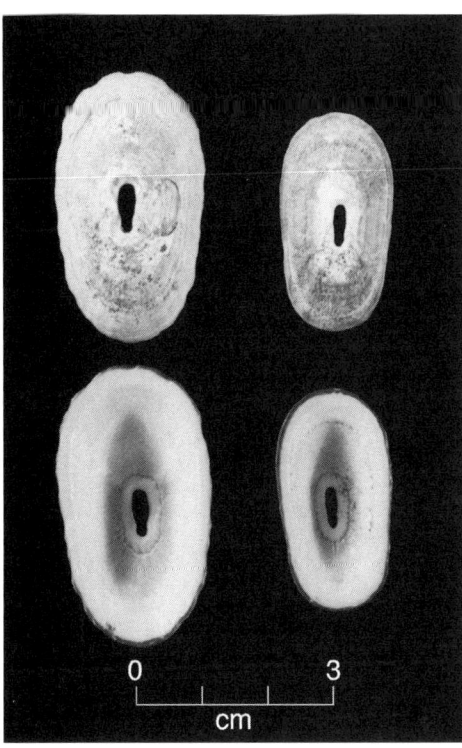

Figure 3.2. Two views of the keyhole limpet *Fissurella crassa* (from archaeological debris).

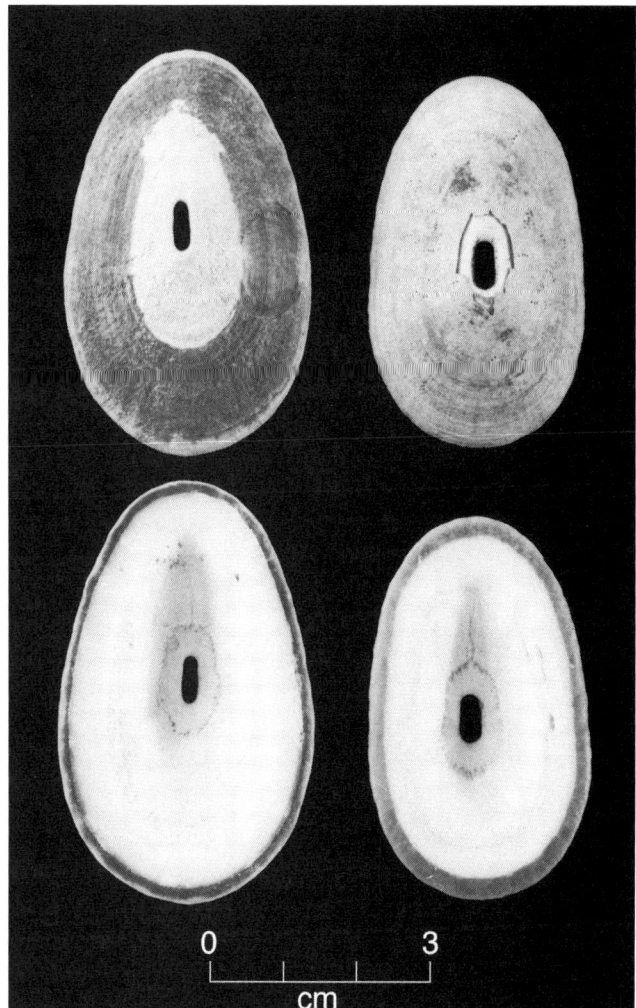

Figure 3.3. Two views of the keyhole limpet *Fissurella limbata* (from archaeological debris).

brown algae, while those higher up tend to graze on algal film. The size of individual limpets, as well as their density per square meter, is affected by such competition.

Keyhole limpets are central-place foragers who remain immobile when dry at low tide; when submerged at high tide they engage in loop-shaped excursions for food, leaving a mucus trail to guide them back to their home base. These creatures were not heavily exploited at Cerro Azul, perhaps because their habitat would have subjected the collectors to pounding from heavy waves.

Let us turn now to the chanque or false abalone, *Concholepas concholepas* (Figs. 3.5, 3.6). This is the largest and meatiest of the local gastropods, and because of its distribution at Cerro Azul we believe that it was a favorite of the elite.

The chanque adheres tightly to rocky surfaces from the tidal zone to depths of 40 m, and can remain in one area for so long that barnacles (*Balanus* sp.) grow on it. *Concholepas* is a true predator that attaches itself to colonies of small mussels (*Semimytilus* and *Perumytilus*) and eats its way through them (Castilla and Durán 1985). A study by Serra et al. (1997) shows that *Concholepas* tends to become a more experienced and efficient predator over time, improving its pursuit of *Perumytilus* while reducing its risk of being eaten by sea stars (starfish). Its predatory behavior is one reason for its name "false abalone," since true abalones are herbivores.

On one occasion at Cerro Azul, at least 26 chanques (and perhaps as many as 34) were served at a sea lion roast (see below). And on August 15, 1984, our workmen Edgar Zavala and Alberto Barraza collected 8 chanques from the cliff face at El Fraile. How did they eat them? Raw, of course, like any lover of ceviche.

Closing out our list of gastropods from Cerro Azul are three species of sea snail: *Tegula atra*, *Calyptraea trochiformis*, and *Thais chocolata*. Taken as a group, they illustrate the diversity of lifeways seen in sea snails. *Tegula* is an herbivore, *Calyptraea* is

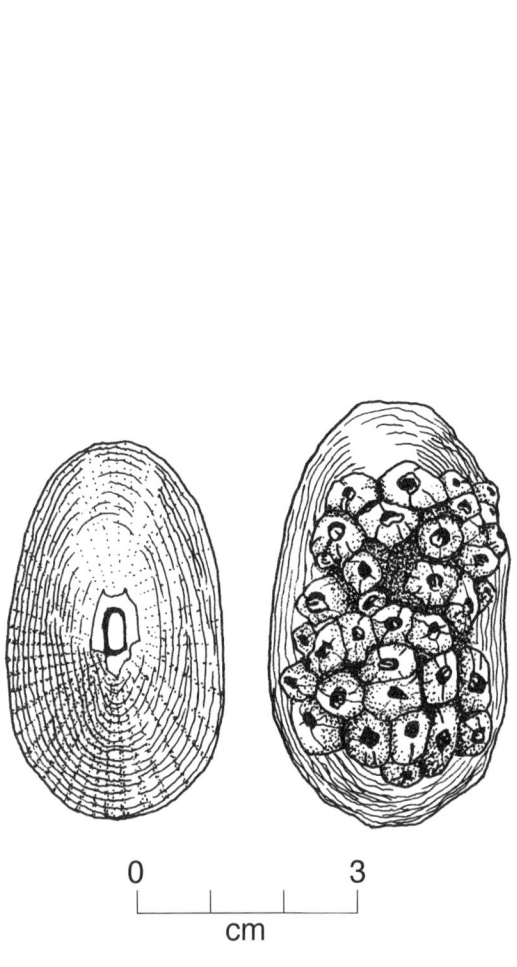

Figure 3.4. Two views of the keyhole limpet *Fissurella limbata*, drawn by Kay Clahassey from living specimens. Note the barnacles (*Balanus* sp.) attached to the limpet on the right.

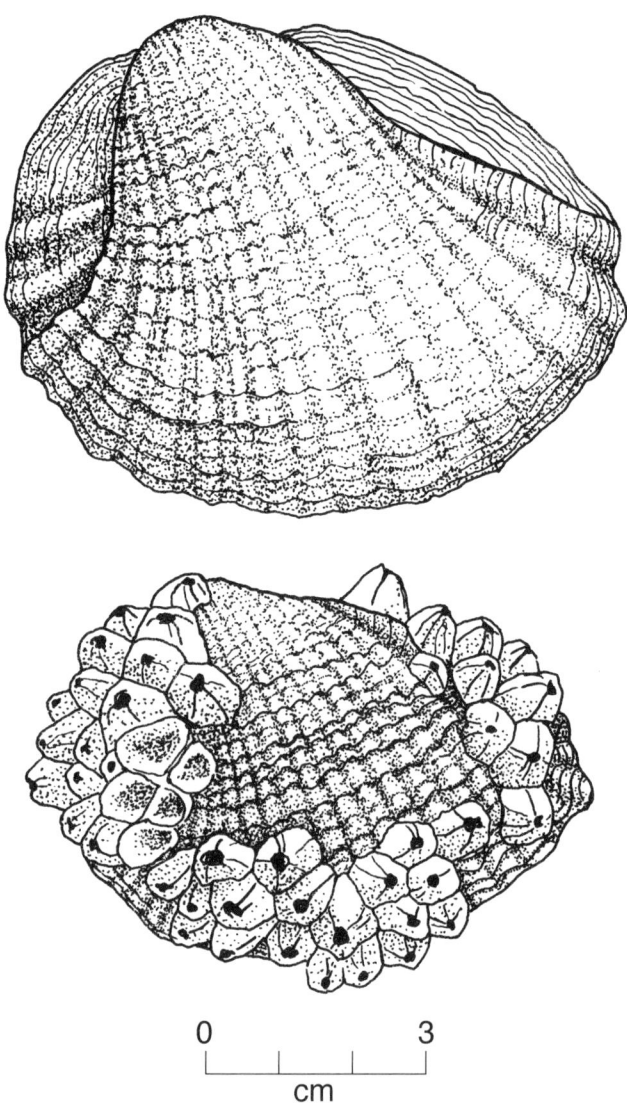

Figure 3.5. Two examples of the chanque (*Concholepas concholepas*), drawn by Kay Clahassey from living specimens. Note the barnacles (*Balanus* sp.) attached to the lower specimen.

a filter feeder, and *Thais* is an active carnivore. All are referred to locally as *caracoles*.

Tegula atra, the black turban shell (Fig. 3.7), is adapted to the lower intertidal zone of rocky coasts, where it coexists successfully with sea urchins. In fact, their relationship is symbiotic; *Tegula* grazes on microalgae while the sea urchins consume macroalgae, thereby improving the substrate for *Tegula* as they eat. According to Ricklefs (1979:629), *Tegula* is an even more efficient herbivore than the common grasshopper, since its algal grazing can net it 750 kcal per square meter per year.

The slipper shell *Calyptraea trochiformis* (Fig. 3.8) survives by filtering plankton and microdetritus from the surrounding water. It lives on rocky substrate just below the littoral zone, and under the right conditions will form vertical stacks in which multiple individuals live atop one another. Such multi-individual stacks are well suited to the hermaphroditic *Calyptraea*, for should the senior female of the stack perish, the largest male will respond by transforming himself into a female.

The rock snail *Thais chocolata* (Fig. 3.9) seems to have been another of the molluscs favored by the elite families of Cerro Azul (see below). Indeed, this large sea snail—which can exceed 5.5 cm in length—is now protected as a food source in parts of Chile (Cantillánez et al. 2011).

The rock snail is a benthic species that can be found living

Figure 3.6. A sample of the 26 chanques recovered from Feature 22, Terrace 16, Quebrada 5. Note that some chanques are encrusted with barnacles (*Balanus* sp.).

Figure 3.7. Two views of the sea snail *Tegula atra* (from archaeological debris).

Figure 3.8. Two views of the sea snail *Calyptraea trochiformis* (from archaeological debris).

Figure 3.9. Two views of the sea snail *Thais chocolata* (from archaeological debris).

Figure 3.10. The mussel *Aulacomya ater* (used as a pigment dish).

on stone or shell substrate at depths of 5–40 m. It is an active predator that tends to live near (and prey upon) natural colonies of mussels such as *Aulacomya ater*. It is known to drill holes in both mussels and barnacles with its radula.

To be sure, *Thais* has moments of vulnerability of its own. At low tide it may be exposed to predators as it crawls about on the rocks. It also tends to aggregate in large colonies during the reproductive season, and these colonies attract both sea otters and humans. During the Late Intermediate, *Thais* was harvested by the shellfish collectors who supplied the elite residents of Structure D at Cerro Azul.

Pelecypods

Seven species of clams and mussels were identified in the refuse at Cerro Azul, but only six of these were common enough to represent the remains of food. One species seems to have been imported for use as an artifact.

Most bivalves are sessile filter feeders or suspension feeders, mobile only during their larval stages. At the same time, many of the species we recovered at Cerro Azul can be dynamic competitors whose behavior plays a role in structuring marine communities.

The relatively large mussel *Aulacomya ater* (Fig. 3.10) lives on rocky coasts from the tide line to depths of 30 m. It may adhere to rocky substrates and form large colonies that take considerable suspended plankton out of the water. In areas of rocky vertical cliffs, *Aulacomya* may form a band between the upper brown algal zone (where its smaller and more air-tolerant individuals live) and the lower red algal zone (where its larger and less air-tolerant individuals live). As mentioned earlier, these mussels are a preferred prey of the rock snail, which means that both *Aulacomya* and *Thais* could have been collected in the same place. *Aulacomya* was rare at Cerro Azul, and there are hints that some of its shells had actually been imported to serve as pigment dishes.

The giant Chilean mussel *Choromytilus chorus* (Fig. 3.11) does not live in the Cerro Azul area today. Its shell makes an even more elegant pigment dish than *Aulacomya*'s, and it appears that it was imported for that purpose by noble women.

Obviously, we would like to identify the area from which Cerro Azul obtained its *Choromytilus*. The chances are good that that area lay to the south. While the range of this mussel is given as Pacasmayo to Tierra del Fuego, its distribution has fluctuated over time; during part of the twentieth century it disappeared from the northern part of its range, only to reestablish itself later (Avendaño and Cantillánez 2011).

Choromytilus adheres to rocky substrates from 4 m to 20 m depth, and its population fluctuations are believed to be due to at least two factors: (1) changes in water masses and circulation in the Pacific over time, and (2) competition with *Aulacomya ater*, a rival that it sometimes displaces.

One place where *Choromytilus* has at times been abundant

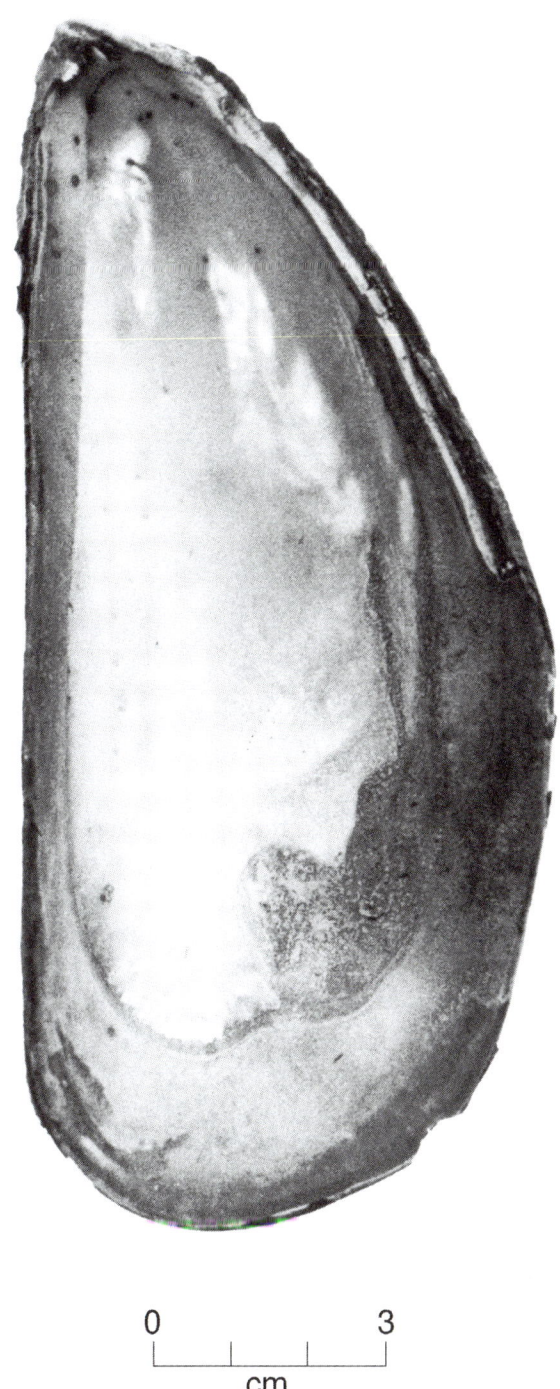

Figure 3.11. The giant Chilean mussel *Choromytilus chorus* (used as a pigment dish).

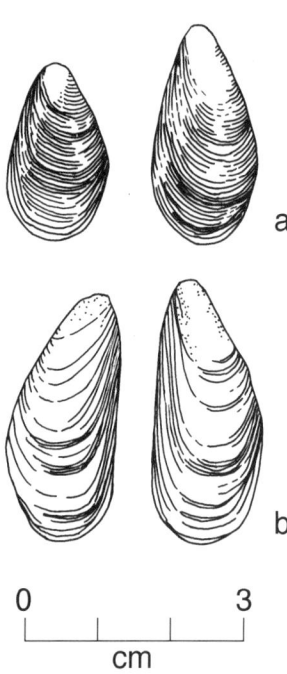

Figure 3.12. Four examples of the mussel *Semimytilus algosus*. a, two modern specimens. b, two Late Intermediate specimens (drawings by Kay Clahassey).

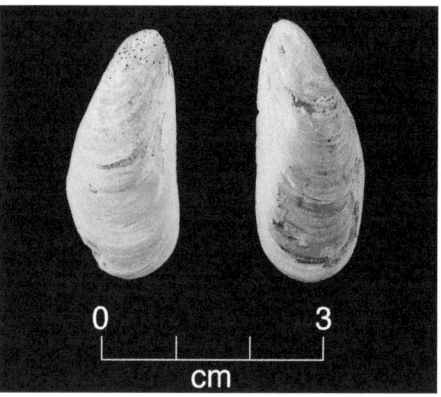

Figure 3.13. The mussel *Semimytilus algosus* (from archaeological debris).

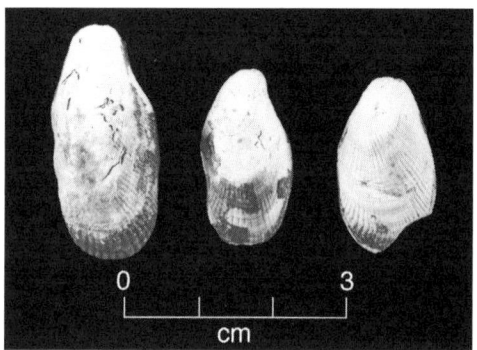

Figure 3.14. The mussel *Perumytilus purpuratus* (from archaeological debris).

Figure 3.15. *Perumytilus* mussels attached to a wooden piling below the pier, Cerro Azul Bay.

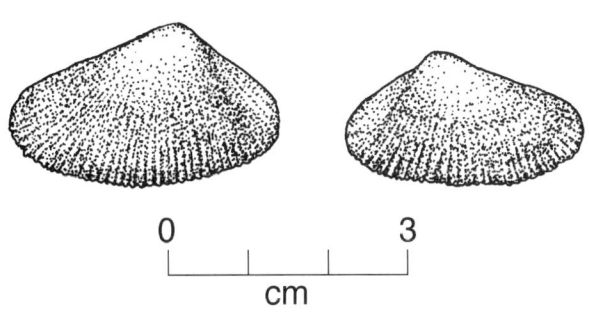

Figure 3.16. Two examples of the coquina clam *Donax obesulus* (= *peruvianus*), drawn by Kay Clahassey from living specimens. This small clam is still abundant at Cerro Azul.

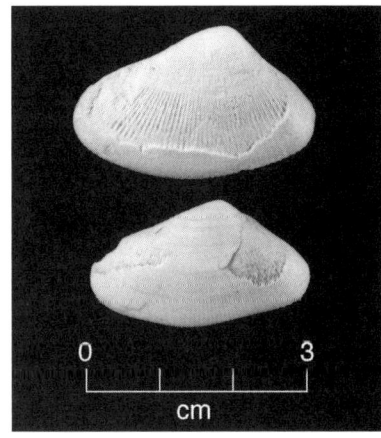

Figure 3.17. The coquina clam *Donax obesulus* (from archaeological debris).

is Cahuachi in the Nasca Valley. There Silverman (1993) reports large numbers of *Choromytilus* shells, some of which contained red pigment (Sandweiss and Rodríguez de Sandweiss 1991:58). The Nasca coast is therefore one possible source for our *Choromytilus*.

Two smaller mussels were important in the diet of Late Intermediate Cerro Azul; these were *Semimytilus algosus* (Figs. 3.12, 3.13) and *Perumytilus purpuratus* (Figs. 3.14, 3.15), both of which can still be found clinging to the posts of the Cerro Azul pier.

Semimytilus algosus, a mussel 2.5–4.0 cm long, inhabits both rocky platforms and emergent, disconnected rocks of the low intertidal zone. These small mussels cement themselves together in colonies, which helps to resist wave shock. Those on rocky platforms tend to have thinner shells, while those adhering to disconnected rocks have thicker shells (Caro and Castilla 2004).

Semimytilus competes with *Perumytilus*, a competition that some investigators feel drives *Semimytilus* to greater depths. As mentioned earlier, both genera are preyed upon by the false abalone, which may actually attach itself to colonies of mussels. In addition, *Semimytilus* falls prey to certain fish species; for example, we collected a freshly caught scaleless blenny (*Scartichthys gigas*) whose stomach contents included dozens of immature *Semimytilus* (Chapter 6).

As for *Semimytilus'* sympatric competitor, *Perumytilus purpuratus*, it has frequently been described as a "bioengineer" or a "key species" in local marine ecosystems. Measuring 2.3–3.2 cm in length, this mussel can form dense beds in the rocky intertidal zone. These beds have been described as three-dimensional matrices that trap both algae and sediment, providing a habitat for as many as 92 other invertebrate taxa (Prado and Castilla 2006). Stated differently, *Perumytilus* engages in "ecological niche construction," strongly influencing biodiversity and helping to structure local ecological communities.

Excavations at Cerro Azul suggest that both *Perumytilus* and its competitor *Semimytilus* were relied on by the Late Intermediate residents, especially in the first half of the period and especially by commoner families. There are also suggestions that the relative frequencies of these two mussels fluctuated over time, the almost predictable result of sympatric competition.

Another mollusc widely available at Cerro Azul was the mariposita or coquina clam, *Donax obesulus* (Figs. 3.16, 3.17). This tiny clam, which provides only a gram of meat per individual, is an occupant of the sandy beach zone. *Donax* burrows only a centimeter or two into the sand, and is so small that we suspect it was probably eaten mainly in soups or stews.

The population dynamics of *Donax* are closely tied to its shallow interment in the sand. Some species of *Donax* can be present at densities of 20,000 per m^2 (McConnaughey and Zottoli 1983:307). On the other hand, coquina clams can easily be dislodged from the sand by rough seas. As a result, *Donax* may be present by the thousands one year and gone the next. Rather than being permanent residents in a given beach, whole communities of *Donax* can move north or south with the alongshore currents, leading to a boom-or-bust presence and a long-term mechanism for colonization of new areas.

We should stress that these fluctuations in the density of coquina clams are not simply the result of El Niño years. Rough seas of any kind are enough to dislodge them and move them, after which they rebury themselves elsewhere. Since these tiny clams are eaten by gulls, their population dynamics would affect the diet of shore birds as well.

When present, coquina clams were eaten by the thousands by the commoners at Cerro Azul. There are also hints in the stratigraphy of the site that their populations may have fluctuated over time.

A larger pelecypod of the *playa* zone is the wedge clam *Mesodesma donacium*, known in Peru as the macha (Fig. 3.18).

Figure 3.18. The clam *Mesodesma donacium* (from archaeological debris).

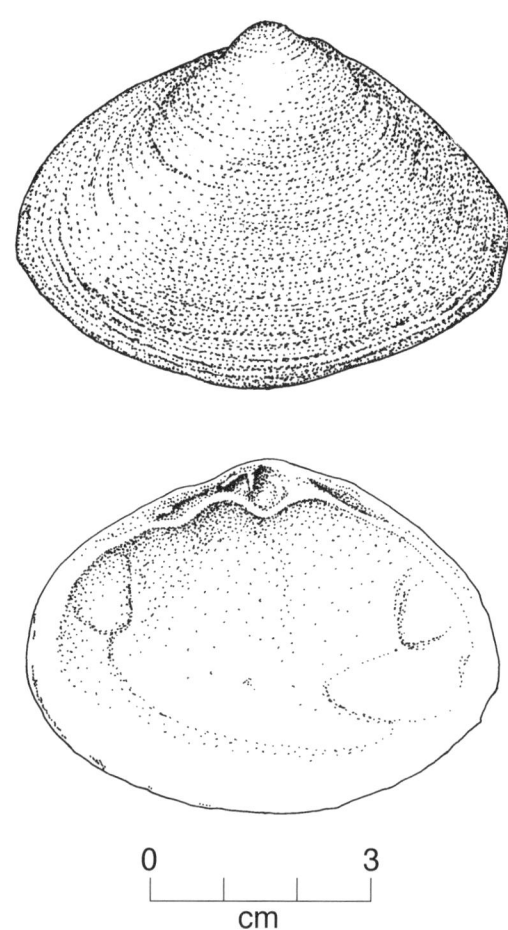

Figure 3.19. Two views of the clam *Mulinia edulis* from Late Intermediate specimens. This white clam, once abundant at Cerro Azul, has changed its geographic range and can no longer be found in the area (drawings by Kay Clahassey).

This filter feeder seeks out soft, sandy beaches where the substrate is loosened by incoming waves (Carstensen 2010). Such softening of the sand allows the wedge clam to dig itself in to depths of 5 cm, much deeper than the coquina clam.

Because it prefers deeper water than *Donax*, the wedge clam is best collected at times of lowest tides. It shares its habitat with the *muy muy* or mole crab (*Emerita analoga*) and like the latter is preyed upon by gulls. The difference is that *muy muy* are taken most often by the smaller garuma or grey gull, *Larus modestus*, while it is the larger kelp gull (*Larus dominicanus*) that usually gets the machas. Eagle rays also get their share of wedge clams.

Mesodesma prefers water temperatures of 14–21° C, which makes it sensitive to the warming of the ocean in ENSO years. During the 1982–83 El Niño, for example, it disappeared from the Tacna coast and did not reestablish itself until 1986 (Ibarcena Fernández et al. 2005). Thus, while *Donax* considers

environmental disturbances a mechanism for migrating to new beaches, *Mesodesma* is temporarily devastated by them.

Wedge clams were eaten in such small numbers at Cerro Azul that we wonder from how far away they came. One possibility is Playa El Hueso, on the coast to the north of Cerro Azul (see Fig. 1.1). Farther away—but still within easy boat travel for today's artisanal fishermen—is the great sandy beach of Asia, where we observed thousands of machas being harvested in 1982.

Finally we come to the white clam *Mulinia edulis* (Figs. 3.19, 3.20), another species that we believe was a favorite of the Cerro Azul elite. This bivalve uses its siphon to circulate water through its filtering system, consuming organic matter and rejecting the inorganic matter in the form of pseudofeces (Garrido et al. 2012). It dislikes turbulence, preferring the mud flats that form in the intertidal zone where rivulets of fresh water enter the sea; its favorite substrate is about 50 percent mud, 50 percent sand.

An unanswered question is where the occupants of Cerro Azul obtained their white clams. According to Dall (1909), *Mulinia* can be found from Callao to Chile, but during the 1980s none of our workmen could tell us where to collect living examples. A few of the older residents of Cerro Azul claimed that white clams had once been available up the coast on Playa El Hueso, but were no longer available.

Two possibilities come to mind. There may once have been mud flats within walking distance of Cerro Azul, either at the mouth of the Río Cañete or the terminus of a major irrigation canal that has since been filled in. Another possibility is that the waters off Cerro Azul are now too polluted by insecticides and sewage to support white clams.

The Ancient Shellfish Environments

As for the sources of molluscs at ancient Cerro Azul, we can offer the following suggestions. The sea cliffs of La Centinela and El Fraile, as well as the adjacent rocky ocean floor, are still replete with limpets, sea snails, false abalone, and small mussels. The sandy beach of Cerro Azul Bay had thousands of coquina clams even in the aftermath of the 1982–83 El Niño. We are not sure, however, where the ancient occupants of the site obtained *Aulacomya*, *Mesodesma*, and *Mulinia*. The clams may have come from Playa El Hueso, north of Cerro Azul, in the era before the cotton-producing haciendas began to overuse pesticides; the *Aulacomya* mussels may have come from Cerro Azul's rocky offshore islands, which we did not visit.

We would be remiss if we did not mention a pelecypod conspicuous for its absence at Cerro Azul: the scallop *Argopecten purpuratus*, not one specimen of which showed up in our excavations. This mollusc is so abundant on the sandy beaches of Asia that its absence at Cerro Azul only underscores how varied the coastal environments in this part of Peru can be.

Finally, let us mention two other molluscs for which we have no archaeological evidence. Both squid and octopus were caught at Cerro Azul in the 1980s, and we have no doubt that they were eaten in the Late Intermediate. Their absence in our faunal remains almost certainly reflects the fact that they do not preserve well, even under the arid conditions of the Cañete Valley.

The Archaeological Evidence for Shellfish

Let us now consider the shellfish remains from our most important archaeological contexts. We begin by considering the number of identified specimens (NISP) from Structure D, Structure 9, and the quebradas of Cerro Camacho. In a later section we choose several of our largest samples, convert NISP to minimum number of individuals (MNI), and estimate the amount of meat contributed by each genus of mollusc.

Some Andeanists have weighed the shells they recovered, in addition to presenting NISP. We decided not to weigh our

Figure 3.20. The clam *Mulinia edulis* (from Late Intermediate archaeological debris).

shell, since that is not the part of the mollusc that the prehistoric inhabitants of the site ate. We satisfied ourselves with calculating NISP and estimating the weight of the meat.

The Shellfish from Structure D

The largest single collection of shellfish remains from Structure D was provided by **Feature 6**, a midden in the Southwest Canchón. This midden, which we passed through two sizes of screen, yielded 1,257 specimens identifiable at least to genus (Table 3.3). The five best represented taxa were the rock snail *Thais chocolata* (320); two genera of small mussels (309 *Perumytilus*, 295 *Semimytilus*); the white clam *Mulinia edulis* (239); and the coquina clam *Donax obesulus* (67). Other species of clam, mussel, sea snail, and limpet were present, but in limited numbers.

In the northern and northeastern parts of the Southwest Canchón, we encountered two structures which we designated the **North and East Platforms**. These platforms had been buried beneath several layers of later fill, each of which produced shellfish remains.

Lying directly on the canchón floor in this area, we found 2 false abalones and 2 white clams. In the lower fill covering the North Platform we found 2 *Fissurella crassa*, 3 *Concholepas*, 2 *Tegula*, one *Semimytilus*, and 3 *Mulinia*. The uppermost, and latest, layer of fill in this area produced 2 *Concholepas*, 8 *Thais*, one *Semimytilus*, 3 *Mesodesma*, and 13 *Mulinia*. Finally, when all of this later fill had been removed, we came upon a *llamkana pata*, or raised work platform.

Table 3.3. Shellfish from the Feature 6 Midden (Structure D).

Fissurella crassa: 3
Fissurella limbata: 6
Fissurella sp.: 2
Concholepas concholepas: 7
Tegula atra: 3
Thais chocolata: 320
Aulacomya ater: 2
Semimytilus algosus: 295
Perumytilus purpuratus: 309
Donax obesulus: 67
Mesodesma donacium: 4
Mulinia edulis: 239

Total NISP: 1,257

Resting on this *pata* were 3 very large false abalones and 3 very large white clams. This food debris may have been left by the commoner-class overseer who occupied the platform.

In the process of clearing fill and exposing the original canchón floor, we discovered a blocked doorway that had once connected the Southwest Canchón to Room 11. Resting on the floor in front of this blocked doorway were 2 *Concholepas*, 2 *Thais*, one *Semimytilus*, and 3 *Mulinia*.

On the floor of the **South Corridor**, which led from the Southwest Canchón to the interior of Structure D, we found a few more molluscs. Included were 2 *Thais*, one *Choromytilus*, one *Semimytilus*, one *Perumytilus*, 3 *Donax*, one *Mulinia*, and an unidentified cone shell. We also discovered one land snail and one small freshwater snail.

The largest open-air work space in the interior of Structure D was the **Northeast Canchón**. This canchón had been used mainly for tasks such as weaving textiles and drying maize, and it did not produce much in the way of shellfish. A few ash-filled hollows in the floor in the northwest corner of this canchón produced one *Fissurella limbata*, one *Semimytilus*, one *Perumytilus*, 10 *Mulinia*, and one land snail.

The most distinctive architectural feature of the Northeast Canchón was **Collca 1**, a tapia-walled storage cell set against the south wall of the canchón. While this *collca* was notable for the weaving equipment stored in it, it also contained the following shellfish: one *Fissurella limbata*, 2 *Concholepas*, 3 *Tegula*, and 6 *Mulinia*.

The **North Central Canchón** of Structure D was a kitchen and chicha brewery. We found only 3 molluscs in this canchón —one *Concholepas* and 2 *Mesodesma*.

Owing to wind erosion, little remained of the **Northwest Canchón** of Structure D. It produced only 3 *Concholepas*.

Room 1

Room 1 had originally been an elite residential unit, but earthquake damage caused it to be converted to fish storage. After its period of use as a storage room was over, a certain amount of refuse, including discarded mollusc shell, was introduced into the room. Included were one *Concholepas*, one *Thais*, 5 *Donax*, and 16 *Mulinia*.

Room 3

Room 3 was another room whose function shifted from residence to storage unit following earthquake damage. The fill in this room could be divided into a lower deposit (resting on the original floor) and an upper deposit that postdated its conversion to a storage room. The shellfish from these two episodes of fill were kept separate.

In her analysis of the pottery from the lower fill (resting on the original floor of Room 3), Marcus (2008:157) observed, "… if I were asked to predict what the ceramic assemblage from an elite residential unit might look like, this collection from just above the floor of Room 3 would fit my prediction." Something similar could be said for the shellfish found in this layer; included were 5 *Concholepas*, 12 *Thais*, one giant *Choromytilus* of the type used for pigment dishes, one *Perumytilus*, 3 *Donax*, and 27 *Mulinia*. In other words, the collection featured four species that we suspect were favored by the elite (white clams, false abalones, rock snails, and the giant Chilean mussel), and had little in the way of limpets, small mussels, and coquina clams.

The upper fill of Room 3 yielded only one fragment of false abalone and 3 white clams. By the time of this later fill the function of the room had changed, and any of the shells it contained were in tertiary context.

Room 4

Room 4 of Structure D had a complex history and a stratigraphy to match. This room's original floor had once been part of the Southwest Canchón floor, and lay at the same depth (3.73 m below datum). At that time, the room seems to have been used for storage. At a later date, 1.5 m of fill were added to the room so that a new floor could be laid in at a depth of 2.23 m below datum. On this later floor Marcus' crew found Feature 5, a cluster of large potsherds that had been created when several whole vessels collapsed in pieces on the floor.

Finally, late in the history of Structure D, Room 4 was filled in to the top of its walls. Each stage in Room 4's fill had its own refuse layer, and these usually included shellfish.

The uppermost (and final) layer of fill in Room 4 yielded only one tiny specimen of *Thais*. Feature 5 was similarly unproductive, containing only two valves of *Mulinia*.

Immediately below the Feature 5 floor, however (2.23 m below datum), Marcus' crew found a refuse deposit with 47 molluscs. As the list below demonstrates, 85 percent of these molluscs were either white clams or false abalones, two species favored by the elite.

Fissurella crassa: 1
Fissurella limbata: 1
Concholepas concholepas: 8
Semimytilus algosus: 1
Perumytilus purpuratus: 2
Donax obesulus: 1
Mesodesma donacium: 1
Mulinia edulis: 32

As Marcus' excavation continued downward, her crew came upon another refuse deposit some 1.5 m below the Feature 5 floor (3.73 m below datum). This refuse lay just above the original floor of Room 4, and yielded a collection of 105 molluscs. As the list below indicates, this collection was about two-thirds white clams (*Mulinia edulis*). The remaining third was divided among eight species, including limpets, false abalones, sea snails, mussels, and clams.

Fissurella crassa: 2
Concholepas concholepas: 4
Thais chocolata: 2
Aulacomya ater: 1
Semimytilus algosus: 12
Perumytilus purpuratus: 7
Donax obesulus: 3
Mesodesma donacium: 4
Mulinia edulis: 70

Having already invested a great deal of effort on the stratigraphy of Room 4, Marcus decided to continue down, removing a portion of the lower floor and exploring the subfloor fill to a depth of 4.73 m below datum. This fill produced a sample of 6 molluscs: 2 *Concholepas*, one *Semimytilus*, one *Perumytilus*, one *Donax*, and one *Mulinia*.

Room 5

Originally designed for storage, Room 5 had been converted to a residential room late in the history of the building. Trapped in the fill between the early floor and the later floor in the west half of the room was a lone *Thais chocolata*. In the upper fill of the room—stratigraphically above the last sleeping bench added, after the room had become residential—Marcus' crew recovered 13 molluscs: 3 *Semimytilus*, one *Perumytilus*, 2 *Mesodesma*, and 7 *Mulinia*. Marcus (2008:167) concluded that this fill dated to sometime around the transition from Late Intermediate to Late Horizon.

Room 7

Room 7 was another that had suffered extensive earthquake damage. Instead of being converted to fish storage like most other damaged rooms, however, it simply had its doorway blocked and became a place for the dumping of refuse. Included in this refuse were 49 molluscs: 6 *Concholepas*, one *Thais*, 3 *Semimytilus*, one *Perumytilus*, one *Donax*, one *Mesodesma*, and 36 *Mulinia*. The fact that 86 percent of the molluscs in this collection were false abalones or white clams gives the impression of elite refuse.

Room 8

Room 8 was devoted to fish storage and did not produce any molluscs. After Structure D was abandoned, however, wind erosion caused disintegrated tapia and sand to drift into the upper part of the room. Found in this drift layer were one limpet (*F. limbata*) and one white clam (*Mulinia edulis*).

Rooms 9 and 10

Rooms 9 and 10 were dedicated to the housing of guinea pigs, their fodder, and their bedding. After the abandonment of the building, however, disintegrated tapia and sand drifted into the upper part of these two rooms. Marcus' crew found some 19 molluscs in this upper fill; included were 4 *Concholepas*, one *Thais*, one *Semimytilus*, one *Perumytilus*, and 12 *Mulinia*. As in the case of the Room 7 refuse, 84 percent of this collection consisted of false abalones and white clams.

A Late Horizon Squatters' House in Structure D

After the abandonment of Structure D, the patio designated "Room 2" became the location of a Late Horizon squatters' hut. The midden associated with this kincha hut was designated **Feature 4**, and it yielded 96 shellfish. As the list below shows, 82 percent of this collection consisted of the small mussels *Perumytilus* and *Semimytilus*; false abalones and white clams made up only 9 percent, and *Thais* was totally lacking. These frequencies suggest commoner-class refuse, which would be appropriate for a kincha house.

Fissurella limbata: 2
Concholepas concholepas: 2
Unidentified sea snails: 2
Aulacomya ater: 3
Semimytilus algosus: 6
Perumytilus purpuratus: 73
Mesodesma donacium: 1
Mulinia edulis: 7

The Shellfish from Structure 9

Our largest sample of shellfish from Structure 9 was provided by the Feature 20 midden. This midden is presumed to represent refuse from the kincha hut occupied by the commoner-class overseer of Structure 9, a fish storage facility (Fig. 2.7). Marcus' crew screened the Feature 20 midden through both 1.5 mm mesh and 0.6 mm mesh, recovering 398 molluscs (Table 3.4).

As an examination of Table 3.4 reveals, Feature 20 yielded a very different mix of mollusc species from Feature 6 (Table 3.3). Some 80 percent of the shellfish in Feature 20 were either small mussels (*Semimytilus* and *Perumytilus*) or coquina clams (*Donax*); *Thais* and *Mulinia* amounted to only 16 percent, and *Concholepas* was not even present. These percentages confirm our in-the-field impression that Feature 20 represented commoner-class debris.

Table 3.4. Shellfish from the Feature 20 Midden (Structure 9).

Fissurella crassa: 2
Fissurella limbata: 3
Tegula atra: 2 (tiny)
Thais chocolata: 36
Semimytilus algosus: 164
Perumytilus purpuratus: 141
Donax obesulus: 15
Mesodesma donacium: 10
Mulinia edulis: 27

Total NISP: 398

Room 5

Room 5 of Structure 9 began as a fish storage room, but late in the history of the building it became a place to dump trash. Included in this trash were one beachworn limpet, 7 *Concholepas*, one *Tegula*, one *Aulacomya*, 7 *Semimytilus*, 3 *Perumytilus*, 2 *Mesodesma*, 28 *Mulinia*, and 2 unidentified snails. The source of the trash is unknown.

Room 6

Room 6 lay adjacent to Room 5, but had a different history; its original use was not for fish storage. Its lowest layer of fill produced 2 *Mulinia* and one *Donax*.

Room 7

Room 7, adjacent to Room 6, was used for fish storage from the beginning. Late in its history it had some trash thrown into it, including one *Thais* and 13 *Mulinia*. The source of the trash is unknown, and it likely dates to a very late stage in the building's use.

Room 8

Room 8 had never been used for fish storage, and contained a number of restorable ceramic vessels. From this room Marcus' crew recovered only one *Thais* and one *Fissurella limbata*.

Room 10

Room 10 was never used for fish storage; one lone valve of *Mulinia* rested on its floor, in a layer of disintegrated tapia.

Room 12

Room 12 was never used for fish storage, but may have been designed for some other kind of domestic storage. It yielded only one *Thais* and one *Mulinia*.

The South Entryway

Finally, one *Mesodesma* and one *Mulinia* were recovered from the sand fill in Structure 9's South Entryway.

Shellfish from the Quebradas of Cerro Camacho

Not surprisingly, excavations in the quebradas of Cerro Camacho supplied the University of Michigan team with excellent samples of shellfish, since many of the terraces in these quebradas had been created by the dumping of trash from one or more of the residential compounds at the site.

Quebrada 5, Terrace 16

Marcus' excavation on Terrace 16 of Quebrada 5 reached a depth of 2.2 m and exposed at least 10 "natural" or "cultural" strata.

Stratigraphic Zone F. This was a 50-cm-thick layer of soft gray ash that represented the first episode of trash dumping on the terrace. It produced one lone *Concholepas*.

Stratigraphic Zone D1. A layer of sterile salitre, or salty crust, filled the excavation at an average depth of 1.10 m below the surface, indicating a hiatus in the use of the terrace. Just above this salitre lay Zone D1, a deposit of ashy midden dumped on the terrace once the aforementioned hiatus had ended. Resting directly on top of this ashy midden was a thick layer of mollusc shell, which provided a cap for Zone D1. This layer consisted of 415 shellfish, 98 percent of which were the small mussels *Semimytilus* and *Perumytilus*. The list is as follows:

Concholepas: 2
Semimytilus: 28
Perumytilus: 377
Donax: 5
Mesodesma: 2
Mulinia: 1

Feature 22. Intrusive into Stratigraphic Zone A2 of the excavation was Feature 22, a basin-shaped hearth or roasting pit filled with fire-cracked rock, ash, and sea lion bones (see Chapter 12). It was clear that this feature had been produced during a *pachamanca* or "cookout" at which sea lions were eaten. Apparently the ceviche course that preceded this feast consisted of false abalones, 26 of which were found in Feature 22.

Stratigraphic Zone A2. In Zone A2, just west of the Feature 22 hearth, Marcus' crew found eight more false abalones that may have been part of the sea lion feast. One white clam and one small *Tegula atra* were also recovered from this area.

Stratigraphic Zone A1. A layer of salitre formed just above Zone A2, indicating a second hiatus in the dumping of trash on Terrace 16. Above this salitre lay Zone A1, the final stratum of midden deposited on the terrace. Five *Mulinia*, one *Aulacomya*, and one unidentified limpet were recovered from this layer.

Quebrada 5-south

One of the notable discoveries in Quebrada 5-south was **Structure 11**, a multiroom storage facility built against the rocky wall of the quebrada. Late in this building's history, someone built a kincha hut in its largest room, Room 3. Associated with this kincha hut were 4 *Semimytilus*, one *Donax*, and 3 *Mulinia*.

Immediately to the west of Structure 11 lay **Structure 12**, a typical Late Intermediate burial cist. This cist had largely been looted, but the looters left behind on the floor two gourd bowls containing food for the afterlife. Among the delicacies left in Gourd Bowl 1 were one *Semimytilus* and one *Perumytilus*. Included in Gourd Bowl 2 were 2 *Semimytilus*.

Quebrada 5a, Terrace 9

Terrace 9 of Quebrada 5a had had a complex history. Its deposits reached a depth of 2.6 m below the surface and could be divided into three general layers of cultural activity, which Marcus designated Zones A, B, and C.

Stratigraphic Zone C was a layer of brown domestic refuse that overlay bedrock at a depth of 2.6 m. Lying atop this refuse were a wooden post and a number of cane-impressed lumps of clay daub from a destroyed kincha structure. The pottery from this stratigraphic zone stood out as the earliest of Marcus' entire excavation; it almost certainly antedated the occupation of Structures D and 9.

Stratigraphic Zone B consisted of a Late Intermediate midden in which the coquina clam *Donax obesulus* was the most common mollusc. This midden layer was full of gray ash, and in places was a meter thick; it appeared to have been created by the dumping of trash from nearby residences.

Stratigraphic Zone A accounted for the uppermost 1.2 m of the Terrace 9 fill. Its matrix was domestic midden, but four Late Intermediate burial cists (Structures 4–7) had been excavated into Zone A. Unfortunately, these stone masonry cists had suffered looting.

Stratigraphic Zone C. Marcus divided Zone C into three approximately equal arbitrary levels, each of which was carefully screened.

Lower Zone C produced 58 shellfish, as follows:
    Semimytilus: 9
    Perumytilus: 18
    Donax: 29
    Mesodesma: 2

Middle Zone C produced 158 shellfish, as follows:
    Concholepas: 1
    Thais: 2
    Semimytilus: 13
    Perumytilus: 53
    Donax: 82
    Mesodesma: 7

Upper Zone C produced 306 shellfish, as follows:
    Concholepas: 1
    Aulacomya: 1
    Semimytilus: 26
    Donax: 214
    Mesodesma: 11

Let us make a few points about Zone C, since it was one of the oldest deposits we found at Cerro Azul. First, we see little or no change in the assemblage of shellfish during the period that Zone C was being redeposited. Second, the species makeup of Zone C is strikingly different from that seen in Structure D (Table 3.3). Of the 522 molluscs found in Zone C, 62 percent were coquina clams (*Donax*) and 33 percent were small mussels (*Semimytilus* and *Perumytilus*). With regard to the species we believe were preferred by the Late Intermediate nobles of Structure D, there were no white clams at all in Zone C, and both *Concholepas* and *Thais* were barely present.

The most parsimonious explanation is that the families responsible for the Zone C refuse were not only commoner class, but also obtained their shellfish from a relatively limited area. We cannot, however, rule out the possibility that the local coastal environments were somewhat different at the time that Zone C was accumulating (see below).

Stratigraphic Zone B. Zone B of Terrace 9 produced the largest shellfish sample from Quebrada 5a. Isolating an area of midden that was free from looters' pits, Marcus' crew screened it through 1.5 cm mesh.

The 1,919 specimens of shellfish (NISP) recovered from this screened sample are listed in Table 3.5. It will be readily seen that the Zone B midden was dominated by coquina clams (*Donax*), which constituted 65 percent of the shellfish. Second in frequency were small mussels (*Semimytilus* and *Perumytilus*), which together made up 31 percent of the sample. No other species supplied as much as 1 percent.

Table 3.5. Shellfish Recovered from Stratigraphic Zone B, Terrace 9, Quebrada 5a.

Fissurella sp.: 1
Concholepas concholepas: 6
Tegula atra: 6
Calyptraea trochiformis: 3
Thais chocolata: 9
Semimytilus algosus: 321
Perumytilus purpuratus: 279
Donax obesulus: 1,250
Mesodesma donacium: 44

Total NISP: 1,919

Zone B contributed our only examples of the slipper shell *Calyptraea trochiformis*; on the other hand, many genera present in Structure D (such as *Aulacomya*, *Choromytilus*, and *Mulinia*) were completely absent. We are not sure how to explain the differences between Features 6 and 20, on the one hand, and Zone

B of Terrace 9 on the other. Did these differences simply reflect elite diet vs. commoner diet, or were there actual differences in coastal environments between the beginning and end of the Late Intermediate period?

The University of Michigan project witnessed several of Cerro Azul's beaches changing from sand to gravel between 1982 and 1986, and there is no telling how many changes might have occurred between A.D. 1100 and 1470. More extensive stretches of sandy beach, for example, would have resulted in higher numbers of *Donax* being present when Zones C and B of Terrace 9 were accumulating.

An Additional Sample from Zone B. While Marcus was excavating **Structure 5**, a stone masonry burial cist on Terrace 9, she noticed that the mortar holding the stones together strongly resembled the gray ashy matrix of Stratigraphic Zone B. These similarities extended to the sherds and mollusc species included in the mortar. To test the possibility that Zone B might be the source of the mortar used for Structure 5, Marcus took a sample from the wall of the burial cist.

The shellfish from this sample of mortar are listed in Table 3.6. It will be noted that, just as in the mollusc sample from Zone B, the collection was dominated by coquina clams (*Donax*), followed by small mussels (*Semimytilus* and *Perumytilus*). No other species contributed as much as 1 percent.

Table 3.6. Shellfish Recovered from the Gray, Ashy Mortar Used in the Building of Structure 5.

Concholepas concholepas: 1
Calyptraea trochiformis: 1
Thais chocolata: 4
Aulacomya ater: 1
Semimytilus algosus: 87
Perumytilus purpuratus: 76
Donax obesulus: 450
Mesodesma donacium: 30
Mulinia edulis: 5

Total NISP: 655

Stratigraphic Zone A. Zone A was composed of gray, ashy midden material, not unlike Zone B. It proved difficult to get a reliable sample of shellfish from this stratum because so much of it had been disturbed by burial cists and looters' pits. By navigating carefully around these intrusive features, Marcus succeeded in isolating and screening enough of Zone A to recover 392 molluscs from this uppermost stratigraphic zone. The list of species is presented in Table 3.7.

In general, the sample of shellfish from Zone A resembled that taken from Zone B. The coquina clam (*Donax*) accounted for 70 percent of the molluscs from Zone A, and small mussels (*Semimytilus* and *Perumytilus*) made up another 26 percent. No other species constituted as much as 1 percent.

Table 3.7. Shellfish Recovered from Stratigraphic Zone A of Terrace 9, Quebrada 5a.

Concholepas concholepas: 1
Unidentified olive shell: 1
Semimytilus algosus: 53
Perumytilus purpuratus: 49
Donax obesulus: 275
Mesodesma donacium: 12
Mulinia edulis: 1

Total NISP: 392

Quebrada 6, Terrace 11

Terrace 11 of Quebrada 6 provided Marcus with additional samples of domestic midden. One plausible source of that midden was Structure H, an elite residential compound that sits directly in front of Quebrada 6.

Terrace 11 was not particularly deep. Marcus' deepest sample came from 30–50 cm below the surface in Squares M8 and N8. This area of midden was screened through 1.5 mm mesh and produced 77 molluscs, as follows:

Concholepas: 1
Thais: 1
Semimytilus: 18
Perumytilus: 6
Donax: 42
Mulinia: 9

At a depth of 30 cm in Square N8, a collection of 29 molluscs came to light. These molluscs had a greater-than-average incrustation of the barnacle *Balanus*, suggesting that they came from an old, relatively undisturbed locality. The genera involved were:

Tegula: 1
Semimytilus: 7
Perumytilus: 4
Mesodesma: 3
Mulinia: 14

In the upper 30 cm of Squares M8 and N8, a layer of brown midden carefully screened by Marcus' crew yielded 219 molluscs. The list of genera was as follows:

Concholepas: 1
Thais: 1
Semimytilus: 81
Perumytilus: 47
Donax: 45
Mesodesma: 6
Mulinia: 38

Finally, at a depth of 0–20 cm below the surface in Squares M6 and N6, Marcus' crew recovered 263 molluscs, as follows:
Semimytilus: 137
Perumytilus: 87
Mesodesma: 5
Mulinia: 24
Small unidentified freshwater snails: 10

The deposits on Terrace 11 showed both changes over time and significant contrasts with Features 6 and 20. The oldest sample (from 30–50 cm below the surface) resembled that from Terrace 9 of Quebrada 5a, in that it was dominated by coquina clams (55 percent) and small mussels (31 percent); on the other hand, it had higher levels of white clam specimens (12 percent) than did Terrace 9 of Quebrada 5a.

Over time, the percentage of *Donax* began to decline and the percentage of *Mulinia* to rise. In the uppermost 20 cm of the terrace *Donax* was actually outnumbered by *Mulinia*, although small mussels continued to make up 85 percent of the sample.

What we do not know is whether these changes over time resulted from changing coastal environments or social factors —for example, a shift from the dumping of commoner trash to the dumping of elite refuse. Either is possible.

Quebrada 6, Terrace 12

Terrace 12 of Quebrada 6—immediately downslope from Terrace 11—had a very different history. It featured a burial cist, a storage cist, and an earth oven. All these features had been set in a supporting matrix of beach gravel, deliberately brought in for that purpose.

There were no molluscs either in the burial cist, the storage cist, the earth oven, or the gravel matrix. Late in the history of the terrace, however, a quantity of brown-to-beige midden had eroded down from Terrace 11 and covered parts of Terrace 12. From this midden deposit Marcus' crew recovered 22 mollusc specimens, as follows:
Concholepas: 3
Aulacomya: 1
Semimytilus: 6
Mulinia: 12

The Meat Contributed by Various Shellfish

Over the course of this chapter we have considered two sources of variation in shellfish. First, noble families seem to have had greater access to false abalones, rock snails, and white clams than did commoner families. Second, there are hints that over the course of the Late Intermediate period, changing coastal environments might have altered the spectrum of shellfish available to the occupants of the site.

We have so far compared our samples in terms of NISP and percentages thereof, but these figures do not take into account the fact that shellfish vary greatly in the amount of meat they provide. False abalones and white clams provide more meat per individual than do coquina clams and small mussels, which may be one reason why they were preferred by nobles. We decided, therefore, that we needed a way of estimating the amount of meat provided by each species of mollusc.

To do this, we needed to convert our data from number of identified specimens (NISP) to minimum numbers of individuals (MNI). This immediately brings up an important difference between gastropods and pelecypods. In the case of a gastropod such as *Thais*, each individual has but one shell; NISP and MNI are identical. In the case of a pelecypod such as *Mulinia*, on the other hand, each individual has two valves, each of which we would have considered an identified specimen; in other words, the MNI is half the NISP.

In addition to converting pelecypod valves to minimum individual molluscs, we had to decide whether to weigh our shellfish meat while it was fresh and moist, or to desiccate it and measure the dry weight. Dry weight is often preferred in malacological research (e.g., Serra et al. 1997), but we decided against using it because we suspect that most of the Cerro Azul molluscs were either eaten raw or made into stews. Their moisture would thus have been included.

Our estimates—given in Table 3.8—therefore come from freshly harvested, still moist shellfish. We obtained most of our estimates by averaging the weight of the meat from a sample of 10 molluscs of each species, either collected at Cerro Azul or purchased in the market at San Vicente Cañete. In the case of *Mulinia*—which was unavailable at Cerro Azul at that time—we were forced to base our estimate on specimens of the white clam *Semele*, whose shells were the same size as our Late Intermediate *Mulinia* specimens.

Finally, it goes without saying that our estimates apply only to shellfish from Cañete, since the same species of mollusc may grow to different sizes in other parts of its range.

Table 3.8. Estimated Weight of the Meat Present in Each Species of Shellfish Found at the Site of Cerro Azul.

Fissurella crassa: 3 g
Fissurella limbata: 4 g
Concholepas concholepas: 40–45 g
Tegula atra: 7–8 g
Calyptraea trochiformis: 6–7 g
Thais chocolata: 7–8 g
Aulacomya ater: 7–8 g
Semimytilus algosus: 3–4 g
Perumytilus purpuratus: 3–4 g
Donax obesulus: 1 g
Mesodesma donacium: 12–15 g
Mulinia edulis: 20–25 g (based on *Semele* of comparable size)

Figure 3.21. Two marisqueros (sharp-edged flakes struck off igneous beach cobbles) found in the upper fill of Room 8, Structure D.

A glance at Table 3.8 makes it clear why certain shellfish were preferred by the elite of Cerro Azul. One would have to eat 40–45 coquina clams to get the same amount of meat provided by one false abalone. For their part, two white clams provide as much meat as 10 or more *Semimytilus*.

Let us now compare some of our larger mollusc collections in terms of the amount of meat they would have provided. We begin with Feature 6 (elite refuse) and Feature 20 (commoner refuse), which are compared in Table 3.9. In Feature 6, which was excavated in its entirety, we have evidence for an estimated 5.9–7.2 kg of shellfish meat. In Feature 20, only a sample of which was excavated, we have evidence for an estimated 1.2–1.4 kg of shellfish meat.

In spite of the fact that one of these middens represents elite refuse and the other commoner refuse, both reveal the dietary importance of white clams. In Feature 6 *Mulinia* provided an estimated 40–41 percent of the shellfish meat. In Feature 20 it provided only 24–25 percent, but that was still the highest percentage provided by any mollusc genus. Thus, although noble families clearly enjoyed more access to white clams, *Mulinia* was important to the commoner family in Structure 9 as well. The head of that family was the storage building's overseer, and his relationship with his noble overlords may have afforded him greater access to white clams than the average commoner.

Table 3.9. A Comparison of Features 6 and 20, Based on the Estimated Amount of Meat Provided by Each Genus of Mollusc (using MNI).

Genus	Feature 6	Feature 20
Fissurella	15–24 g	18 g
Concholepas	280–315 g	0 g
Tegula	21–24 g	42 g
Thais	2240–2560 g	252–288 g
Aulacomya	7–8 g	0 g
Semimytilus	444–592 g	246–328 g
Perumytilus	465–620 g	213–284 g
Donax	34 g	8 g
Mesodesma	24–30 g	60–75 g
Mulinia	2400–3000 g	280–350 g
Totals	**5,930–7,207 g**	**1,179–1,393 g**

A clear difference between elite and commoners can be seen in the genus that provided the second highest amount of meat. For the elite who produced Feature 6 that genus was the rock snail *Thais*, which provided 36–38 percent of their shellfish meat. For the commoners who produced Feature 20, *Thais* provided only about 21 percent of their shellfish meat; they relied more heavily on small mussels (*Semimytilus* and *Perumytilus*), which together provided 40–44 percent of their shellfish meat. For the elite who produced Feature 6, in contrast, *Semimytilus* and *Perumytilus* together provided only 15–17 percent of their shellfish meat.

To place Features 6 and 20 further in perspective, let us now look at our meat estimates for Stratigraphic Zone B of Terrace 9, Quebrada 5a (Table 3.10). This gray, ashy midden layer produced enough shell to account for 2.2–2.6 kg of meat. One is immediately struck by the fact that in this midden, small mussels (*Semimytilus* and *Perumytilus*) provided the bulk of the meat (42–47 percent). In second place was the tiny coquina clam (*Donax*), which provided some 24–29 percent.

The households whose refuse ended up in Zone B, in other words, relied heavily on three small pelecypods that are still readily available at Cerro Azul. They appear to have had little or no access to white clams or large mussels and ate only modest amounts of false abalones and sea snails. Almost certainly, the families whose refuse produced Zone B were commoners. However, since Zone B also dates to an earlier part of the Late Intermediate than Structures D and 9, there is also a chance that the coastal environments available to them were different.

Table 3.10. Estimates of the Weight of Meat Provided by Each Genus of Mollusc Present in Stratigraphic Zone B of Terrace 9, Quebrada 5a (using MNI).

Fissurella: 3–4 g
Concholepas: 240–270 g
Tegula: 42–48 g
Calyptraea: 18–21 g
Thais: 63 72 g
Semimytilus: 483–644 g
Perumytilus: 420–560 g
Donax: 625 g
Mesodesma: 264–330 g

Totals: 2,158–2,574 g

The Use of Marisqueros

Many mussels attach themselves tightly to cliffs, rocks, or stony sections of sea floors. Removing them can be difficult. When our workmen went out to collect *Concholepas*, for example, they often took with them a stone tool they called a marisquero. This is a thin but sturdy flake struck from an igneous beach cobble. It can be wedged between the mollusc and its substrate and used to pry it free.

Stone flakes of this type were not uncommon at Cerro Azul (Fig. 3.21). In some cases they showed sufficient edge wear to convince us that they had been used to pry molluscs free; in other cases we had trouble convincing ourselves that they were anything but fortuitous flakes. Like so many of the stone tools at Cerro Azul, they were rough-and-ready implements that took only a few seconds to prepare, and were so easy to make that they were probably discarded more often than curated.

4

The Collection of Crustaceans

Kent V. Flannery and Jeffrey D. Sommer

The waters of the Kingdom of Huarco would have teemed with crustaceans. Despite the variety of species available, the Late Intermediate occupants of Cerro Azul seem to have concentrated on three that were easily collected. These were the cangrejo moro or purple stone crab, *Platyxanthus orbignyi*; a large crayfish, *Cryphiops caementarius*; and the *muy muy* or mole crab, *Emerita analoga*. All three of these crustaceans were still being collected in 1982 when the University of Michigan began work at Cerro Azul.

Crustaceans at Cerro Azul in the 1980s

The Purple Stone Crab

The purple stone crab, *Platyxanthus orbignyi* (Figs. 4.1*a*, 4.2), was the most frequently captured crab at Cerro Azul during the 1980s. So many of these crabs occupy Cerro Azul Bay that when conditions are right, dozens can be caught in less than half an hour. One day in 1984 Edgar Zavala, one of Marcus' archaeological workmen, waded into Cerro Azul Bay and began feeling for crabs with his bare feet (Fig. 4.3). Within 15 minutes he had caught a dozen stone crabs of different sizes and ages (Fig. 4.4).

The purple stone crab has an interesting life history. During its larval stage—when it is called a nauplius—it spends most of its time in shallow water, consuming plankton. Once having metamorphosed into a juvenile crab it becomes a scavenger and detritus feeder, teaming up with other small crabs to consume the carcasses of dead fish. Eventually, after molting several times, it reaches its adult size; at this point, *Platyxanthus* becomes a true benthic predator (Farias et al. 2014).

Adult stone crabs like the one shown in Figure 4.1*a* are described as having dimorphic chelae, meaning that they have one big claw and one small claw. The large claw is called a crusher, and its role is to crack the shells of bivalve molluscs (Fig. 4.2 shows a crusher from Late Intermediate refuse). The smaller claw is called a cutter, and its role is to slice up softer material such as carrion or polychaete worms.

Each claw consists of a dactylus or moveable finger, and a fixed finger that is an extension of the crab's manus or "hand." These fingers of the crab claw were by far the most common anatomical parts discarded during the Late Intermediate, which suggests that they were removed before the crab was cooked.

Despite their strong claws, stone crabs seek out molluscs with thin shells. For example, Farias et al. (2014) found that *Platyxanthus* usually did not try to crack *Tegula*, the black turban shell; it concentrated instead on thin-shelled mussels such as *Semimytilus*, or wedge clams such as *Mesodesma*. The high frequency of purple stone crabs in Cerro Azul Bay in 1984 makes it likely that they were preying upon the coquina clam *Donax obesulus*, which was present by the thousands and only shallowly buried in the sand (see Chapter 3).

The Jaiba or Arched Blue Crab

During the years 1983–1984 the artisanal fishermen of Cerro Azul were catching a second delicious crab, *Callinectes arcuatus* (Fig. 4.1b). This swimming crab—known locally as a jaiba—is a distant relative of the Chesapeake Bay blue crab, *Callinectes sapidus*. Known as the arched blue crab, it is usually found in warm water and in most years would not be captured off Cerro Azul. According to Fischer and Wolff (2006:301), however, *Callinectes* "extends its range to Northern Chile in times of El Niño." The availability of this blue crab at Cerro Azul in 1984 shows us that the effects of the 1982–83 ENSO were still being felt.

Callinectes is another crab that changes its foraging habits during life. As a nauplius it dines on zooplankton in the 45–80 micron range; as a juvenile crab it consumes detritus and feeds on the corpses of catfish and mullet. Finally, as an adult crab it becomes a predator on sessile benthic invertebrates. A study by Paul (1981) showed that *Callinectes* stomach contents were 28 percent molluscs by dry weight and 22 percent crustaceans (including other crabs and ghost shrimp).

It is significant that not a single specimen of the arched blue crab was found in our Late Intermediate refuse at Cerro Azul. This lack of evidence for jaiba supports our conclusion that Features 6 and 20—our two most productive middens—did not accumulate during El Niño years.

Unidentified Crabs

Bits of other crab species showed up in our screened refuse deposits, but they were too fragmentary to identify. The variety of crabs known from the central coast of Peru is very high (Sánchez Romero 1973:326–405), but only a limited number of species are eaten today.

Spider crabs (Superfamily Oxyrhyncha) are caught on Peru's continental shelf, often as the unintended casualties of shrimp fishing. These long-legged, omnivorous scavengers search for decaying matter at considerable depth. They like both sandy and rocky ocean floors and sometimes confuse predators by camouflaging themselves with sponges.

Porcelain crabs (Family Porcellanidae) usually live in pairs, hiding under rocks and in crevices in the Cerro Azul area. These filter feeders are one of the favorite foods of the local drums (Family Sciaenidae) and when necessary, are said to leave some of their limbs behind to escape predators. It is this brittle aspect of their limbs that has led Porcellanids to be named for a delicate type of pottery.

The Sally Lightfoot crab (*Grapsus grapsus*) swarms on the sea cliffs of the *peña* zone at Cerro Azul, often just above the

Figure 4.1. Two species of crab caught at Cerro Azul in 1984. *a*, the cangrejo moro or purple stone crab (*Platyxanthus orbignyi*). *b*, the jaiba or arched blue crab (*Callinectes arcuatus*).

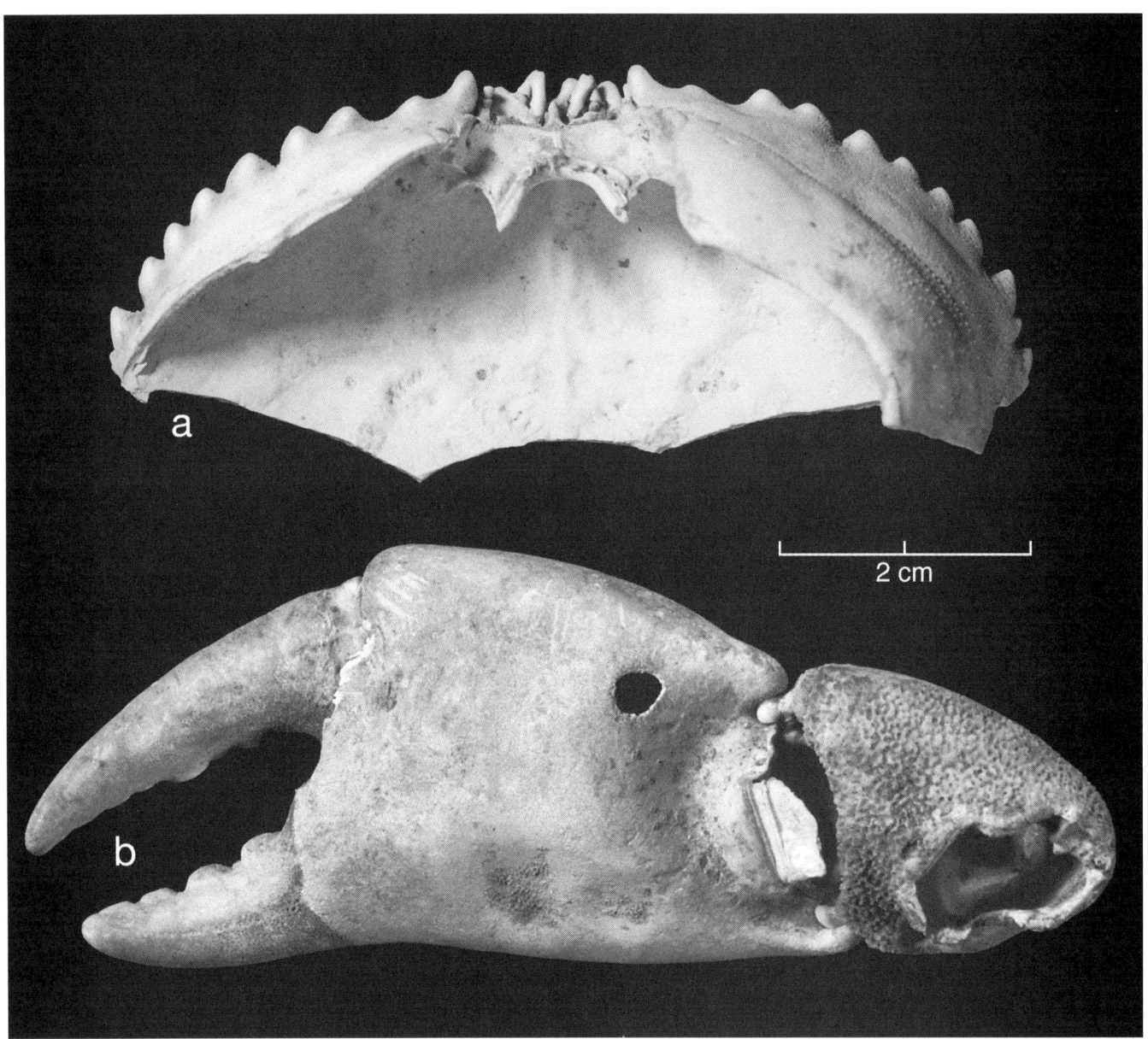

Figure 4.2. Fragments of stone crab from Structure D. *a*, a carapace fragment from Feature 6. *b*, a large "crusher" cheliped from the fill of the North Central Canchón. From left to right, we see the dactylus (above) and the fixed finger; the manus; and the carpus.

Figure 4.3. Edgar Zavala and two friends stand thigh-deep in Cerro Azul Bay, searching for stone crabs with their bare feet.

Figure 4.4 (below). Edgar's harvest: a dozen stone crabs, the product of only 15 minutes of barefoot searching.

tide line. This agile little crab divides its time between grazing on algae and scavenging dead fish, protecting itself from predators by scampering away on its tiptoes.

Despite the relatively large numbers of the aforementioned crabs, they are not pursued by Cerro Azul's population today, and we failed to identify them in our Late Intermediate refuse.

Crayfish

One of the most important crustaceans collected in the Cañete region is the freshwater crayfish, *Cryphiops caementarius*. This crayfish, which can weigh as much as 80 g, is still abundant in the Río Cañete (Méndez G. 1981). It would also have entered the large irrigation canals that carried water westward from the Cañete River toward Cerro Azul.

Cryphiops is another crustacean with dimorphic chelae. The dactylus and fixed finger of the larger claw were among its most frequently discovered anatomical parts in Late Intermediate refuse.

For more information we refer the reader to Chapter 5 of this volume, where Marcus and Matos illustrate this crayfish and describe the traditional traps used to capture it.

Mole Crabs

Despite its vernacular name, the mole crab *Emerita analoga* is not actually a true crab; it belongs to a separate superfamily, the Hippoidea. In Peru its indigenous name is *muy muy*, a gloss that comes to us not from Quechua but from a language the Spaniards called "Pescadora." This tongue was spoken by the indigenous fishermen of the north coast (Torero 1986; Rabinowitz 1980:142; see also Rabinowitz 1983; Netherly 2009; Rostworowski de Diez Canseco 1975, 1977a, 1977b).

Muy muy are seasonally available by the thousands in the *playa* environment at Cerro Azul. Mole crabs live in the shifting sands of wave-swept beaches; they bury themselves just below the surface, facing seaward with only their antennae and eye stalks out of the sand (Haig 1980). As waves retreat from the beach, the mole crab extends its long feathery antennae farther out of the sand so that it can catch small suspended food particles. At times—as when a gentle current of water passes over it—the *muy muy* may even emerge partially from the sand.

At such times the mole crab becomes vulnerable to human or animal predators. At Cerro Azul Bay, for example, the garuma or grey gull (*Larus modestus*) can be seen running along the tide line, plucking mole crabs from the sand with its beak (see Chapter 11). Should it emerge in somewhat deeper water, the *muy muy* becomes prey for the corvina (*Cilus gilberti*) and other members of the family Sciaenidae.

Muy muy were eaten in ancient times, and we found specimens in our Late Intermediate refuse (Fig. 4.5). The fishermen of the 1980s used *Emerita analoga* mainly as fish bait, but could still remember the times when their mother added *muy muy* to a stew.

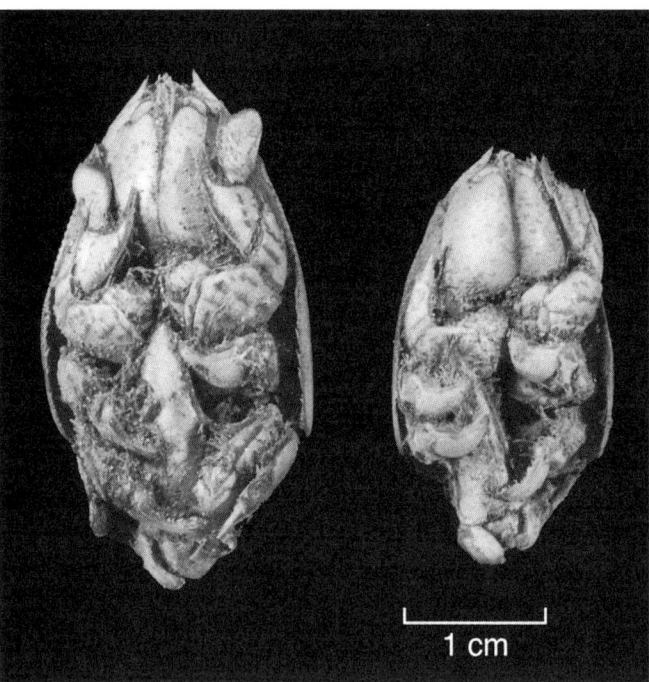

Figure 4.5. Two desiccated specimens of *muy muy* or mole crab (*Emerita analoga*) from the fill of Room 7, Structure D.

Ghost Shrimp

Another crustacean used as fish bait today is the marucha or ghost shrimp, *Callianassa islagrande* (Figs. 4.6, 4.7). Ghost shrimp dig elaborate burrows in sandy bays and tidal flats, including both the shores of Cerro Azul Bay and the Playa El Hueso that stretches north from Cerro Azul. In the course of their burrowing, ghost shrimp have been known to turn over 30 inches of sediment every 240 days (Brusca 1980:258); this activity has a definite impact on the *playa* environment.

Ghost shrimp rely on deposit feeding and suspension feeding, meaning that they extract nourishment both from particles in the sand and particles suspended in water. In the course of their activities they leave a number of uneven, closely spaced, telltale holes on the surface. These holes provide seabirds and fish with clues to the ghost shrimp's whereabouts, and are especially noticeable during seasons when large colonies of *Callianassa* are present.

In the Cerro Azul area, fishermen in search of maruchas use a large metal piston to extract these crustaceans from their burrows (Fig. 4.7). The piston is pushed deep into the wet sand when a wave passes over, and when the plunger is pulled upward the marucha follows.

We find it difficult to believe that these periodically abundant creatures were not exploited in the Late Intermediate. Unfortunately, only a few of our unidentified crustacean remains reminded us of the marucha's long, spindly chelipeds.

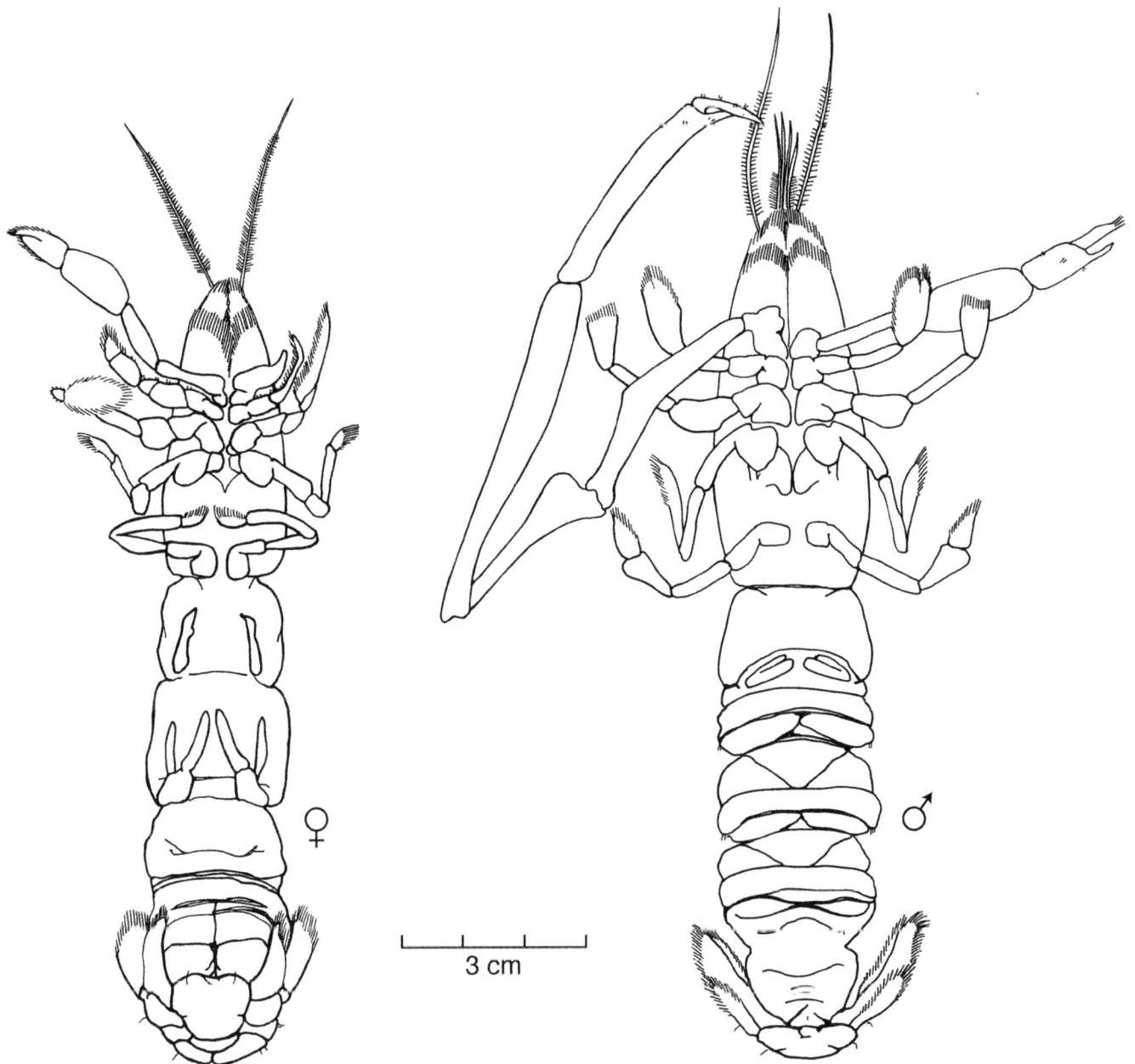

Figure 4.6. A pair of maruchas or ghost shrimp (*Callianassa islagrande*) from a sandy beach habitat at Cerro Azul.

Figure 4.7. A fisherman uses a metal piston to extract ghost shrimp from a sandy beach at Cerro Azul. Note the traditional *chiwa*, or net collecting bag, hanging beside his left leg.

The Late Intermediate Remains of Crustaceans

More than 1000 specimens of crustaceans were recovered from Late Intermediate refuse at Cerro Azul; all but 124 fragments could be identified to species. The vast majority of identified specimens were fragments from the chelipeds of stone crabs or crayfish.

Structure D

By far the greatest number of identified crustacean remains came from **Feature 6**, a midden in the Southwest Canchón that Marcus' workmen screened through 1.5 mm and 0.6 mm mesh. This midden produced 902 fragments of crustaceans, weighing a total of 357.34 grams (Table 4.1).

Table 4.1. Crustaceans from Feature 6 of Structure D.

Purple stone crabs
  Fragments of cheliped
    90 R. dactyli
    85 L. dactyli
    82 R. fixed fingers
    66 L. fixed fingers
    36 other manus fragments
    23 R. carpi
    15 L. carpi
    14 R. meri
    8 L. meri
    2 R. bases
    1 L. base
    5 R. bases + ischia
    4 L. bases + ischia
    2 R. coxae
    1 L. coxa
  Fragments of maxillipeds
    1 L. merus
    5 R. ischia
    3 L. ischia
    1 L. merus + carpus + ischium
    1 L. merus + carpus + ischium + exognath
    1 R. ischium + exognath
    1 R. exognath
  Fragments of walking legs
    29 dactyli
    15 other segments
  Other body parts
    65 carapace fragments (see Fig. 4.2)
    2 abdomen fragments
    7 sternum fragments
    34 unidentified
Totals:
  NISP = 599
  MNI = 90
  Weight = 338.74 g

Crayfish
  Fragments of cheliped
    21 R. dactyli
    21 L. dactyli
    1 dactylus (side unknown)
    1 R. dactylus + manus
    21 R. mani
    14 L. mani
    26 other fragments

Fragments of first walking leg
- 7 R. dactyli
- 8 L. dactyli
- 3 R. dactyli + propodi
- 3 L. dactyli + propodi
- 5 R. propodi
- 4 L. propodi
- 25 other leg segments
- 60 other body parts

Totals:
NISP = 220
MNI = 22
Weight = 14.50 g

Unidentified fragments of crustaceans
NISP = 83
Weight = 4.10 g

The large number of specimens recovered from Feature 6 allows us to draw a few conclusions about the way the occupants of Structure D prepared their crabs and crayfish. When it comes to the chelipeds of the purple stone crab, for example, it appears that the farther from the carapace a segment was, the greater the likelihood that it would wind up in a midden. Of the 434 identifiable cheliped fragments, 40 percent were pieces of dactylus. Only 9 percent were pieces of carpus; 5 percent were pieces of merus, and less than 1 percent were pieces of coxa. In other words, lots of people removed the claw itself before cooking, but only a tenth as many removed the cheliped at mid-limb, and almost no one bothered to remove the entire cheliped at its point of attachment to the body of the crab.

Similar conclusions can be drawn about the preparation of crayfish. Of the 220 fragments of crayfish recovered from Feature 6, nearly half (48 percent) were pieces of the cheliped; another 25 percent were pieces of the first walking leg. Abdominal somites were virtually absent from the midden debris, presumably because they belonged to the part of the crayfish that got eaten. In other words, the occupants of Structure D often removed the large claw; sometimes removed the first walking leg; and almost always left the abdomen intact for eating.

The North Platform

Feature 6 was a midden on the floor of Structure D's Southwest Canchón. Not far to the north were a series of platforms, one of which was likely used by the overseer of that canchón. Late in the occupation of Structure D, these platforms were buried under refuse and earthen fill. Included in the refuse were the following crustacean remains:
Purple stone crab:
  1 large R. dactylus of cheliped (4.63 g)
Crayfish:
  1 fragment of cheliped (0.39 g)

The South Corridor

The South Corridor was the principal access route from the Southwest Canchón into the interior of Structure D. Late in the occupation of the building, plant and animal refuse was discarded on the floor of the South Corridor. Included were these remains:
Purple stone crab:
  3 fragments of carapace (7.48 g)
  1 R. fixed finger of cheliped (0.66 g)

The Northeast Canchón

The Northeast Canchón was a large, unroofed work space in Structure D. In the floor of its northwest corner there were a number of ash-filled hollows with organic debris. Included was the following specimen:
Purple stone crab:
  1 large L. dactylus of cheliped (2.34 g)

Collca 1

Collca 1 was a storage unit set against the south wall of the Northeast Canchón. It produced the following:
Purple stone crab:
  1 L. dactylus of cheliped (0.50 g)

The North Central Canchón

The North Central Canchón was Structure D's kitchen and chicha brewery. The following crab remains were found in the fill of the canchón:
Purple stone crab:
  Still-articulated dactylus, fixed finger, manus and carpus of R. cheliped (12.01 g); see Fig. 4.2

Room 1

Room 1 of Structure D was originally part of an elite residential unit, but after suffering earthquake damage it was converted to a fish storage room. Marcus' crew recovered crab remains from both periods in the room's history. A sand layer just above the room's original floor produced the following:
Purple stone crab:
  1 R. dactylus
  1 L. fixed finger from the chelipeds of 2 different crabs (2.28 g)

The fill in the upper part of Room 1, postdating its conversion to storage, yielded the following:
Purple stone crab:
  1 L. fixed finger of cheliped (1.46 g)

Room 3

Room 3 had originally been part of an elite residential unit, but like Room 1 it had been converted to storage after suffering earthquake damage. Marcus' crew found crab remains only in the later fill of the room. The 17 specimens, from a minimum of 7 crabs, were as follows:

Purple stone crab:
>7 fragments of R. dactylus, cheliped
>2 fragments of L. dactylus, cheliped
>4 fragments of R. fixed finger, cheliped
>4 fragments of L. fixed finger, cheliped

Totals:
NISP=17
MNI = 7 crabs
Weight = 32.36 g

Room 4

Room 4 had a complex stratigraphic history. Its original floor lay 3.73 m below the datum point established for Structure D. At a later date, however, it was partially filled in, and a new floor was constructed at a depth of 2.23 m below datum. On this upper floor Marcus discovered Feature 5, a group of pottery vessels that had collapsed under the weight of later overburden. Her crew found crab remains at three depths within Room 4.

The oldest crab remains, found on the room's original floor at a depth of 3.73 m below datum, were as follows:

Purple stone crab:
>2 L. dactyli from the chelipeds of 2 different crabs (3.82 g)

The second oldest crab remains came from just below the room's upper floor, as follows:

Purple stone crab:
>1 R. dactylus of cheliped
>1 L. fixed finger of cheliped
>(6.61 g)

The most recent crab fragment in Room 4 was found among the broken vessels of Feature 5, as follows:

Purple stone crab:
>1 R. dactylus of cheliped (3.63 g)

Room 5

Room 5 of Structure D, like Room 4, had a complex history. Originally built as a storage room, it was converted to a residential room later on. In the fill between its lower and upper floors was one right dactylus from a cheliped of stone crab (0.99 g). In the fill above the upper floor was a right dactylus from the cheliped of an even larger stone crab (3.91 g).

Room 7

Room 7 was a small chamber adjacent to the kitchen/brewery of Structure D. After it suffered earthquake damage, this room became a convenient place to dump refuse. Included in that refuse was the following:

Purple stone crab:
>3 R. dactyli of chelipeds
>1 L. dactylus of cheliped
>1 L. fixed finger of cheliped

Totals:
NISP = 5
MNI = 3
Weight = 6.74 g

Mole crab: 2 complete individuals (3.01 g); see Fig. 4.5

Because of Room 7's location, the stone crab and *muy muy* remains may be kitchen debris from the North Central Canchón.

Room 8

Room 8 was a sand-filled facility for drying fish. In the uppermost part of this room there was a layer of debris that appeared to have eroded downslope and into Room 8 from the Southwest Canchón. This eroded layer contained the following crab remains:

Purple stone crab:
>1 R. dactylus of cheliped
>2 R. fixed fingers of chelipeds

Totals:
NISP = 3
MNI = 2
Weight = 2.62 g

Rooms 9 and 10

Rooms 9 and 10 of Structure D were devoted to the raising of guinea pigs. In the upper (post-abandonment) fill of this room pair, there was a layer of debris that appeared to have eroded down from the Southwest Canchón. Included in that layer of eroded material were two left dactyli from the chelipeds of two different stone crabs.

Structure 9

The largest collection of crustacean remains from Structure 9 came from Feature 20, a midden resting against the east side of the building (Marcus 2008: Fig. 8.3). This midden was screened through both 1.5 mm and 0.6 mm mesh; its contents are given in Table 4.2.

Table 4.2. Crustaceans from Feature 20 of Structure 9.

Purple stone crabs

    Fragments of cheliped

        17 R. dactyli

        15 L. dactyli

        7 R. fixed fingers

        8 L. fixed fingers

        1 manus fragment

        3 L. carpi

        1 fragment of L. maxilliped

        1 dactylus of walking leg

        2 carapace fragments

    Totals:

        NISP = 55

        MNI = 17

        Weight = 35.17 g

Crayfish

        2 fragments of chelate hand

        1 fragment of walking leg

    Totals:

        NISP = 3

        MNI = 1

        Weight = 0.05 g

Unidentified fragments of crustaceans

        NISP = 29

        Weight = 1.09 g

Although fewer in number than our sample from Feature 6, the crustacean remains from Feature 20 were broadly similar, and did not provide evidence that Cerro Azul's commoners and elite had significant differences in their access to crabs or crayfish.

Room 4

Room 4 of Structure 9 was a small storage unit in the northeast corner of the building. This room produced one carapace of a juvenile stone crab (1.37 g).

Room 5

Room 5 was a small unit along the west side of Structure 9; it had originally been used for the storage of fish. Late in its history, this room had been used for the dumping of domestic trash, perhaps from the nearby wattle-and-daub house of the building's commoner-class overseer. Included were the following remains:

Purple stone crabs:

    1 R. dactylus

    1 L. dactylus

    2 L. fixed fingers

    6 small cheliped fragments

    1 carapace fragment

Totals:

    NISP = 11

    MNI = 2

    Weight = 10.69 g

Room 6

Room 6 of Structure 9 was a small storage unit immediately to the south of Room 5. Late in its history, this room was used as a place to dump trash. That trash included one left dactylus from the cheliped of a purple stone crab (1.11 g).

Room 8

Room 8 was a small unit in the northeast quadrant of Structure 9. The debris in this room—which included large sherds from a number of partially restorable vessels—had probably not been transported far. Included in that debris was one right dactylus from the cheliped of a purple stone crab (1.65 g).

Crustacean Remains from the Quebradas of Cerro Camacho

Marcus' crew recovered additional crustaceans from the quebradas of Cerro Camacho. Some of these remains were found in stratified midden deposits on the terraces in the quebradas; other traces were identified by John G. Jones in human and dog coprolites (see Chapter 20).

Quebrada 5, Terrace 16

Terrace 16 of Quebrada 5 had 2.2 m of stratified deposits. Remains of purple stone crab were found in both Zone F—an early midden layer that may predate the occupation of Structure D—and Zone A1, a Late Intermediate midden layer. The remains were as follows:

    Zone F: 3 fragments from 1 carapace (4.02 g)

    Zone A1: 1 R. fixed finger of cheliped (3.92 g)

Quebrada 5-south

Our Quebrada 5-south specimens came from **Structure 11** (a multiroom storage facility) and **Structure 12** (a partially looted burial cist).

Room 3 of Structure 11 contained traces of a wattle-and-daub house. The associated debris produced the following:

Purple stone crab:

    1 R. dactylus of cheliped

    1 fragment of walking leg

Totals:
    NISP = 2
    MNI = 1
    Weight = 0.90 g

Crayfish:
    1 R. dactylus
    1 L. dactylus
    1 R. fixed finger
    4 other fragments

Totals:
    NISP = 7
    MNI = 2
    Weight = 2.5 g

Structure 12 of Quebrada 5-south was a stone-lined burial cist, found to be partially looted. This cist produced the merus, ischium, basis, and coxa from the left cheliped of a purple stone crab, all still in articulation (1.67 g). We do not know whether this crab was part of the food offering for a burial, or refuse accidentally included in the fill of the cist.

Quebrada 5a, Terrace 9

Terrace 9 of Quebrada 5a was 2.6 m deep and had a complex stratigraphic history. Its oldest stratum (Zone C) was a refuse midden that antedated Structure D. A second midden (Zone B) accumulated on this terrace during the early occupation of Structure D. Finally, when the midden deposits were deep enough, Zone A of Terrace 9 became a convenient place to construct subterranean burial cists. Marcus' crew recovered crustacean remains from all three stages in the history of the terrace.

The oldest crab remains came from Zone C and consisted of one left dactylus from the cheliped of a purple stone crab.

The second oldest crab specimen from this terrace came from the gray ashy mortar of **Structure 5**, a burial cist. This mortar appeared to contain reused ashy midden from Stratigraphic Zone B. The crab remains consisted of a left dactylus from the cheliped of a purple stone crab.

The most recent sample of crustacean remains came from the vicinity of **Burials 6 and 7**, found in the upper portion of Zone B. They were as follows:
Purple stone crab:
    1 L. dactylus of cheliped (0.66 g)

Crayfish:
    1 fragment of R. chelate hand (0.62 g)

Quebrada 6, Terrace 11

Terrace 11 of Quebrada 6 provided the University of Michigan project with an excellent sample of highly organic midden, dated to the Late Intermediate period. The crustacean remains from this midden all came from a depth of 30–50 cm and showed up in the 1.5 mm mesh screen. They were as follows:

Purple stone crabs:
    1 R. fixed finger of cheliped, large
    1 L. fixed finger of cheliped, small
    1 fragment of carapace: 1

Totals:
    NISP = 3
    MNI = 2
    Weight = 6.01 g

Unidentified fragments of crustaceans:
    NISP = 12
    Weight = 0.50 g

Crayfish Remains in Coprolites

John G. Jones (Chapter 20) discovered microscopic traces of crayfish in four coprolites from the quebradas. Three of these coprolites were human and were found either (1) in the abdominal cavities of intact burials or (2) in the debris from looted burials. One coprolite was canine, and came from a looted burial cist; this is not surprising, since dogs had been placed with several Late Intermediate burials (see Chapter 18). The following coprolites included traces of crayfish:

Coprolite 1 (human, Burial 19)
Coprolite 2 (human, Structure 5 burial cist)
Coprolite 5 (dog, Structure 12 burial cist)
Coprolite 6 (human, Structure 12 burial cist)

We find it interesting that so many coprolites associated with burials contained crayfish remains. It raises the possibility that individuals who were seriously ill (in some cases, terminally ill) were fed crayfish in some form, perhaps as a soup or stew.

Conclusions

Out of the dozens of available species of crustaceans, the occupants of Cerro Azul chose to concentrate on three that were easy to collect. The purple stone crab could be collected in large numbers simply by wading barefoot into Cerro Azul Bay. Crayfish were also readily available, both in the Río Cañete and in the main irrigation canals to which its waters were diverted; in Chapter 5, Marcus and Matos illustrate part of a Late Intermediate crayfish trap from Cerro Azul. Finally, mole crabs would have been seasonally available by the thousands on sandy beaches in the Cerro Azul area. Collectors would only have had to observe the behavior of the local grey gulls, which would have alerted them to the presence of the shallowly buried mole crabs.

5

Crayfish Trapping

Joyce Marcus and Ramiro Matos

One of the most important crustaceans of the south-central Peruvian coast is the crayfish *Cryphiops caementarius* (Fig. 5.1). According to a study by Viacava et al. (1978), this is Peru's most commercially significant crayfish. Those authors estimated that during the period 1975–1976, an estimated 400–500 metric tons of crayfish reached the main market in Lima.

The importance of crayfish for the Late Intermediate occupants of Cerro Azul can be confirmed by three lines of evidence. First, we found actual fragments of the "chelate hands" (pincers) of *Cryphiops* in middens such as Feature 6 of Structure D (Fig. 5.2). Second, John G. Jones discovered remains of crayfish in the coprolites associated with Late Intermediate burial cists in the quebradas (Chapter 20). Third, we recovered part of a damaged crayfish trap from a Late Intermediate storage bin, Collca 1 of Structure D (Fig. 5.3).

The Behavior of *Cryphiops*

Cryphiops caementarius is a resident of cobble-floored coastal rivers. A female with a cephalothorax 50 mm long can weigh 50 grams; a male with a cephalothorax 60 mm long can weigh 80 grams.

Adult crayfish tend to live upriver in the *chaupi yunga*, but their larvae migrate downstream to river mouths and coastal estuaries. Once the young crayfish have grown to 15 mm or longer, they return upriver. For their part, the pregnant adult females move downriver between November and January, months corresponding to the Andean spring and summer. All these migrations expose the crayfish to predation from humans and sea otters.

For Cerro Azul, the nearest source of crayfish would probably have been the irrigation canal known as the Acequia de la Quebrada de Hualcará or "María Angola" (Fig. 1.2). In ancient times this large canal took off from the Río Cañete near the Fortress of Ungará and ran west through the Kingdom of Huarco until it ended near Cerro Azul Bay (Chapter 1). Crayfish are said to have entered the canal and followed the current all the way to Cerro Azul.

Today that ancient acequia has fallen into disuse, as have several later versions of it. Dagoberto Sánchez, a longtime resident of Cerro Azul, took us to see the vegetation-clogged remains of a canal that once reached Cerro Azul (Fig. 5.4). According to don Dagoberto, the water in the canal shown in Fig. 5.4 was once as "crystalline" as that of the Río Cañete, and served as a ready source of crayfish and catfish.

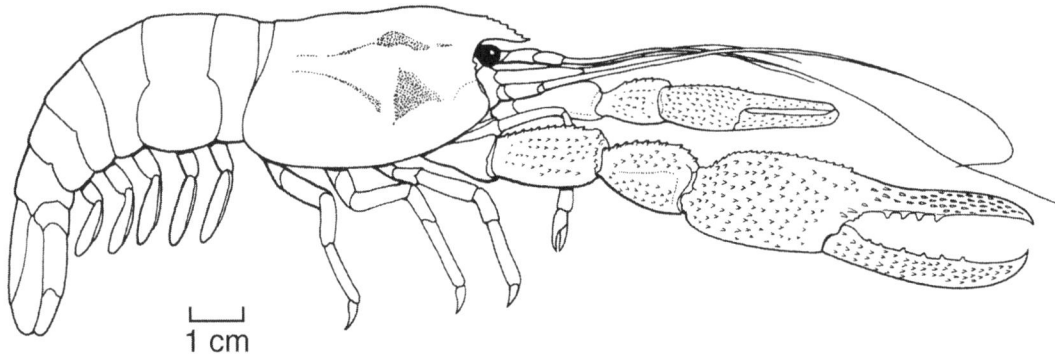

Figure 5.1. *Cryphiops caementarius*, the crayfish most frequently eaten at ancient Cerro Azul. (Drawing by Kay Clahassey, based on an illustration in Méndez 1981: Fig. 247).

Figure 5.2. Fragments of the large chelate hands (pincers) of *Cryphiops caementarius* from Feature 6, Structure D.

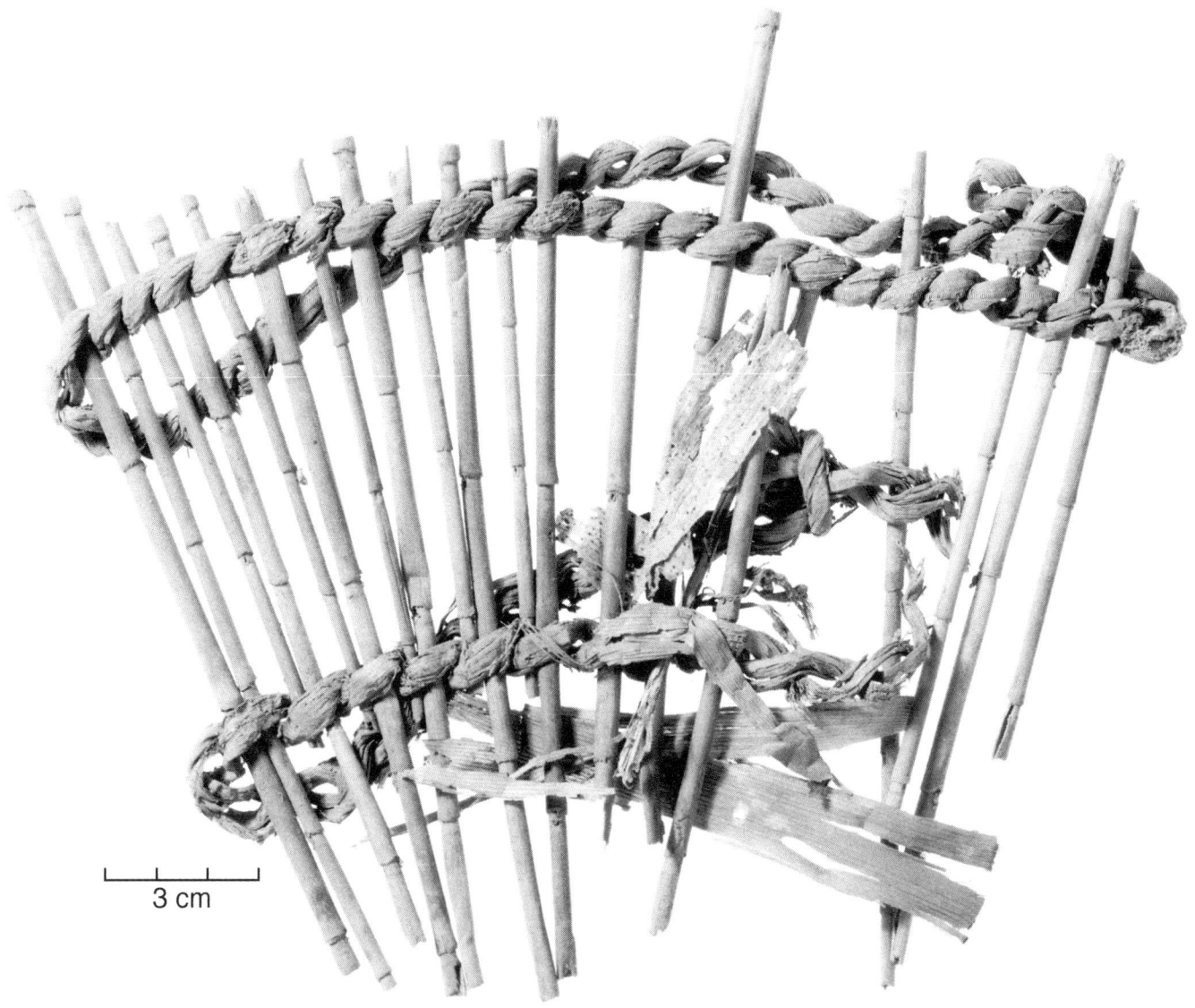

Figure 5.3. Portion of damaged *chauchu* (crayfish trap) from Collca 1, Structure D (Late Intermediate period).

Two human decisions doomed the canals reaching Cerro Azul. First, the cotton boom of the 1930s and 1940s led to an overuse of pesticides, which inevitably poisoned the waters of the canals. Finally, around 1970 the citizens of Cerro Azul filled in the town's main canal to prevent it from sanding in the new Cerro Azul pier. Dagoberto lamented both changes, which he felt had had unanticipated and environmentally unfortunate consequences.

Crayfish Trapping in 1986

Once we began to recover both crayfish remains and fragments of crayfish traps, we naturally became interested in seeing how the latter were used. Given the fact that the relevant canals at Cerro Azul are now no longer in use, we knew that we would have to visit Boca del Río on the Cañete River. There we found a farmer named Carlos Torres who was an experienced crayfish trapper. He agreed to demonstrate his craft for us.

On August 13, 1986—Andean midwinter—we joined Carlos on the right bank of the Río Cañete. He brought with him three indigenous pieces of equipment and one concession to modern technology. His three indigenous artifacts were a *chauchu* or multicomponent crayfish trap; a canasta or simple basket trap; and a *chiwa* or net bag in which the captured crayfish would be carried home. Carlos' one modern piece of equipment was a pair of scuba-diving goggles, which allowed him to submerge his face in water while searching for individual crayfish.

Figure 5.4. The vegetation-clogged remains of an ancient canal near Cerro Azul.

The Chauchu

The *chauchu* is a conical trap made from reeds or canes; it measures 80 cm to one meter in length (Fig. 5.5). The trap consists of two cones, one inside the other (Fig. 5.6). The smaller interior cone, called an embudo or funnel, is made of thicker cane than the larger exterior cone (Fig. 5.7). The embudo fits inside the mouth of the larger cone, which measures 40–50 cm in diameter. A circular hoop called an armazón, made from the flexible branch of a guava tree, forms the border of the trap's mouth. To it are affixed the lower ends of the 90 or more slender reeds or canes that form the large outer cone of the trap. Their upper ends are tied together in such a way that the *chauchu* can be opened when it comes time to remove the crayfish.

The outer reeds or canes are bound to one another by a series of three hoops that encircle the *chauchu*. The lower hoop—roughly one-third of the distance from the mouth to the top—consists of a vegetal strap that passes over the first reed, under the second, over the third, and so on. The middle hoop, about halfway up the trap from the mouth, consists of a strap that passes over two reeds, under two, over two, and so on. The uppermost hoop—two-thirds of the distance from the mouth to the top—consists of a strap that passes over three reeds, under three, over three, and so on. The smaller funnel is also affixed to the armazón with vegetal straps.

The *chauchu* is always laid on its side in the river with its mouth facing downstream. To prevent the current from moving the trap, Carlos piled heavy cobbles alongside it (Figs. 5.8, 5.9). If all goes well, crayfish seeking a hiding place enter the mouth of the embudo and continue on until they have passed through the small opening at its upper end. This opening leads the crayfish into the larger cone of the *chauchu*, from which they are unable to escape. Once a trap is relatively full, it can be removed from the river and the upper end untied to allow access to the crayfish.

According to Carlos, the best months for crayfish trapping were January, February, and March. During the Andean summer, melting glaciers swell the Andean rivers and crayfish are abundant. Carlos also revealed that the best crayfishing territory was upstream in the *chaupi yunga* between Lunahuaná and Zúñiga. Even downstream at Boca del Río, however, a *chauchu* could trap up to a kilogram of crayfish during one session; Carlos frequently used four or five traps simultaneously to increase his yield. He considered 8–10 kg of crayfish to be the maximum that a single fisherman could collect during one trip to the river.

During July and August, Carlos reported, the water level tends to drop and the crayfish become smaller and scarcer. This is the season for repairing traps that have been damaged; a common problem is that the strong current can tear the embudo out of the *chauchu* and wash it downstream. We are not sure whether that happened to the trap we found in Collca 1 of Structure D (Fig. 5.3). The struts on the surviving part of that trap appear to be the slender kind used for the larger cone of a *chauchu*.

Figure 5.5 (right). Informant Carlos Torres poses with his *chauchu* crayfish trap, 1986.

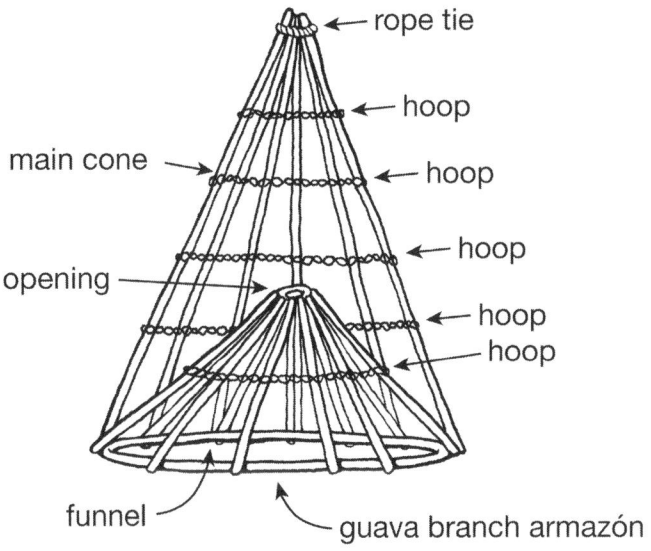

Figure 5.6 (above). Simplified diagram of a *chauchu*. This traditional crayfish trap consists of a smaller embudo, or funnel, inside a larger and more elongated cone of reeds or canes. When the *chauchu* is placed with its mouth facing downriver, crayfish seeking shelter enter the funnel, pass through the small opening into the main cone, and are unable to exit.

Figure 5.7 (right). The embudo or funnel from a *chauchu*, damaged and repaired in 1985.

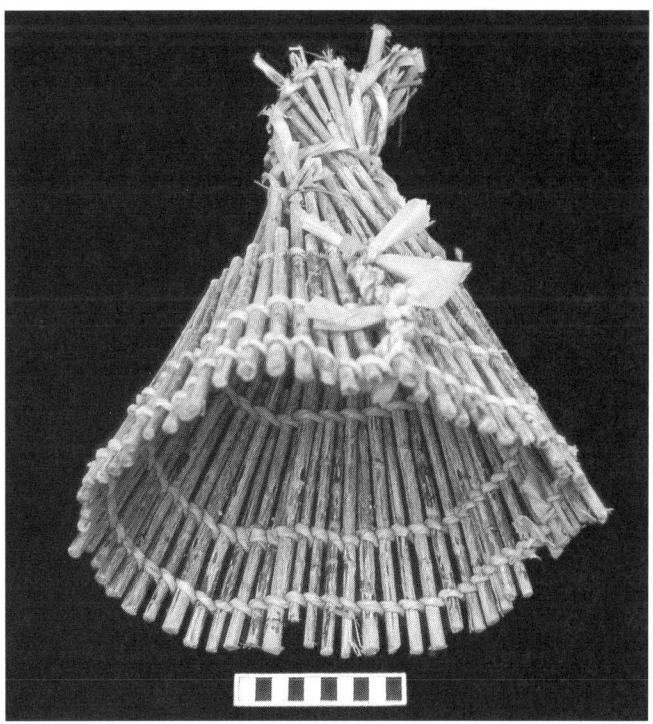

Figure 5.8. Carlos places heavy cobblestones alongside his *chauchu*, ensuring that it will not be moved by the current.

Figure 5.9. The *chauchu* is now secure and can be left unattended for hours while Carlos pursues other tasks.

Figure 5.10. Donning scuba goggles, Carlos lowers his face underwater and lifts cobbles to search for crayfish. Any crayfish caught by hand will be stored in his *chiwa*, or net collecting bag, which can be seen hanging down (and to the right) from his belt.

Figure 5.11. Two live crayfish, caught by Carlos in a very short time while wearing scuba goggles.

The Canasta

A second and less complex type of crayfish trap is the canasta, which lacks the double funnel construction of the *chauchu*. Canastas are made from caña brava or carrizo and are taller than a *chauchu*, reaching a height of 2–2.5 m. Like the *chauchu*, they are conical. The circular base, one meter in diameter, is held open by an armazón of flexible wood; the top of the trap forms a narrow peak, held shut by a bulrush tie.

Each canasta has a small rectangular door, cut into the wall just above the base. Five or six flexible wooden branches are made into the hoops that encircle the canasta at different heights, decreasing in diameter as one approaches the peak of the cone.

In contrast to the *chauchu*, the canasta is always placed on its side with its open base facing upstream. Most crayfish that are on the move get carried into the canasta by the current, while others enter through the small rectangular door in search of shelter. We were unable to witness the use of a canasta because Carlos chose not to use his during our trip to Boca del Río.

The Chiwa

The *chiwa* is a small net collecting bag, put to a variety of uses. Carlos suspended the *chiwa* from his belt by means of the same drawstring that he used to close the bag when it was full of crayfish.

Catching Crayfish by Hand

In the process of turning over cobblestones from the bed of the river and using them to stabilize the *chauchu*, fishermen almost always uncover hidden crayfish. In the past, they simply placed their faces under water to detect these crayfish and captured them by hand. Carlos donned modern scuba-diving goggles to make the process even easier (Fig. 5.10). On the day we accompanied him to the river, Carlos caught two crayfish by hand while selecting cobbles to stabilize his *chauchu* (Fig. 5.11). The short length of time required for this capture gave us some idea how many thousands of crayfish were hiding among the cobbles of the Río Cañete.

Archaeological Remains of Crayfish

More than 200 specimens of crayfish were recovered from Late Intermediate deposits at Cerro Azul. Most of the identifiable specimens were fragments of chelate hands (pincers), followed by the dactyli of walking legs. Virtually no abdominal somites were recovered, presumably because that part of the crayfish had been consumed. Even crayfish that had been eaten, however, were sometimes detectable from microscopic traces in coprolites. In the case of at least three Late Intermediate burials, coprolite analysis suggests that their last meals had included crayfish (Chapter 20).

Our archaeological specimens of crayfish have already been described in Chapter 4; we recapitulate the data here in Table 5.1.

Table 5.1. Archaeological Occurrences of Crayfish at Cerro Azul.

Structure D

Feature 6: NISP =220, MNI =22, 14.5 g (includes 79 fragments of chelate hands; 21 dactyli of walking legs; dozens of limb segments; virtually no abdominal somites)

Southwest Canchón, Fill above North Platform: NISP =1, MNI = 1, 0.39 g (one fragment of chelate hand)

Structure 9

Feature 20: NISP =3, MNI = 1, 0.05 g (2 fragments of chelate hand; 1 fragment of walking leg)

Quebrada 5-south

Structure 11, Room 3: NISP =7, MNI = 2, 2.5 g (fragments of chelate hands from 2 different crayfish)

Quebrada 5a

Terrace 9, Zone B, near Burials 6 & 7: NISP =1, MNI = 1, 0.62 g (1 fragment of chelate hand)

Traces of Crayfish Found in Coprolites

Coprolite 1 (human, Burial 19)
Coprolite 2 (human, Structure 5 burial cist)
Coprolite 5 (dog, Structure 12 burial cist)
Coprolite 6 (human, Structure 12 burial cist)

(Data from John G. Jones. See Chapter 20.)

6

The Fish Resources of Cerro Azul in the 1980s

Kent V. Flannery and Joyce Marcus

The coastline of the Cañete Valley, as pointed out in Chapter 1, features a multitude of marine habitats: sea cliffs, rocky sea floor, offshore islands, cobble beach, sandy beach, and the delta of the Río Cañete. It is therefore no surprise that Cañete's fish resources are rich and varied in "normal" (i.e., non-ENSO) years. Those resources oscillated between 1982 and 1986, a period that included the strong 1982–1983 El Niño and the two mild La Niñas that followed it. In 1982 and 1983 the fishermen at Cerro Azul were catching arched blue crab, shrimp, pompano, and dolphinfish, four warm-water species that visit Cañete only in ENSO years. By 1986 those species were gone, and the fishermen of Cerro Azul were celebrating the return of cool-water species like the silversides.

Establishing a Comparative Collection

We knew from the start of the Cerro Azul project that we would be recovering fish bones from Late Intermediate refuse, and lacked a good comparative collection with which to identify them. We also wanted to get an idea of the fish typical of the Cañete coast. In 1983, therefore, we began collecting two fish of each species captured by the artisanal fishermen of Cerro Azul (Fig. 6.1). Many of Marcus' archaeological workmen were experienced fishermen and participated enthusiastically in this endeavor, once she had explained it to them.

In this chapter we describe what we learned by collecting and skeletonizing fish from 1983 to 1986. During the last three years of her project (1984–1986) Marcus recorded the catch of every boat returning to the Capitanía del Puerto (Appendix A), and we obtained further data on fishing from Peru's Instituto del Mar. We ended up observing 36 species of fish, but this was just the tip of the iceberg; the real number of species caught at Cerro Azul could be 50 or more.

Table 6.1 lists the 36 species we observed. They came from 21 families and included such diverse forms as sharks, rays, sea catfish, sea basses, mackerels, grunts, drums, blennies, clingfish, flounders, and hakes. Nine species of drum alone were caught at Cerro Azul during our stay; there were also 3 species of the mackerel-tuna family and 3 species of the jack-pompano family. We assigned each specimen a catalogue number, recorded its length and weight, and measured the diameter of its largest vertebra for comparison with archaeological remains of the same species.

Table 6.1. Fish Species Observed by the University of Michigan Project, 1982–1986.

Triakidae (dogfish sharks)
    Mustelus mento (smoothhound)
Myliobatidae (eagle rays)
    Myliobatis peruvianus (eagle ray)
Callorhinchidae (plow-nosed chimaeras)
    Callorhinchus callorynchus (plow-nosed chimaera)
Clupeidae (herrings)
    Sardinops sagax (Pacific sardine)
Engraulidae (anchovetas)
    Engraulis ringens (anchoveta negra)
    Anchoa nasus (anchoveta blanca)
Ariidae (sea catfish)
    Galeichthys peruvianus (sea catfish)
Mugilidae (mullets)
    Mugil cephalus (common mullet)
    Mugil curema (white mullet)
Polynemidae (threadfins)
    Polynemus approximans (Pacific threadfin)
Atherinidae (silversides)
    Odontesthes regia regia (silversides)
Serranidae (sea basses)
    Hemilutjanus macrophthalmos (grape-eye sea bass)
    Paralabrax humeralis (Peruvian rock bass)
Carangidae (jacks and pompanos)
    Trachurus symmetricus murphyi (Chilean jack mackerel)
    Trachinotus paitensis (pompano)
    Seriolella violacea (blackruff)
Coryphaenidae (dolphinfish)
    Coryphaena hippurus (dolphinfish)

Haemulidae (grunts)
    Anisotremus scapularis (grunt)
Sciaenidae (drums)
    Menticirrhus ophicephalus (mismis)
    Menticirrhus rostratus (zorro)
    Paralonchurus peruanus (coco)
    Cynoscion analis (ayanque)
    Sciaena fasciata (burro)
    Sciaena deliciosa (lorna)
    Sciaena starksi (róbalo)
    Cilus gilberti (corvina)
    Stellifer minor (mojarrilla)
Cheilodactylidae (morwongs)
    Cheilodactylus variegatus (pintadilla)
Blennidae (blennies)
    Scartichthys gigas (scaleless blenny)
Clinidae (clinids)
    Labrisomus philippii (scaled blenny)
Scombridae (mackerels and tunas)
    Scomber japonicus peruanus (Pacific mackerel)
    Sarda sarda chiliensis (bonito)
    Scomberomorus maculatus sierra (sierra mackerel)
Gobiesocidae (clingfish)
    Sicyases sanguineus (pejesapo)
Bothidae (left-eye flounders)
    Paralichthys adspersus (left-eye flounder)
Merluccidae (hakes)
    Merluccius gayi peruanus (Peruvian hake)

Figure 6.1. The Capitanía del Puerto, Cerro Azul Bay. In the foreground are ten of the small boats used by the artisanal fishermen who provided us with our comparative collection.

The Cartilaginous Fish: Sharks, Rays, and Chimaeras

Even in regions as arid as the Cañete coast, cartilaginous fish pose real problems for the zooarchaeologist. Most elasmobranchs begin to spoil in a matter of days, leaving no archaeological evidence except for an occasional shark tooth or the crusher plate of a ray.

Older sharks and rays may undergo ossification of their vertebral centra, and we did find a number of ossified centra at Cerro Azul. Younger individuals, however, are predominantly cartilage, and will always be underrepresented in archaeological refuse. We are confident that the occupants of Cerro Azul ate sharks and rays, but Marcus found no shark teeth or crusher plates in her excavations.

The sharks most frequently caught at Cerro Azul in the 1980s were smoothhounds or dogfish sharks (*Mustelus* spp.), known locally as tollos (Fig. 6.2). *Mustelus mento*, the most common, ranges from 75 to 150 cm in length and can be taken by gill net when it ventures into the shallow waters of the continental shelf. It is less easily caught when farther out at sea, where it can descend to 16–50 m.

Mustelus mento loves to eat both shellfish and mobile benthic crustaceans. Its cousin, *M. whitneyi*, prefers the rocky substrate around offshore islands, where it eats crabs, ghost shrimp, and small fish.

Our efforts to preserve any part of a young *M. mento* by sun-drying were unsuccessful, so we simply saved some of its cartilaginous vertebrae in alcohol to see if they resembled the ossified vertebrae that we were recovering archaeologically.

Among the other sharks captured by artisanal fishermen at Cerro Azul were the tiburón azul (*Prionace glauca*), the pichirrata or thresher shark (*Alopias* spp.), and the tiburón martillo or hammerhead shark (*Sphyrna* spp.). While we recovered few archaeological traces of these species, we believe that we have some evidence for requiem sharks (*Carcharhinus* sp.).

The ray most commonly captured at Cerro Azul was the raya águila or eagle ray, *Myliobatis peruvianus* (Figs. 6.3, 6.4). Averaging 2.4 kg in weight, this ray is often dried in the sun, after which its meat is stripped off and sold as *ch'arki de raya*. We were able to preserve its cartilaginous vertebrae in alcohol, but found that even the *ch'arki* eventually spoiled.

Eagle rays are bentho-pelagic; they like to hang out near the coast, but their depth range is from 1–50 m. They do not like rocky substrate, but do well on muddy, sandy, or even gravelly ocean floors. There they eat an eclectic mix of mobile benthic molluscs, sea urchins, sea cucumbers, starfish, shrimp, crabs, polychaete worms, squid, and even small octopus. In Cerro Azul Bay they can be seen gliding over the sandy bottom, searching

The Fish Resources of Cerro Azul in the 1980s

Figure 6.2. The tollo or smoothhound shark (*Mustelus* sp.). These small sharks, usually in the 75–150 cm range, are among the most common cartilaginous fish caught at Cerro Azul today. Unfortunately, except for an occasional ossified vertebral centrum, virtually no trace of the tollo was preserved in the Late Intermediate archaeological debris.

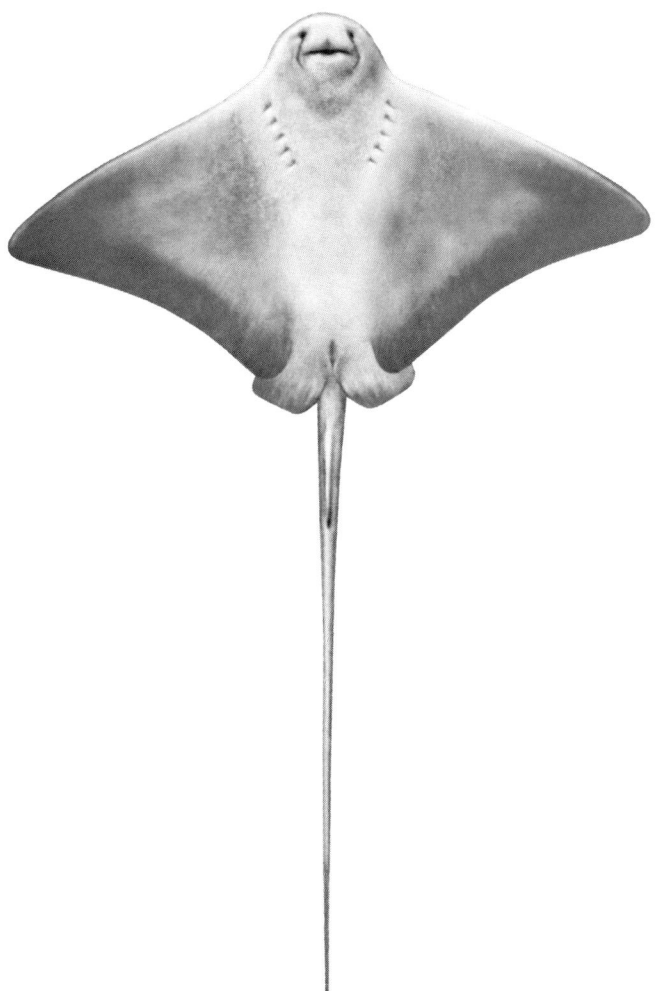

Figure 6.3. The eagle ray (*Myliobatis peruvianus*) is perhaps the most common of the rays caught at Cerro Azul today. Like the tollo, it leaves no archaeological trace beyond an occasional ossified vertebral centrum.

Figure 6.4. Artisanal fisherman drying more than 100 eagle rays (*Myliobatis peruvianus*) on a concrete slab near the Capitanía del Puerto, 1984.

for currents rising from the siphons of buried pelecypods. Once they detect such currents, they scour away the sand with powerful strokes of their mantles and consume the exposed molluscs (Marcus et al. in press).

Many other rays, including the angelote (*Squatina* spp.) and the temblador (*Discopyge* spp.), are caught near Cerro Azul, but we found no convincing Late Intermediate specimens of either. Most disappointing is our lack of prehistoric evidence for the guitarfish (*Rhinobatos planiceps*), for which Peru has a considerable archaeological record. This ray is caught regularly by artisanal fishermen at Cerro Azul, some of whom sail as far as Asia to get it.

Another cartilaginous fish seen at the Cerro Azul dock in the 1980s was the pejegallo or plow-nosed chimaera, *Callorhinchus callorynchus* (Fig. 6.5). Sometimes it was captured off Cerro Azul Bay, sometimes off Asia. It was one of those species that fishermen turn to in El Niño years, when many of the usual bony fish are unavailable. We find it likely that pejegallo was eaten at Cerro Azul, but you couldn't prove it from the archaeological remains.

As suggested by its abundance at Asia today, the plow-nosed chimaera likes sandy or muddy ocean floors. There it uses its fleshy, hook-like snout to probe the sand, searching for small molluscs like the coquina clam. This cartilaginous fish has an operculum over its gill slits and can even filter food from sediment. It will consume small octopus, but does not have the speed and body design to pursue bony fish.

Menhaden

The fishermen of Cerro Azul are familiar with the machete or Pacific menhaden, *Brevoortia* [= *Ethmidium*] *maculata chilcae*. In the past it appeared in great schools in the waters of Peru's continental shelf, where it filtered phytoplankton (mostly diatoms) and ate zooplankton (mainly copepods). It preferred shallow coastal water, less than 10 m in depth, and deposited its eggs on sandy ocean floors near the shore.

A few bones of menhaden appeared in the Late Intermediate refuse of the Cerro Azul site, but we failed to collect a single living specimen. There are three likely reasons for this. One reason is that *Brevoortia* today is overfished by commercial trawlers, not as food for humans but as a source of fish oil, fish meal, and dog food. A second reason is that menhaden are sensitive to El Niño years like 1982–83, and take a long time to recover. The third reason is that most artisanal fishermen consider the machete a "trash fish" and are content to leave it for the cormorants, boobies, pelicans, and seagulls.

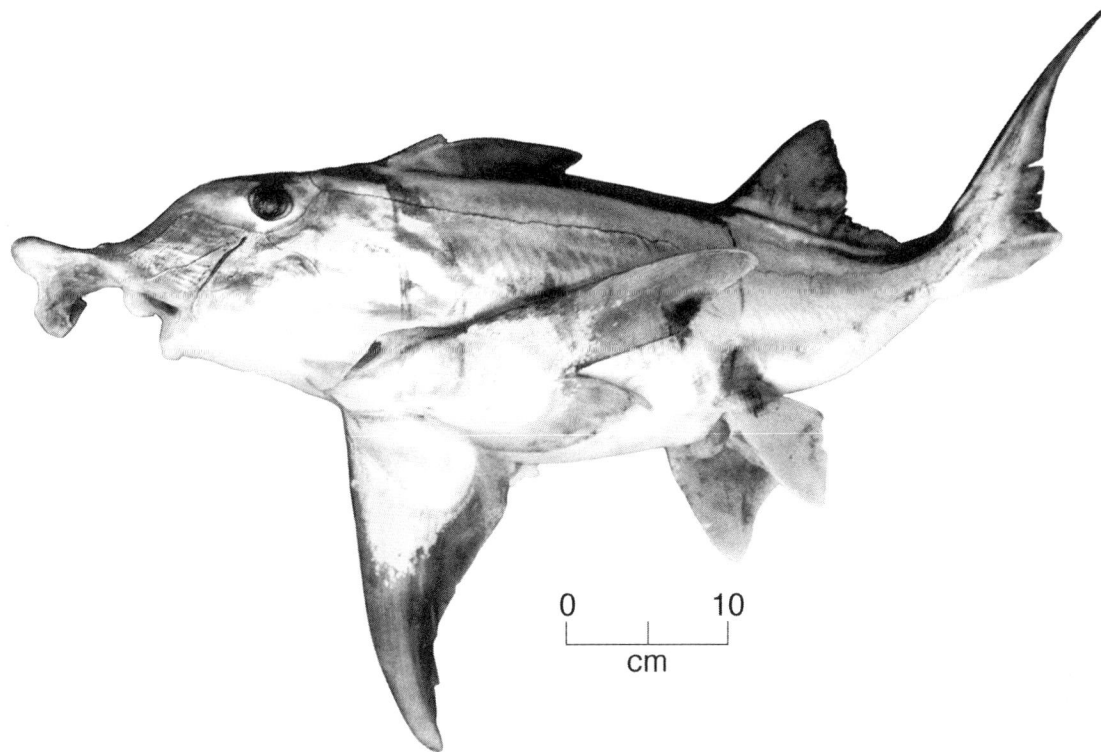

Figure 6.5. The pejegallo or plow-nosed chimaera (*Callorhinchus callorynchus*), another cartilaginous fish, caught often at Asia and Cerro Azul during the 1980s but difficult to detect archaeologically.

The Alternating Regimes of Sardines and Anchovetas

Let us now turn to the teleosts, or bony fishes, that we were able to collect during 1983–1986. We will begin, in phylogenetic order, with the sardines and anchovetas, two small fish caught by the tens of thousands during Late Intermediate times.

Recent oceanographic studies have revealed that Pacific sardine *Sardinops sagax* (Fig. 6.6) and the black anchoveta *Engraulis ringens* (Fig. 6.7) are "ecological neighbors but not ecological analogues" (Swartzman et al. 2008:236). Their populations respond to a series of oscillations much longer and more significant than those of El Niño. As Alheit and Ñiquen (2004:201) have put it,

> The long-term dynamics of the [Humboldt Current] ecosystem are controlled by shifts between alternating anchovy and sardine regimes that restructure the entire ecosystem from phytoplankton to the top predators.

To understand why these two species are "not ecological analogues," we must look first at the phytoplankton and zooplankton of the Humboldt Current. The two key categories of phytoplankton are diatoms (unicellular algae that live within a protective silica capsule) and dinoflagellates (unicellular algae with a cellulose membrane and a whip-like flagellum). The two key categories of zooplankton are copepods (crustaceans 0.2 mm–2 cm in length) and euphausiids (shrimplike crustaceans 1–6 mm in size, sometimes referred to as krill). Other categories of zooplankton are nauplii (tiny crustacean larvae), ostracods (small crustaceans, covered by a hinged carapace), and foraminifera (shell-covered protozoa with pseudopods).

Sardinops and *Engraulis* compete for, but do not eat identical mixes of, this plankton. Sardines tolerate warm water better than the anchoveta negra and like to visit bays and other continental shelf regions, where they consume the zooplankton and filter the phytoplankton. They tend to eat organisms that are closer to the base of the food chain than those eaten by anchovetas, and they also consume *Engraulis* eggs. Compared to anchovetas, sardines tend to eat smaller copepods and fewer euphausiids; in terms of phytoplankton, they eat more dinoflagellates and fewer diatoms than anchovetas (Espinoza et al. 2009). Finally, when conditions become unfavorable for sardines in Peru, they engage in extended migrations to Chile or Ecuador (Alheit and Ñiquen 2004:204).

Anchovetas negras, on the other hand, are described as omnivores or opportunists whose dietary flexibility allows them to adapt to changing conditions while performing only limited migrations (Espinoza and Bertrand 2008, 2014). Rojas

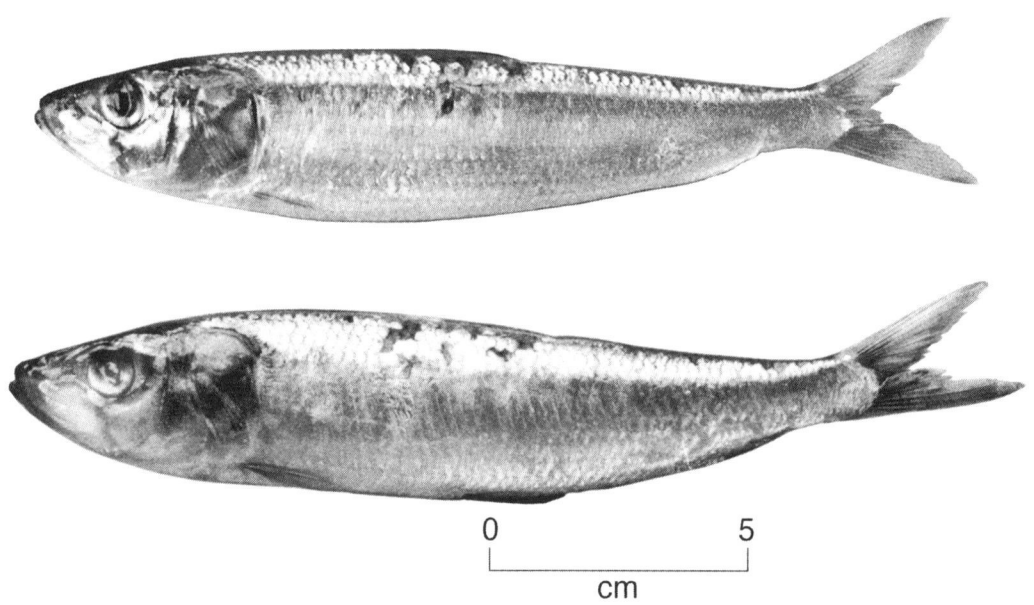

Figure 6.6. The Pacific sardine (*Sardinops sagax*), one of the small fish dried for export at ancient Cerro Azul.

Figure 6.7. The anchoveta negra (*Engraulis ringens*), a small fish dried for export at ancient Cerro Azul.

de Mendiola et al. (1969) examined the stomach contents of anchovetas off Tambo de Mora (Chincha Valley) and found them to contain 16 species of diatoms; 8 species of dinoflagellates; 5 types of crustaceans, including copepods, euphausiids, and nauplii; 3 types of protozoa; 1 type of tunicate; and miscellaneous fish eggs. The phytoplankton that anchovetas eat is dominated by diatoms, with dinoflagellates second; the zooplankton they eat is dominated by euphausiids, with copepods second (Espinoza and Bertrand 2008). However, it can be shown that when food resources change (for example, when copepods increased relative to euphausiids during 1996–2003), anchovetas adapt (Espinoza and Bertrand 2014).

We now know that the Humboldt Current was cooler on average from 1950 to 1970 and from 1985 to 2004. From 1970 to 1985, on the other hand, it was warmer on average. This warm period saw a dominance of sardine over anchoveta, presumably because warmer water encourages the smaller zooplankton that sardines prefer (Alheit and Ñiquen 2004, Swartzman et al. 2008). The two cooler periods (1950–1970 and 1985–2004) saw a dominance of anchoveta over sardine.

Note that these cycles have nothing to do with El Niño years. Indeed, Alheit and Ñiquen (2004:218) describe ENSO as "only a short-term perturbation for the [Humboldt Current] anchovy," a minor setback from which it recovers in a few years. It is the longer, 15–20 year cycles that have the capacity to restructure the ecosystem.

Alheit and Ñiquen give an example of that restructuring. Two of the anchoveta negra's main predators are the Pacific mackerel (*Scomber japonicus peruanus*) and the Chilean jack mackerel (*Trachurus murphyi*). Both fish feed on the anchoveta when it is abundant, but tend to pursue it in the warmer water of the open sea rather than in the cool Humboldt Current. During episodes of warmer sea water, however, the anchoveta swims shoreward in search of cooler upwelling water. When that happens, both mackerel and jack mackerel follow the anchoveta shoreward into the cooler Humboldt Current, exposing themselves to artisanal fishermen.

Elderly fishermen at Cerro Azul told us of years in which one could "fill a basket with anchoveta simply by wading into the bay." We suspect that these years occurred during warm, sardine-dominated water regimes, when both the anchoveta and its predators came closer to shore. At such times, the same elderly fishermen claim, "the jurel (Chilean jack mackerel) was so abundant we couldn't even sell it — we were giving it away." Clearly, the jurel had followed the anchovetas to the coast.

In Chapter 9 Sommer and Flannery argue that elite Late Intermediate compounds at Cerro Azul were catching sardines and anchovetas by the tens of thousands and drying them for shipment inland. They also conclude that in terms of their own meals, the Cerro Azul nobles preferred sardines to anchovetas. Obviously, to support this argument Sommer and Flannery have to show that the differences seen in nobles' and commoners' food debris reflect human choice, rather than natural changes in sardine and anchoveta regimes. Their evidence is given in Chapter 9.

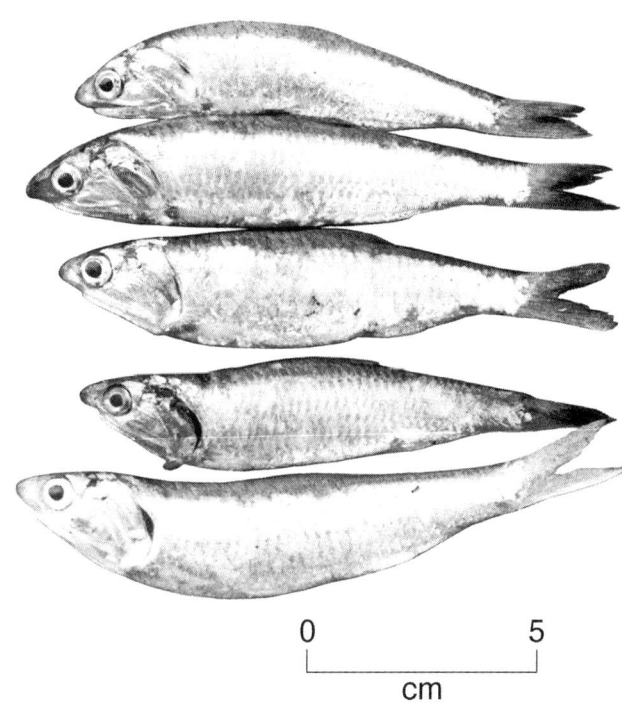

Figure 6.8. The anchoveta blanca (*Anchoa nasus*) visited Cerro Azul during the 1980s.

Finally, let us mention a second anchoveta species caught at Cerro Azul. This is *Anchoa nasus*, known locally as the anchoveta blanca (Fig. 6.8). We know less about this fish, which normally does not range south of Callao; under the warmer conditions of the 1982–83 El Niño, however, it occasionally reached Cerro Azul. The anchoveta blanca likes the shallower waters of the continental shelf, seeks out bays, and tolerates the lower salinity of estuaries. This presumably helps to reduce competition with its cousin, the anchoveta negra.

We collected the following specimens of sardines and anchovetas at Cerro Azul in 1984, and list them with their catalogue numbers, lengths, weights, and maximum vertebral diameters.

Sardinops sagax
[C054] Length 20 cm, weight 60 g, vertebral diameter 2.5 mm
[C055] Length 20 cm, weight 60 g, vertebral diameter 2.5 mm

Engraulis ringens
[C039] Length 16 cm, weight 25 g, vertebral diameter 1.85 mm
[C040] Length 15 cm, weight 25 g, vertebral diameter 1.84 mm

Anchoa nasus
[C047] Length 13 cm, weight 20 g, vertebral diameter 1.15 mm
[C048] Length 13 cm, weight 20 g, vertebral diameter 1.15 mm

Figure 6.9. The sea catfish (*Galeichthys peruvianus*), a fish eaten mainly by commoner families at Late Intermediate Cerro Azul.

Sea Catfish

The bagre or sea catfish (*Galeichthys peruvianus*) is attracted to the cool Humboldt Current and the soft sandy floor of Cerro Azul Bay (Fig. 6.9). Abundant in "normal" years, it disperses during ENSO events.

Local fishermen report that *Galeichthys* comes closer to the surface at night and seeks deeper water during daylight hours. It also protects itself with a venomous, serrated dorsal spine that is readily identifiable archaeologically. Many of Marcus' workmen considered the bagre a "trash fish" of little commercial value. In Chapter 9 Sommer and Flannery present evidence that *Galeichthys* was eaten mainly by commoners during the Late Intermediate, suggesting that elite families had as little interest in it as Marcus' workmen.

The young sea catfish consumes juvenile polychaete worms. As an adult it hangs around nurseries of anchoveta, which constitute half its food in the Andean summer. In the Andean spring and fall it increases its reliance on crustaceans.

In 1984 and 1985 we collected two *Galeichthys*, as follows:
[C056] Length 32 cm, weight 250 g, vertebral diameter 4 mm
[C081] Length 26 cm, weight 200 g, vertebral diameter 3.5 mm

Mullets

Mullets are prototypic scavengers and bottom feeders. Distributed worldwide in coastal waters, lagoons, and estuaries, they filter organic detritus from the sediment of the ocean floor. Among the 130 microscopic forms they are known to eat are diatoms, amphipods, nematode worms, foraminifera, copepods, ostracods, small molluscs, eggs, and larvae.

Two species live off the coast of Peru: the common or gray mullet *Mugil cephalus* (Fig. 6.10) and the white mullet *Mugil curema*. Both are referred to locally as lisas, and their archaeological remains are difficult to tell apart. The white mullet is better adapted to warm tropical environments than the gray mullet (Ibáñez and Colín 2014:946), but during El Niño years both lisas reach Cerro Azul.

Both *M. cephalus* and *M. curema* compete for essentially the same invertebrates, but where their ranges overlap they spawn in different seasons of the year (Ibáñez and Colín 2014). This could be seen as a strategy for reducing competition.

We believe that most of the mullet bones from Cerro Azul belonged to *Mugil cephalus*, which is common in "normal" years, and we do not believe that Cerro Azul Bay was the only local habitat for the gray mullet. Our elderly informants spoke often of a large, water-filled depression called Huaca Chola, just inland from Cerro Azul, which had existed in their youth. Huaca Chola, whose water was said to be "mineral but not warm," was considered a source of mullet, waterfowl, bulrush, cattails, and other "industrial" plants. Unfortunately, this depression was deliberately filled in during the cotton boom of the 1930s.

In 1983 we collected two *Mugil cephalus*, as follows:
[C015] Length 44 cm, weight 650 g, vertebral diameter 9.2 mm
[C016] Length 41 cm, weight 600 g, vertebral diameter 7.5 mm

Threadfins

The Pacific threadfin (*Polynemus* [=*Polydactylus*] *approximans*) inhabits continental shelf waters, estuaries, and lagoons (Fig. 6.11). It prefers sandy or muddy ocean floors and is an omnivore, consuming crabs, shrimp, clams, and marine worms.

The larval forms of *P. approximans* migrate to the open sea, but they return as adults to bays and other nearshore habitats. The threadfin's name derives from the fact that some of its pectoral fin rays extend outward in the form of threadlike filaments; the resemblance of these filaments to a wispy beard leads Peruvian fishermen to call this fish the barbudo.

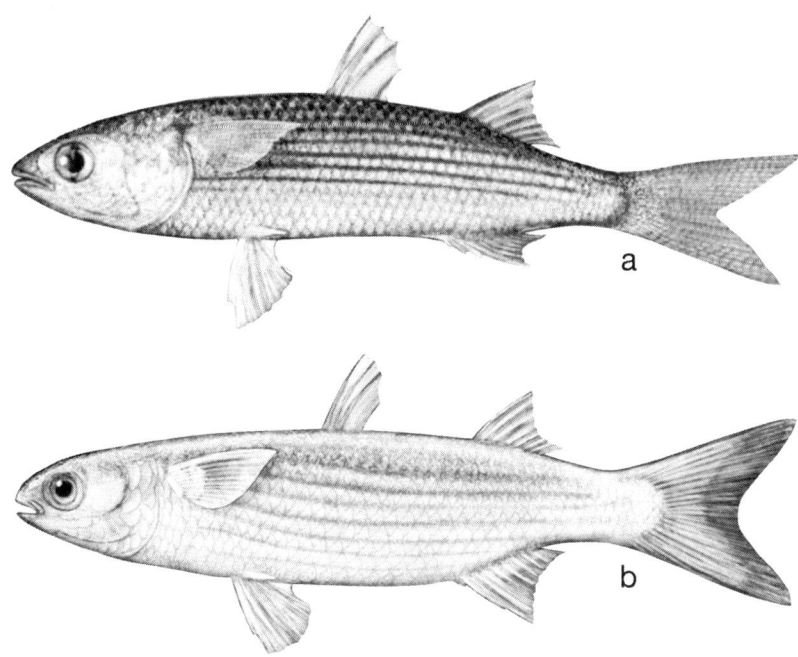

Figure 6.10. Two mullets visit Cerro Azul. *a* shows the lisa or common mullet (*Mugil cephalus*). *b* shows the white mullet (*Mugil curema*).

Figure 6.11. The barbudo or threadfin (*Polynemus approximans*) ranges south of Callao primarily in ENSO years. This specimen was purchased in the San Vicente Cañete market in 1984, during the aftermath of the 1982–83 El Niño. The filamentous pectoral fin rays that give the barbudo its Spanish name are not visible in this photograph; they are stuck to the creature's wet belly.

Figure 6.12. The pejerrey or silversides (*Odontesthes regia regia*) virtually disappeared from Cerro Azul during the 1982–83 El Niño, but returned in force in 1986.

In "normal" years the Pacific threadfin does not range as far south as Cerro Azul, but during 1984—in the aftermath of the 1982–83 El Niño—we were able to purchase one in the fish market of San Vicente Cañete. Its measurements were as follows:

[C066] Length 30 cm, weight 300 g, vertebral diameter 6.1 mm

Silversides

Despite its small size, the pejerrey or silversides (*Odontesthes regia regia*) is an active hunter (Fig. 6.12). Anchoveta eggs are one of its favorite foods, which explains why it virtually disappears in ENSO years along with the anchoveta (Llompart et al. 2013). It also eats foraminifera and the larval forms of crustaceans and molluscs (Vargas et al. 1999).

The pejerrey was one of the fish notably absent from Cerro Azul during the El Niño of 1982–83. Through the grapevine, Marcus' fishermen informants were convinced that it had gone south to the cooler waters of Chile. By 1985 they were hearing from other fishermen that silversides were slowly returning to the north. First they were "in Nasca," then "in Pisco," then "in Chincha." They finally returned to Cerro Azul in force in 1986.

Silversides pose a problem for zooarchaeologists because their bones are tiny and fragile. Left on the floor of a tapia-walled room, they tend not to survive a vigorous sweeping of the floor. That is presumably why we found no evidence of *Odontesthes* in the Feature 6 midden of Structure D (Chapter 9). That midden appeared to be largely debris swept up from the floor of an open work area. We are confident that pejerrey were eaten at Cerro Azul, but they were not preserved in any of the Late Intermediate refuse. Almost certainly, this fish is archaeologically underrepresented because of its fragility.

In 1985 we collected two *Odontesthes* that had been netted off Pisco. Their measurements were:

[C071] Length 16 cm, weight 30 g, vertebral diameter 1.15 mm
[C072] Length 16 cm, weight 30 g, vertebral diameter 1.15 mm

In 1986 we collected a third *Odontesthes*, caught near the lighthouse at Cerro Azul:

[C082] Length 18.5 cm, weight 49 g, vertebral diameter 1.3 mm

Sea Basses

Two members of the sea bass family were taken at Cerro Azul in the 1980s. These were the ojo de uva or grape-eye sea bass (*Hemilutjanus macrophthalmos*) and the cabrilla or Peruvian rock bass (*Paralabrax humeralis*). The grape-eye sea bass (Fig. 6.13) prefers areas of rocky substrate like the sea cliffs of La Centinela and El Fraile; there it feeds on small fish and crustaceans. The Peruvian rock bass (Fig. 6.14) also prefers rocky substrate and ranges widely in depth, from near the surface to the ocean floor. It tends to travel in schools, eating octopus, squid, stone crabs, opossum shrimp, and small fish (Vargas et al. 1999).

We found sea bass virtually unavailable at Cerro Azul during the 1982–83 El Niño. By late 1984 and early 1985 they were available again, and we collected the following specimens:

Hemilutjanus macrophthalmos
[C062] Length 35 cm, weight 700 g, vertebral diameter 6.9 mm
[C073] Length 24 cm, weight 225 g, vertebral diameter 4.6 mm
[C074] Length 24.5 cm, weight 250 g, vertebral diameter 4.7 mm

Paralabrax humeralis
[C078] Length 28 cm, weight 300 g, vertebral diameter 4.9 mm
[C079] Length 27 cm, weight 280 g, vertebral diameter 4.6 mm

Jacks and Pompanos

Two members of the jack-and-pompano family frequent Cerro Azul in "normal" years; these are the jurel or Chilean jack mackerel (*Trachurus symmetricus murphyi*) and the cojinova or blackruff (*Seriolella violacea*). A third member of the family—the pámpano or "paloma pompano" (*Trachinotus paitensis*)—is a warm-water fish that visits Cerro Azul only during ENSO years.

The Chilean jack mackerel (Fig. 6.15) is a wide-ranging, commercially important fish that travels in schools and can be found throughout the South Pacific from New Zealand to the coasts of Chile and Peru. It loves open water and can plunge to depths of 70 m, but seems just as at home in bays or the shallow waters of the continental shelf. Adult jack mackerels can grow to be 45–75 cm in length.

A study carried out in northern Chile (Medina and Arancibia 2002) showed the jack mackerel's favorite food to be euphausiid crustaceans, followed by small fish like the anchoveta and the lightfish or bristlemouth (*Vinciguerria* sp.). They also eat copepods, mole crabs, porcelain crabs, squid, and fish larvae.

The jurel, as mentioned earlier, will follow anchovetas to the Humboldt Current in El Niño years, when the open waters of the Pacific are too warm. For its part, jack mackerels are preyed upon by sharks, swordfish, and tuna, especially in the open sea.

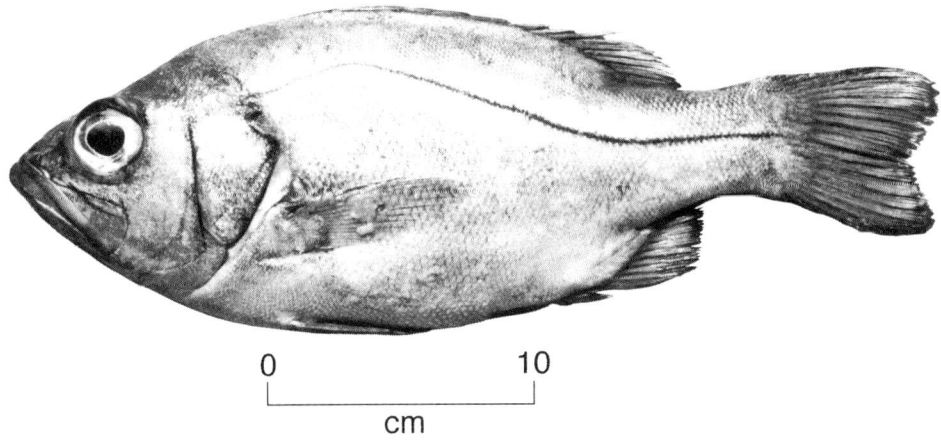

Figure 6.13 (top). The ojo de uva (*Hemilutjanus macrophthalmos*) is a member of the sea bass family.
Figure 6.14 (middle). The cabrilla or Peruvian rock bass (*Paralabrax humeralis*).
Figure 6.15 (bottom). The jurel or Chilean jack mackerel (*Trachurus symmetricus*) was a popular fish at Late Intermediate Cerro Azul.

Figure 6.16. The paloma pompano (*Trachinotus paitensis*) is a warm-water fish that visited Cerro Azul during the 1982–83 El Niño; its numbers dwindled during the 1984–1986 transition to cooler water. This was a smaller-than-average specimen.

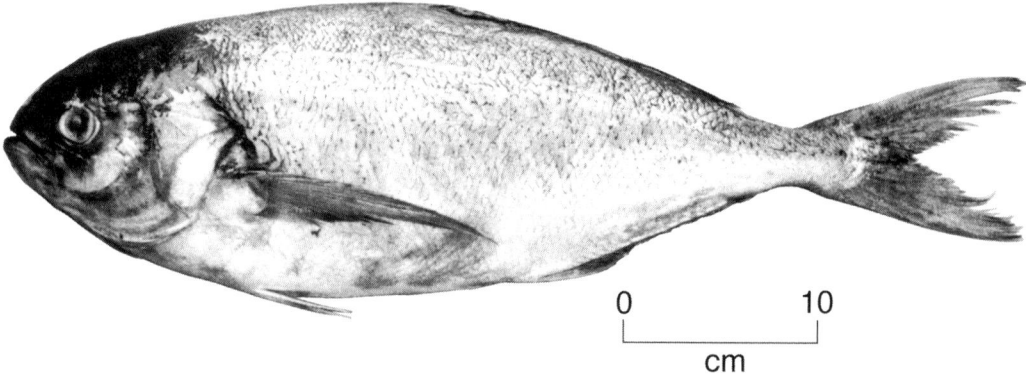

Figure 6.17. The cojinova or blackruff (*Seriolella violacea*) was an occasional visitor to Cerro Azul during the 1980s.

Figure 6.18. The dorado or dolphinfish (*Coryphaena hippurus*) is a warm-water species that visited Cerro Azul during the 1982–83 El Niño.

We collected two specimens of *Trachurus symmetricus murphyi* at Cerro Azul in 1983, toward the end of an El Niño episode. Both were near the lower end of the size range for adults, as follows:

[C025] Length 40 cm, weight 400 g, vertebral diameter 6.7 mm
[C026] Length 43 cm, weight 490 g, vertebral diameter 6.8 mm

During the 1982–83 El Niño, the paloma pompano (Fig. 6.16) spread south from the Gulf of Guayaquil, taking advantage of the abnormally warm waters off Cerro Azul. With the cooling of the ocean in 1984, pompanos withdrew again to the north.

Pompanos swim higher in the water column when out to sea. Nearer to the shore, they school over the sandy ocean floor in search of crustaceans and molluscs. Adults of the species range in length from 30 to 50 cm.

We collected four *Trachinotus paitensis* during the 1982–83 El Niño, as follows:

[C001] Length 28 cm, weight 250 g, vertebral diameter 5.16 mm
[C002] Length 28 cm, weight 350 g, vertebral diameter 5.15 mm
[C003] Length 26 cm, weight 200 g, vertebral diameter 5.10 mm
[C004] Length 24 cm, weight 200 g, vertebral diameter 5.10 mm

The cojinova (Fig. 6.17) is a fish that likes the shallower waters of the continental shelf. There it may, on occasion, form small schools with its cousin the Chilean jack mackerel. Cojinovas will eat planktonic invertebrates such as amphipods and copepods, small fish such as sardines and anchovetas, and even juvenile Chilean jack mackerels.

Unlike the pompano, *Seriolella violacea* became scarce during the ENSO years of 1982–83. It returned in 1984–85, allowing us to collect the following two specimens:

[C049] Length 55 cm, weight 2000 g, vertebral diameter 12.4 mm
[C075] Length 59 cm, weight 2600 g, vertebral diameter 13.8 mm

Dolphinfish

Another fish that visits Cerro Azul in El Niño years is *Coryphaena hippurus*, the dorado or dolphinfish (Fig. 6.18). The dorado prefers its water warmer than 21° C, and usually frequents tropical seas from Panama to Ecuador. This is the same fish known in Hawaii as the mahi mahi.

Coryphaena is a migratory pelagic fish whose adult weight can reach 40 kg. Its populations fluctuate seasonally, usually peaking between November and May. When it expands its range to central Peru in ENSO years it is probably following two of its favorite foods—shrimp and flying fish—both of which undertake the same southward migration. The dorado also feeds on squid, crabs, and smaller pelagic fish, including the Pacific sardine.

During the El Niño of 1982–83 we came upon three partial skeletons of *Coryphaena hippurus*; all were part of the refuse behind small seafood restaurants. Our workmen informed us that local restaurants were willing to buy as many dorado as they could get, knowing that this tasty fish would no longer be available when cooler ocean temperatures returned.

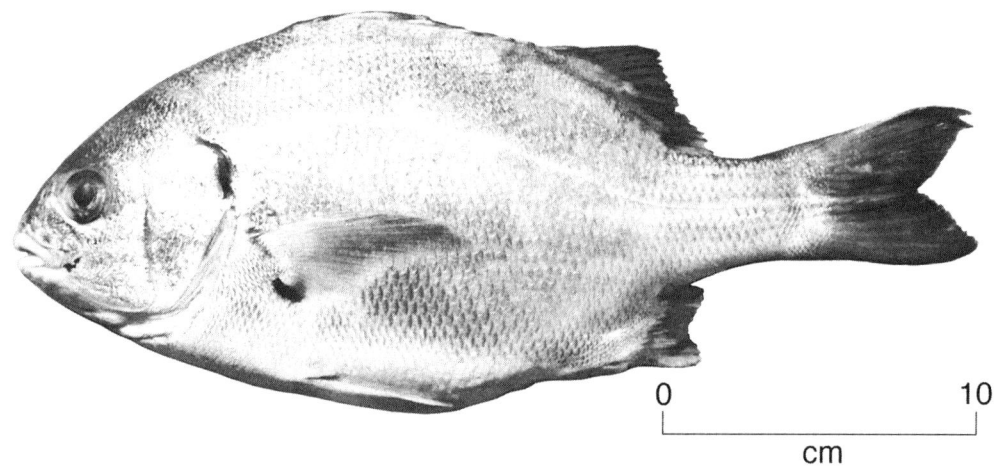

Figure 6.19. The chita (*Anisotremus scapularis*) is the most common member of the grunt family caught at Cerro Azul today.

Grunts

At least two species of grunt frequent the central coast of Peru: the chita (*Anisotremus scapularis*) and the cabinza (*Isacia conceptionis*). The chita (Fig. 6.19) was so common during our stay that we could have collected many more specimens than we did. We were told that cabinza were present at Cerro Azul during April and May of 1985, but we were not in the field at that time and failed to collect one.

The genus *Anisotremus* is believed to have evolved in an environment of rocky reefs with sandy or muddy bottoms; later, some species went on to colonize coral reefs (Bernardi et al. 2008). The environment in which the chita is found today at Cerro Azul is the sea cliff of La Centinela and the small rocky islands just offshore. The ocean bottom there varies from rocky to sandy or gravelly. The waves crash hard against La Centinela, but the chita comes so close to the cliffs that it can be caught in cast nets by fishermen standing on rocky ledges (see Chapter 7).

Grunts tend to school during the day, dispersing at night to forage individually (Bernardi et al. 2008). At Cerro Azul, chitas come to the sea cliffs to eat kelp. They are omnivores who also eat copepods, amphipods, ostracods, small crabs, small fish, larvae, and polychaete worms (Berrios and Vargas F. 2004). One need only look at their pharyngeal teeth (Chapter 9) to realize that chitas spend a lot of their time crushing shellfish. When threatened by predators, the chita may seek shelter among the sea urchin colonies of the ocean floor, relying on the urchins' sharp spines for protection from larger fish.

Local fishermen report that Isla Corriente, a rocky island some 10 km north of La Centinela, is another excellent fishing locality for chita. During the 1982–83 El Niño and its aftermath, many artisanal fishermen from Cerro Azul made the trip to Isla Corriente when the fishing was poor at Cerro Azul.

We collected three *Anisotremus scapularis* in 1983 and 1984, as follows:

[C006] Length 40 cm, weight 1000g, vertebral diameter 7.6 mm

[C009] Length 35 cm, weight 750 g, vertebral diameter 7.5 mm

[C063] Length 31 cm, weight 600g, vertebral diameter 6.2 mm

In addition we were allowed to measure the following specimens, which we did not skeletonize:

Length 40 cm, weight 1500 g

Length 49 cm, weight 2100 g

Length 43 cm, weight 1900 g

Length 32 cm, weight 600 g

Drums

No family of fish is better represented on the Cañete coast than the Sciaenidae, referred to as drums or croakers because of the vocalizations amplified by their air bladder. Sciaenids are small to medium-sized carnivores of warm temperate or tropical waters, grazing the ocean floor for invertebrates and pursuing sardines and anchovetas. No fewer than nine series of drum were caught at Cerro Azul between 1983 and 1986. Included were two of the most sought-after fish on the menus of Lima's seafood restaurants, the corvina and róbalo.

When nine species of fish from the same family occupy the same stretch of ocean, one of the first questions asked is: How do they keep from competing? Do they eat different foods, or different sizes of foods? Do some feed in the morning, and others at night? Do they take turns, coming to a particular locality in different seasons? Unfortunately, we can give only partial answers to these questions.

The nine species of Sciaenids caught at Cerro Azul between 1983 and 1986 do, in fact, show substantial size variation;

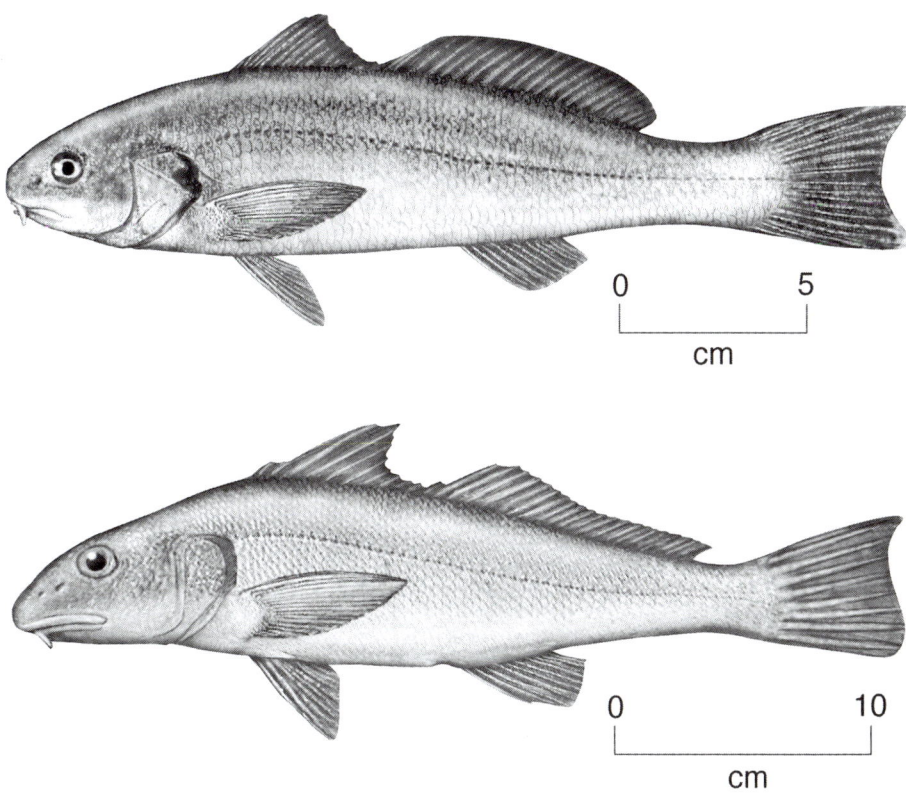

Figure 6.20 (top). The mismis (*Menticirrhus ophicephalus*) is one of the common species of the drum family (Sciaenidae) taken at Cerro Azul.
Figure 6.21 (bottom). The zorro (*Menticirrhus rostratus*), a relative of the mismis.

included are differences in mouth size that could affect their choice of prey. Both corvina (*Cilus gilberti*) and róbalo (*Sciaena starksi*) would be considered large drums (greater than 50 cm length as adults). At the other end of the size spectrum, both the mismis (*Menticirrhus ophicephalus*) and the mojarrilla (*Stellifer minor*) would be considered small drums (less than 25 cm length as adults). The latter two species consume similar invertebrate prey, but are thought to use temporal differences in activity to reduce competition.

Unfortunately that still leaves us with five species, all of which would be considered medium-sized drums (roughly 26–40 cm in adult length). These species are the zorro (*Menticirrhus rostratus*), the coco (*Paralonchurus peruanus*), the ayanque (*Cynoscion analis*), the burro (*Sciaena fasciata*), and the lorna (*Sciaena deliciosa*). All feed on the invertebrate fauna of sandy or gravelly ocean floors, and we have incomplete knowledge of the subtle differences in their behavior that may reduce competition. There is evidence in the daily records of the Capitanía del Puerto that some of these fish undergo cycles of abundance and scarcity at Cerro Azul (Appendix A).

Let us now consider these drums in phylogenetic order. The mismis or snakehead croaker (Fig. 6.20) was one of the most frequently caught Sciaenids at Cerro Azul during 1984–86, and its peaks of abundance seemed to be correlated with those of the lorna (Appendix A). The mismis is a small fish of coastal waters and bays, where it feeds on polychaete worms, molluscs, mole crabs, and other small crustaceans. Although its diet resembles that of the lorna (Vargas et al. 1999), those two drums coexist well, perhaps because the larger lorna can capture prey that would be too large for the mouth of a mismis.

We collected two specimens of *Menticirrhus ophicephalus* in 1984, after the 1982–83 El Niño had ended. Their measurements were as follows:

[C037] Length 24 cm, weight 150 g, vertebral diameter 4 mm
[C038] Length 24 cm, weight 150 g, vertebral diameter 4 mm

The zorro (*Menticirrhus rostratus*), although a member of the same genus as the *mismis*, is more than half again as large (Fig. 6.21). It was abundant at certain moments, but then might disappear from Cerro Azul for as much as six months at a time.

We collected three specimens of *Menticirrhus rostratus* in 1983, which indicated to us that this medium-sized drum did not necessarily vanish during ENSO conditions. Their measurements were as follows:

[C013] Length 36 cm, weight 650 g, vertebral diameter 7.9 mm
[C014] Length 36 cm, weight 650 g, vertebral diameter 7.9 mm

Figure 6.22 (left). The coco (*Paralonchurus peruanus*), a medium-sized member of the drum family.

Figure 6.23. The ayanque (*Cynoscion analis*), another medium-sized member of the drum family.

Figure 6.24. The burro (*Sciaena fasciata*), another medium-sized member of the drum family.

Figure 6.25. The lorna (*Sciaena deliciosa*). This medium-sized drum was one of the most frequently caught fish at Late Intermediate Cerro Azul.

[C024] Length 43 cm, weight 750 g, vertebral diameter 8.4 mm

The *coco* or Peruvian banded croaker (Fig. 6.22) inhabits the waters of sandy beaches and bays from Panama to Peru. There it scours the ocean floor for mobile crustaceans such as crabs and shrimp, and consumes molluscs and marine worms. This medium-sized drum can descend to more than 10 m depth if necessary. After the 1982–83 El Niño, *Paralonchurus peruanus* did not return to Cerro Azul in large numbers until 1985. We collected one specimen in late 1984 and two more in 1985, as follows:

[C065] Length 26 cm, weight 250 g, vertebral diameter 4.6 mm
[C069] Length 33 cm, weight 400 g, vertebral diameter 5 mm
[C070] Length 29 cm, weight 300 g, vertebral diameter 5 mm

The ayanque or Peruvian weakfish (Fig. 6.23) is another medium-sized drum that likes sandy or gravelly ocean floors. In its juvenile stage it enters estuaries and shallow bays from Ecuador to Chile; as an adult it tends to move farther out to sea, where it can descend to 50 m below the surface. Depending on its size and stage of development, the ayanque may eat shrimp and smaller crustaceans or pursue octopus, squid, and small fish. It may therefore compete with small Sciaenids during its juvenile stage and with medium-sized Sciaenids as an adult.

We collected two specimens of *Cynoscion analis* in 1984, after the 1982–83 El Niño had ended. Their measurements were as follows:

[C044] Length 32 cm, weight 370 g, vertebral diameter 5 mm
[C045] Length 29 cm, weight 270 g, vertebral diameter 4.9 mm

Artisanal fishermen at Cerro Azul spoke of having captured the burro (Fig. 6.24), but it was so scarce during our stay that we failed to collect one locally. Finally, in 1985, we purchased two specimens of *Sciaena fasciata* in the fish market of Lima's Chorrillos district. Their measurements were as follows:

[C076] Length 35 cm, weight 850 g, vertebral diameter 13 mm
[C077] Length 30 cm, weight 500 g, vertebral diameter 7 mm

By far the most common drum caught at Cero Azul—both during the 1980s and the Late Intermediate—is the "delicious drum" or lorna (Fig. 6.25). *Sciaena deliciosa* prowls sandy or gravelly ocean floors, from the sunlit upper zone to depths of 37 m. It seeks out mole crabs and other benthic crustaceans, pelecypod molluscs, opossum shrimp, crabs, and polychaete worms. Lornas were often present in huge schools during the 1980s, during which times they were netted by the thousands. Between the 13th and 25th of July, 1984, some artisanal boats reported catches of 50 to 140 dozen lorna. Their numbers then dwindled, but lornas returned in force between January 3rd and 8th, 1985 (Marcus et al. in press). Such cycles of abundance and scarcity continued throughout our stay. Often, as mentioned earlier, peaks of lorna coincided with peaks of the smaller mismis (Appendix A).

We collected two lorna in 1984, and could have had dozens more. Their measurements were as follows:

[C042] Length 32 cm, weight 400 g, vertebral diameter 6 mm
[C043] Length 32 cm, weight 400 g, vertebral diameter 6 mm

We come now to the two largest and most commercially important drums of the Cañete coast, the róbalo and the corvina. Once considered to belong to the same genus, they have now been placed in different genera.

The róbalo, *Sciaena starksi* (Fig. 6.26) is the largest drum caught at Cerro Azul today, and the archaeological fish bones indicate that those caught in the Late Intermediate were larger still. For example, in 1984 we collected a 61 cm-long róbalo whose maxilla measured 4.5 cm in length. From Late Intermediate refuse, however, we recovered a róbalo maxilla measuring 9.7 cm (Chapter 9). A róbalo twice as big as the one we collected in 1984 would have been an impressive fish indeed.

The two róbalos we collected at Cerro Azul were caught off the *costa*, or cobble beach, immediately southeast of the archaeological site. Both were caught with the espinel, a long hand-held line to which multiple hooks are attached at regular intervals (see Chapter 7). In one case, the hooks were baited with Pacific sardines; in the other case, they were baited with white chicken feathers. Apparently, róbalos mistake the white feathers for sardines or anchovetas.

Róbalos and corvinas have been observed congregating near the mouths of rivers on the Peruvian coast during times of flooding, waiting for crayfish and smaller crustaceans to be carried from the river to the ocean (Murphy 1920, 1923). This could have happened seasonally at the mouth of the Río Cañete, but along the cobble floor of the ocean near Cerro Azul we suspect that róbalo were eating the polychaete worms so abundant in that zone.

The dimensions of the róbalo we collected were as follows:

[C052] Length 56 cm, weight 2200 g, vertebral diameter 13 mm
[C059] Length 61 cm, weight 2700 g, vertebral diameter 14.8 mm

The corvina, *Cilus gilberti* (Fig. 6.27), is a predatory carnivore of temperate and tropical coasts from Panama to Chile; it seeks out areas of sandy or gravelly ocean floor at depths of 5 to 50 m. While the corvina's abundance can be temporarily affected by El Niño conditions, it cares little whether the ocean is in a sardine-dominant or anchoveta-dominant phase, since it eats both. In fact, the corvina is described as heterotrophic because it dines on so many levels of the food chain (Chong et al. 2000, Fernández and Oyarzún 2001).

In an area of *playa* and *costa* like the one fronting Cerro Azul, corvinas eat tiny amphipods, opossum shrimp, and floating larvae; take polychaetes and other marine worms when they emerge from their cobble retreats; pluck molluscs from the bottom; grab mole crabs (*Emerita*) and ghost shrimp (*Callianassa*) from the sand when they expose themselves; and pursue schools of anchoveta and sardine.

We had trouble collecting corvina at Cerro Azul because we

Figure 6.26. The róbalo (*Sciaena starksi*) is the largest of the drums caught at Cerro Azul today, and our Late Intermediate refuse included róbalos even larger than those caught in the 1980s.

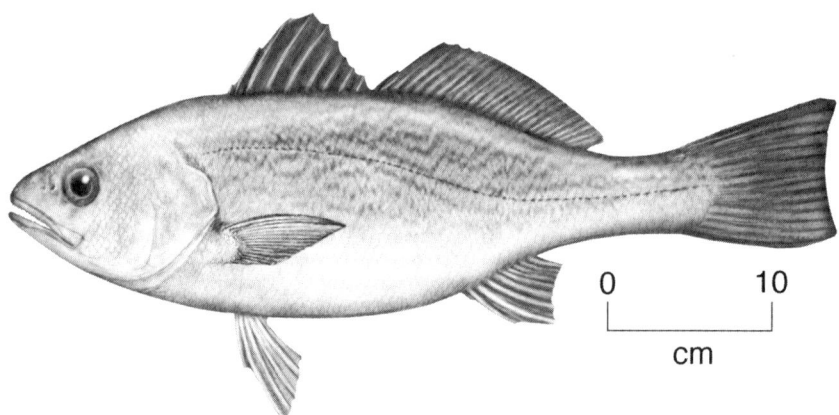

Figure 6.27. The corvina (*Cilus gilberti*) is so desirable a drum that most specimens caught today at Cerro Azul go directly to the restaurants of Lima.

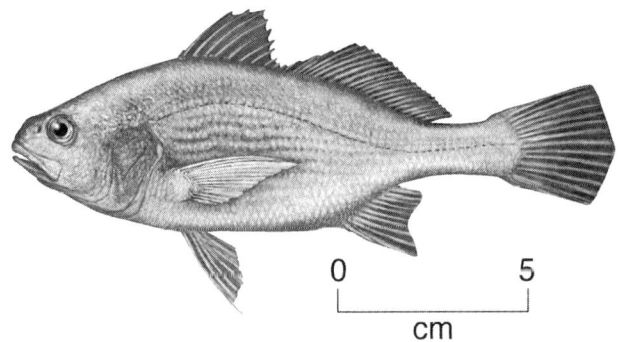

Figure 6.28 (right). The mojarrilla (*Stellifer minor*) is the smallest of the drum species captured at Cerro Azul today.

Figure 6.29. The pintadilla (*Cheilodactylus variegatus*), a medium-sized fish eaten mostly by elite families at Cerro Azul during the Late Intermediate.

were competing with the best restaurants on the central coast. Only by paying top dollar did we collect two in 1983, as follows:
  [C007] Length 50 cm, weight 1125 g, vertebral diameter 11.2 mm
  [C008] Length 49 cm, weight 1100 g, vertebral diameter 11 mm

In addition, we were allowed to measure an undersized specimen that we did not skeletonize:
  Length 35 cm, 600 g, vertebral diameter 8.4 mm

Finally we come to the mojarrilla or minor stardrum, *Stellifer minor* (Fig. 6.28), the smallest of the drums caught at Cerro Azul during our stay. The mojarrilla likes to live in warm, shallow water over sandy bottoms, where the bulk of its diet consists of small crustaceans. The latter range in size from planktonic copepods and small mysids (opossum shrimp) to larger shrimp, crabs, and their free-swimming larvae.

Given its size, the mojarrilla's main Sciaenid competitor would be the mismis. One difference between the two is that *Stellifer* actually increases in abundance under ENSO conditions (Iannacone 2004), while the mismis becomes scarce. Two specimens of *Stellifer minor* that we collected in 1983 had the following dimensions:
  [C030] Length 16 cm, weight 68 g, vertebral diameter 3 mm
  [C008] Length 15 cm, weight 55 g, vertebral diameter 2.7 mm

Morwongs

The Cheilodactylidae are a family of South Pacific fishes, adapted to the rocky coasts of Australia, New Zealand, Easter Island, Peru, and Chile. These fish are so well represented on the coast of Australia that we call them by their Australian name, morwong.

The pintadilla or Peruvian morwong, *Cheilodactylus variegatus* (Fig. 6.29), swims in small groups off the sea cliffs and stony offshore islands of the Cañete coast. It hides from its predators in rocky outcrops and beds of seaweed, venturing

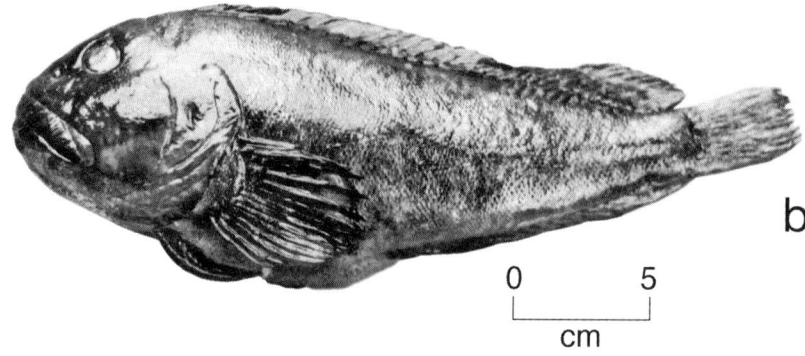

Figure 6.30. Two species of blenny caught at Cerro Azul. *a*, the borracho or scaleless blenny (*Scartichthys gigas*). *b*, the trambollo or scaled blenny (*Labrisomus philippii*). *Scartichthys* was more common in Late Intermediate refuse than *Labrisomus*.

out to pick sea snails off the igneous rock of the cliffs. The pintadilla also preys on copepods, amphipods, porcelain crabs, and polychaete worms (Berrios and Vargas F. 2004). The local fishermen catch it with gill nets, maintaining that it has little interest in baited hooks.

The distribution of morwong in the archaeological deposits at Cerro Azul suggests that it was another fish which (like the corvina and róbalo) was preferred by the Late Intermediate elite.

In 1984 (after the 1982–83 El Niño had ended) we collected two *Cheilodactylus variegatus*, as follows:

[C050] Length 28 cm, weight 300 g, vertebral diameter 5 mm
[C051] Length 27 cm, weight 320 g, vertebral diameter 5.5 mm

Blennies

Two blennies are taken by the fishermen of Cerro Azul. One is primarily a vegetarian, the other a carnivore/generalist. Both hang around the kelp beds of the intertidal zone, although they have been known to descend to 20 m depth.

The borracho or scaleless blenny, *Scartichthys gigas* (Fig. 6.30 *a*), is largely herbivorous; more than 90 percent of its food by weight is seaweed (Berrios and Vargas F. 2004). The remaining 10 percent includes molluscs, amphipods, and porcelain crabs.

In 1984 we collected two *Scartichthys gigas*, whose measurements are given below. The stomach contents of both specimens included tiny juvenile mussels (*Semimytilus algosus*). The presence of these mussels confirms the observations of Díaz and Muñoz (2010:296) that *Scartichthys*—in spite of its largely herbivorous diet—will eat invertebrates, although the latter are "small prey and are always in a low number."

[C060] Length 14.5 cm, weight 30 g, vertebral diameter 2 mm
[C061] Length 14.5 cm, weight 30 g, vertebral diameter 2 mm

The trambollo or scaled blenny, *Labrisomus philippii* (Fig. 6.30 *b*), also frequents kelp beds on the rocky coasts of Peru and Chile. Unlike the borracho, however, the trambollo is mainly looking for small game such as polychaete worms, amphipod crustaceans, and even small fish (Biffi and Iannacone 2010). One reason it likes the kelp beds is that they provide it with places to hide from its main predator, the sea otter (*Lutra felina*).

In 1984 we collected two *Labrisomus philippii*, as follows:
[C057] Length 28 cm, weight 450 g, vertebral diameter 5 mm
[C058] Length 30 cm, weight 450 g, vertebral diameter 6 mm

Figure 6.31 (right). The caballa or Pacific mackerel (*Scomber japonicus peruanus*).

Figure 6.32. The Pacific bonito (*Sarda sarda chiliensis*). This elegant fish was both consumed and featured in the art of Late Intermediate Cerro Azul.

The Tuna-Mackerel Family

The family Scombridae (tunas and mackerels) includes some of the more important game fish of the South Pacific. Three species that visit Cerro Azul are the caballa or Pacific mackerel (*Scomber japonicus peruanus*), the Pacific bonito (*Sarda sarda chiliensis*), and the sierra or sierra mackerel (*Scomberomorus maculatus sierra*).

Like most members of its family, the Pacific mackerel (Fig. 6.31) is a wide-ranging migratory fish. It forms schools of thousands of individuals, sometimes even schooling jointly with the very sardines that it preys upon.

Since these fish swim relatively high in the water column, they are vulnerable to the dive-bombing attacks of birds like the Peruvian booby (Chapter 11). To protect it from such attacks, the upper surface of the caballa has evolved a protective coloration that confuses birds, making it difficult to identify a mackerel in the sunlight-dappled water.

Because the Pacific mackerel has been extensively studied, we know that its diet tends to change as it grows older. As a larval fish it feeds voraciously on copepods and rotifers. This reliance on zooplankton continues so long as the herring is a juvenile. As an adult, the mackerel begins to eat opossum shrimp, euphausiids, tunicates, and sardines. Juvenile mackerel are attracted to kelp beds, sandy beaches, and open bays; as adults they move to open water and expand their depth range to 300 m. Schools of mackerel attract tuna, swordfish, sharks, dolphins, and sea lions.

In 1983 (an El Niño year) we collected two specimens of *Scomber japonicus peruanus*, as follows:

[C020] Length 34.5 cm, weight 325 g, vertebral diameter 5.1 mm
[C021] Length 32.5 cm, weight 325 g, vertebral diameter 5 mm

In addition, we were able to measure a third specimen that we did not skeletonize:

Length 33 cm, weight 300g

The Pacific bonito (Fig. 6.32) ranges from Ecuador in the north to Chile in the south. Owing to its tolerance of warm water, it may increase in abundance during El Niño years; for example, bonito was more available than usual at Cerro Azul during the 1982–83 El Niño and its aftermath. Its numbers declined after cooler water temperatures were reestablished, but it never really became unavailable (Appendix A).

Like their cousins the mackerels, bonito are speedy denizens of the upper water column. They tend to form schools by developmental stages, juveniles hanging out with other juveniles and adults with adults. In the Cerro Azul area they eat squid,

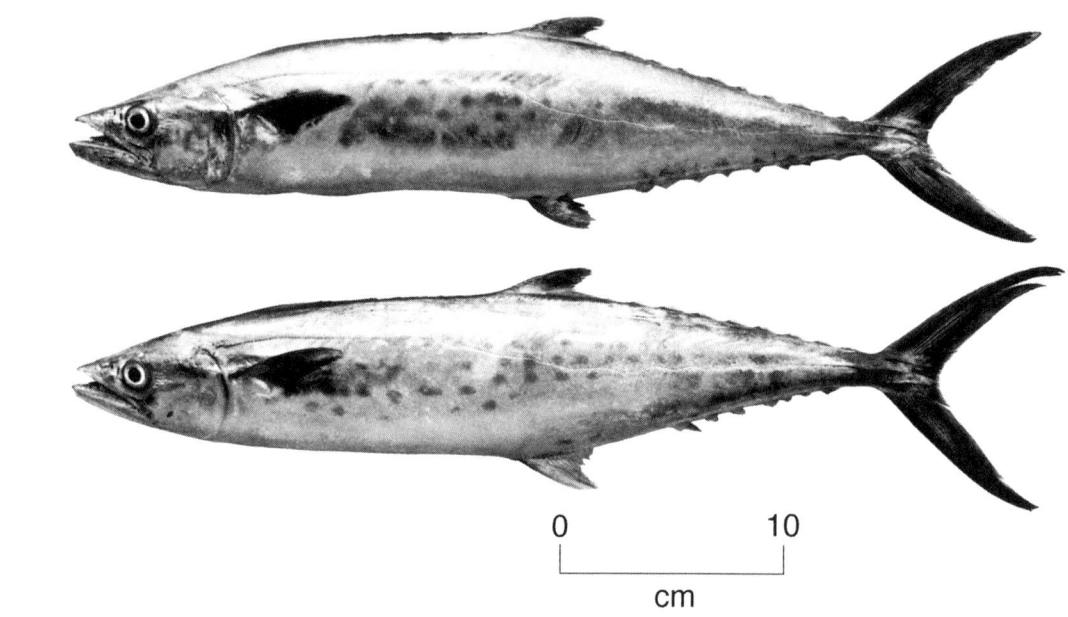

Figure 6.33 (above). The sierra (*Scomberomorus maculatus sierra*), a member of the mackerel family.

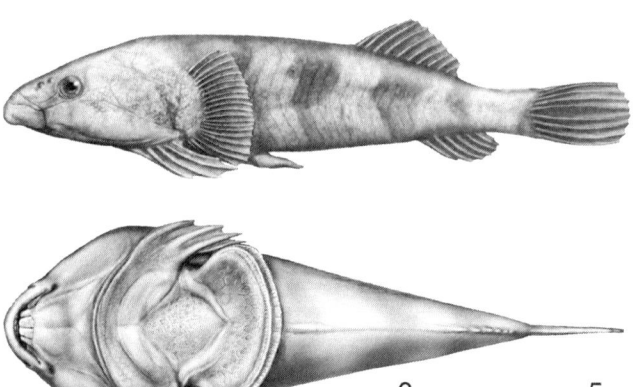

Figure 6.34 (left). Lateral and ventral views of the pejesapo or clingfish (*Sicyases sanguineus*), which attaches itself to rocks and sea cliffs by means of a sucker disc.

shrimp, and small schooling fish such as sardines and anchovetas.

According to Ramiro Matos, the fact that bonito tends to last a long time without spoiling led to its being one of the fish that the Inca were most interested in drying and salting. Not only was it eaten at Cerro Azul in Late Intermediate times, it was featured in the art of that period (Marcus 1987a: Fig. 63). In today's Peru, however, many people complain that bonitos have too gamey a flavor and smell unpleasant while being cooked.

We collected one *Sarda sarda chiliensis* in 1983 and another in 1984, as follows:

[C019] Length 41 cm, weight 600 g, vertebral diameter 8 mm
[C020] Length 58 cm, weight 1700 g, vertebral diameter 10.9 mm

Finally we come to the sierra mackerel (Fig. 6.33). This is another wide-ranging migratory fish, seen as far north as California and as far south as Chile. The sierra tends to spawn in large schools near the shore before heading out to open sea. It eats sardines and anchovetas and (in contrast to the bonito) is prized as ceviche because of its excellent, delicate flavor.

We collected two *Scomberomorus maculatus sierra* during the El Niño year of 1983. Their measurements were as follows:

[C032] Length 44 cm, weight 450 g, vertebral diameter 6.2 mm
[C033] Length 41 cm, weight 400 g, vertebral diameter 6.1 mm

Clingfish

One of the most interesting fish of the Cerro Azul area is the pejesapo or clingfish, *Sicyases sanguineus* (Fig. 6.34), whose range extends from central Peru to southern Chile. Its unique adaptation is to attach itself to vertical rock walls in the intertidal zone, where it consumes seaweed, chitons, limpets, sea snails, mussels, barnacles, and small crabs. Never mind that it is exposed to the air when the tide goes out; the pejesapo tolerates periodic exposure so well that some authors consider it "virtually amphibious."

Natural selection has fused the pejesapo's pelvic fins and converted them into a sucker disk bearing dermal papillae. By combining the suction of the disk and the adhesion of the papillae, the pejesapo clings to the sheer rock face despite taking a pounding from the waves. Its upper teeth are incisor like and its lower teeth chisel-like, allowing it either to scrape algal film or to seize prey, depending on circumstances. Clingfish are so good at what they do that they may live on cliffs at densities of up to 10 fish per m^2 (Cancino and Castilla 1988).

Like so many fish we have considered here, pejesapos display different behavior and diet preferences at different stages of their life. They are more gregarious when young, tend to occur higher on the rock face, and tend to eat more algae. They are more quiescent as adults, tend to occur lower on the rock face, and tend to be more carnivorous. Clingfish are alert to movement around them, and when threatened by a predator they quickly release their grip on the cliff and swim to safety.

Because colonies of pejesapo do such an impressive job of consuming *Enoplochiton*, *Fissurella*, *Concholepas*, *Tegula*, *Semimytilus*, and *Perumytilus* (see Chapter 3), Cancino and Castilla (1988) consider them to play a role in structuring invertebrate communities on vertical rock faces. Clingfish not only eat herbivorous mussels, they also eat the carnivorous whelks that eat the mussels.

The Late Intermediate occupants of Cerro Azul ate *Sicyases* (Chapter 9), which made us eager to collect and skeletonize a specimen. Our failure to collect one largely reflects the fact that our artisanal fishermen did not see them as commercially important. "If you want a pejesapo," one grizzled veteran advised us, "try the restaurants in Lima's barrio chino." We later learned that *chifas*, or Chinese restaurants, know how to convert the "toadfish" into a fabulous dish.

Flounders

The lenguado or left-eye flounder, *Paralichthys adspersus* (Fig. 6.35), is another fish with a distinctive evolutionary adaptation. It hatches as a normal, bilaterally symmetrical larval fish, feeding on copepods and other zooplankton. As the lenguado grows from larva to juvenile, however, its right eye begins to migrate to the left side of its head. As an adult it lies motionless on its right side on the sandy ocean floor, camouflaged by protective coloration and partially covered with sand. In this setting it becomes an ambush predator, waiting for its prey to swim within reach of its mouth. Then, with unexpected speed, it lunges upward and grasps its prey with needle-like teeth.

Paralichthys adspersus is a fish of temperate, sandy floored continental shelves from Paita (Peru) to Chile (Oliva et al. 1996);

it is particularly fond of shallow bays like the one at Cerro Azul. As a juvenile it eats copepods and polychaete worms, but as it grows it begins to add juvenile fish to its diet. By the time it has reached its full adult length of 70 cm, it is dining on silversides, anchovetas, and crustaceans such as *muy muy*.

The lenguado's predilection for shallow bays makes it vulnerable to predation by both humans and guanay cormorants (Chapter 11). Many fishermen of Cerro Azul capture flounders by wading into the bay with a trammel net or red de cortina, a device like a tennis net, whose weights hold it to the sandy ocean floor (Chapter 7). Owing to the lenguado's tendency to lie flat in the sand, a red de cortina stretched between two men may be more effective than fishing from a small boat.

We collected two specimens of *Paralichthys adspersus* in 1983, as follows:

[C034] Length 40 cm, weight 750 g, vertebral diameter 8.8 mm
[C035] Length 39 cm, weight 700 g, vertebral diameter 8.4 mm

Hake

One of the most commercially important fish of Peruvian coastal waters is the merluza or Peruvian hake, *Merluccius gayi peruanus* (Fig. 6.36). It is, however, a demersal fish that is more likely to be taken by deepwater trawlers than by artisanal fishermen in rowboats. Not a single hake bone was found in Late Intermediate refuse at Cerro Azul.

The merluza breeds in the Paita-Chimbote section of the northern Peruvian coast, and tends to be larger in size on the north coast than on the south. After leaving their spawning grounds, hake forage to depths of 50 m over the continental shelf and up to 500 m in the open sea. This would have put them largely out of range for Late Intermediate fishermen in caballitos de totora (Chapter 7).

Between 1995 and 1997, Orrego and Mendo (2012) examined the stomach contents of 2,836 merluza. They found that these fish adapted to the El Niño of 1997–98 by broadening the list of species they ate. In years of cooler water temperatures the hake eats mostly small fish, supplemented by squid and crustaceans (crabs, mantis shrimp, and euphausiids). One of the most surprising discoveries of Orrego and Mendo's study was the level of cannibalism in merluza; other hake were the most common fish found in their stomachs, followed by anchoveta negra, anchoveta blanca, sardines, and mojarrillas.

During the El Niño year of 1997, hakes settled for smaller prey on average, and expanded their diet to include Pacific mackerel and Chilean jack mackerel. They also broadened the range of crustacean species they ate.

In 1983 (under El Niño conditions) we purchased two *Merluccius gayi peruanus*, as follows:

[C027] Length 45 cm, weight 500 g, vertebral diameter 6 mm
[C028] Length 42 cm, weight 500 g, vertebral diameter 6 mm

Figure 6.35. The lenguado or left-eye flounder (*Paralichthys adspersus*). This was the flounder most frequently caught at Cerro Azul in the 1980s.

Figure 6.36. The merluza or Pacific hake (*Merluccius gayi peruanus*). This demersal fish is taken mostly by deep-sea trawlers; no specimens were found in Late Intermediate middens at Cerro Azul.

Summary

Within a kilometer of the site of Cerro Azul one can find a shallow bay, sandy beaches, cobble beaches, sea cliffs, and rocky offshore islands. The result is an extremely varied set of fish resources. Some species, like those of the mackerel family, are wide-ranging and migratory; others, like the clingfish, attach themselves to a specific cliff. Many drum species graze the rocky sea floors and cliff faces; the flounder lies half buried in the sand. Sardines and anchovetas display 15–20 year cycles in which they take turns being the dominant small fish.

Once having stored our comparative fish collection at the Universidad Nacional Mayor de San Marcos for safekeeping, we considered the following research questions: How comparable would the fish resources of the Late Intermediate period be? How much richer was the sea before there were commercial trawlers, refrigerator ships, and artisanal boats with outboard motors? How limiting was the fact that caballitos de totora were relatively unseaworthy? Could El Niño years, or the longer sardine-anchoveta cycles, be detected in our archaeological refuse? Would we be able to recognize differences in fish preference between the nobles and commoners of Late Intermediate Cerro Azul? Chapter 9 provides partial answers to these questions.

7

Fishing Strategies and Fishing Gear

Joyce Marcus

Chapters 1 and 6 have prepared us for the fact that Cerro Azul has at least three major marine environments, featuring dozens of species of fish. In the 1980s, however, fishing was being done with nylon nets and watercraft unavailable in Late Intermediate times. I knew that once we began to excavate, we would almost certainly find that both the Late Intermediate fishing technology and the mix of species differed from today's. To assess the differences, I expected to rely on ancient fish bones and fishing nets, both of which appeared to be well preserved at Cerro Azul. I also knew that Ramiro Matos and I could interview living fishermen at Cerro Azul, and that María Rostworowski and I could consult Spanish Colonial documents concerning the fishermen of the Peruvian coast.

Fishing at Cerro Azul in the 1980s

Matos and I began our interviews with fishermen on Cerro Centinela, a sea cliff that belongs to the *peña* zone described in Chapter 1. Boats are of no use here, as the waves would dash them to pieces against the cliff. Instead, the fishermen of the *peña* scramble down the sheer rock face until they find a ledge they can stand on. From that vantage point they throw an atarraya, or traditional circular cast net, into the sea below (Fig. 7.1). They are hoping that this net will close around some of the fish that approach the cliff, either to graze on algae or to pick mussels off the rock. One of these fish is the chita or grunt (Fig. 7.2).

Peña fishermen have given names to each part of the cliff, and they described to us the different fish and molluscs that could be caught or collected there. Each day the fisherman decides where he wants to go and which net he takes with him. Among the factors affecting his decision are today's weather, how well his fishing went the previous day, and the time of day he plans to leave. Fishermen emphasized to me that the type of net they take to the *peña* is different from that taken to other environments; some men own as many as four different types (Fig. 7.3).

In 1982 I spoke to two elderly fishermen after they had ascended the cliff face, each still holding his cast net. They explained that the use of the atarraya requires a lot of skill, practice, and patience. On that day they were using it to catch mullet, because the lisa will not bite a hook. As our conversation continued, they began to talk about the strategies they used with some of the nets shown in Fig. 7.3. "We use the atarraya on the cliffs of El Fraile and Centinela to catch lisa; we use the espinel on the *costa* (the gravel and cobble beach to the south of Cerro Centinela); and we use the red de cortina (curtain or trammel net) in Cerro Azul Bay."

Each of the fishermen I interviewed owned several types of nets, and the burials at Cerro Azul suggest that this was also true in ancient times. On any given day, however, each fisherman said that he would leave home carrying just one kind of net.

Elderly fishermen talked about the importance of studying the tides, since when the tide was high, the water was deeper near the cliffs. When waters near the cliff face were calm and tranquil, with little foam and turbulence, a lot of fish came directly to the

Figure 7.1 (above). A fisherman, standing on a rocky ledge of Cerro Centinela, throws his atarraya (circular cast net) into the Pacific. His goal is to catch the fish that approach the sea cliff in search of algae or molluscs.

Figure 7.2 (right). A fisherman of the *peña* zone retrieves a grunt (*Anisotremus scapularis*) from his atarraya.

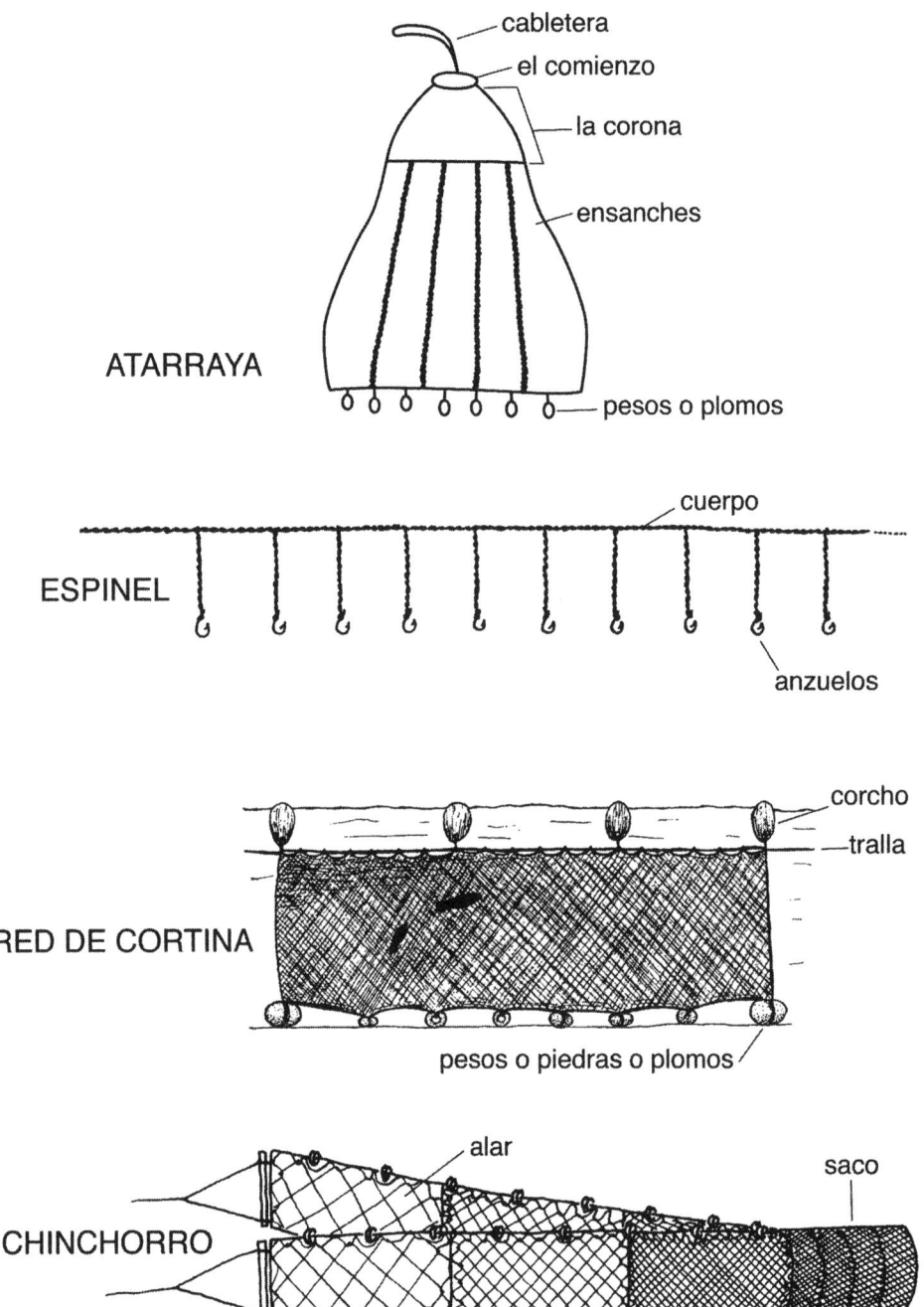

Figure 7.3. Four types of fishing gear used by traditional fishermen on the Peruvian coast: the atarraya or cast net, the espinel or multi-hook line, the red de cortina or trammel net, and the chinchorro, a longer version of the trammel net that can be stretched between two watercraft.

Figure 7.4. A caballito de totora, kept in the home of an elderly Cerro Azul fisherman out of nostalgia for the olden days.

cliffs—even, sometimes, the highly prized corvina. High tide, called "mar llena," alternated with low tide or "mar seca." On that particular day in 1982, the period from 6 pm to midnight was "mar seca"; midnight to 6 am was "mar llena"; 6 am to noon was "mar seca"; and noon to 6 pm was "mar llena."

On the cobble beach stretching south from Cerro Centinela, the fishermen of the 1980s used the espinel (Fig. 7.3), a line up to 100 m long and bearing 40–50 hooks. They said that they could catch a wide variety of fish on these hooks, from drums to rays to dogfish sharks. Since we found no fishhooks at Cerro Azul, we suspect that the espinel was not part of the Late Intermediate technology. In other parts of the Peruvian coast, however, it seems to have been present from early times. For example, at the preceramic site of Asia, Engel (1963:97–98) reports that Grave 15 had "5 (at least) mussel shell fishhooks, possibly on the same fish line, and 1 bone fishhook." My guess is that Late Intermediate fishermen at Cerro Azul used nets instead of espineles to exploit the *costa*.

On the *playa*—the sandy beach encircling Cerro Azul Bay and stretching north along Playa el Hueso—the red de cortina or "curtain net" can be stretched between two fishermen; this technology captures rays, dogfish sharks, left-eye flounders, and a number of species of schooling fish.

Ethnohistoric Data on Ancient Watercraft

In addition to interviewing living fishermen at Cerro Azul, I looked at 16th-century ethnohistoric documents to learn as much as I could about indigenous fishermen and their technology. Unfortunately, few of these documents refer specifically to the Cañete Valley. I therefore had to look to nearby valleys, bearing in mind that data from other parts of the coast might not be identical to the fishing strategies at Cerro Azul. Admittedly, these documents describe 16th-century fishermen rather than Late Intermediate fishermen. The ethnohistoric documents, nevertheless, stimulated me to develop a set of questions that could be addressed at Cerro Azul.

At least three kinds of watercraft are mentioned in 16th-century documents: (1) wooden rafts made of large tree trunks (balsas); (2) kayak-like boats made of bulrushes (caballitos de totora); and (3) inflated watercraft made from sea lion skins sewn together (cueros de lobomarino). Of these three kinds of watercraft, the only one I actually saw with my own eyes at Cerro Azul was a 20th-century caballito de totora, nostalgically displayed at the home of a retired fisherman (Fig. 7.4).

We found no ancient remains of caballitos de totora in our excavations. Recovering such evidence would have required an incredible stroke of luck, even though bulrushes themselves were preserved at Cerro Azul (Chapter 16). Since clay models and painted depictions of caballitos de totora are known from Nasca, Moche, and Chimu pottery (e.g., Buse de la Guerra et al. 1977:160, 203, 204, 244–245; Donnan and McClelland 1999; McClelland et al. 2007; Proulx 2006: Plates 32, 33) and since elderly residents of Cerro Azul remember such boats, I assume that our Late Intermediate fishermen also used them. We were simply not lucky enough to find one.

In places where caballitos de totora are still used, such as Huanchaco in the Moche Valley (Banack et al. 2004, Murphy 1923), the bulrushes are not simply harvested in the wild; they are cultivated in sunken beds called totorales, wachaques balsares, or pukios. Approximately 200 bulrush beds have been found in the area called Los Balsares de Huanchaco, situated just 1 km north of the town. These lands are owned by family groups, who keep multiple beds with plants at different growth stages so that some bulrushes will always be available to harvest (Banack et al. 2004:12–13).

Huanchaco residents once collected bulrushes for boat building from an area near Chan Chan, 6 km to the south, but had to stop when Chan Chan was declared a protected archaeological site. It was then that they decided to collect totora rhizomes and transplant them near Huanchaco. With careful transplanting and cultivation, a bed can produce bulrushes suitable for caballitos for five to seven years. After that, it begins to produce inferior plants that are too short for boat-building.

One of the first beds began as a pit excavated to the subsurface water table, 1 km north of Huanchaco. Its initial success encouraged other fishermen to dig their own beds. The area set aside for bulrushes now covers 46.7 ha. The beds themselves lie between 50 and 150 m from the high tide line, and are placed in areas of saltgrass (*Distichlis spicata*) so that this native plant can slow the movement of sand into the beds. Totora achieves its maximum height within 9 months.

A 22–25 m² planting produces enough totora to build a single caballito. Once fully dried (a process that takes 2–3 weeks), totora is classified by size: first-class bulrushes are those greater than 3.0 m in length; second-class specimens are 2.5–3.0 m in length; and third-class specimens are less than 2.5 m in length.

One caballito observed by Banack et al. (2004:15) required 2200 plants to construct and was usable for only 14–20 days. As a result of this short life, it was necessary for each fisherman to own two or three caballitos in order to have backups. Little effort was expended to prolong the life of a caballito, other than propping it upright after use to drain off excess water and expedite drying.

Given the amount of labor required to construct a caballito, and the short time they last, it is fortunate that the Cerro Azul fishermen did not have to travel far in them. As we will see in Chapter 9, all the Late Intermediate fish bones at Cerro Azul were from species that could have been caught within 100 m of the coast.

There is also reason to believe that Cerro Azul turned inflatable sea lion skins into watercraft. As we learn in Chapter 12, there is evidence that the occupants of Structure D spent time trimming the flippers off such skins. Since we recovered no sea lion garments, it seems likely that the skins were being prepared for use at sea.

Ethnohistoric Data on the Neighboring Chincha Valley

According to one 1558 document the señorío of Chincha, like those of Cañete and Ica, had its own lord and separate government. The powerful Chincha lord was said to have administered a valley with a dense population and to have exported dried fish, as well as decorated gourds, to the highlands in exchange for other items. He is also said to have owned many balsas and to have participated in long-distance trade for *mullu* (*Spondylus*) and copper.

Pedro Cieza de León (1932[1550]: Chapter LVIII) says that the Chincha polity alone had 10,000 fishermen. Other sources describe the ruler as having a total of 30,000 tribute-payers, including 10,000 farmers, 10,000 traders, and the aforementioned 10,000 fishermen (see Rostworowski de Diez Canseco 1977a:99; Biblioteca del Palacio Real de Madrid, Miscelánea de Ayala, Tomo XXII, folios 261–273v; Lizárraga 1946[1605]: Chapter XLVII).

Chincha fishermen are said to have lived on a single road that ran for two leagues before it reached Chincha; the road then was said to have continued on to Lurinchincha. Fishermen lived in a barrio separate from others "en gran orden y concierto" (*Aviso* 1570). They were seen as a separate economic and cultural unit and considered members of a type of guild (un gremio de pescadores). Whether the fishermen of Cerro Azul were considered to be a guild-like unit within the Kingdom of Huarco is an interesting question.

Fishermen were exempt from labor demands by the state; when they were not fishing, they passed their time on land dancing and drinking. "Cada día o los más de la semana entravan [sic] en la mar, cada uno con su balsa y redes y salían y entraban en sus puertos señalados y conocidos, sin … tener competencia los unos con los otros" (*Aviso* 1570).

Ethnohistoric Data on Lurín Fishermen

During his 18th-century travels in the Lurín Valley, Hipólito Ruiz (1952, Vol. I, page 52) reported seeing two groups of fishermen—one group residing on the hill near Mamacona and the other residing in Quilcay. At the time of his visit, these Lurín Valley fishermen had filed complaints against the owner of a neighboring hacienda who was trying to drive them out of the region by drying up the local swamps, allowing him to convert that land to agricultural use. The fishermen declared that they owned no fields and were solely dedicated to fishing. This court document indicates that fishing took place in coastal swamps as well as in the ocean. This fact may have relevance for the use of Huaca Chola near Cerro Azul, which is said to have produced mullet.

From Ruiz we also learn that there were five roads crossing the Lurín Valley. One of these roads was reserved for fishermen needing to transport their fish, while another was reserved for *chaskis* or professional messengers. This dedication of certain roads to specific professions is an example of the economic specialization and division of labor that Rostworowski de Diez Canseco (1970, 1975, 1977a, 1977b) has documented from 16th-century sources. The extent to which Cerro Azul was involved in such specialization remains to be seen.

Late Intermediate Nets from Cerro Azul

More than 100 fragments of nets were preserved at Cerro Azul. The two most common proveniences in which we found nets were burials and storage bins. Unfortunately, most of our burials with nets had been disturbed by looters. By combining our data with those of Kroeber (1937), however, we can at a minimum say this: fishing nets tended to occur with men's burials, and a given man was likely to have nets of several types buried with him. What this suggests is that, like the fishermen of the 1980s, Late Intermediate men were prepared to go wherever the fish were biting. They may have been specialized in the sense of being fishermen (as opposed to farmers), but they were evidently not further specialized into categories such as *peña* fishermen, *playa* fishermen, and so on.

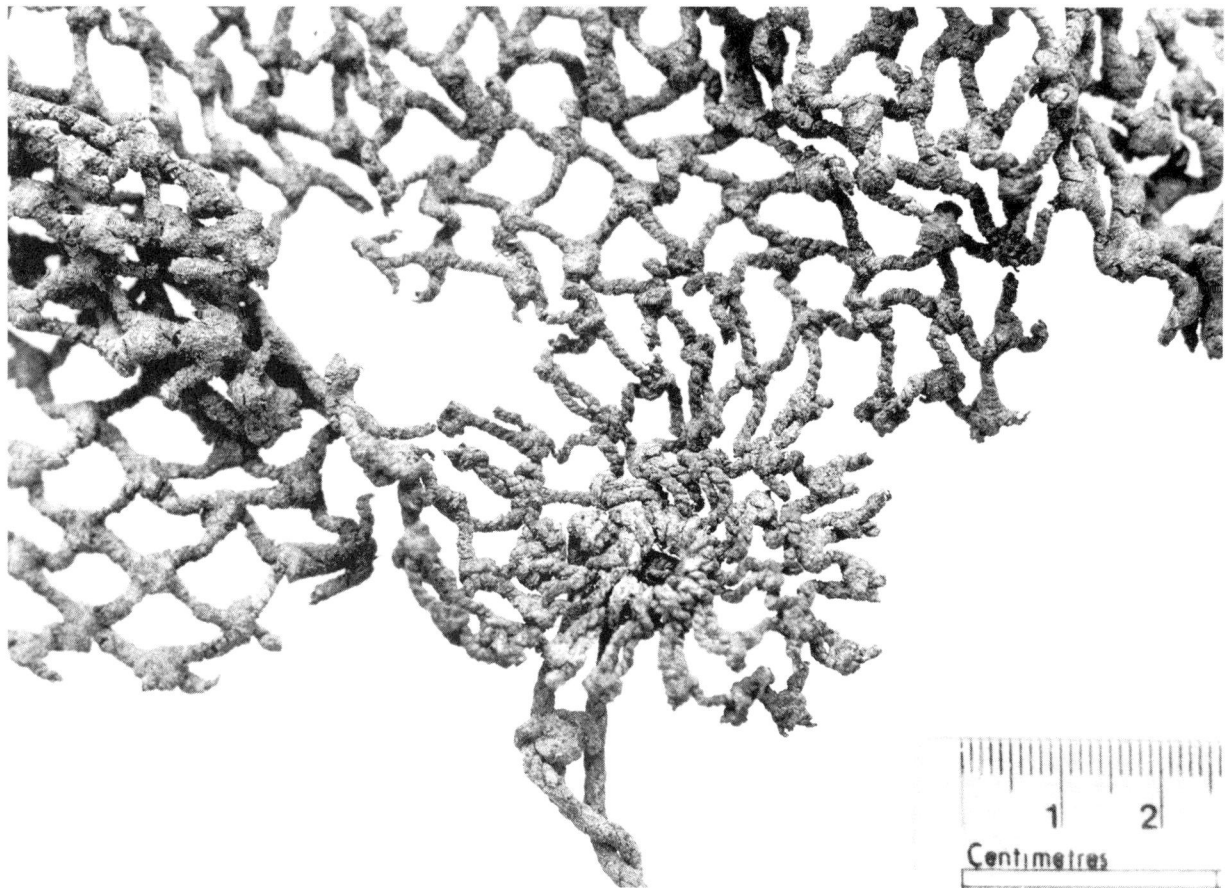

Figure 7.5. Close-up of the comienzo of a Late Intermediate atarraya, consisting of 11 pairs of cords tied with cow-hitch knots to form a central ring. The loop hanging down from the comienzo is the cabletera to which the fisherman tied his retrieval line. This net was found in Structure 5, a looted burial cist in Quebrada 5a.

Some net fragments found in the storage bins, however, opened our eyes to the fact that not all nets at Cerro Azul were for fishing. Some appeared to be what Egyptologists call "carrier nets"—essentially harnesses that made it easier to carry large jars and amphorae. Carrier nets thus became the third major type of net we found at Cerro Azul, along with the cast nets and trammel nets.

An Atarraya from Structure 5

Figure 7.5 presents the photographic close-up of an atarraya, or cast net, from a looted burial cist in Quebrada 5a. Shown in the photo are the comienzo, or central starting point, and the cabletera to which the retrieval line would have been attached. Figure 7.6 provides two diagrammatic sketches of this same net fragment.

This net consisted of 11 paired cords (for a total of 22), tied with asymmetrical overhand knots or cow-hitches to form the comienzo. When the cabletera was pulled, weights attached to the bottom row of mesh came together to turn the net into a bell-shaped bag, thus preventing any trapped fish from escaping. Fishermen to whom we showed this Late Intermediate net fragment said that the mesh was appropriate for catching grunt, morwong, or scaled blenny.

A Trammel Net from Structure 4

Figure 7.7 shows the lower border of a red de cortina, or trammel net, from a looted burial cist at Cerro Azul. C. Earle Smith, Jr. identified this particular net as having been constructed of cotton fiber.

At the bottom of the net we see five pairs of dangling cords. Each of these pairs of cords would have been attached to one of the weights that held the red de cortina to the ocean floor and kept it taut like a tennis net. We recovered no weights with this particular net, but several appropriate stone weights were recovered at Cerro Azul in 1925 by Kroeber (see Fig. 7.8). As the photograph shows, some weights still had cords attached to each end.

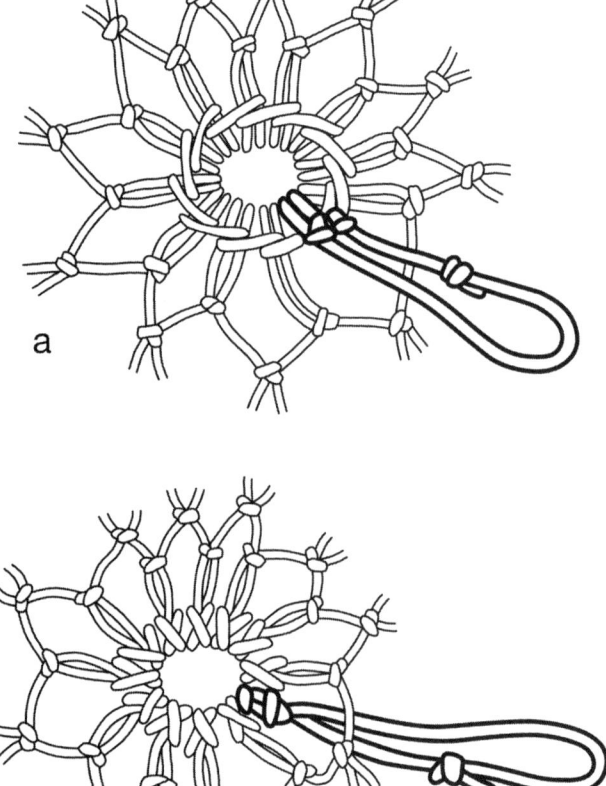

Figure 7.6. Two sketches of the comienzo de atarraya shown in Fig. 7.5. *a* is a view from above (outside the closed net). *b* is a view from below (inside the closed net). The interior diameter of the central ring is 4 mm; the spacing between knots is 8–9 mm.

Figure 7.9 presents an 18th-century drawing of Peruvian fishermen stretching a chinchorro between two balsa rafts. The net is kept open by a series of wooden (or gourd) floats on its upper edge, and a series of weights on its lower edge (Martínez de Compañón 1936).

We found no wooden floats with our specimens of trammel nets, but Kroeber did (Kroeber 1937:52 and Plate LXXXVII, Fig. 2). In Quebrada 1, which Kroeber (1937:249) describes as Cerro Azul's "most seaward gully," he found a shallowly buried mesh bag that contained what may have been a fisherman's kit. This bag, designated Parcel B1, had its own drawstring and was part of a cache that (Kroeber believed) the fisherman had hidden temporarily for safe-keeping. Inside the bag were 43 wooden floats still strung on a line, skeins of string and agave fiber that could have been used to repair nets, and a *tumi* or lunate knife.

One of the numerous burials disturbed by looters in Quebrada 5a of Cerro Azul may have been provided with a trammel net. Figure 7.10 shows a photographic close-up of this net, which displays dangling cords with loops that we suspect were for weights.

While excavating Structure 12, a looted burial cist in Quebrada 5-south, we learned that not all net weights at Cerro Azul were stones. In the looters' debris we found two examples of pot sherds used as weights (Figs. 7.11 and 7.12). Whether these sherds were attached to trammel nets or cast nets cannot always be determined, but we suspect the former in most cases. The mesh of the net shown in Fig. 7.11 would easily stop a fish the size of a mismis or mojarrilla.

A Carrier Net from Collca 1

Figure 7.13 shows a carrier net found in Collca 1, a storage bin set against the south wall of Structure D's Northeast Canchón. This net has the shape of a conical bag, and would fit comfortably around a typical Late Intermediate amphora (e.g., Marcus 2008: Fig. 3.8). The net is made of cotton, 2 ply z-s, and features half-hitch knots (e.g., Emery 1966). Because this carrier net was stored in Collca 1, we consider it possible that it was used to carry jars of water or chicha to the weavers in the Northeast Canchón.

Peru, of course, had a long history of using net bags for transport (Bird 1985:207, 210; Engel 1963:47). So, too, did Egypt, and carrier nets like ours at Cerro Azul evolved there as well. Figure 7.14 shows a pottery vessel from Qurna, Egypt, which is still inside the remains of its carrier net. Veldmeijer and Bourriau (2009) studied 10 such vessels from an 18th Dynasty tomb at Qurna; these large pots were still inside their carrier nets, and suspended from a wooden pole. A wall painting in the tomb of Rekh-mi-re' at Thebes depicts two individuals using just such a pole to transport heavy vessels inside carrier nets (Wendrich 1989:182–183; Davies 1943: Plate L). In that part of the world, of course, nets were made of flax fiber rather than cotton.

Many of the nets we found at Cerro Azul were only fragments, and we have no sure-fire method for separating

Fishing Strategies and Fishing Gear 105

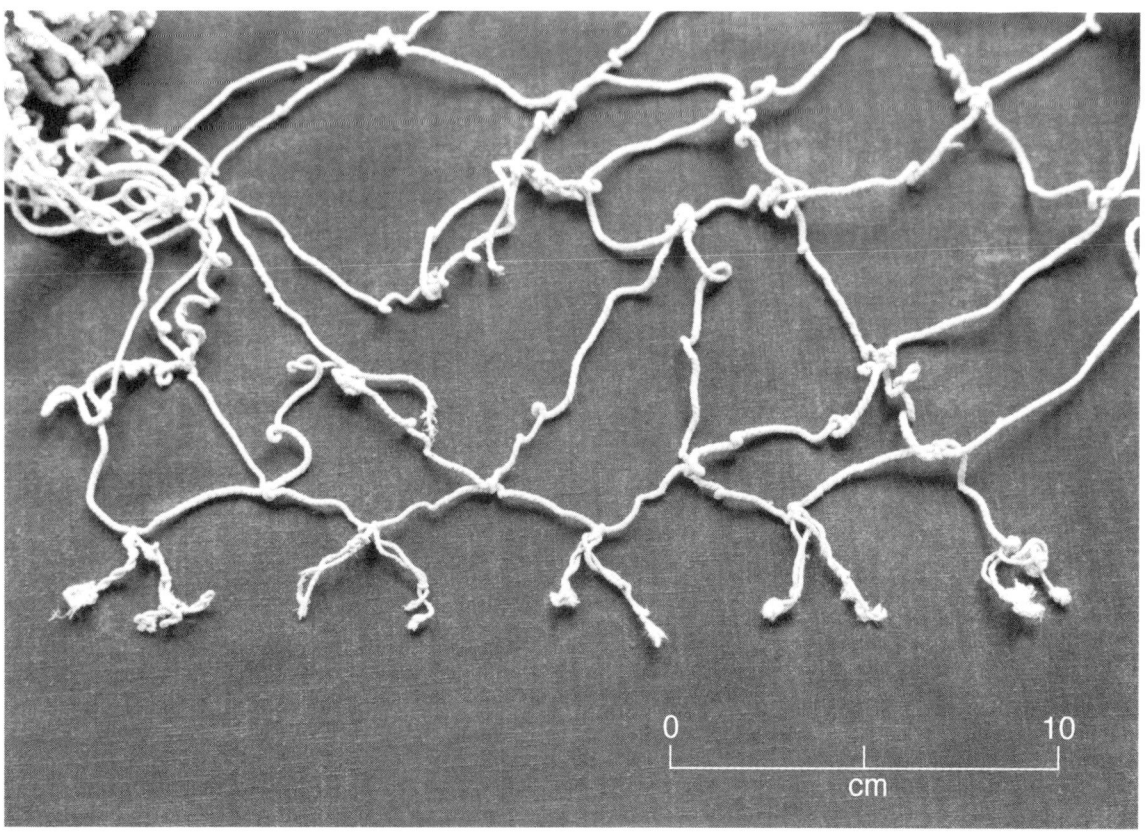

Figure 7.7. Lower border of a cotton trammel net from Structure 4 (a looted burial cist in Quebrada 5a), showing the paired sets of dangling cords that were once attached to weights.

Figure 7.8. Stone weight from a trammel net, with cords still attached to both ends. This weight was found at Cerro Azul in 1925. (Photo courtesy of the Field Museum of Natural History and Robert Feldman)

Figure 7.9. Eighteenth-century Peruvian fishermen using a chinchorro stretched between two log rafts. Note the gourd floats along the upper margin of the net, and the weights along the lower margin (redrawn from Martínez de Compañón 1936).

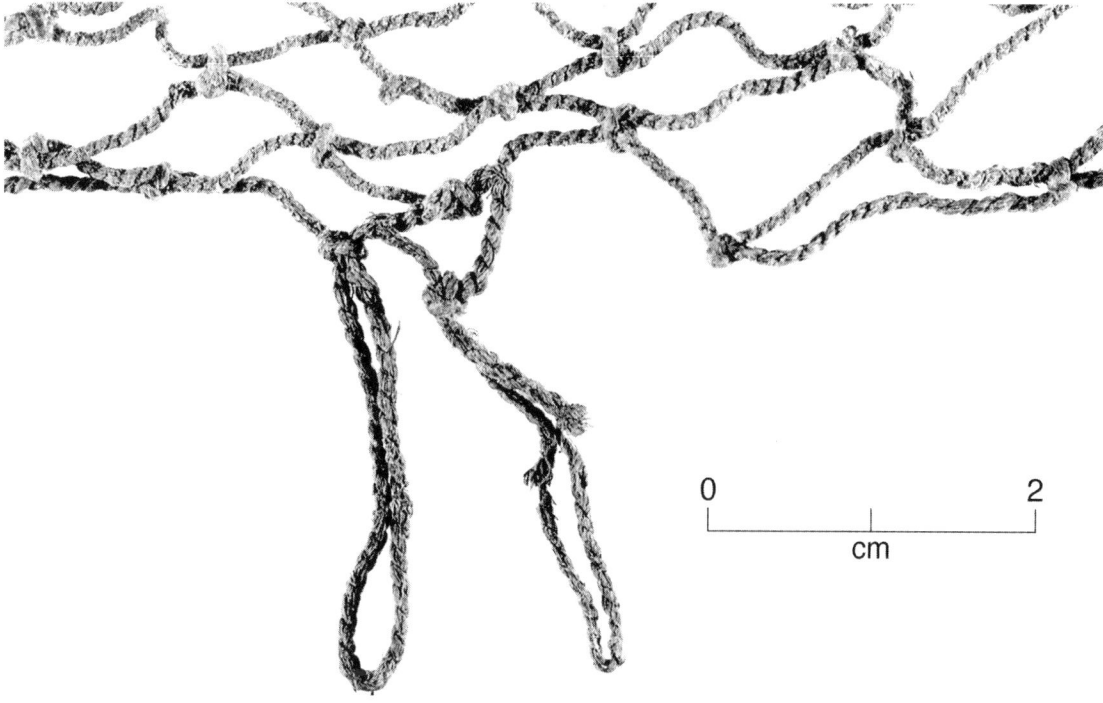

Figure 7.10. Lower border of a trammel net from looters' debris on Terrace 9 of Quebrada 5a. Two loops for weights dangle from the net.

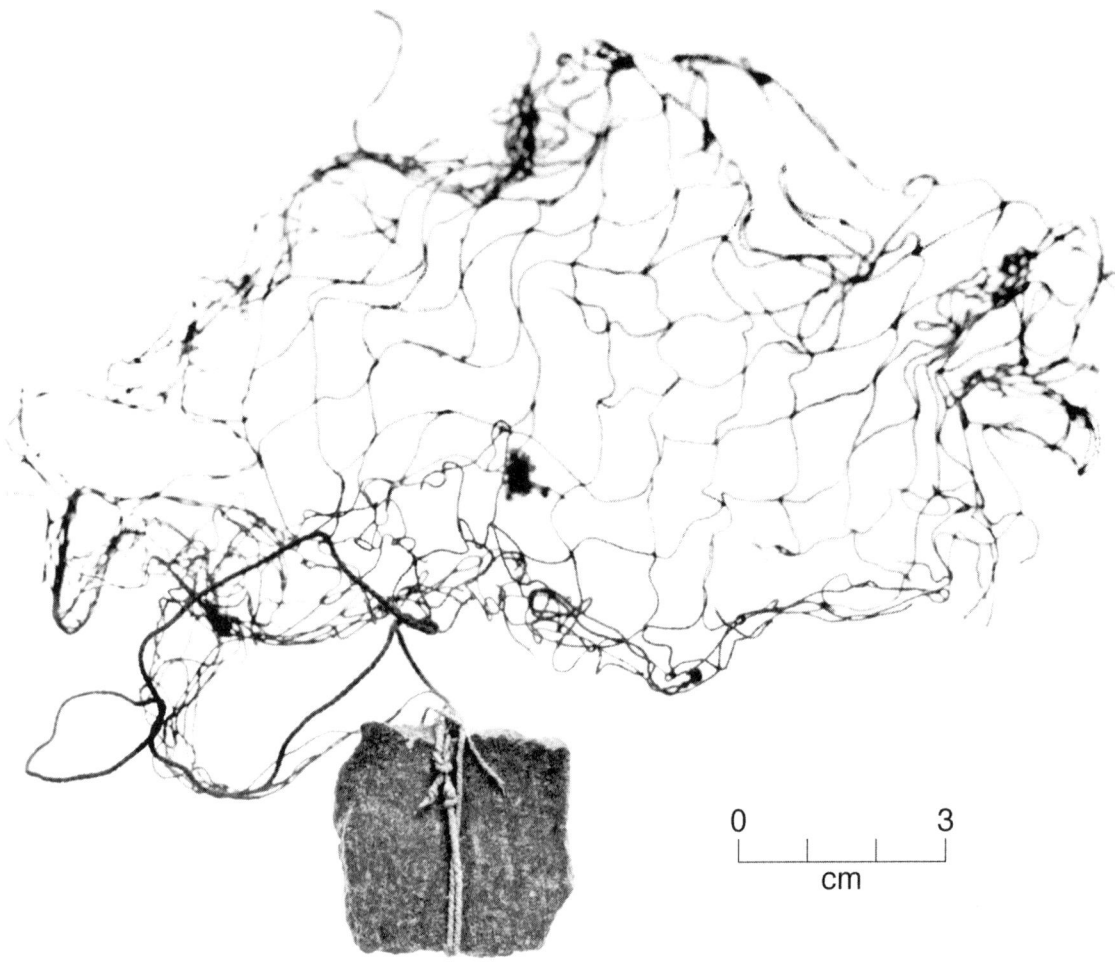

Figure 7.11. Fragment of fine-yarn trammel net with a pot sherd attached as a weight. Found in looters' debris in Structure 12, a burial cist in Quebrada 5-south.

Figure 7.12. Pot sherd used as a weight for a trammel net. Found in looters' debris in Structure 12, a burial cist in Quebrada 5-south.

Fishing Strategies and Fishing Gear

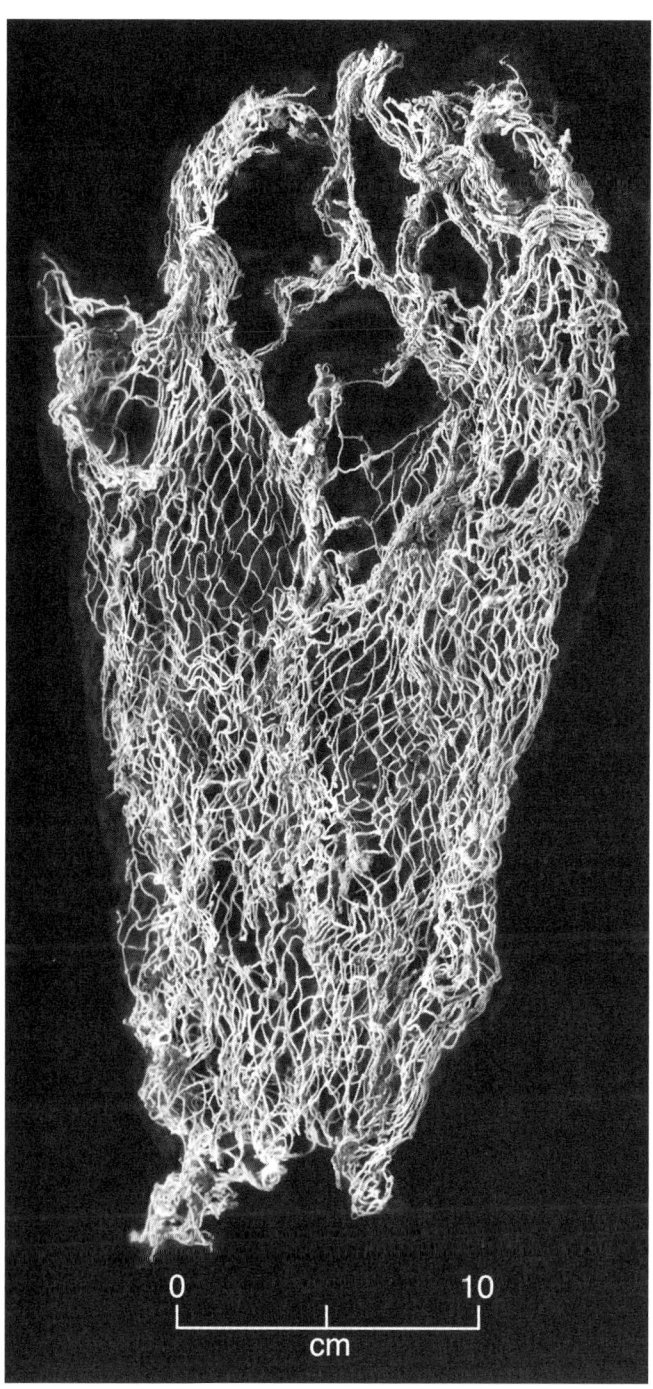

Figure 7.13. Carrier net from Collca 1, Structure D, Cerro Azul.

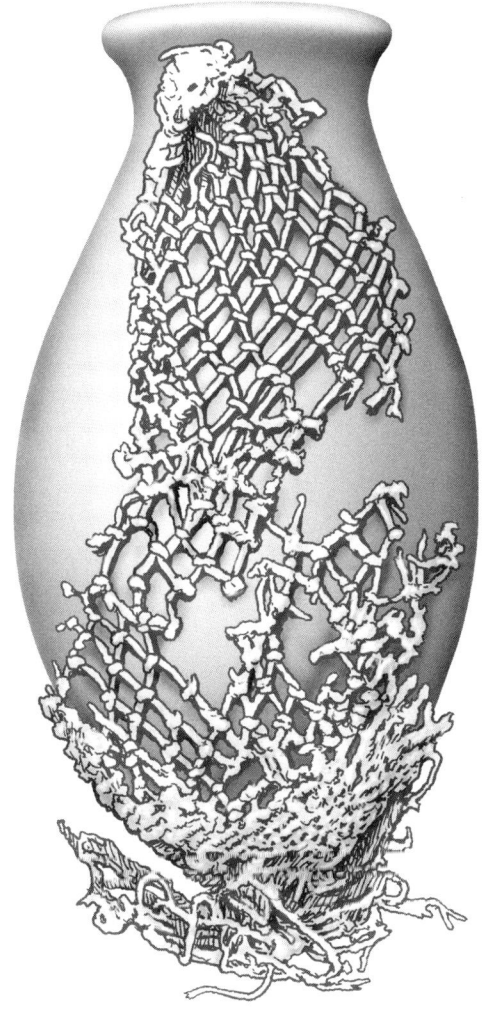

Figure 7.14. Pottery vessel from a burial at Qurna, Egypt, inside its carrier net. Drawing by John Klausmeyer, based on a photograph in Veldmeijer and Bourriau (2009: Fig. 7).

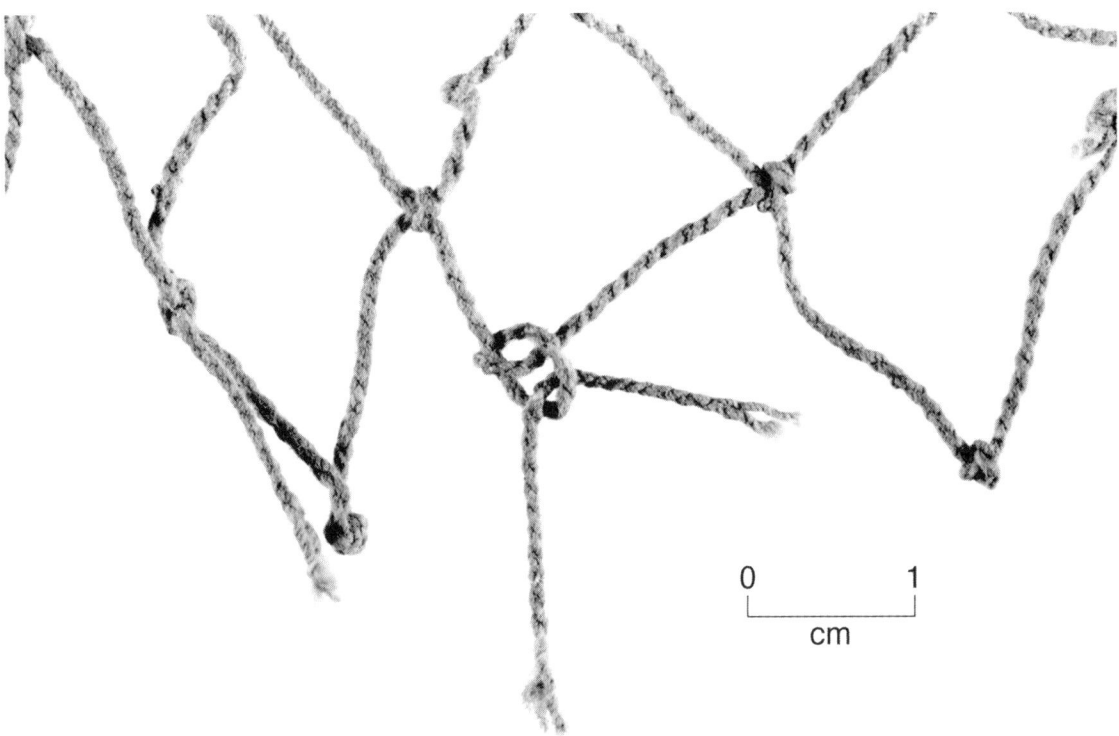

Figure 7.15. Close-up of a net fragment found in Room 7 of Structure D. Its relatively loose, open mesh and its use of the lark's head knot suggest that it may have been a carrier net.

bits of carrier nets from bits of fishing nets. Two attributes of our carrier nets, however, may be significant. First, the mesh size of many carrier nets is too large to prevent most fish from escaping. Second, several obvious carrier nets have lark's head knots, rather than square knots or half-hitches. Let us look at two possible examples.

Figure 7.15 displays a net fragment from Room 7 of Structure D. Its relatively loose, open mesh and its use of the lark's head knot suggest that it may have been a carrier net.

Figure 7.16 gives two views of a possible carrier net found with a looted burial on Terrace 9 in Quebrada 5a, just southeast of the Structure 4 burial cist. This net has the very large mesh typical of a carrier net and employs the lark's head knot. Unfortunately, only time (and a larger sample) will determine whether these two criteria will serve to identify Andean carrier nets.

Examples of Late Intermediate Men Buried with More Than One Type of Net

We mentioned earlier the possibility that Late Intermediate fishermen owned several types of nets, allowing them to fish in whichever habitat they chose. Let us now look at a few examples of burials with multiple nets.

Not far from Structure G of Cerro Azul, we came across the looted burial of a fisherman. Looters had made off with most of his burial offerings, but evidently considered his two nets to be of no commercial value. Net 1 of this burial is shown in Figures 7.17–7.19. Judging by its border (Fig. 7.19) this net had probably been an atarraya. Unfortunately, it was damaged and incomplete; it may once have been 2 m in diameter, but now its largest fragment is 120 x 65 cm (Fig. 7.17). The mesh size was 5 x 5 mm—small enough to catch anchovetas and sardines—and appears to have employed sheet bend knots (Fig. 7.18).

Net 2 from this same burial was designed to catch larger fish, such as grunts or drums. Its mesh was roughly 20 x 20 mm and it utilized square knots (Fig. 7.20). Unfortunately, this net was so damaged that we could not determine whether it was an atarraya or a trammel net.

We discovered another looted burial with multiple nets on Terrace 9 of Quebrada 5a, just southeast of Structure 4. This individual was found with several fishing nets and at least one carrier net. The carrier net (shown in Fig. 7.16) had likely been found covering one of the pottery vessels the looters made off with.

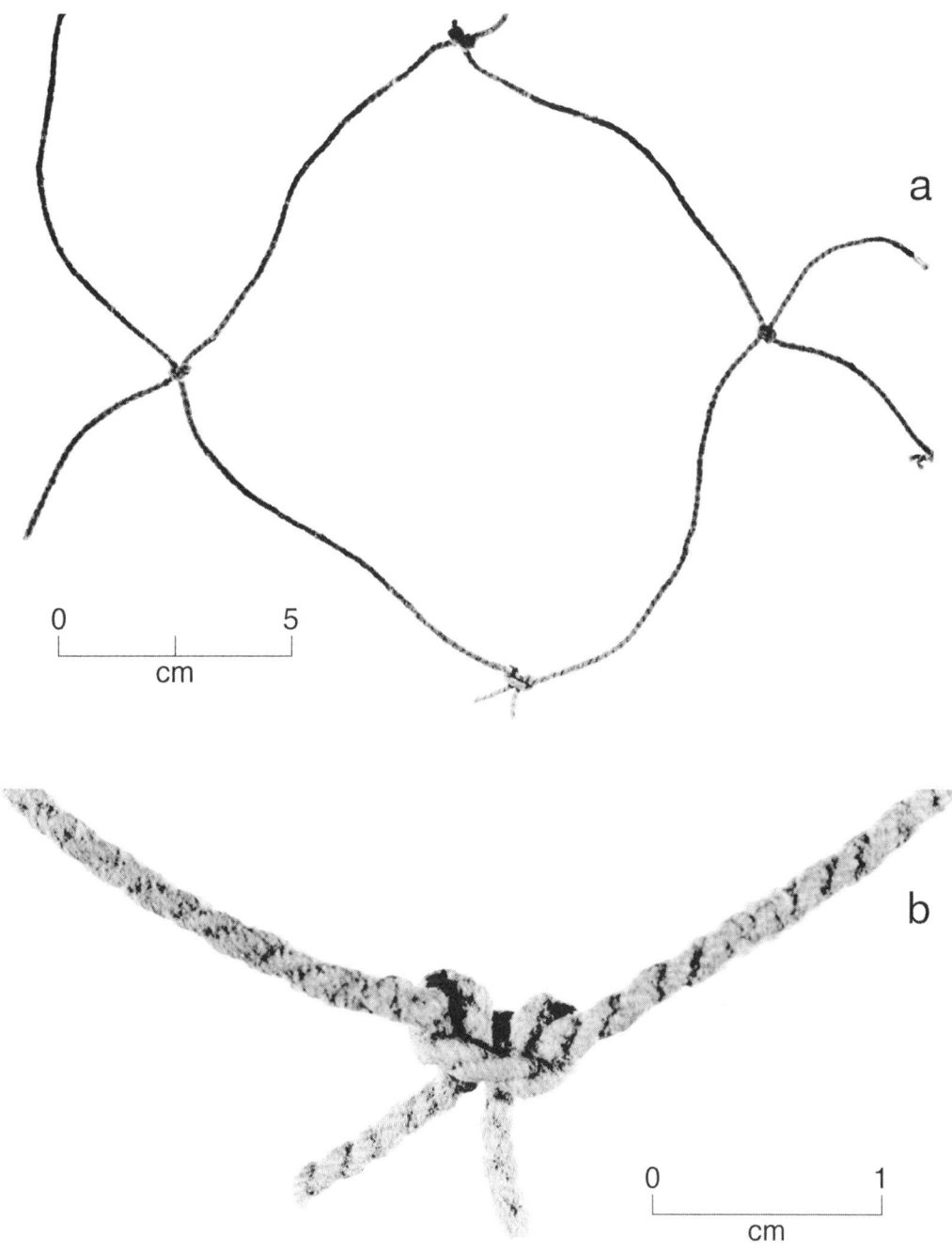

Figure 7.16. Two views of Net 1 from a looted burial on Terrace 9 of Quebrada 5a. *a*, a view showing the roughly 10 by 10 cm mesh. *b*, close-up photograph of a lark's head knot. This was almost certainly a carrier net.

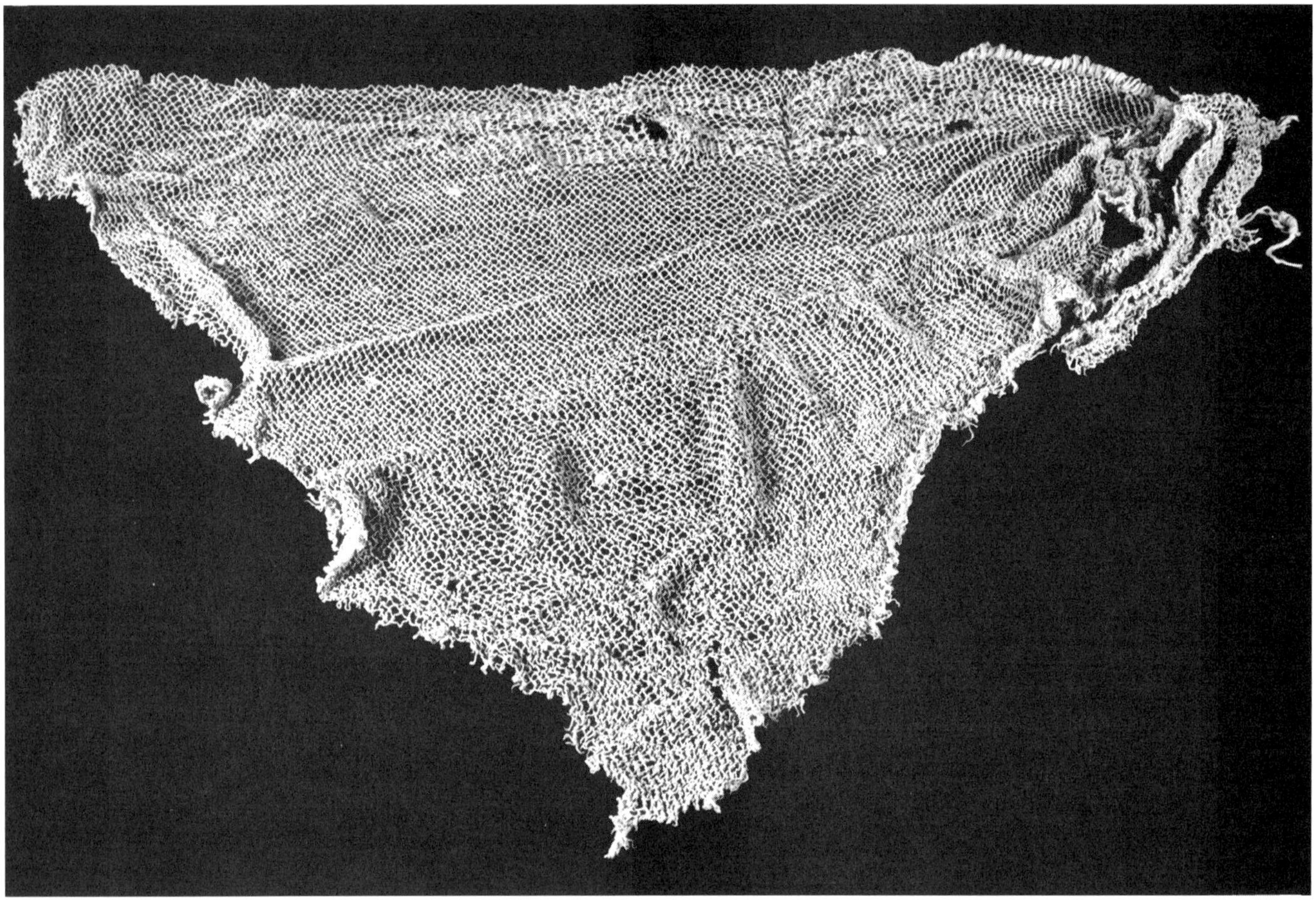

Figure 7.17. Net 1 from a looted burial near Structure G of Cerro Azul. This fragment measures 120 x 65 cm, and its mesh size is 5 x 5 mm. Before being damaged by looters, the net may originally have had a diameter of 2 m.

Figure 7.18. Close-up photograph of the sheet bend knots used in the net seen in Fig. 7.17. The mesh of this net was roughly 5 x 5 mm.

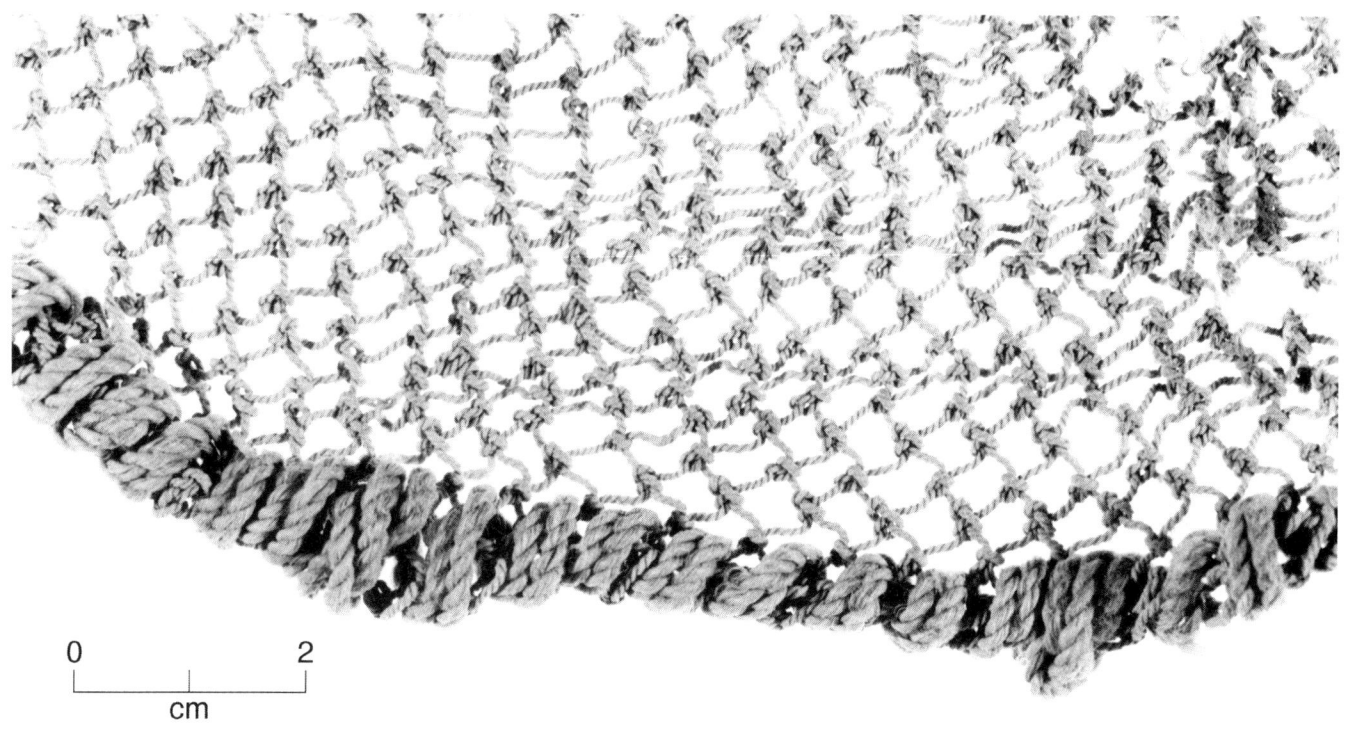

Figure 7.19. Close-up photograph of the border of the net seen in Fig. 7.17. Such a border is typical of an atarraya.

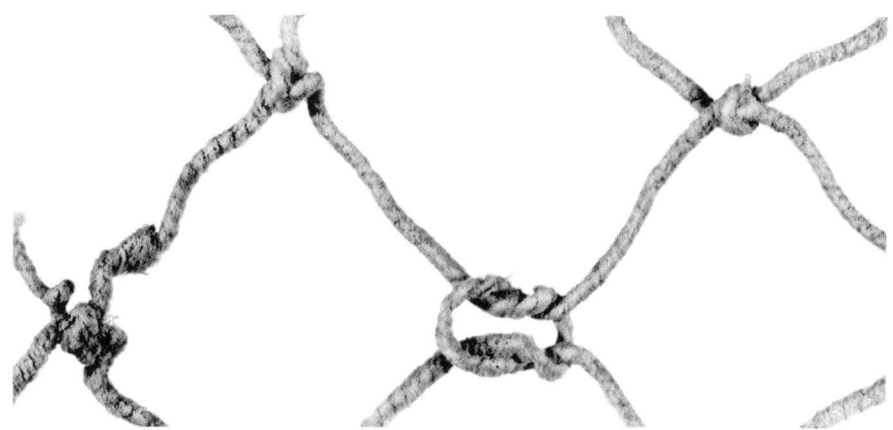

Figure 7.20. Close-up photograph of the square knots used in Net 2 from a looted burial near Structure G. The mesh of this net was roughly 20 x 20 mm.

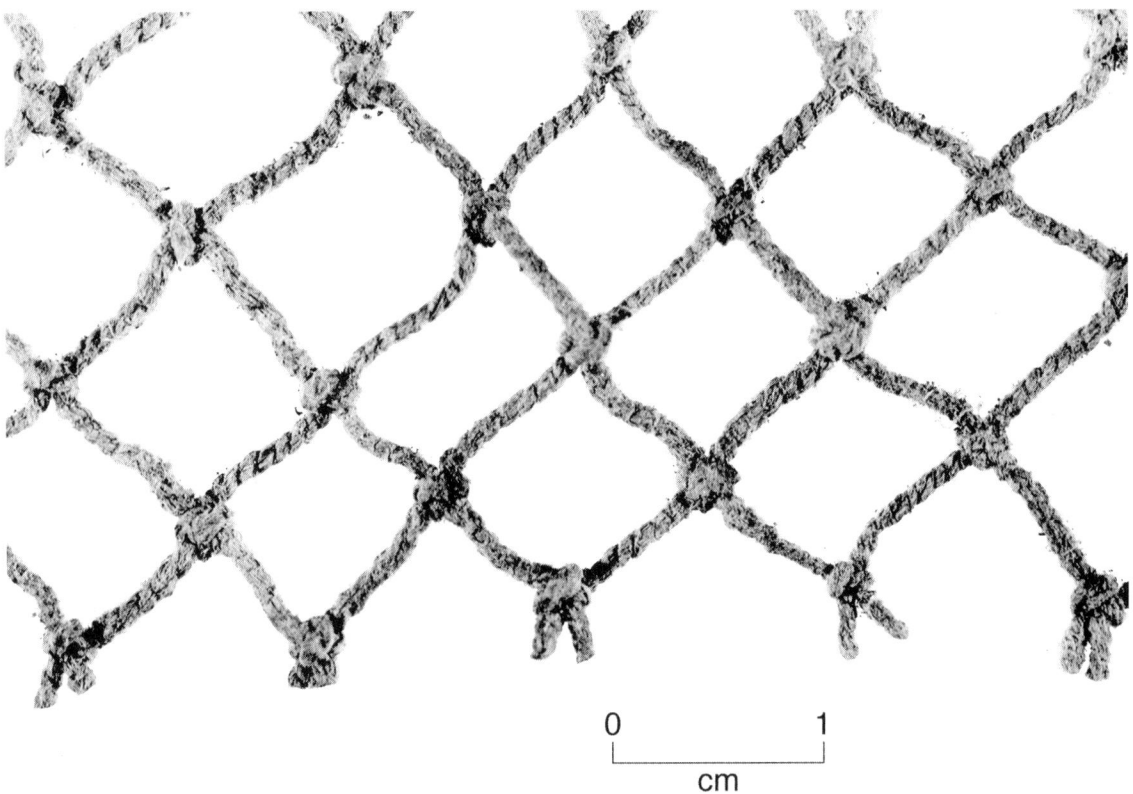

Figure 7.21. Close-up photograph of Net 2 from another looted burial on Terrace 9 of Quebrada 5a. This net was almost certainly used to catch small fish.

Figure 7.21 shows a photographic close-up of another net from the Terrace 9 burial just described. This appears to have been a net for catching small fish, since its knot spacing was only 10 mm. The other nets from this burial— all fragmentary— have different mesh sizes.

Structure 4 of Quebrada 5a produced a large sample of net fragments, including both cast nets and trammel nets. Unfortunately, the fact that the looters beat us to this burial cist makes it impossible for us to confirm the variety of nets buried with any individual fisherman.

Creating Evenly Spaced Mesh

Late Intermediate fishing nets displayed a variety of mesh sizes. One attribute that nets of all mesh sizes shared was uniform spacing between knots. This uniformity was achieved by the use of a mallero, or wooden template.

Figure 7.22 shows two templates found at Cerro Azul. Our workmen were convinced that both were made of wood from the chonta palm, but this identification has not been verified by an ethnobotanist.

Mallero *a*—roughly 9.0 x 2.5 cm in size—was found in looters' debris northeast of Structure 4, a burial cist that produced many fishing nets. Our fishermen informants believed that this mallero was for making nets to catch lisa, or mullet.

Mallero *b*—roughly 7 x 5 cm in size—was found in looters' debris from Structure 12, a looted burial cist in Quebrada 5-south. This template was suitable for creating very large mesh, either for a carrier net or a net for catching róbalo.

Sandweiss (1992:74) found a mallero not unlike our Fig. 7.22 *a* at the Late Horizon site of Lo Demás in the neighboring Chincha Valley. His mallero measures 12.75 x 2.39 cm, longer than ours but similar in width. We are not sure how many sizes of templates there were in Late Intermediate times, but we have reason to suspect that the number is high. At the north coast site of Pacatnamú, Donnan reports that templates were the most common fishing implements

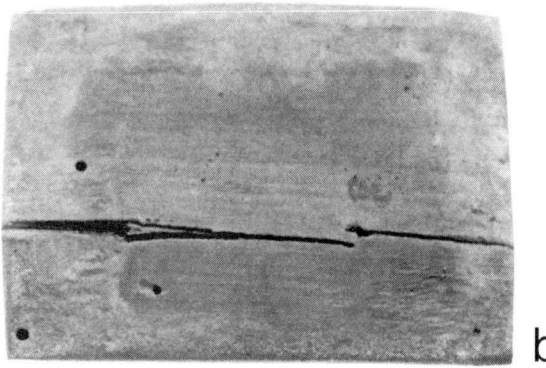

Figure 7.22. Two malleros (templates for standardizing the spacing of knots in netmaking). *a*, roughly 9.0 x 2.5 cm in size, was found in looters' debris immediately northeast of Structure 4. This tool could be used to create the proper knot spacing for a mullet net. *b*, roughly 7 x 5 cm in size, was found in looters' debris from Structure 12.

found in the men's burials. He adds, "These [malleros] appear to have been as exclusive to male graves as weaving implements were to women's graves" (Donnan 1995:150).

Conclusions

Based on the data presented in this chapter, we now have an idea how Late Intermediate fishermen would have pursued the fish listed in Chapter 6. Within our sample of more than 100 net fragments from Cerro Azul, elderly fishermen of the 1980s believed that they could recognize meshes suitable for anchoveta and sardine; mullet; grunt and drum; and larger fish such as bonito and róbalo. Many nets were made from cotton yarn, using knot-spacing templates made of imported wood.

We have indisputable evidence for the atarraya, or cast net, used today to fish in the *peña* environment. We also have evidence for the red de cortina, or trammel net, including the necessary wooden floats and stone or sherd weights. Such nets are typically used to fish in Cerro Azul Bay. We have no fishhooks or evidence for the espinel, however, and we are not confident that we could tell fragments of the chinchorro from fragments of trammel net.

As for the watercraft used in the Late Intermediate, we have circumstantial evidence for the working of sea lion skins. We also suspect that caballitos de totora were used at Cerro Azul, but did not find an archaeological example. (Few archaeologists do.)

In Chapter 9 we will learn which species of fish were exploited at Late Intermediate Cerro Azul. We do not expect the prehistoric species mix to closely resemble that of the 1980s. On the one hand, the Late Intermediate fishermen did not have to compete for their fish with commercial trawlers and the fish meal industry. On the other hand, they did not have outboard motors or nylon nets. The bad news is that their fragile watercraft probably kept them within 100 m of the shore. The good news is that they lived in an era when one could get plenty of anchovetas simply by wading into Cerro Azul Bay with a basket.

8

The Drying of Fish for Export

Joyce Marcus

Catching fish was only the first step in Cerro Azul's economic specialization. The second step was drying massive amounts of certain fish for shipment inland. It appears that it was the noble families at Cerro Azul who directed this economic activity, although the actual harvesting of small fish was presumably done for them by commoner-class fishermen. Intervening between the nobles and fishermen was a group of overseers who supervised the filling and emptying of fish storage rooms.

We found little evidence at Cerro Azul to suggest that ordinary fishermen, unsupervised by nobles, harvested thousands of anchovetas on their own. On Terrace 9 of Quebrada 5a, for example, we found extensive commoner-class middens in which anchovetas and sardines were sparsely represented (see Chapter 9). The commoner households that produced these middens seem to have concentrated on a series of medium-sized fish such as grunt, drum, mullet, and mackerel.

Our highest concentrations of anchovetas were found in specialized, sand-filled storage rooms such as Room 8 of Structure D. Such rooms occurred either in the large elite residential compounds that surrounded Cerro Azul's central patio, or in a series of smaller storage facilities adjacent to them (e.g., Structure 9).

The sand that filled the fish storage rooms did not come from the beach adjacent to Cerro Azul. That sand would have been contaminated with wave-borne flotsam, dead fish, salt, and the droppings of shore birds. The sand in the storage rooms at Cerro Azul was green to blue-green in color. The sand was clean and uniform in grain size. Our workmen were confident that it had been brought to the site from an inland source, but were not sure where that source lay.

The technique of fish drying appears to have been as follows. First, thousands of anchovetas and sardines would be dried in the sun, presumably on a stretch of cobble beach. Then a layer of fine clean sand would be laid on the floor of a storage room. Over this went a single layer of partially dried anchovetas and/or sardines. A second layer of sand would be placed over the fish;

then came a second layer of dried fish; then a third layer of sand, and so on. The hygroscopic qualities of the sand drew the last of the moisture out of the fish, completing the drying process.

In the case of Room 8 of Structure D, the dried fish would have been periodically removed from the sand and loaded onto the backs of llamas that had entered the nearby Southwest Canchón. This activity was likely overseen by a commoner-class supervisor stationed on the North Platform of the Southwest Canchón. Off the fish would then go to inland communities.

Not only was Structure D involved in the export of fish, its capacity for fish storage was increased over time. At the outset, Room 8—a doorless unit, clearly designed for storage—appears to have been the only room involved in this activity. Whenever another room was badly damaged by an earthquake, however, it was filled with sand and repurposed as a fish storage unit. Such conversions took place in Rooms 1 and 3 of Structure D.

Structure 9, which lay only 22 m from Structure D, was a smaller tapia compound devoted even more fully to fish storage. Of its 13 rooms, roughly half either began as fish storage units or were converted to fish storage units later in the building's history. The filling and emptying of these rooms was likely overseen by a commoner-class overseer whose kincha house occupied a central location in the building.

The fact that both Structure D and Structure 9 saw more and more of their rooms filled with sand during their occupation suggests that Cerro Azul felt pressure to increase its capacity for fish storage over time. Whether that pressure came from the nobles running each elite compound, or from the ruler of Huarco himself, is an important question that will require more research.

The Fish Remains in Storage Rooms

While the fish in the center of a storage room normally dried thoroughly in the sand, there were cases in which some fish came into direct contact with the clay floor or tapia walls of the room. These surfaces never fully dried. They were kept moist by periodic episodes of salt fog, and when the fish accidentally touched them, patches of skin and scales stuck to the clay. We also recovered bones from disintegrated fish when we screened the sand in the storage rooms through carburetor mesh.

Figures 8.1 through 8.4 illustrate several examples of fish body parts that stuck to the walls or floors of storage rooms in Structure 9. Patches of skin and scales had adhered to the walls of Room 1 (Fig. 8.1 *a*); strings of still-articulated vertebrae had stuck to the floor (Fig. 8.1 *b*). Heads and tails of anchoveta littered the floor of Room 11 as well (Figs. 8.2, 8.3). On rare occasions, the bones of anchovetas and sardines were accompanied by vertebrae of larger fish like the corvina (Fig. 8.4). It is not clear whether such large fish had been deliberately dried in the storage rooms, or whether some of their bones were introduced accidentally while the rooms were being filled or emptied.

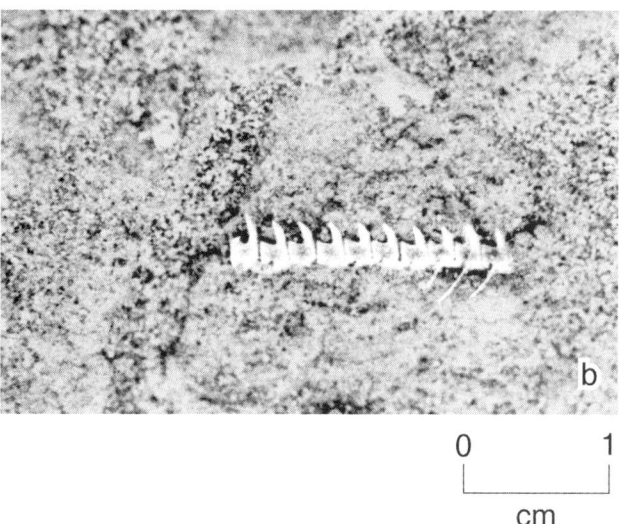

Figure 8.1. Remains of dried fish from Room 1, Structure 9. *a*, patches of skin and scales sticking to the tapia wall. *b*, ten still-articulated vertebrae (and other body parts) sticking to the clay floor.

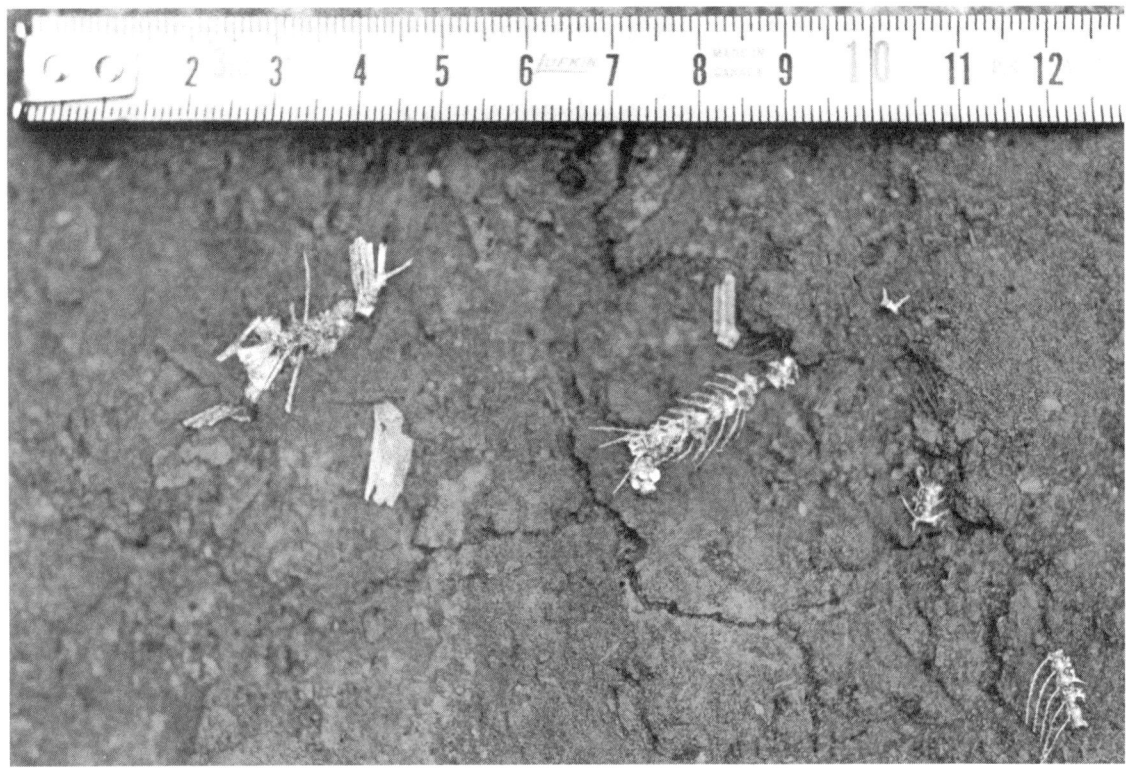

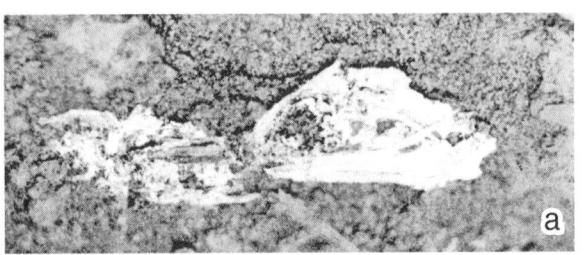

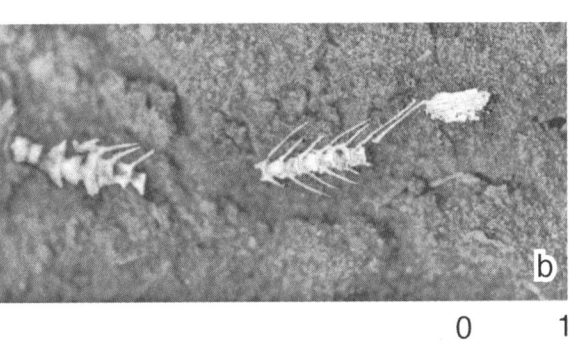

Figure 8.2 (above). The clay floor of Room 11, Structure 9, showing the remains of fish the size of anchovetas and sardines (scale in centimeters).

Figure 8.3 (left). Remains of dried fish sticking to the floor of Room 11, Structure 9. *a*, head of anchoveta (Family Engraulidae). *b*, sections of vertebral column.

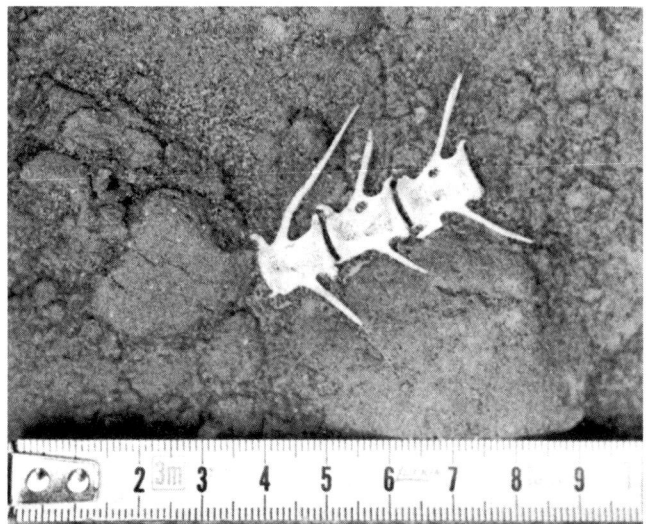

Figure 8.4. Vertebrae from a corvina resting on the clay floor of Room 11, Structure 9 (scale in centimeters).

Comparisons with Other Regions

The drying of fish at Cerro Azul strengthens the position of the Andes as an epicenter for drying food of all kinds. Andean peoples made llama meat into *ch'arki* and potatoes into *chuño* (e.g., Duviols 1973, Flannery et al. 1989, Miller 1979, Rowe 1946); they dried fish, molluscs, and crayfish for shipment inland; and they desiccated or freeze-dried many plant foods.

Other parts of the world had similar traditions of drying fish for shipment to other regions. For example, ElMahi (2000) describes the drying of fish by coastal communities of the Sultanate of Oman. Drying was performed on sardines and other small pelagic fish caught off sandy coastlines, and to tuna and shark caught off rocky coastlines.

The fishermen of Oman used three traditional methods for preserving fish: drying, grilling, and salting. Among the small fish dried in the sun were sardines and anchovies. Coastal Omani communities also dried larger fish such as sharks, guitarfish, and eagle rays, three categories of marine life that Cerro Azul's fishermen still convert into *ch'arki*.

In Oman, cartilaginous fish are cut into pieces and soaked in salt for a day or two; then the pieces are laid on pebbles to dry for days, or hung from wooden beams. In other cases, fish are stored in the sand (ElMahi 2000:103–104).

According to ElMahi (2000:108), the Oman exchange system is one "in which fish protein plays a dynamic role. It is now clear that fishing communities along the coasts of Oman were part of a complementary network of goods exchange with nomad groups, oasis farmers and other groups in the interior parts of the country. It is not a simple exchange of commodities between two variable economic systems or a coastal adaptation and an inland adaptation. It is the dependency of the inland system or tradition on the coastal and *vice versa*." The analogies with the Andes—where dried fish moved inland and *ch'arki* moved to the coast—should be clear.

9

The Archaeological Fish Remains from Cerro Azul

Jeffrey D. Sommer and Kent V. Flannery

The University of Michigan excavations at Cerro Azul produced more than 25,000 fish bones. Somewhere between two-thirds and three-quarters of these bones could be identified at least to the level of family, and thousands could be identified to species. All identified taxa, listed in phylogenetic order, can be found in Table 9.1.

We note that our list of Late Intermediate species is only two-thirds as long as the list of fish Marcus and Flannery collected at Cerro Azul in the 1980s (Table 6.1). There are a number of reasons for this. First—as pointed out in Chapter 6—sharks, rays, and chimaeras will always be underrepresented archaeologically, because their skeletons are largely cartilage. Second—as explained in Chapter 7—many species caught in the 1980s with rowboats or outboard motors would have been beyond the reach of Late Intermediate caballitos de totora. Finally, some deep-water species like the Pacific hake (*Merluccius gayi*) are more likely to be caught by commercial trawlers than by small boats of any kind. What we see in Table 9.1 is essentially a list of fish that could be caught by wading in Cerro Azul Bay with baskets or trammel nets, hurling cast nets off Cerro Centinela, or keeping your caballito within 100 m of the shore.

Obstacles to Identification

Before presenting the archaeological fish bone from Cerro Azul, let us discuss some of the obstacles we faced during our analysis. Most have to do with the presence of closely related species from the same family.

1. The specimens we have identified as Clupeids are overwhelmingly *Sardinops sagax*; perhaps no more than 1 percent are the machete, *Brevoortia maculatum*. There are problems, however, in distinguishing the vertebrae of these two Clupeids.
2. The specimens we have identified as Engraulids are overwhelmingly *Engraulis ringens*, the anchoveta negra. Since the anchoveta blanca (*Anchoa nasus*) occasionally shows up at Cerro Azul, however, its presence cannot always be ruled out. Distinguishing the vertebrae of these two species is particularly difficult.
3. All the grunts that could be identified to species turned out to be the chita, *Anisotremus scapularis*. Since the cabinza (*Isacia conceptionis*) occasionally shows up at Cerro Azul, however, it could not always be ruled out.

4. The drums (Family Sciaenidae) presented special problems. Seven members of this family were caught by Late Intermediate fishermen, and many of these species are about the same size. As a result, some drums simply had to be listed as "small-to-medium Sciaenids" or "medium-to-large Sciaenids." Fortunately, there was little doubt about the most frequently captured Sciaenid at Cerro Azul: more than 90 percent of the identified drum remains belonged to the lorna (*Sciaena deliciosa*).
5. Most *Mugil* specimens were probably the common mullet, *Mugil cephalus*, whose teeth Medina Chauca (1982) classifies as villiform (resembling shaggy hair). Since the white mullet *M. curema* occasionally shows up at Cerro Azul, however, it could not always be ruled out.
6. The left-eye flounder *Paralichthys adspersus* was the Bothid most frequently caught at Cerro Azul in the 1980s, and all identified flounders were consistent with this species. Related flounders, however, could not always be ruled out, especially in the case of identifications made from vertebrae.
7. Calculating the NISP (number of identified specimens) for the Cerro Azul fish was not difficult. Calculating the MNI (minimum number of individuals) was often problematic, especially when (as in the case of anchovetas) the most abundant elements were vertebrae. According to a study of 14,156 anchovetas by Jordán (1963), the number of vertebrae in *Engraulis ringens* can vary from 43 to 49 (mean = 46.75). Jordán concluded that lower numbers of vertebrae could be a phenotypic response to years with higher-than-normal ocean temperatures. He also detected a slight trend toward higher numbers of vertebrae with increasing latitude (traveling south from Chimbote to Ilo).

Structure D

Our largest sample of fish bones came from **Feature 6**, a midden in the Southwest Canchón of Structure D. This midden was screened through both 1.5 mm and 0.6 mm mesh and produced more than 13,400 identified specimens.

Why so many thousands of fish bones ended up in this midden is an interesting question. Feature 6 is thought to represent debris swept up from the floor of the Southwest Canchón and piled against its east wall. Under normal circumstances, this debris would eventually have been carried to one of the quebradas of Cerro Camacho and left on a terrace. Since this never happened, we suspect that Feature 6 accumulated not long before A.D. 1470, when the Inca conquered Huarco and Structure D was abandoned.

Why were so many fish bones present on the floor of the Southwest Canchón? Did some initial cleaning of fish take place there? If fish heads were removed there, for example, it would explain the large number of bones from the feeding apparatus and gill covers found in Feature 6. It would not, however, explain the thousands of vertebrae in the midden; these look more like the refuse from meals. It is possible, therefore, that fish refuse

Table 9.1. Taxa of Identified Fish from Archaeological Contexts at Cerro Azul.

Cartilaginous fish	sharks, rays, or chimaeras
Carcharhinus sp.	requiem shark
Clupeidae	sardine or machete
cf. *Brevoortia maculatum*	likely machete
Sardinops sagax	Pacific sardine
Engraulidae	black or white anchoveta
Engraulis ringens	anchoveta negra
Galeichthys peruvianus	sea catfish
Mugil sp.	common or white mullet
Paralabrax humeralis	Peruvian rock bass
Trachurus symmetricus	Chilean jack mackerel
cf. *Anisotremus scapularis*	grunt; likely chita
Menticirrhus ophicephalus	mismis
Paralonchurus peruanus	coco
Cynoscion analis	ayanque
Sciaena fasciata	burro
Sciaena deliciosa	lorna
Sciaena starksi	róbalo
Cilus gilberti	corvina
Stellifer minor	mojarrilla
Small-to-medium Sciaenids	drums (small/medium)
Medium-to-large Sciaenids	drums (medium/large)
Cheilodactylus variegatus	pintadilla
Scartichthys gigas	scaleless blenny
Labrisomus philippii	scaled blenny
Scomber japonicus	Pacific mackerel
Sarda sarda	bonito
Sicyases sanguineus	clingfish
cf. *Paralichthys adspersus*	left-eye flounder

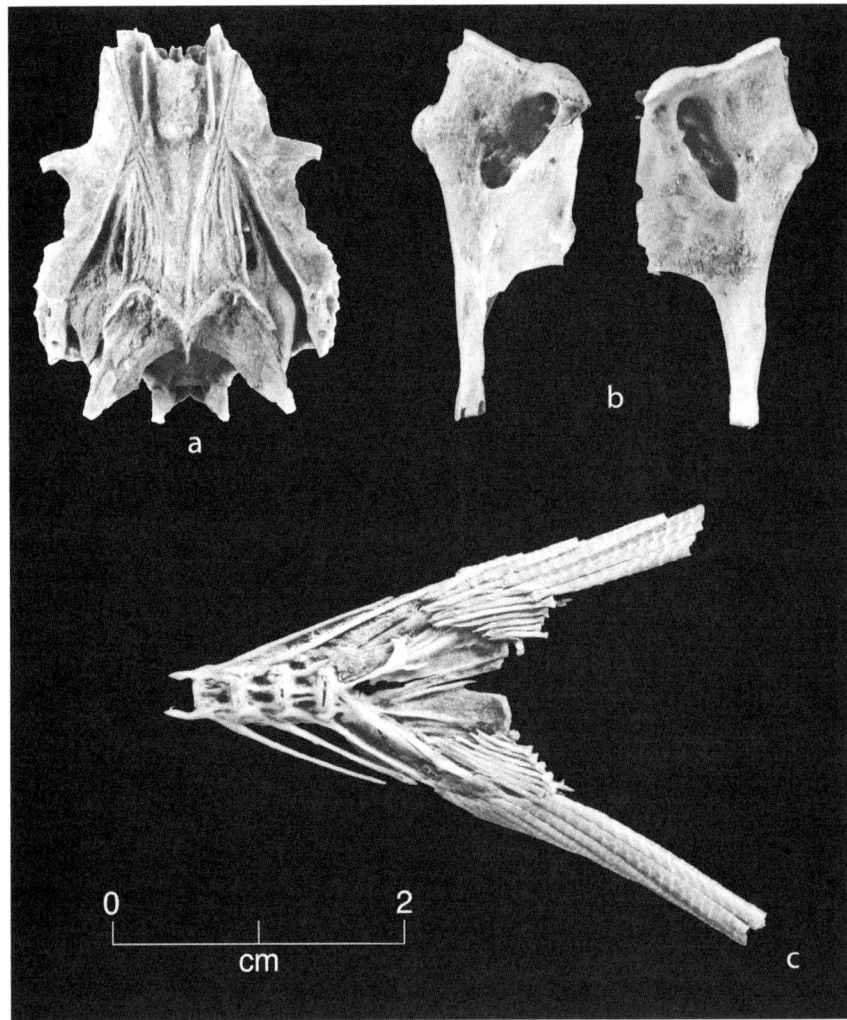

Figure 9.1. Bones of Pacific sardine (*Sardinops sagax*) from Feature 6, Structure D. *a*, neurocranium. *b*, two hyomandibulars. *c*, desiccated tail.

from Structure D was left in the Southwest Canchón for a while before being carried off to a quebrada.

Figures 9.1 through 9.26 illustrate some of the skeletal parts that allowed us to identify the osteichthys (or bony fish) from Feature 6. We have published these bones at two times natural size to make their landmarks easier to see. Sadly, the 25 ossified vertebral centra from cartilaginous fish in Feature 6 probably reflect only a small percentage of the sharks, rays, and chimaeras that were consumed.

Pacific sardines were well represented in Feature 6; we recovered their distinctive neurocrania and tails (Fig. 9.1), many of the bones from the mouth and gill covers (Fig. 9.2), and even parts of the pectoral girdle (Fig. 9.3).

In the case of the anchoveta, two of the most recognizable items we recovered were desiccated heads and tails (Figs. 9.4, 9.5). To our surprise, even the lenses from the eyes of anchovetas were sometimes preserved (Fig. 9.6).

The chita (*Anisotremus scapularis*) was represented in Feature 6 by hundreds of bones. Among its most diagnostic characteristics were its supraoccipital crest, its pharyngeal teeth, and the bones making up its jaws (Figs. 9.7, 9.8).

As mentioned earlier, so many species of drums were captured at Cerro Azul that hundreds of their bones could be identified no further than family. An unexpected discovery was that both Late Intermediate and modern drums display occasional exostoses (abnormal bony swellings) of the vertebrae (Fig. 9.9).

The bones of the feeding apparatus helped us to distinguish among the various species of drums; the articular proved especially useful (Figs. 9.10–9.12). The fact that Feature 6 produced more than 1000 bones of the lorna (*Sciaena deliciosa*) gave us a number of reliable landmarks for that species (Figs. 9.12, 9.13). Especially diagnostic were its teeth, which Medina Chauca (1982) classifies as cardiform (resembling the wire bristles on a wool-carding tool).

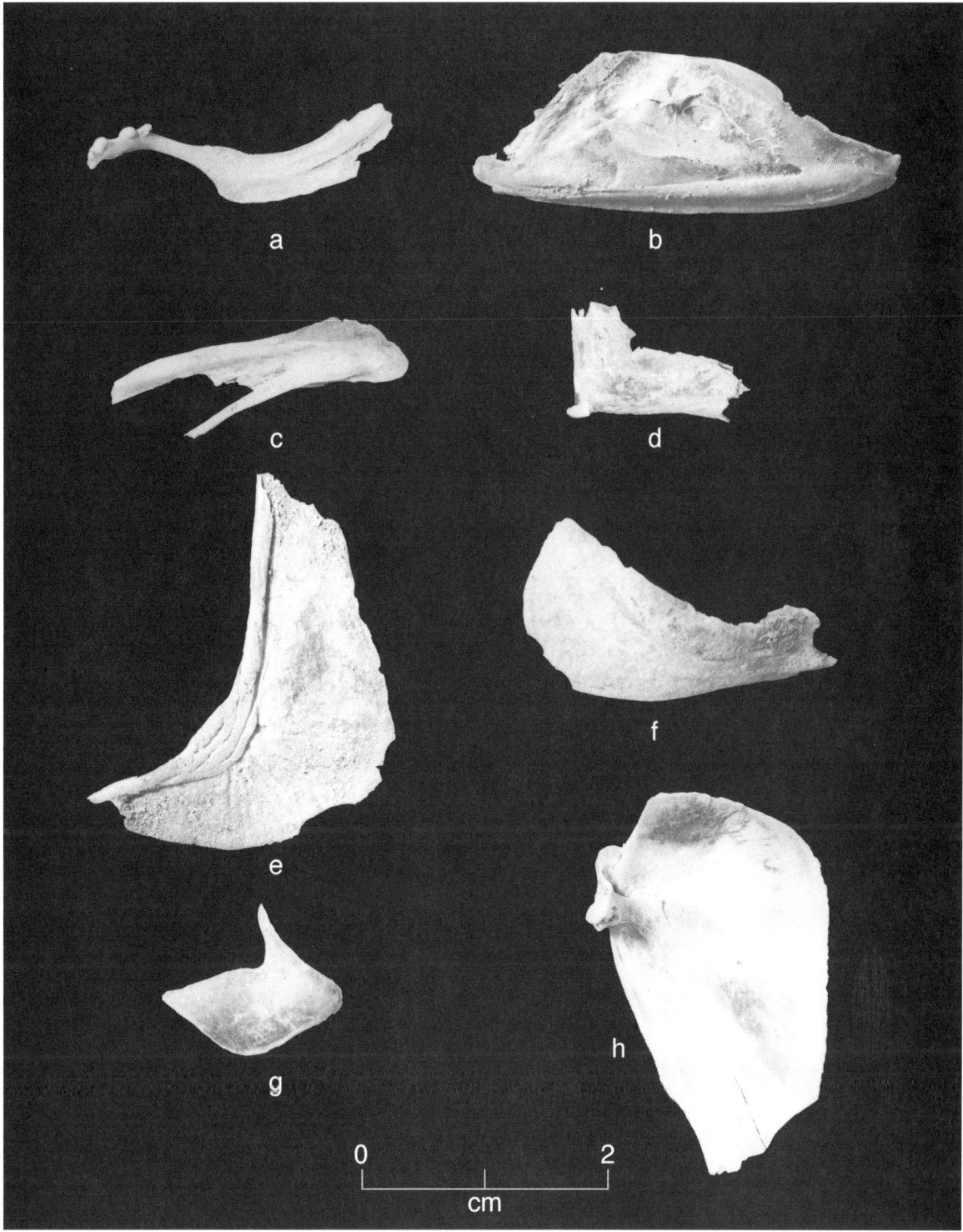

Figure 9.2. Bones of Pacific sardine from Feature 6, Structure D. *a*, maxilla. *b*, still-articulated dentary and articular. *c*, posttemporal. *d*, quadrate. *e*, preopercular. *f*, interopercular. *g*, subopercular. *h*, opercular.

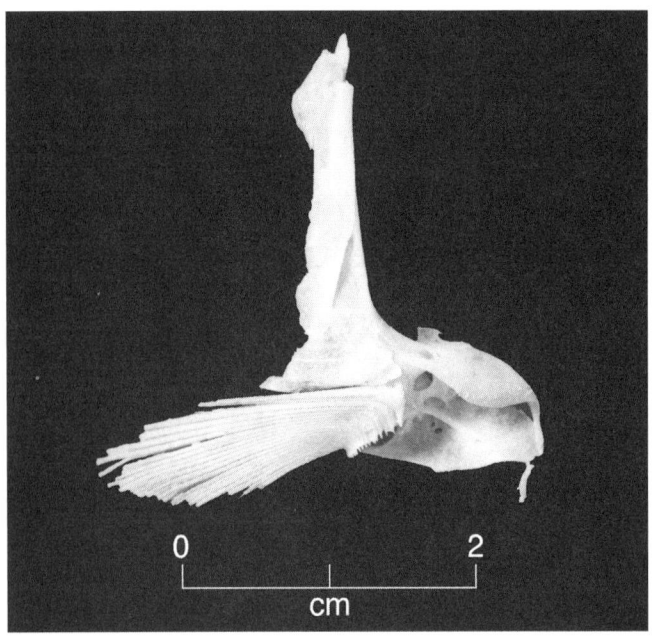

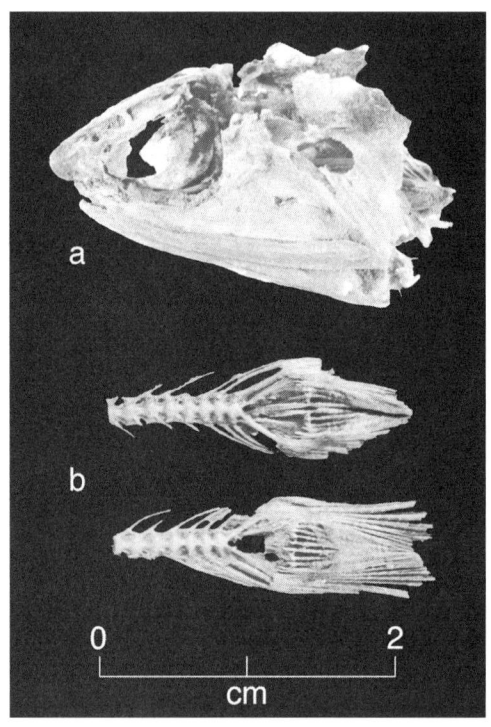

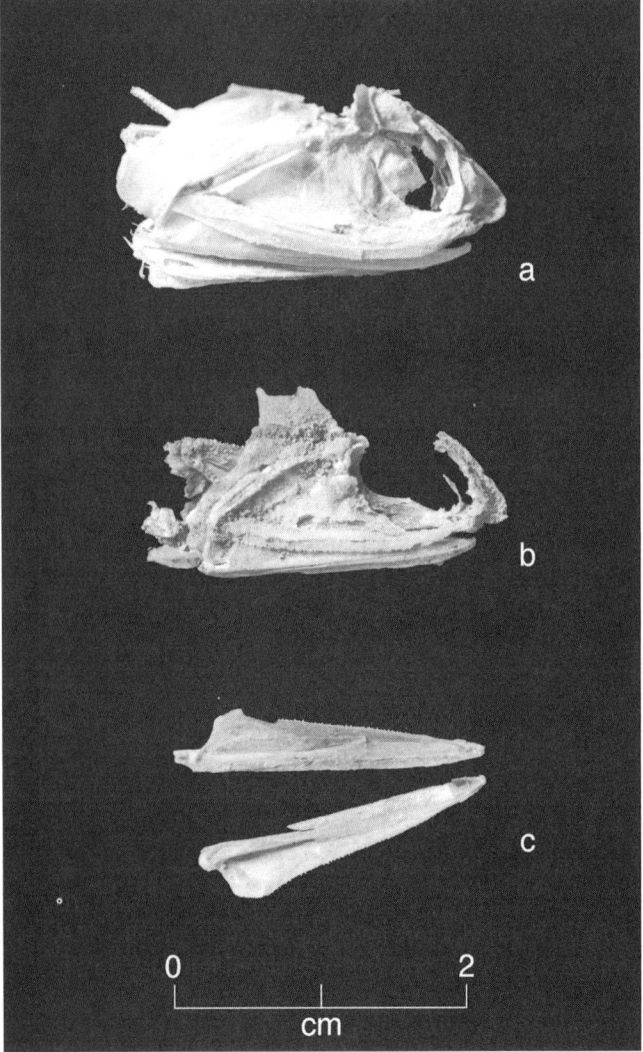

Figure 9.3 (above left). Cleithrum and pectoral fin rays of Pacific sardine from Feature 6, Structure D.

Figure 9.4 (above). Remains of anchoveta from Feature 6, Structure D. *a*, desiccated head. *b*, two desiccated tails.

Figure 9.5 (left). Remains of dried anchovetas from Feature 6, Structure D. *a, b*, desiccated heads. *c*, a pair of still-articulated dentaries and articulars.

Figure 9.6 (below). Lenses from the eyes of anchovetas, Feature 6, Structure D.

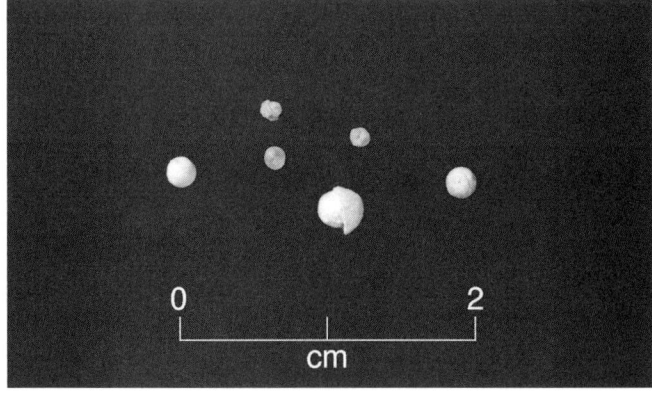

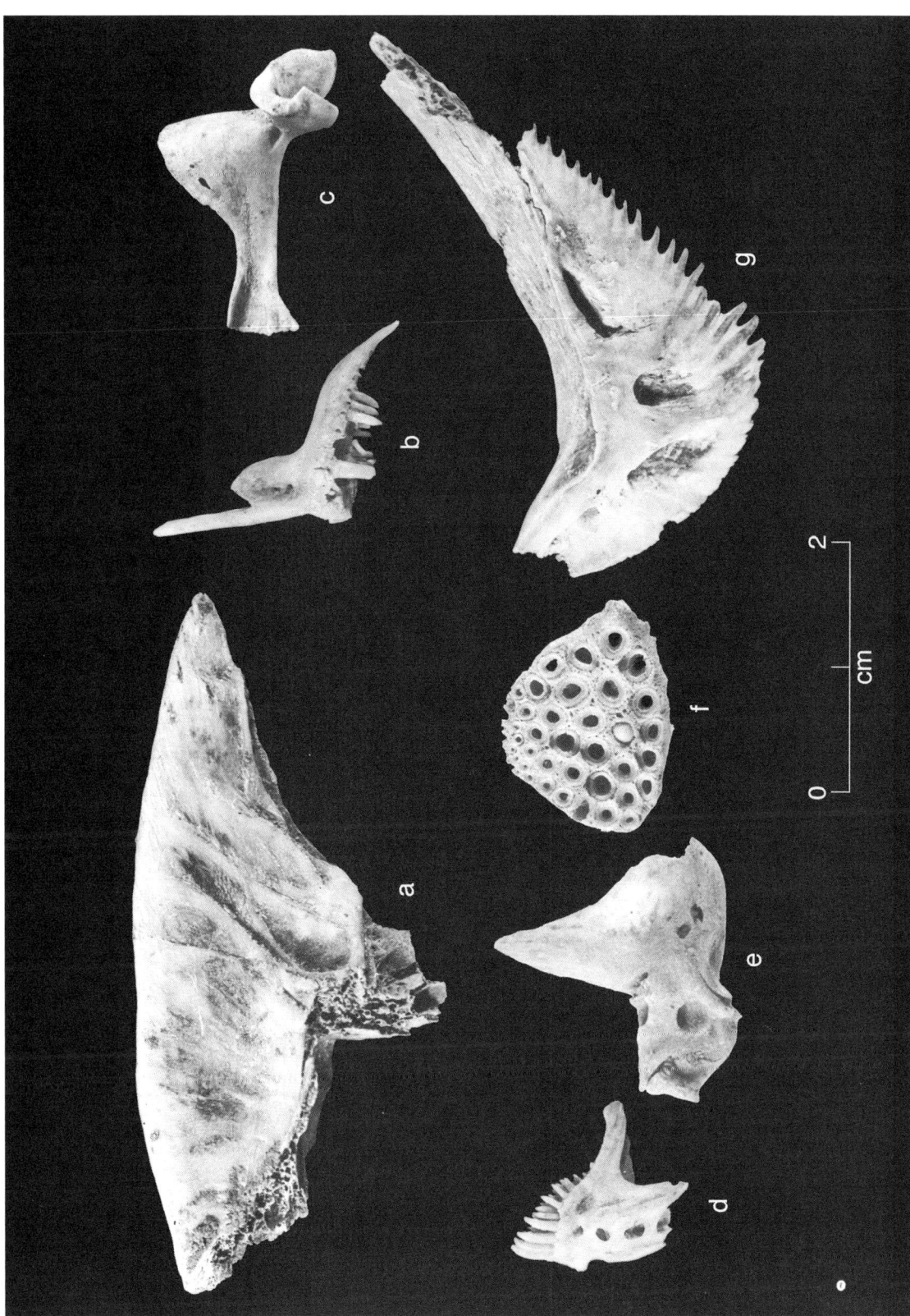

Figure 9.7. Bones of grunt (*Anisotremus scapularis*) from Feature 6, Structure D. *a*, supraoccipital crest. *b*, premaxilla. *c*, maxilla. *d*, dentary. *e*, articular. *f*, pharyngeal teeth. *g*, preopercular.

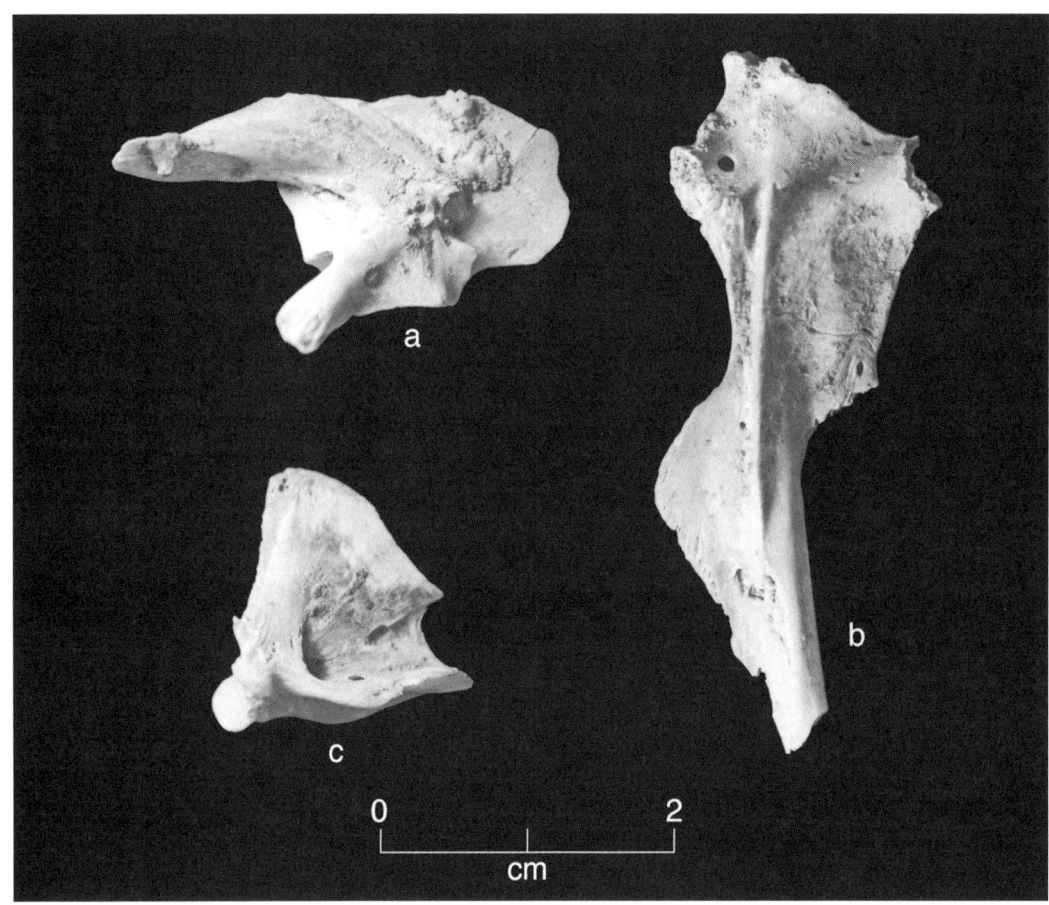

Figure 9.8. Bones of grunt from Feature 6, Structure D. *a*, posttemporal. *b*, hyomandibular. *c*, quadrate.

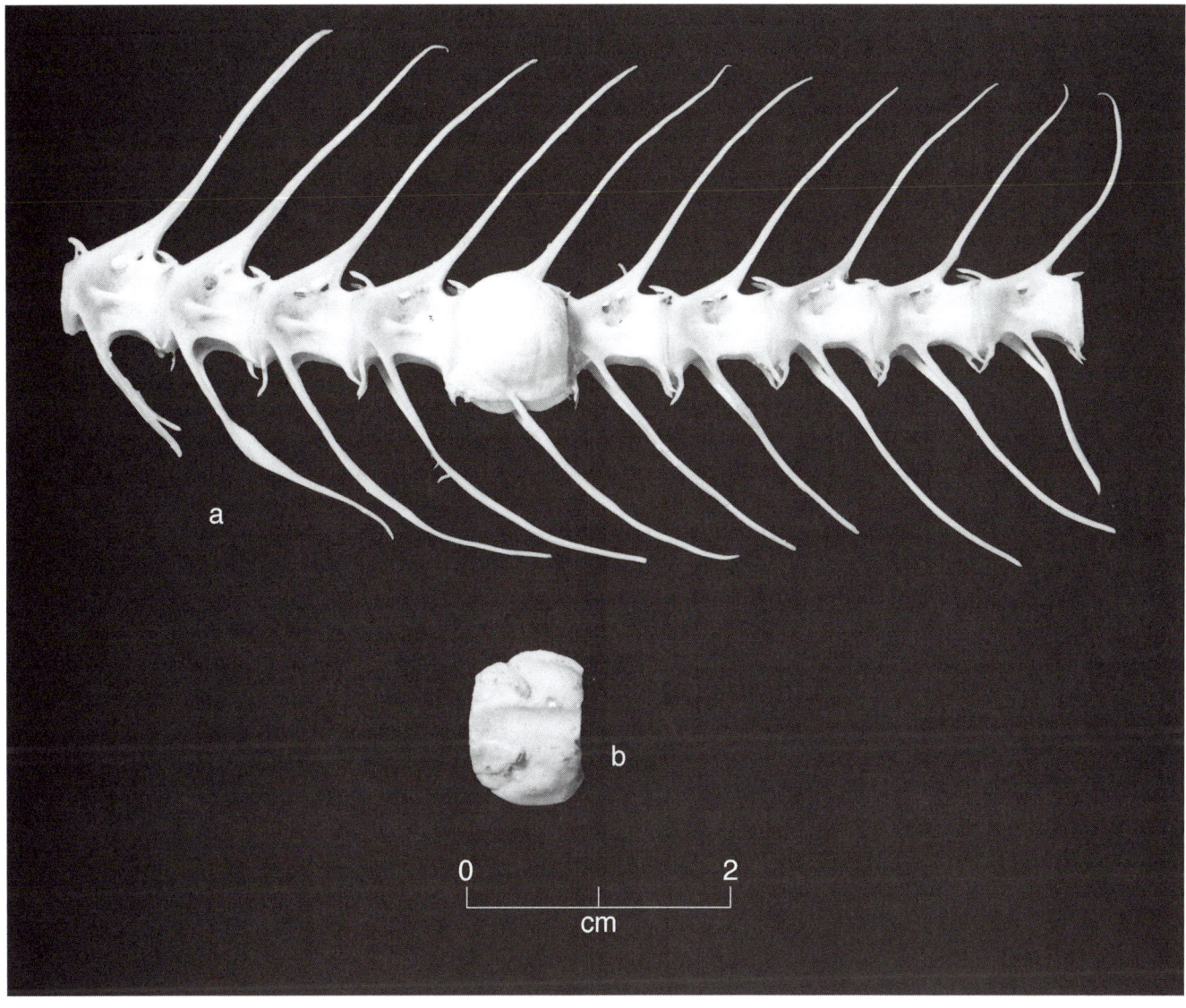

Figure 9.9. Both ancient and modern fish of the drum family caught at Cerro Azul display occasional exostoses (abnormal swellings) of the vertebral centra. *a*, spinal column of coco (*Paralonchurus peruanus*) caught in 1985, featuring an exostosis. *b*, exostosis on a vertebral centrum in Zone C of Terrace 9, Quebrada 5a.

0
2
cm

0
2
cm

a
b
c
d
e
g
h
f
i
0
2
cm
j
k

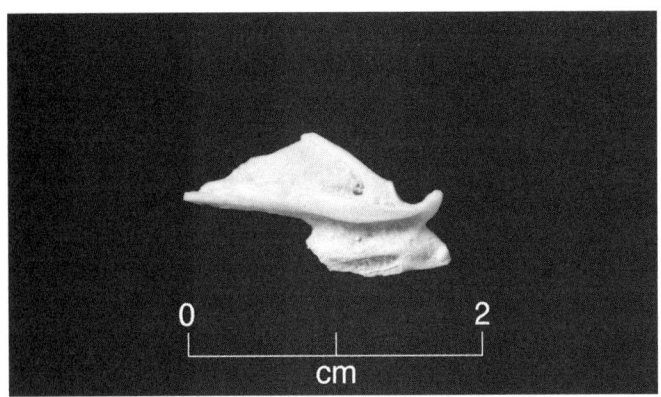

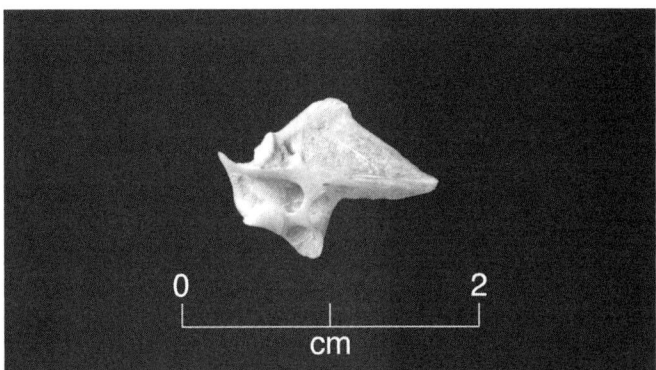

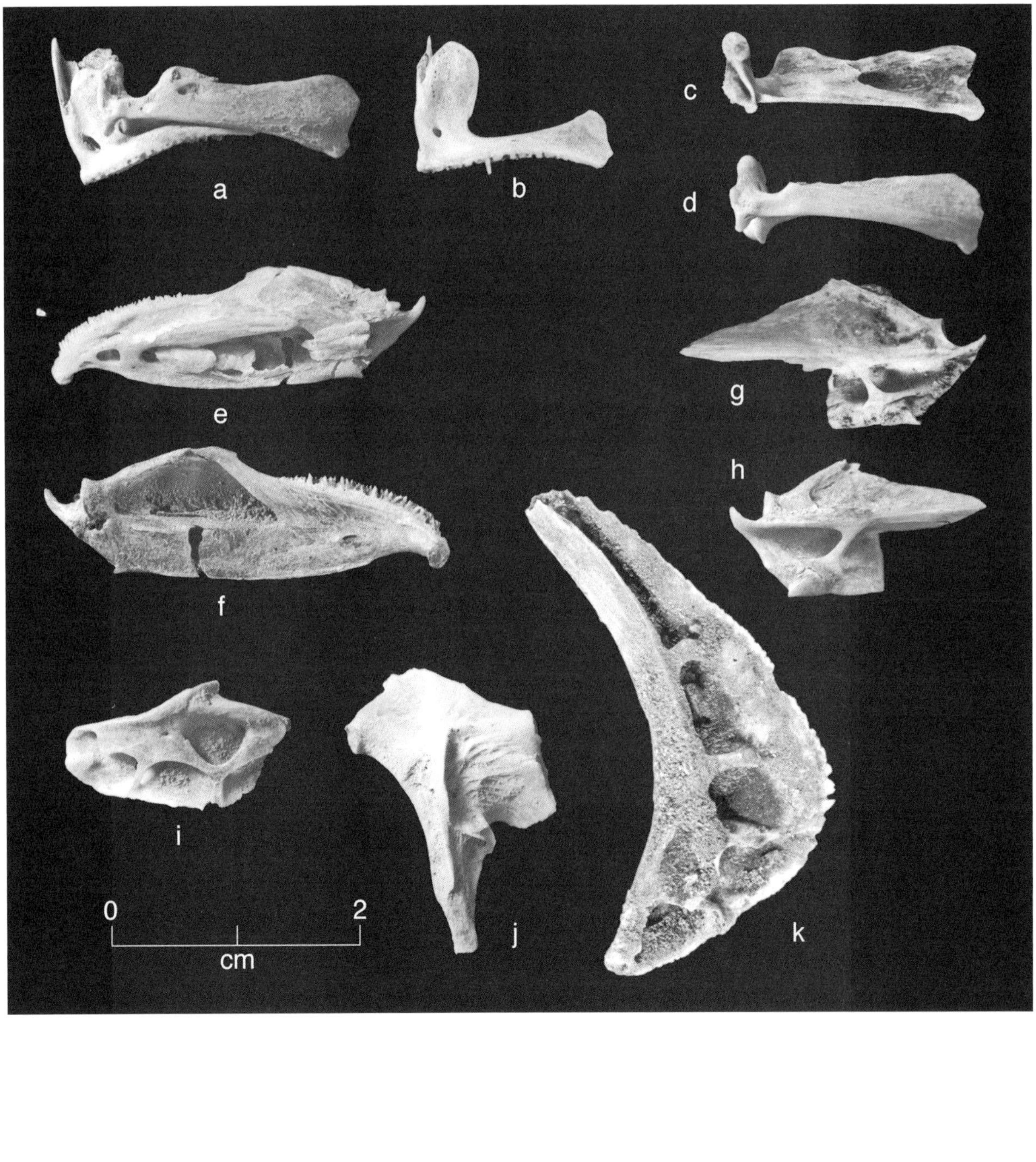

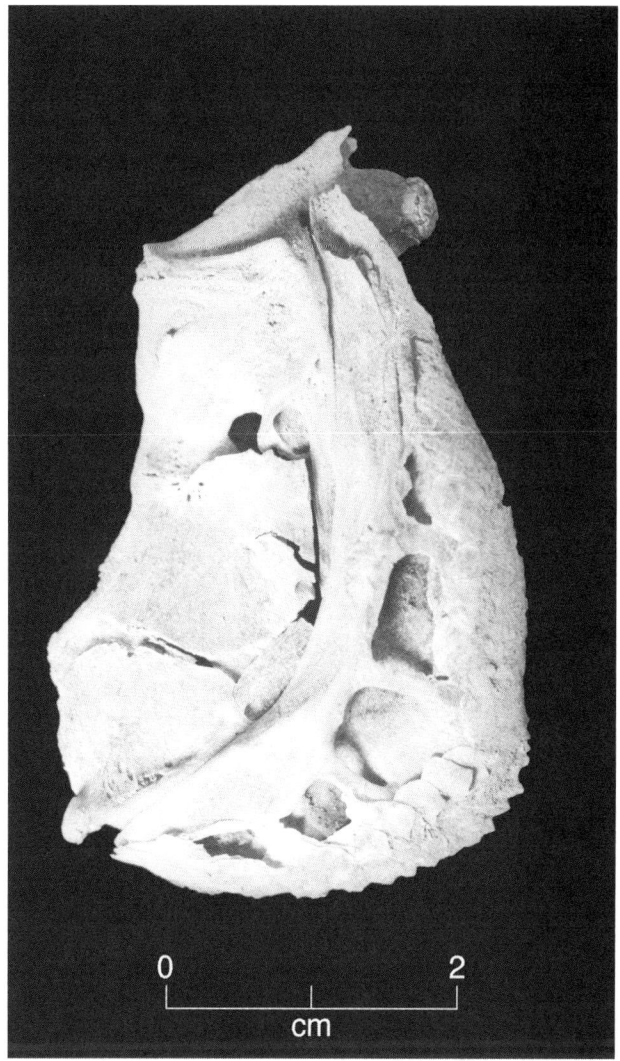

Opposite page:
Figure 9.10 (top left). Articular of ayanque (*Cynoscion analis*) from Feature 6, Structure D.

Figure 9.11 (top right). Articular of burro (*Sciaena fasciata*) from Feature 6, Structure D.

Figure 9.12 (below left). Bones of the lorna (*Sciaena deliciosa*) from Feature 6, Structure D. *a*, still-articulated premaxilla and maxilla. *b*, premaxilla. *c, d*, maxillae. *e, f*, two examples of still-articulated dentary and articular. *g, h*, articulars. *i*, lachrymal. *j*, hyomandibular. *k*, preopercular.

Figure 9.13 (right). Still-articulated preopercular, hyomandibular, quadrate, symplectic, and metapterygoid of lorna from Feature 6, Structure D.

The largest of the drums taken at Cerro Azul was the róbalo (*Sciaena starksi*), and some of the mouth parts recovered from Structure D were larger, on average, than those of róbalos collected in the 1980s (Figs. 9.14, 9.15). It was also the bones of the feeding apparatus that allowed us to identify the corvina (Fig. 9.16).

Smallest of the local drums was the mojarrilla (*Stellifer minor*), also recognizable by its mouth parts and cardiform teeth (Fig. 9.17).

Drums are known for their otoliths, a number of which are presented in Fig. 9.18. Given the large number of lornas recovered from Feature 6, this species was well represented in our otolith collection.

The pintadilla (*Cheilodactylus variegatus*) was also found in Feature 6, and its feeding apparatus (including its cardiform teeth) was clearly identifiable (Fig. 9.19).

Both scaled and scaleless blennies were caught during the Late Intermediate, though not in equal numbers; in Feature 6 there were five times as many bones of the scaleless blenny (*Scartichthys gigas*) as bones of the scaled blenny (*Labrisomus philippii*). Both blennies could be identified by their feeding apparatus (Figs. 9.20, 9.21). For the distinctive premaxillae of the scaleless blenny, see also Medina Chauca (1982: Fig. 21 *b*).

The Pacific bonito (*Sarda sarda chiliensis*) has needlelike teeth in its jaws and an unusual hypural plate in its tail (Fig. 9.22), making it easily recognizable.

It appears that the Late Intermediate residents of Cerro Azul successfully pried the pejesapo or clingfish (*Sicyases sanguineus*) from the sea cliffs. This fish's unique adaptation and incisiform teeth make its feeding apparatus distinctive (Fig. 9.23; see also Medina Chauca 1982: Fig. 21 *c*).

Because both its eyes are on one side of its head, the left-eye flounder (*Paralichthys adspersus*) has a distinctive neurocranium. Its stiletto-like teeth proved diagnostic as well (Figs. 9.24, 9.25). Figure 9.24 also makes it clear that Late Intermediate flounders came in a variety of sizes.

In spite of the large number of fish bones found in Feature 6, only a few showed cut marks. One of those was the ossified vertebral centrum from a cartilaginous fish (Fig. 9.26). These cut marks may have been made when the animal was being sliced into strips for drying.

In Table 9.2 we present all of the fish bones recovered from Feature 6 of Structure D.

Table 9.2. Fish Bones from Feature 6, Structure D.

Cartilaginous fish	25 ossified vertebral centra (from numerous fish)		1 cleithrum/coracoid/pectoral fin/scapula/tripus
Likely machete	1 hyomandibular		1 branchiostegal + interopercular joined
	1 posttemporal		28 tripuses
	2 operculars (MNI = 1)		9 ventral scales
Pacific sardine	9 neurocrania		43 final vertebrae
	31 maxillae	Clupeidae, cf. *Sardinops*	7,202 vertebrae
	3 maxilla + premaxilla joined		
	2 maxilla + premaxilla + supramaxilla1	Engraulidae, cf. *E. ringens*	
	1 maxilla + premaxilla + supramaxilla1 + supramaxilla2		7 complete heads
			9 dentaries
	1 maxilla + premaxilla + supramaxilla2		15 articulars
			9 cleithra
	9 maxilla + supramaxilla1 + supramaxilla2		3 maxillae
			1 quadrate
	1 maxilla + supramaxilla2		8 hyomandibulars
	1 supramaxilla1 + supramaxilla2		1 hyomandibular + lense
	6 supramaxillae 2		6 hyal + ceratohyal joined
	39 dentaries		11 articular + quadrate
	10 articular + dentary joined		7 articular/dentary/quadrate
	5 articular + dentary + quadrate		7 articular/dentary/maxilla
	47 articulars		22 articular/dentary/maxilla/quadrate
	20 basioccipitals		
	75 cleithra		2 articular/dentary/hyomandibular/quadrate
	5 ceratohyals		
	2 ceratohyal + epihyal joined		2 articular/dentary/hyomandibular/maxilla/quadrate
	21 quadrates		
	6 prootics		
	9 ethmoids		15 articular + dentary joined
	10 frontals		69 articular/dentary/ceratohyal (MNI = 46)
	1 ectopterygoid + quadrate		
	182 hyomandibulars (MNI = 97)		2,663 vertebrae (MNI = 54 to 62)
	1 hyomandibular + opercular joined	Mullet, cf. *M. cephalus*	1 articular (MNI = 1)
	1 hyomandibular + preopercular		1 interopercular
	1 hyomandibular/opercular/subopercular		5 vertebrae
	13 interoperculars	Peruvian rock bass	1 articular (MNI = 1)
	1 interopercular/opercular/preopercular		1 dentary
			1 supracleithrum
	145 operculars	Chilean jack mackerel	1 premaxilla
	2 opercular + subopercular joined		1 palatine
	32 preoperculars		2 operculars (MNI = 1)
	25 posttemporals		
	12 suboperculars		

Table 9.2. Fish Bones from Feature 6, Structure D.

	1 posttemporal 8 vertebrae	Burro	1 articular (MNI = 1) 1 prevomer

Grunt, cf. *Anisotremus*	4 supraoccipital crests
	12 premaxillae
	17 maxillae
	13 dentaries (MNI = 8)
	11 articulars
	7 quadrates
	2 prevomers
	1 frontal
	8 upper pharyngeals
	6 lower pharyngeals
	5 other pharyngeals
	10 preoperculars
	3 operculars
	9 hyomandibulars
	5 fragments basioccipital
	6 cleithra
	2 supracleithra
	3 postcleithra
	4 scapulae
	10 posttemporals
	1 prootic + sphenotic joined
	6 palatines
	6 parasphenoids
	1 interopercular
	9 hyomandibulars
	1 epihyal
	6 urohyals
	3 ceratohyals
	5 basipterigia
	1 coracoid
	1 opisthotic + pterotic joined
	1 undetermined
	82 vertebrae (7 gnawed, possibly by dog?)
		Lorna	27 premaxillae (MNI = 17)
			1 premaxilla + maxilla joined
			27 maxillae
			19 dentaries
			5 dentary + articular
			31 articulars (MNI = 17)
			18 hyomandibulars
			1 hyomandibular/metapterygoid/
			  preopercular/quadrate/
			  symplectic
			3 frontals
			3 palatines
			13 quadrates
			1 quadrate + preopercular joined
			14 preoperculars
			2 prootics
			4 prootic + sphenotic
			1 sphenotic
			5 posttemporals
			8 lachrymals
			8 interoperculars
			2 epihyals
			7 ceratohyals
			3 ceratohyal + epihyal joined
			2 ceratohyal + hypohyal
			3 ceratohyal/epihyal/hypohyal
			9 cleithra
			790 vertebrae

Mismis	57 vertebrae (MNI = 2)	Róbalo	1 maxilla (MNI = 1)
Coco	1 dentary (MNI = 1) 1 otolith	Corvina	6 premaxillae (MNI = 4) 5 maxillae 1 dentary 3 articulars 4 quadrates 2 hyomandibulars 2 otoliths
Ayanque	1 articular 1 hyomandibular 1 ceratohyal + epihyal joined 3 vertebrae	Mojarrilla	1 premaxilla 3 dentaries 7 preoperculars (MNI = 4) 1 prevomer 1 ceratohyal + epihyal 3 otoliths

Table 9.2. Fish Bones from Feature 6, Structure D.

Small-to-medium Sciaenids	8 premaxillae		3 articulars
	10 maxillae		4 quadrates
	13 dentaries		1 palatine
	10 hyomandibulars		1 prevomer
	13 operculars		5 hyomandibulars
	1 subopercular		1 preopercular
	3 interoperculars		3 operculars
	1 prootic		2 interoperculars
	2 prootic + sphenotic joined		1 lachrymal ?
	1 parasphenoid		1 ceratohyal
	2 exoccipitals		1 ceratohyal + epihyal joined
	2 posttemporals		2 cleithra
	4 cleithrum + scapula joined		2 supercleithra
	3 basipterygium		1 postcleithrum ?
	43 otoliths		47 vertebrae
	9 urohyals	Bonito	6 premaxillae (MNI = 4)
	19 tails		4 dentaries
	6 final vertebrae		2 fragments dentary or premaxilla
Medium-to-large Sciaenids			1 articular
	4 premaxillae		1 hyomandibular
	1 otolith		2 palatines
	239 vertebrae		2 prevomers
Pintadilla	2 maxillae (MNI = 2)		1 quadrate + ectopterygoid joined
	1 dentary		1 preopercular
	2 articulars (MNI = 2)		3 ceratohyals
	2 quadrates		2 epihyals
	1 palatine		1 hypohyal
	2 interoperculars		1 cleithrum
	1 ceratohyal		1 scapula
	1 epihyal		60 vertebrae
	1 posttemporal	Clingfish	4 premaxillae
	17 vertebrae		1 maxilla
Scaleless blenny	2 premaxillae		2 dentaries
	20 maxillae (MNI = 13)		1 articular
	17 dentaries		1 quadrate
	3 articulars		3 operculars (MNI = 3)
	12 quadrates		3 cleithra (MNI = 3)
	18 preoperculars		10 vertebrae
	5 operculars	Left-eye flounder, cf. *Paralichthys*	4 premaxillae (MNI = 3)
	9 cleithra		1 maxilla
	345 vertebrae		1 dentary
Scaled blenny	6 maxillae (MNI = 4)		2 articulars
	3 dentaries		2 dentary + articular joined
			3 hyomandibulars

Table 9.2. Fish Bones from Feature 6, Structure D.

Unidentified bony fish	3 preoperculars 1 parasphenoid 1 prevomer 1 urohyal 50 vertebrae 3 maxillae 3 articulars 2 quadrates 17 hyomandibulars 2 preoperculars 1 opercular + subopercular joined 3 interoperculars 9 pharyngeal teeth 2 palatines 3 prevomers 2 posttemporals 21 cleithra 19 supracleithra 12 scapulae 17 basioccipitals 4 exoccipitals	2 ceratohyals 7 epihyals 1 sphenotic 4 basipterigia 12 radials 16 hypurals 2 urohyals 9 pectoral spines 89 pterygiophores 11 pterygiophore + spine 134 miscellaneous spines 69 fine rays 1 complete tail 1871 vertebrae (5 with exostoses) 1596 miscellaneous unidentified bones 7.95 grams of scales

Totals:
Cartilaginous fish, NISP = 25
Bony fish, NISP = 13, 393
Unidentified = 3, 944

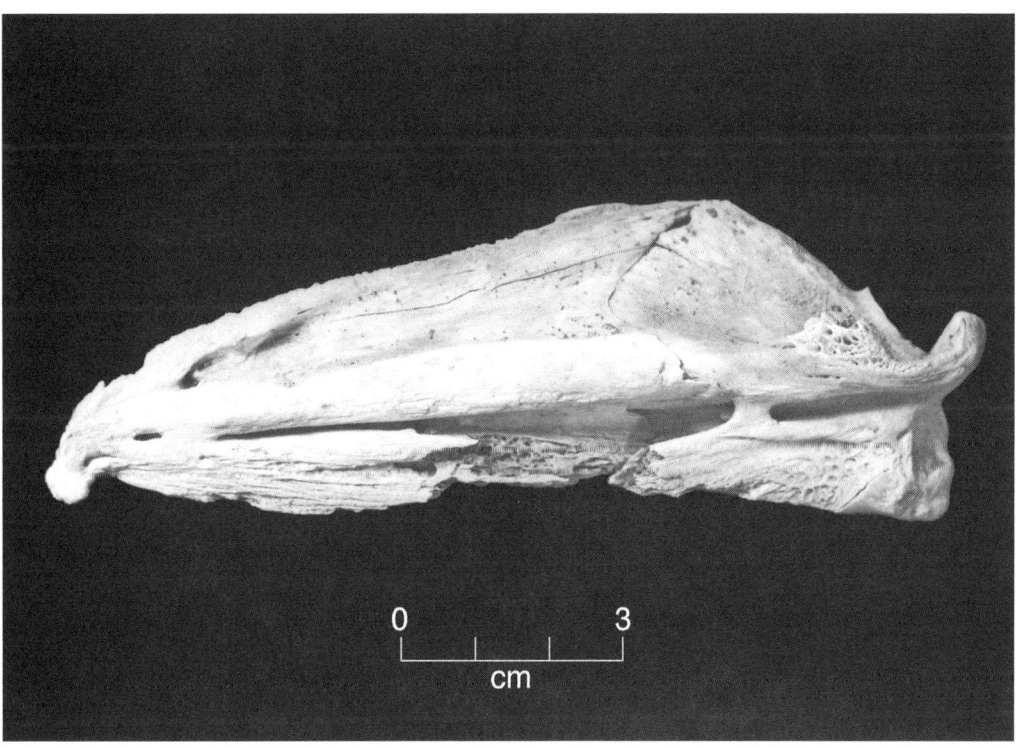

Figure 9.14. Articular and dentary of large róbalo (*Sciaena starksi*), damaged but still articulated. These bones were found in the upper fill of Room 3, Structure D.

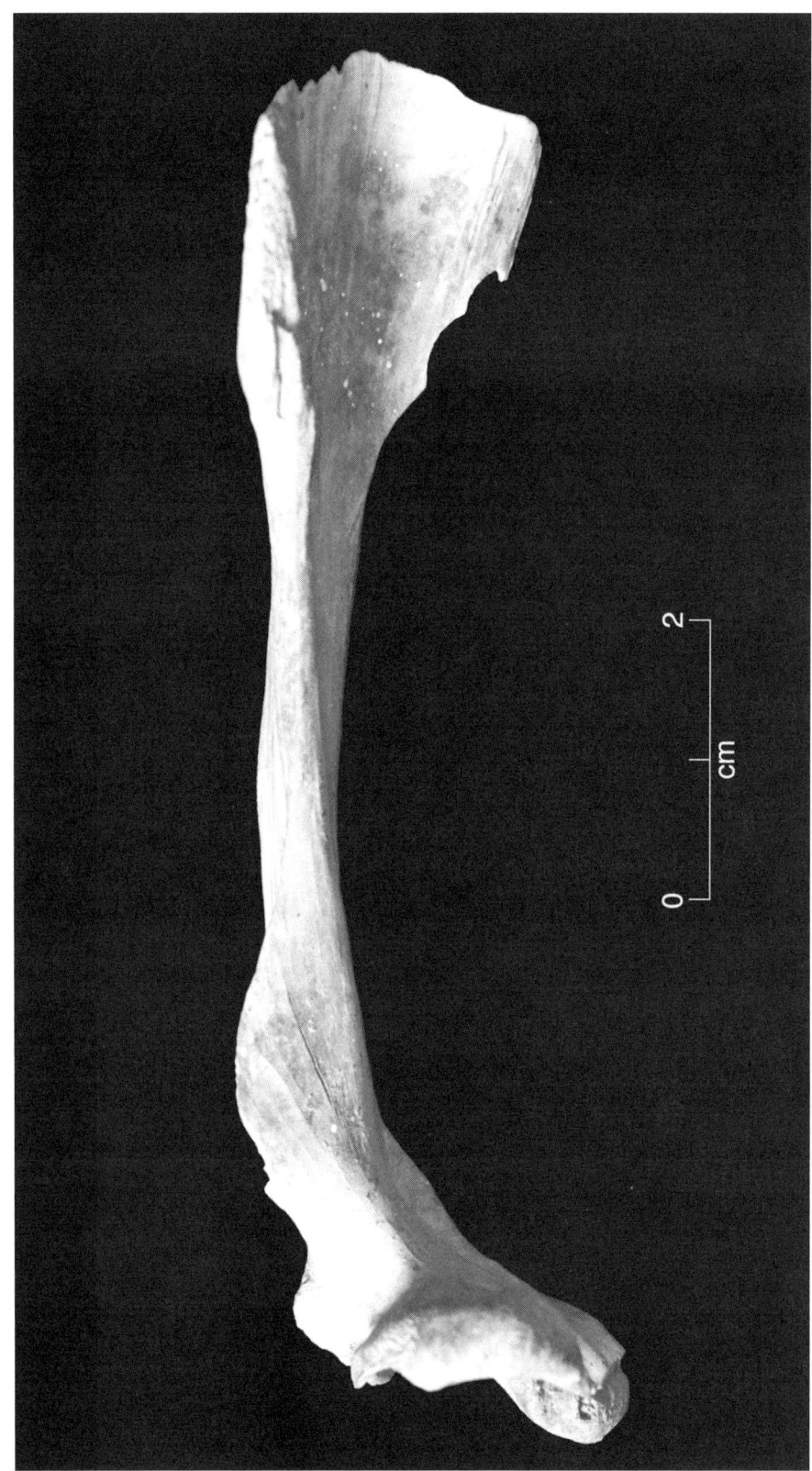

Figure 9.15. Maxilla of large róbalo from Late Intermediate refuse, Structure D.

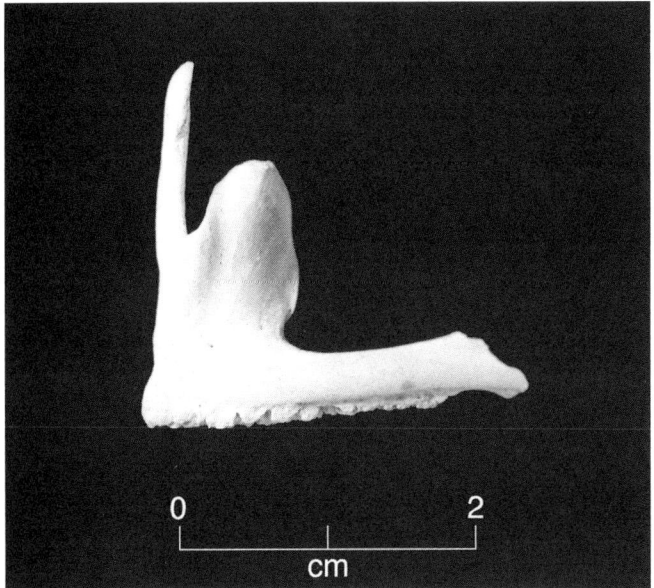

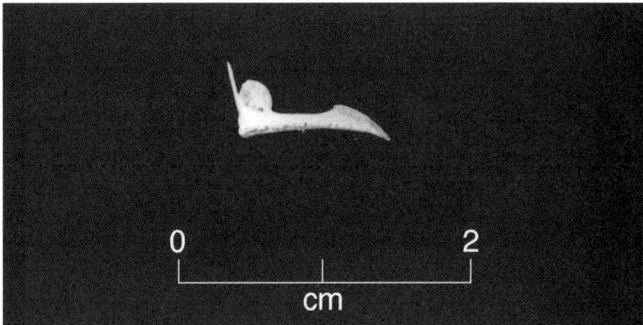

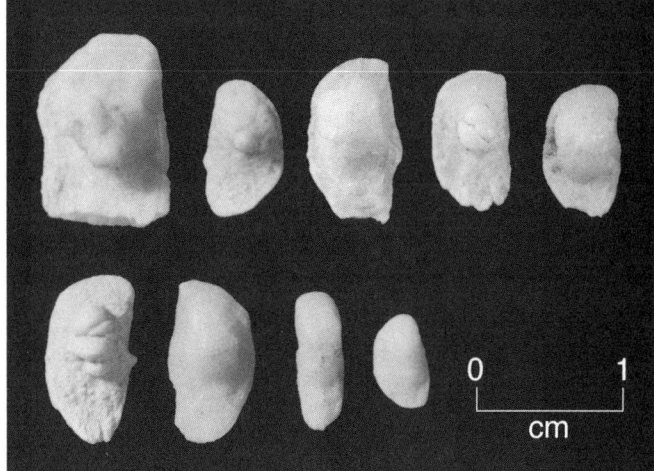

Figure 9.16 (above). Premaxilla of corvina (*Cilus gilberti*) from Feature 6, Structure D.

Figure 9.17 (upper right). Premaxilla of mojarrilla (*Stellifer minor*) from Feature 6, Structure D.

Figure 9.18 (right). Drum otoliths from Feature 6, Structure D.

The Southwest Canchón

A few fish bones on the floor of the Southwest Canchón had escaped being swept up into the Feature 6 midden. Those closest to the midden were one vertebra from a corvina and one vertebra from a left-eye flounder. Farther from the midden, Marcus' crew found the pharyngeal teeth of a grunt, one vertebra from a medium-to-large Sciaenid, the right premaxilla of a left-eye flounder, and one ossified vertebral centrum from a large cartilaginous fish.

In the northern portion of the canchón, Marcus' crew found two tapia platforms. One of these, the North Platform, is believed to be the vantage point from which a commoner-class overseer supervised the movement of commodities in and out of Structure D. One vertebra of chita was found on the overseer's *pata*, or sleeping area; scattered around it, on the surface of the North Platform, were the head of an anchoveta negra, the premaxilla of a medium-to-large Sciaenid, and the vertebra of an unidentified fish. On the floor of the Southwest Canchón, near the base of the North Platform, were (1) sardine remains, including 5 vertebrae and a still-articulated hyomandibular + preopercular, and (2) bonito remains, including a dentary.

Late in the occupation of Structure D, the North Platform was buried under a layer of fill. The fish bones from this fill were as follows:

Pacific sardine	3 vertebrae (MNI = 1)
Grunt, cf. *Anisotremus*	1 articular (MNI =1)
	1 lachrymal?
	4 vertebrae
Scaleless blenny	1 quadrate (MNI = 1)
Left-eye flounder	1 hyomandibular
	5 vertebrae (MNI = 2, based on size)
Unidentified fish	6 spines
	9 vertebrae
	8 unknown elements

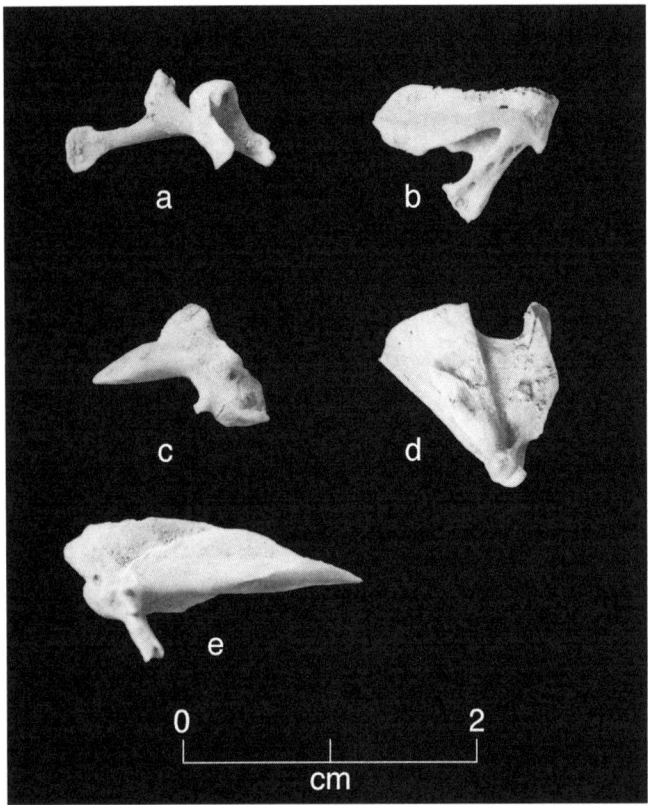

Figure 9.19. Bones of the pintadilla (*Cheilodactylus variegatus*) from Feature 6, Structure D. *a*, maxilla. *b*, dentary. *c*, articular. *d*, quadrate. *e*, posttemporal.

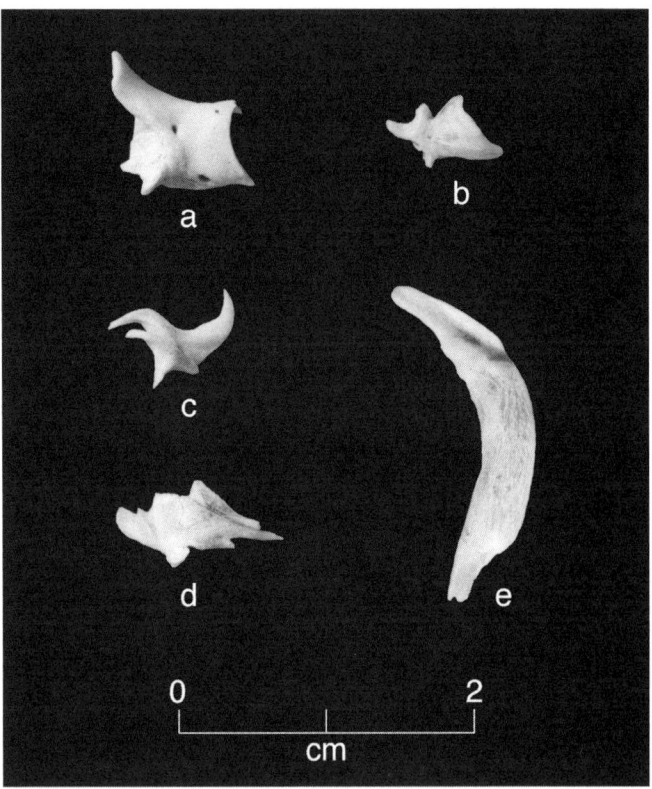

Figure 9.20. Bones of the scaleless blenny (*Scartichthys gigas*), some incomplete. *a*, dentary. *b*, articular. *c*, maxilla. *d*, quadrate. *e*, preopercular.

The South Corridor

The South Corridor was a single-file passageway that led from the Southwest Canchón to the interior of Structure D. Late in the occupation of Structure D, someone dumped a considerable amount of refuse on the floor of the corridor. While this refuse was dominated by maize cobs, it also included one opercular of sardine, one vertebra of grunt, five vertebrae of lorna, one vertebra from a larger Sciaenid, and two premaxillae of bonito.

The Northeast Canchón

The Northeast Canchón was a large, unroofed work area where (among other things) weaving was done. The floor in the northwest corner of the canchón featured some ash-filled hollows containing small amounts of refuse. One of these hollows produced the vertebra of a large róbalo.

Not far from these ash-filled hollows were traces of a campfire made very late in the building's history. This campfire produced a fish bone with an exostosis like those occasionally seen on drum vertebrae.

Collca 1

Collca 1 was a storage bin built against the south wall of the Northeast Canchón. Although originally designed for storage, this bin had some debris tossed into it late in its history. Included in this debris were three heads of anchoveta negra, the vertebra of a large corvina, and the ossified vertebral centra from two different cartilaginous fish.

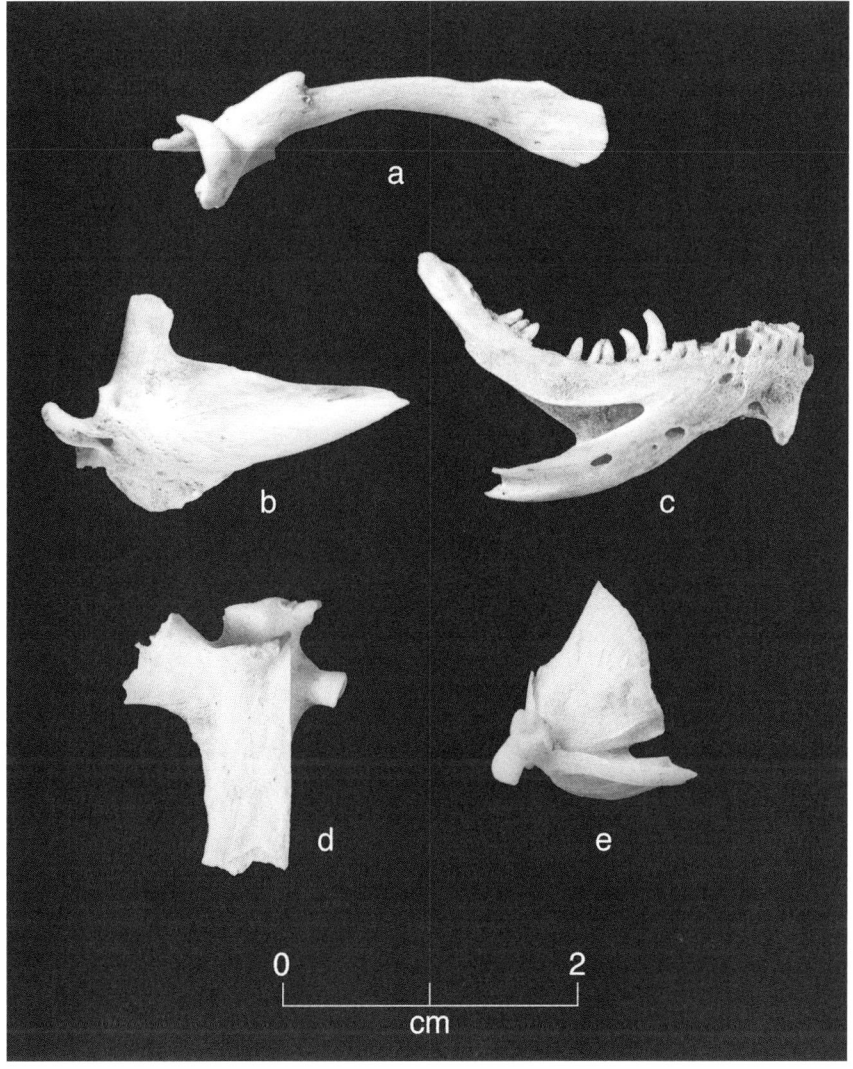

Figure 9.21. Bones of the scaled blenny (*Labrisomus philippii*) from Feature 6, Structure D. *a*, maxilla. *b*, articular. *c*, dentary. *d*, hyomandibular. *e*, quadrate.

Figure 9.22. Bones of the Pacific bonito (*Sarda sarda chiliensis*) from Feature 6, Structure D. *a*, dentary. *b*, articular. *c*, prevomer. *d*, caudal vertebrae. *e*, hypural plate.

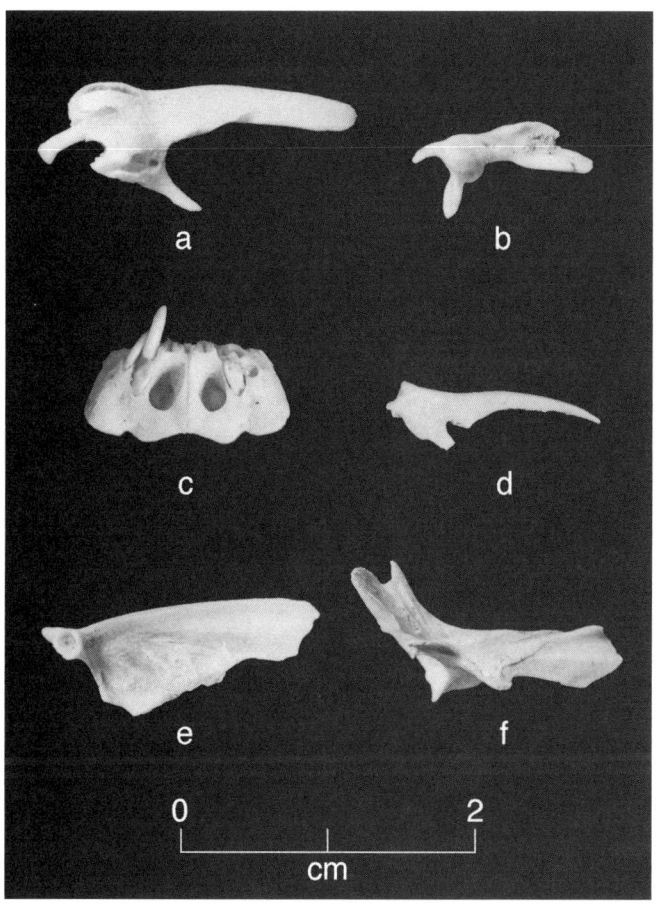

Figure 9.23. Bones of the pejesapo or clingfish (*Sicyases sanguineus*) from Feature 6, Structure D. *a*, premaxilla. *b*, maxilla. *c*, still-articulated left and right dentaries. *d*, quadrate. *e*, opercular. *f*, cleithrum.

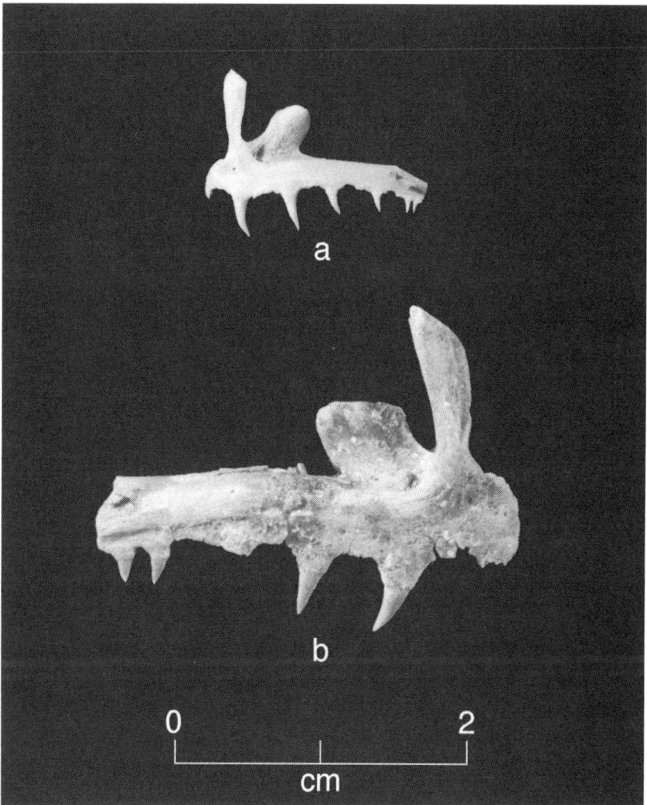

Figure 9.24. Premaxillae of left-eye flounders (*Paralichthys adspersus*), showing the size range of specimens caught at Late Intermediate Cerro Azul. *a*, left premaxilla from Feature 6, Structure D. *b*, right premaxilla from the Southwest Canchón, Structure D.

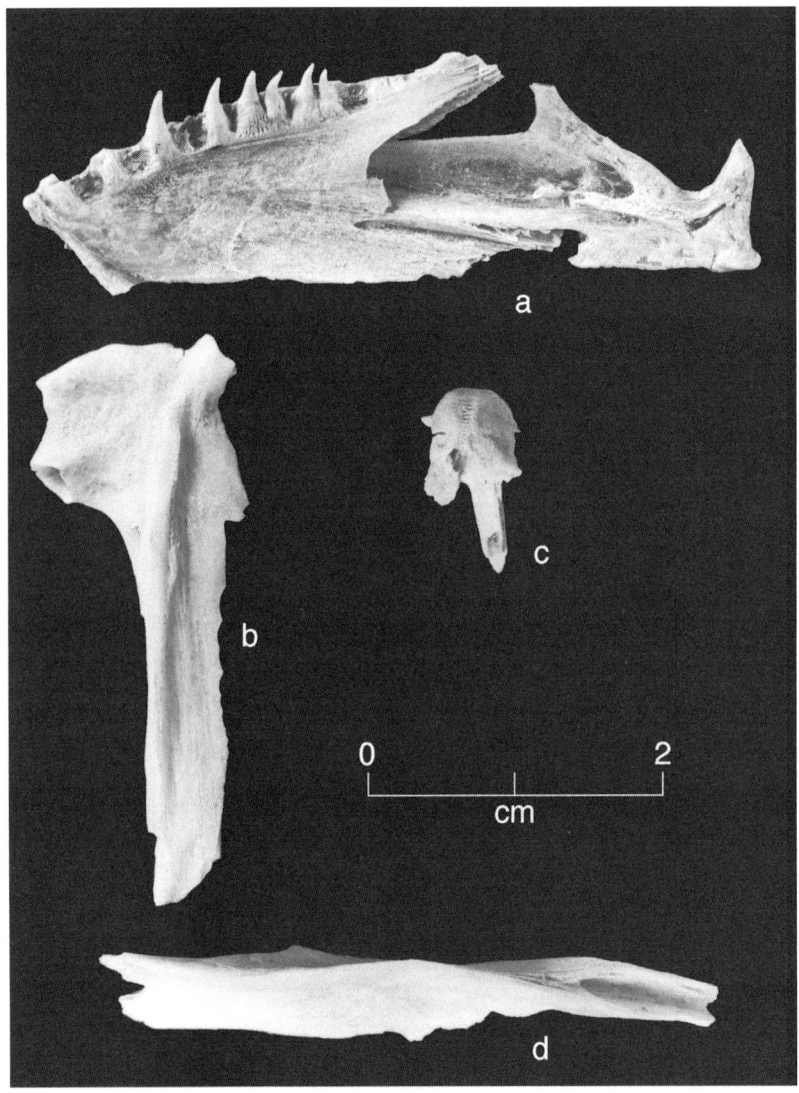

Figure 9.25. Bones of the left-eye flounder from Feature 6, Structure D. *a*, still-articulated dentary and articular. *b*, hyomandibular. *c*, prevomer. *d*, parasphenoid.

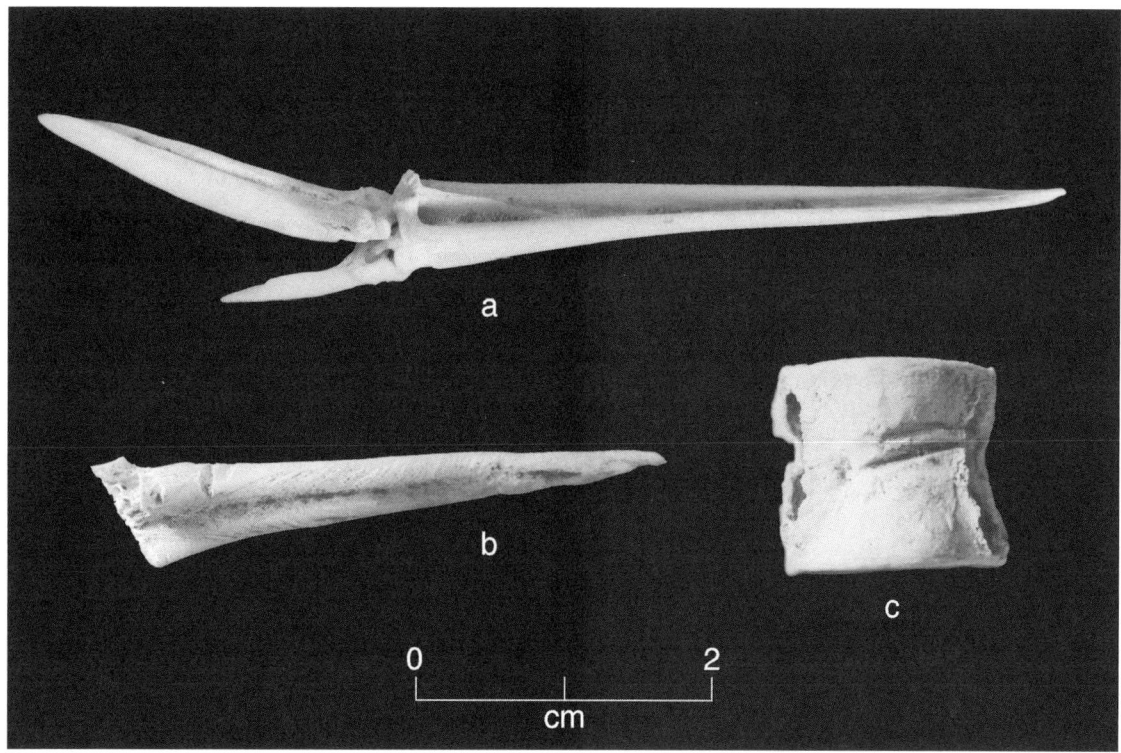

Figure 9.26. Fish bones displaying cut marks, all from Feature 6 of Structure D. *a*, still-articulated pterygiophore and spine from the pelvic girdle. *b*, fin ray of bony fish. *c*, ossified vertebral centrum from shark, ray, or chimaera.

The North Central Canchón

The North Central Canchón was a large kitchen/brewery for Structure D. Among the debris scattered on the floor of the canchón were the maxilla of a lorna, two vertebrae from a very large corvina, and two ossified vertebral centra from large cartilaginous fish. The two vertebral centra lay in ash deposits, and both had been burned.

Room 1

Room 1 of Structure D was originally part of a two-room elite residential apartment. After suffering earthquake damage, it was converted to a fish storage room. One articular of grunt had been left behind on the original floor. Six vertebrae from a bonito were found in the upper fill, postdating the room's conversion to fish storage.

"Room 2"

After excavation, "Room 2" turned out to be a private interior patio adjacent to Rooms 1 and 3. An earthen ramp against its west wall led up toward Rooms 5 and 6. Included in the fill of the ramp was one vertebra from a grunt, probably *Anisotremus*.

Room 3

Room 3, immediately adjacent to Room 1, had originally been part of the same two-room residential apartment mentioned above. It, too, had suffered earthquake damage, and was later converted to fish storage. The clean sand added for this purpose filled Room 3 to the depth of its original sleeping bench. The fish bones discovered in this sand layer are listed in Table 9.3.

Table 9.3. Fish Bones From the Sand Layer Added to Room 3 after Earthquake Damage.

Shark, cf. *Carcharhinus*	1 ossified vertebral centrum (MNI =1)
Pacific sardine	18 vertebrae (MNI =1)
Anchoveta negra	1 dentary + articular
	3 articulars (MNI = 3)
	1 cleithrum
	1 hyomandibular
	151 vertebrae (MNI = 3)
Grunt, cf. *Anisotremus*	1 articular (MNI = 1)
	1 palatine
	1 preopercular
	3 vertebrae
Ayanque	1 vertebra (MNI = 1)
Lorna	7 vertebrae (MNI = 1)
Large Sciaenid, cf. *S. starksi*	1 dentary + articular joined (MNI = 1)
Bonito	6 vertebrae (MNI = 1)
Left-eye flounder	1 vertebra (MNI = 1)
Unidentified fish	3 branchials
	4 pterygiophores
	10 spines
	23 vertebrae
	5 unidentified elements

Totals:
NISP = 198
Unidentified = 45

We note that while 80 percent of the identified specimens from the sand layer turned out to be anchovetas, at least 8 other species were present in Room 3. We consider the evidence for a possible requiem shark to be surprising, given its size and weight relative to a caballito de totora. Perhaps the shark came into shallow water, or accidentally beached itself.

Room 4

Room 4 had a complex stratigraphic history. Its original floor lay 3.73 m below the datum point established for Structure D. Lying on this floor were two ossified vertebral centra from large cartilaginous fish, and five vertebrae from unidentified bony fish. One of the latter vertebrae featured an exostosis like those seen on members of the drum family.

At a later date Room 4 was partially filled in, and a later floor was created at a point 2.23 m below datum. No fish bones were associated with the later floor.

Room 5

Room 5 also had a complex stratigraphic history. Originally designed as a doorless storage room, it was converted to a residential unit after other rooms in Structure D had suffered earthquake damage. No fish bones or sand were associated with the room's early construction stage, indicating that it was used to store something other than fish.

During the conversion of Room 5 to residential quarters, a sleeping bench was added at a depth of 1.03 m below datum. Incorporated into fill of this bench were fragments of the neurocranium and first vertebra of a chita. Resting on the surface of the bench were three vertebrae from a bonito.

Finally, in the uppermost fill of Room 5 Marcus' crew recovered one vertebra from a grunt, two vertebrae from a bonito, one vertebra from a left-eye flounder, and two elements from unidentified fish.

Room 7

Room 7 was a small unit whose access doorway was deliberately blocked after the room suffered earthquake damage. Late in its history Room 7 became a convenient place to dump refuse. This refuse included one vertebra from a Chilean jack mackerel and 12 elements from unidentified fish.

Room 8

Room 8 of Structure D was a doorless unit dedicated to the storing of dried fish (Marcus 2008:128–130). When discovered this room still had more than a meter of sand inside it, and there were patches of fish skin and scales stuck to its tapia walls.

Marcus' crew screened the Room 8 sand through both 1.5 mm and 0.6 mm mesh, recovering 339 fish bones. The inventory from Room 8 is given in Table 9.4.

Table 9.4. Fish Bones Recovered From the Sand Deposit in Room 8, Structure D.

Cartilaginous fish	1 ossified vertebral centrum (MNI = 1)
cf. Pacific sardine	4 vertebrae (MNI = 1)
Anchoveta negra	17 maxillae
	31 dentaries
	12 dentary + articular joined
	36 articulars
	9 quadrates
	3 ethmoids
	55 hyomandibulars (MNI = 34)
	2 hyomandibular + preopercular
	18 preoperculars
	2 operculars
	15 cleithra
	3 parasphenoids
	6 frontals
	2 prootics
	12 ceratohyals
	3 basioccipitals
	37 vertebrae
	2 unspecified elements
Peruvian rock bass	1 articular (MNI = 1)
Chilean jack mackerel	1 maxilla (MNI = 1)
Mismis	2 vertebrae (MNI = 1)
Lorna	1 maxilla (MNI = 1)
	1 basioccipital
	2 vertebrae
Pacific mackerel	1 hyomandibular (MNI = 1)
Bonito	1 ceratohyal (MNI = 1)
Unidentified fish	5 pharyngeal teeth
	1 basioccipital
	1 otolith
	4 branchials
	20 pterygiophores
	11 vertebrae
	17 unspecified elements

Totals:
  NISP = 280
  Unidentified = 59

Since 96 percent of the identified fish bones from Room 8 could be assigned to *Engraulis ringens*, it would seem that the drying of anchovetas for export was the main purpose of that room. The proximity of Room 8 to the Southwest Canchón—where we know that llama caravans remained long enough to leave their dung pellets—reinforces that possibility. Whether the drums, rock bass, mackerels, jack mackerels, and other fish we found were actually dried in Room 8, or represent stray bones introduced into the room when it was being filled or emptied, remains uncertain.

Rooms 9 and 10

Rooms 9 and 10 were dedicated to the raising of guinea pigs. The rodents themselves were kept in Room 9; their food was kept in Room 10. No fish remains were found on the floors of these rooms. During the centuries following the abandonment of Structure D, however, wind erosion caused part of the fill covering the North Platform of the Southwest Canchón to drift downward into Rooms 9 and 10. Included in this fill was one vertebra from a Chilean jack mackerel.

Late Horizon Squatters' Debris

At some point after Structure D had ceased to function as an elite residential compound, commoner-class squatters built a kincha house in the "Room 2" patio. Pottery from this house dated it to the Late Horizon.

Feature 4 was a Late Horizon midden associated with the kincha house. Included in this midden were the following fish bones:

Cartilaginous fish	2 large, ossified vertebral centra
Chita	1 maxilla
	2 preoperculars
	2 vertebrae
Lorna	1 vertebra
Unidentified fish	1 unspecified element

Structure 9

The largest collection of fish bone from Structure 9 came from **Feature 20**, a midden piled up against the east wall of the building. It is believed that this midden contained domestic refuse from the household of Structure 9's commoner-class overseer.

Marcus' crew screened Feature 20 through both 1.5 mm and 0.6 mm mesh. The fish bones from this midden are listed in Table 9.5.

Table 9.5. Fish Bones from Feature 20, Structure 9.

Cartilaginous fish	13 ossified vertebral centra (MNI = 2, based on size)		2 supraoccipitals 3 basioccipitals 3 metapterygoids
Pacific sardine	6 maxillae 7 dentaries 12 articulars (MNI = 7) 3 quadrates 1 ethmoid 5 hyomandibulars 8 operculars 1 posttemporal 2 cleithra 4 basioccipitals 6 tripuses 928 vertebrae (MNI = 19)		11 branchiostegals 1 dorsal spine 3 pectoral spines 1 dorsal or pectoral spine 6 pterygiophores (MNI = 4) 1 hypural 1 final vertebra 59 other vertebrae
		Grunt, cf. *Anisotremus*	3 premaxillae 4 maxillae (MNI = 4) 2 dentaries 3 articulars
Engraulidae, cf. *E. ringens*	21 maxillae 32 dentaries 4 dentary + articular joined 41 articulars 3 quadrates 3 frontals 24 hyomandibulars 53 prootics (MNI = 27) 30 cleithra 1 ceratohyal 3 basioccipitals 1 basioccipital + supraoccipital 3 final vertebrae 60 first vertebrae (MNI = 60) 1981 other vertebrae		3 quadrates 1 upper pharyngeal 1 lower pharyngeal 1 palatine 2 operculars 3 ceratohyals 1 ceratohyal + epihyal joined 1 urohyal 1 cleithrum 1 opisthotic + pterotic joined 1 spenotic + prootic 1 parasphenoid 17 vertebrae
		Mismis	11 vertebrae (MNI = 2, based on size)
		Coco	1 maxilla 2 operculars (MNI = 2) 1 otolith
Sea catfish	1 dentary 1 quadrate 1 quadrate + preopercular joined 2 operculars 4 hyomandibulars 3 parasphenoids 2 cleithra 1 coracoid 2 pterotics 2 ceratohyals 1 ceratohyal + hypohyal joined 2 otoliths	Ayanque Lorna	2 premaxillae (MNI = 1) 1 premaxilla 3 maxillae 9 dentaries (MNI = 5) 4 articulars 1 quadrate 2 palatines 2 preoperculars

Table 9.5. Fish Bones from Feature 20, Structure 9.

	1 opercular	Clingfish	1 quadrate (MNI = 1)
	1 interopercular		4 vertebrae
	1 lachrymal	Left-eye flounder	1 premaxilla, large
	5 hyomandibulars		7 vertebrae (MNI = 2, based on size)
	5 posttemporals		
	84 vertebrae (MNI = 4)		
Corvina	2 otoliths (MNI =2)	Unidentified fish	4 premaxillae
	1 vertebra		1 maxilla
			6 prevomers
Mojarrilla	4 premaxillae		1 dentary
	3 maxillae		5 articulars
	4 dentaries		3 quadrates
	3 articulars		2 loose teeth
	1 preopercular		9 pharyngeals
	2 ceratohyals		15 hyomandibulars
	8 otoliths (MNI = 6)		2 preoperculars
	7 vertebrae		4 operculars
			5 ceratohyals
Small-to-medium Sciaenids	7 premaxillae		1 ceratohyal + epihyal joined
	1 maxilla		
	4 dentaries		9 epihyals
	1 quadrate		6 cleithra
	3 interoperculars		8 supracleithra
	3 posttemporals		12 parasphenoids
	19 otoliths		1 posttemporal
	1 urohyal		11 basioccipitals
	11 first vertebrae		4 exoccipitals
	45 other vertebrae		8 scapulae
			3 basipterygia
Scaleless blenny	6 maxillae?		14 pterygiophores
	2 articulars		10 final vertebrae
	6 quadrates (MNI = 5)		675 other vertebrae (4 with exostoses)
	5 preoperculars		
	3 operculars		7 scales
	1 ceratohyal + hypohyal joined		1280 unspecified elements
	2 ceratohyal + epihyal		
	1 epihyal		
	1 cleithrum		
	102 vertebrae		
Scaled blenny	1 hyomandibular		
	11 vertebrae (MNI = 2, based on size)		
Bonito	1 maxilla (MNI = 1)		
	1 final vertebra		
	24 other vertebrae		

Totals:
Cartilaginous fish, NISP = 13
Bony fish, NISP= 3,848
Unidentified = 2,106

The South Entryway

The main entryway into Structure 9 was from the south. This entryway led directly to a small interior patio, then to the tapia platform on which the building's overseer had built his kincha house.

During the course of the building's occupation, the South Entryway had been renovated. Its original floor lay 1.77 m below the datum point established for Structure 9. Found on this early floor were three fish vertebrae; one was the ossified vertebral centrum from a cartilaginous fish, one was the vertebra from a Chilean jack mackerel, and one was the vertebra from a bonito. It is possible that these bones were from fish eaten by the overseer's household.

Later in the history of the South Entryway, a new floor was laid down at a depth of 99 cm below datum. No fish bones were found on this later floor.

Room 1

Whatever its original function may have been, Room 1 had suffered so much earthquake damage that it was eventually converted to fish storage. Marcus' crew found the room filled with sand, which they fine-screened in its entirety. The list of fish was as follows:

Cartilaginous fish	3 ossified vertebral centra (multiple fish)
Engraulidae, cf. *E. ringens*	3 maxillae
	6 dentaries
	4 dentary + articular joined
	5 articulars
	6 quadrates
	5 preopeculars
	4 operculars
	10 hyomandibulars
	14 ceratohyals (MNI = 8)
	6 epihyals
	2 parasphenoids
	3 frontals
	16 prootics (MNI = 8)
	2 posttemporals
	4 basioccipitals
	3 otoliths
	39 vertebrae
	0.32 g of scales
Lorna	1 basioccipital + exoccipital (MNI = 1)
	4 vertebrae
Bonito	1 dentary (MNI = 1)
	1 parasphenoid
	1 basioccipital + exoccipital
	3 vertebrae
Unidentified fish	10 unspecified elements

Totals:
 NISP = 146
 Unidentified = 10

In addition to the bones recovered by screening the sand, Marcus' crew found crushed anchovy heads, patches of fish scales, and sections of vertebral column adhering to the clay floor of the room (Chapter 8).

The fact that 90 percent of the identified fish bone could be assigned to anchoveta indicates that Room 1 was dedicated to the storage of that small fish. Whether the lorna and bonito had also been dried in the sand was unclear; their bones might have been introduced accidentally when workers were filling or emptying the room.

Room 4

Room 4 was a small storage unit containing a bit of greenish gray sand. It yielded only one vertebra of lorna.

Room 5

Room 5 contained roughly 1.9 m³ of greenish gray sand, which Marcus' crew screened through both 1.5 mm and 0.6 mm mesh. The fish remains were as follows:

Cartilaginous fish	1 ossified vertebral centrum (MNI = 1)
Pacific sardine	1 dentary (MNI = 1)
	1 hyomandibular
	1 cleithrum
	10 vertebrae
Engraulidae, cf. *E. ringens*	5 maxillae
	14 dentaries (MNI = 9)
	2 dentary + articular joined
	14 articulars
	1 quadrate
	4 ceratohyals
	3 prootics
	3 cleithra
	2 first vertebrae
	2 final vertebrae
	36 other vertebrae
Lorna	1 preopercular
	1 palatine (MNI = 1)
Mojarrilla	1 preopercular (MNI = 1)
Unidentified fish	1 hyomandibular
	2 vertebrae
	2 scales
	159 unspecified elements

Room 6

Room 6 lay immediately adjacent to Room 5, but was not converted to fish storage until relatively late in its history. The following fish bones were recovered:

Cartilaginous fish	1 ossified vertebral centrum (MNI = 1)
Pacific sardine	1 dentary (MNI = 1)
	1 quadrate
	1 hyomandibular
	1 preopercular
	1 subopercular
	30 vertebrae
Engraulidae, cf. *E. ringens*	10 maxillae
	11 dentaries
	28 articulars
	5 quadrates
	7 frontals
	11 hyomandibulars
	6 ceratohyals
	5 epihyals
	53 prootics (MNI = 27)
	1 posttemporal
	9 cleithra
	11 basioccipitals
	3 otoliths
	22 first vertebrae
	352 other vertebrae
Chilean jack mackerel	1 maxilla (MNI = 1)
	1 quadrate
	1 opercular
	1 hyomandibular
	1 cleithrum
	1 supracleithrum
	1 supraoccipital
	11 vertebrae
Ayanque	1 prevomer (MNI = 1)
	2 vertebrae
Mismis	10 vertebrae (MNI = 1)
Small-to-medium Sciaenids	2 premaxillae
	1 articular
	2 prevomers (MNI = 2)
	1 frontal
	1 vertebra
Scaleless blenny	1 maxilla?
	1 articular (MNI = 1)
	1 opercular
	3 vertebrae
Bonito	2 vertebrae (MNI = 1)
Unidentified fish	3 hyomandibulars
	1 scapula
	27 vertebrae
	5 scales
	656 unspecified elements

Totals:
NISP = 616
Unidentified = 692

Among the identified fish bones from the sand in Room 6, 87 percent can be assigned to anchoveta. This is comparable to the situation in Room 1, and indicates that anchovetas were the principal fish being stored in the room. Sardines were a minor element. Once again, we cannot be sure how many of the bones of larger fish were accidentally introduced, and how many came from fish deliberately dried in the room.

Room 7

Room 7, adjacent to the South Entryway, appeared to have been used for fish storage from the outset. Its sand layer yielded the following fish bones:

Anchoveta negra	1 articular
	3 quadrates (MNI = 2)
Chilean jack mackerel	5 vertebrae (MNI = 2, based on size)
Bonito	1 vertebra (MNI = 1)
Unidentified fish	12 unspecified elements

Room 8

Room 8 contained no sand; rather, it yielded domestic refuse that appeared not to have been brought far from its place of original use. It may, in other words, represent refuse from the nearby kincha house of the building's overseer. Scattered among large fragments of utilitarian pottery, Marcus' crew recovered one vertebra from a grunt, one vertebra from a bonito, 5 vertebrae from a left-eye flounder, and one ossified vertebral centrum from a cartilaginous fish.

Room 10

Room 10, although small enough to be a storage unit, appears never to have been filled with sand. It yielded only one vertebra from a large Sciaenid.

Room 11

Room 11 was a small storage unit with a tapia bench along one wall. Late in its history, it had been filled with sand and converted to fish storage. Found in the sand were the following fish bones:

Pacific sardine	3 vertebrae (MNI = 1)
Engraulidae	13.11 grams of anchoveta (the bones of multiple individuals, too fragile to be separated and counted)
Corvina	3 articulated vertebrae (MNI = 1) (Fig. 8.5)
Bonito	1 vertebra (MNI = 1)
Unidentified fish	1 opercular
	2 vertebrae
	14 unspecified elements

In addition to the fish recovered from the sand layer, dozens of anchoveta heads and tails were stuck to the sand-covered tapia floor of Room 11. These anchoveta remains proved too fragile to remove from the clay.

Room 12

Room 12 was another storage unit with a tapia bench against one wall. It had never been converted to fish storage, but a number of fish bones lay scattered among the sherds and other debris in the room. These bones were as follows:

Pacific sardine	1 quadrate
	2 hyomandibulars
	(MNI = 1)
Chilean jack mackerel	1 vertebra (MNI = 1)
Small-to-medium Sciaenids	3 vertebrae (MNI = 1)
Bonito	1 final vertebra (MNI = 1)
	8 other vertebrae
Unidentified fish	91 unspecified elements

Differences in the Diet of Commoners and Nobles: Comparing Features 6 and 20

Before proceeding to the fish remains from the quebradas of Cerro Camacho, let us consider what Structures D and 9 can tell us about dietary differences between commoners and elites at Cerro Azul. Structure D was a large residential compound in which a noble extended family lived, presumably accompanied by their servants; during the day, the North Platform of the Southwest Canchón may have been occupied by a commoner-class overseer. The fish remains from Structure D should therefore reflect the diet of an elite family and a handful of trusted commoners. Structure 9, on the other hand, was a small storage facility, overseen by a commoner-class overseer who lived in a kincha house. Its fish remains should therefore reflect the diet of a commoner who was the equivalent of a civil servant.

Two of the fish we expect to have been prized by elite families were the corvina and róbalo. As we have already seen, Structure D produced bones of these two fish that exceeded in size our comparative skeletal material. This is, of course, only anecdotal data. Perhaps the most rigorous way to compare the noble and commoner diets would be to compare Feature 6 (a midden in Structure D) and Feature 20 (a midden in Structure 9). In Table 9.6 we have done this for all the categories of bony fish.

One of the first differences we see is that while the nobles of Structure D were committed to drying anchovetas for export, they preferred eating sardines. This is reinforced by the fact that some nobles buried at Cerro Azul were given sardines as food for the afterlife. Overall, it appears that while the Structure D nobles consumed more sardines (8,013) than anchovetas (2,857), the Structure 9 commoners ate more anchovetas (2,260) than sardines (983). We do not see this as a taste preference, but as a reflection of privilege; given a choice, noble families simply took most of the sardines.

We should now bring up one possible caveat to our interpretation: we have seen, in Chapter 1, that sardines and anchovetas have 15- to 20-year cycles of relative abundance. In other words, if it could be shown that Structures D and 9 dated to different periods, any differences between them might be chronological.

Two factors mitigate against a chronological explanation. First, our available paleoenvironmental data suggest that during the period A.D. 1400–1800 (the "Little Ice Age"), neither sardines nor anchovetas were particularly abundant (Gutiérrez et al. 2009). Second, the ceramics from Structures D and 9 indicate that both buildings were occupied at the same time (Marcus 2008). Thus, the differences seen in Table 9.6 most likely reflect elite preference rather than chronology.

Table 9.6 provides further evidence for elite privilege when it came to desirable fish species. Feature 6 contained 244 bones of the category "medium-to-large Sciaenids"; Feature 20, on the other hand, contained only "small-to-medium Sciaenids." The elite of Structure D consumed corvina at twice the frequency (0.17) as the commoners of Structure 9 (0.08). Feature 6 produced

Table 9.6. A Comparison of Features 6 and 20 at Cerro Azul

Bony Fish	Feature 6 NISP	percent of total	Feature 20 NISP	percent of total
cf. Machete	4	0.03	0	0.00
Pacific sardine	8,013	59.83	983	25.55
cf. Anchoveta negra	2,857	21.33	2,260	58.73
Sea catfish	0	0.00	113	2.94
Mullet	7	0.05	0	0.00
Peruvian rock bass	3	0.02	0	0.00
Chilean jack mackerel	13	0.10	0	0.00
cf. Chita	261	1.95	46	1.20
Mismis	57	0.43	11	0.29
Coco	2	0.01	4	0.10
Ayanque	6	0.04	2	0.05
Burro	2	0.01	0	0.00
Lorna	1,007	7.52	119	3.09
Róbalo	1	0.01	0	0.00
Corvina	23	0.17	3	0.08
Mojarrilla	16	0.12	32	0.83
Small/medium Sciaenids	150	1.12	95	2.47
Medium/large Sciaenids	244	1.82	0	0.00
Pintadilla	30	0.22	0	0.00
Scaleless blenny	431	3.22	129	3.35
Scaled blenny	84	0.63	12	0.31
Bonito	88	0.66	26	0.68
Clingfish	25	0.19	5	0.13
Left-eye flounder	69	0.52	8	0.21
Totals	13,393		3,848	

modest amounts of mullet, Peruvian rock bass, Chilean jack mackerel, burro, róbalo, and pintadilla; Feature 20 produced none of those species, and they were rare to absent elsewhere in Structure 9.

Just as revealing are the frequencies of sea catfish, a notoriously low-prestige fish. There were 113 bones of this species in Feature 20 (almost 3 percent of the total identified specimens); Feature 6 had none.

To be sure, there were some similarities between the two middens as well. The occupants of both buildings ate a lot of lorna, chita, and scaleless blenny, three of the most common fish of the Cerro Azul region. Fish of the drum family were important to the occupants of both buildings, but the elite seem to have eaten more of the large drums while the commoners settled for smaller drums like mismis and mojarrilla.

Fish Remains from the Quebradas of Cerro Camacho

Many terraces in the quebradas of Cerro Camacho had been created by the dumping of refuse over a long period of time. These midden-filled terraces produced hundreds of fish bones, albeit not as well preserved as the bones in Features 6 and 20. It is likely that each time refuse was moved from a residential compound to one of the quebradas, the bones suffered some damage.

Quebrada 5, Terrace 16

Terrace 16 of Quebrada 5 was excavated to a depth of 2.2 m and featured 9 "cultural" or "natural" strata. The only levels producing fish bone, however, were Stratigraphic Zone F (a layer of soft gray ash, dating to an early stage of the Late Intermediate) and Stratigraphic Zones A1 and A2 (two layers dating to a later stage of the Late Intermediate).

Zone F, the earliest stratum with midden deposits, yielded 48 ossified vertebral centra from a minimum of two cartilaginous fish; two operculars from a minimum of two cocos (*Paralonchurus peruanus*); and two bones (one exoccipital and one vertebra) from unidentified fish.

Zones A1 and A2 were separated by a thin layer of salitre or salty crust, probably the result of exposure to years of salt fog. Zone A2, below the salitre layer, produced one vertebra from a Pacific mackerel (*Scomber japonicus*). Zone A1, a soft midden layer above the salitre, produced one ossified vertebral centrum from a cartilaginous fish and one vertebra from a bonito (*Sarda sarda*).

Quebrada 5-south

Quebrada 5-south produced a multiroom tapia storage facility (Structure 11), a looted burial cist (Structure 12), and a domestic midden that filled the space between the two structures.

Structure 11

Owing to time constraints, Marcus was only able to expose three of the rooms in Structure 11. Her crew was able to excavate roughly 4 m^2 of Room 3, the largest. Rooms 1 and 2, two small storage units, were excavated in their entirety. Room 1 produced no fish bone. Room 2 contained the following fish bone:

Anchoveta negra	1 nearly complete desiccated head
Chilean jack mackerel	1 preopercular
	1 opercular
	1 quadrate + ectopterygoid joined
	1 interopercular/epihyal/ branchiostegal
Unidentified fish	1 vertebra

Room 3 had a somewhat more complex history. Artificial fill had been used to level its original clay floor. Then, late in its history, Room 3 was given a second floor, 10 cm above the first. The narrow space between the two floors was packed with sand containing fish bone. After having this deposit fine-screened, Marcus concluded that the sand used to level the upper floor contained so many anchovetas that it had probably been brought from a fish storage room in one of the nearby tapia compounds. Unfortunately, the fish bone seemed badly fragmented, as if it had been moved several times. The list of species from the sand is given in Table 9.7.

We note that 82 percent of the identified bones from the sand layer between Floors 1 and 2 could be assigned to the anchoveta. This fact suggests that the sand might well have been "borrowed" from a fish storage room similar to Room 8 of Structure D.

Late in its history, Room 3 of Structure 11 served briefly as the venue for a kincha house; the evidence consists of eight postmolds and several chunks of clay daub with cane impressions (Chapter 16). This wattle-and-daub house may have been occupied by a commoner-class overseer who supervised the activities in Structure 11. The following fish bones were associated with the kincha house:

Table 9.7. Fish Bones from the Sand Layer between Floors 1 and 2 of Room 3, Structure 11.

Pacific sardine	1 dentary		9 cleithra
	1 hyomandibular		3 posttemporals
	1 epihyal		10 parasphenoids
	1 preopercular		20 prootics
	3 operculars (MNI = 2)		9 basioccipitals
	1 interopercular		12 otoliths
	1 subopercular		6 complete tails
	2 cleithra (MNI = 2)		1 final vertebra
	1 tripus		81 other vertebrae
	1 final vertebra		3.96 g of other fragments
	51 other vertebrae		
		Chilean jack mackerel	5 vertebrae (MNI = 1)
Engraulidae, cf. *E. ringens*	14 desiccated heads		
	25 maxillae	Grunt, cf. *Anisotremus*	1 lachrymal
	9 dentaries		1 preopercular
	16 dentary + articular joined		1 opercular (MNI = 1)
	10 articulars		1 coracoid
	1 dentary + articular + quadrate		1 cleithrum
	18 quadrates		1 prootic + pterotic + sphenotic
	12 frontals		
	13 hyomandibulars	Lorna	1 vertebra (MNI = 1)
	10 hyomandibular + preopercular		
		Left-eye flounder	1 dentary (MNI = 1)
	2 hyomandibular/ preopercular/quadrate	Unidentified fish	1 vertebra
	5 preoperculars		16 unspecified elements
	12 operculars		
	32 ceratohyals (MNI = 16)	**Totals:**	
	1 ceratohyal + hypohyal joined	**NISP = 434**	
	26 epihyals	**Unidentified = 17**	

Pacific sardine	1 supramaxilla2 3 dentaries 1 dentary + articular 2 articulars 1 parasphenoid 2 hyomandibulars 4 preoperculars 10 operculars (MNI = 7) 3 interoperculars 1 posttemporal 1 cleithrum		2 parasphenoids 1 epihyal 3 epihyal + ceratohyal (MNI = 3) 60 unspecified elements

Anchoveta negra	3 complete heads & 2 maxillae
	  (MNI = 4)
	1 quadrate
	1 hyomandibular + preopercular
	  joined
	1 ceratohyal
	2 epihyals
	28 unspecified elements
		Chilean jack mackerel	1 interopercular/ceratohyal/
			  epihyal/branchiostegal
			  joined (MNI = 1)

Lorna	2 operculars (MNI = 1) 1 cleithrum 1 vertebra		
Unidentified fish	28 unspecified elements 4 scales		

While our sample from the kincha house is only a little more than 100 specimens, and thus could be susceptible to sampling bias, it would appear that the overseer in Structure 11 had somewhat greater access to sardines than the overseer in Structure 9.

Structure 12

Structure 12 of Quebrada 5-south was a stone-lined burial cist that had been looted prior to Marcus' arrival at Cerro Azul. There were about 90 fish bones in the earthen fill of the cist, but since most of that fill postdated the looting activity, we cannot be sure of the source of the bones. The collection was dominated by anchoveta, as follows:

Engraulidae, cf. *E. ringens*	1 complete head
	1 maxilla
	1 dentary
	2 dentary + articular joined
	1 dentary + articular +
	  quadrate
	2 hyomandibulars
	1 hyomandibular +
	  preopercular

Quebrada 5a, Terrace 9

Terrace 9 of Quebrada 5a was a relatively deep terrace with a complex stratigraphic history. The earliest level consisted of brown domestic refuse, topped by the remains of a destroyed kincha house. This midden was designated Stratigraphic Zone C, and the associated ceramics suggest that it accumulated before Structures D and 9 were built.

Above the canes, posts, and burnt clay daub of the kincha house, a Late Intermediate midden accumulated on Terrace 9. This gray ashy midden, assigned to Stratigraphic Zone B, featured hundreds of shells of the coquina clam *Donax obesulus*. Zone B was contemporaneous with the initial occupation of Structures D and 9.

Once the midden deposits on Terrace 9 had accumulated to a depth of 2.6 m, the terrace was considered a good place to install stone masonry burial cists. These cists—many of which had been looted prior to Marcus' arrival—originated in Stratigraphic Zone A, but a few intruded into Zone B. The ceramics from Zone A equate it chronologically with the final occupational stages of Structures D and 9.

Fish bones were present at all stratigraphic levels of the terrace. Let us consider them in chronological order, from earliest to latest.

Zone C

The brown domestic refuse of Zone C yielded one ossified vertebral centrum from a cartilaginous fish and one vertebra (with exostosis) of a medium-sized drum (Fig. 9.9 *b*).

Zone B

The lowermost 20 cm of Stratigraphic Zone B (2.20–2.40 m below the surface of the terrace) were intact, unlooted, and unpenetrated by burial cists. A fine-screened sample of these midden deposits produced one vertebra of mullet, one fragment of the pharyngeal teeth from a chita, and one vertebra from an unidentified fish.

At a depth of 1.90–2.20 m below the surface of the terrace, Zone B was intact in some places and penetrated by burials in others. Marcus found an undisturbed area at 1.90–2.20 m depth and fine-screened it. The fish remains were as follows:

Cartilaginous fish	2 ossified vertebral centra
Mullet	1 vertebra
Bonito	1 vertebra
Unidentified fish	1 basipterygium
	4 vertebrae

In the process of excavating Structure 5, one of the intrusive burial cists, Marcus noted that the mortar used in the construction of the cist's cobblestone walls appeared to have been created by mixing the ashy matrix of Zone B with water. She had her crew fine-screen a sample of this gray ashy mortar, which produced the following fish remains:

Cartilaginous fish	2 ossified vertebral centra
Pacific sardine	1 opercular
Lorna	1 hyomandibular
	11 vertebrae
Bonito	4 vertebrae
Unidentified fish	11 vertebrae
	2 scales

In looking at all the evidence from Zone B, we are struck by the total lack of anchovetas. The fish remains in the deposits of Zone B could not be more different from those in Features 6 and 20. Whatever building this midden debris came from, it does not appear to have been a place devoted to the drying and storing of anchovetas. Instead, the Zone B midden seems to have come from a place where people were subsisting on a variety of medium-sized fishes. It was also a place, as we saw in Chapter 3, where coquina clams were by far the most common shellfish consumed. We suspect, therefore, that the source of the Zone B midden was a commoner-class neighborhood.

The uppermost part of Stratigraphic Zone B had been penetrated by Burials 6, 7, and 9. Some of these interments had evidently been missed by the local looters, and lay undiscovered beneath the looters' own backdirt pile. Marcus recovered fish bones from this part of Zone B but did not screen the deposits, in view of the fact that they had been disturbed by burial pits.

Fish bones recovered from the unscreened portion of Zone B were as follows:

Cartilaginous fish	4 ossified vertebral centra
Mullet	8 vertebrae (MNI = 2, based on size)
Chilean jack mackerel	2 vertebrae
Chita	1 posttemporal
	8 vertebrae
Coco	3 vertebrae (one with exostosis)
Lorna	4 vertebrae
Small-to-medium Sciaenids	2 first vertebrae (MNI = 2)
Medium-to-large Sciaenids	1 maxilla
Pacific mackerel	1 vertebra
Bonito	1 vertebra
Unidentified fish	1 pterygiophore
	12 vertebrae
	5 scales
	3 unspecified elements

We note that this unscreened sample of fish bone from Zone B does nothing to change our opinion of the midden. No sardines or anchovetas at all were recovered, and the medium-sized fish are the same species seen in the screened samples from Zone B.

Zone A

The uppermost 1.2 m of Terrace 9 consisted of midden deposits used as a matrix for stone masonry burial cists. Most of these burial cists had been looted, but Marcus managed to find some burials that the looters had missed. Both the burials and the midden layer in which they had been interred could be dated to the second half of the Late Intermediate period. A litter pole from one of the burials was radiocarbon dated, producing a date with a calibrated two-sigma range of A.D. 1276–1431 (Marcus 2008:285–286). Fish bones were abundant in Stratigraphic Zone A, but owing to the disturbance by looters their context was less than ideal.

From the gray ashy Zone A midden surrounding Structure 4, one of the burial cists, the following fish bones were recovered:

Cartilaginous fish	2 ossified vertebral centra
Pacific sardine	1 hyomandibular
Anchoveta negra	1 complete head
Mullet	2 vertebrae
Coco	1 vertebra
Lorna	1 vertebra
Small-to-medium Sciaenid	1 scapula
Bonito	1 vertebra
Unidentified fish	1 vertebra
	1 spine
	2 unspecified elements

In the area below the floor of Structure 4, this same Zone A midden produced 16 still-articulated vertebrae from an anchoveta.

Structure 4

Structure 4 was a stone masonry burial cist that had been extensively looted. When the University of Michigan crew arrived, they found it filled with a combination of windblown dust and churned-up looters' debris. Aware that the looters might well have missed a number of small objects, Marcus asked her crew to screen the looters' debris. The following fish bones were recovered:

Pacific sardine	1 preopercular
	1 preopercular + hyomandibular
	1 opercular
	1 opercular + subopercular
	1 opercular/subopercular/ interopercular
	9 vertebrae
Engraulidae, cf. *E. ringens*	3 maxillae
	2 dentary + articular joined
	3 quadrates
	1 hyomandibular
	2 hyomandibular + preopercular
	1 preopercular
	1 interopercular
	1 ceratohyal
Bonito	2 vertebrae
Unidentified fish	3 vertebrae
	6 scales

The fact that Structure 4 had been looted makes it difficult to draw firm conclusions about its contents. We are struck, nevertheless, by the fact that these fish remains do not look as if they had come from the Zone A or Zone B middens. Essentially, we are dealing with the fragmentary remains of 2–3 sardines and 2–3 anchovetas. This fact is intriguing, because several of the burials at Cerro Azul had been supplied with gourd bowls containing sardines and/or anchovetas as food for the afterlife (Marcus 1987a: Fig. 43). Tentatively, therefore, the fish remains from Structure 4 look more like disturbed burial offerings than debris from the surrounding midden deposits.

Structure 5

Structure 5, a second looted burial cist, produced one opercular of Pacific sardine.

Structure 6

Structure 6, a third looted burial cist, produced another opercular of Pacific sardine.

Burial 3

In the fill surrounding Burial 3, an interment on Terrace 9, Marcus' crew found one ossified vertebral centrum from a cartilaginous fish. There is no evidence that this bone was associated with the burial.

Burial 9, Individual 3

Individual 3 was a woman aged 35 to 39 years of age (Guillén n.d.). Her food for the afterlife, which had been placed in a gourd vessel, included corn, lúcuma, three large sardines, two mussels, two clams, one false abalone, and two guinea pigs. The fact that this woman had been supplied with sardines for the afterlife reinforces our suspicion that the small fish found in Structure 4 were more likely burial offerings than intrusive midden debris.

Quebrada 6, Terrace 11

Quebrada 6 of Cerro Camacho sits directly behind the large tapia compound designated Structure H. Marcus' attention was drawn to Terrace 11 of this quebrada because its striking black color reminded her of Huaca Prieta. Excavation of the terrace led to the conclusion that the black color was due, at least in part, to the effects of salt fog on the large quantities of organic material (including fish oil) in this midden. Structure H was the nearest compound from which this midden material might have been brought.

The deposits on Terrace 11 were not particularly deep and, for all practical purposes, could be considered a single cultural deposit, contemporaneous with the latest stages in the occupation of Structures D and 9. Bedrock was reached at a depth of 30–50 cm below the surface of the terrace. Marcus' crew screened all deposits from this depth through both 1.5 mm and 0.6 mm mesh, and the fish bones recovered are listed in Table 9.8.

As we look over Table 9.8, we are struck by its similarity to the list of fish from Feature 6 of Structure D. The high frequencies of sardines and anchovetas (with sardines being the most abundant) suggest to us that the source of the Terrace 11 midden was probably an elite residential compound that participated in the drying of small fish for export. Whether Structure H was that compound is undetermined, since that building remains unexcavated.

Owing to the steplike nature of the bedrock, there were parts of the Terrace 11 midden that only reached a depth of 20–30 cm below the surface of the terrace. Marcus' crew fine-screened the deposits at this depth and recovered the following bones:

Cartilaginous fish	2 ossified vertebral centra
Pacific sardine	2 operculars (MNI = 2)
	5 articulated vertebrae
Anchoveta negra	1 maxilla/dentary/articular/ quadrate (R.)
	1 maxilla/dentary/articular/ quadrate/ceratohyal/ hypohyal (L.)
	1 epihyal (MNI = 1)
	1 epihyal + ceratohyal
	1 hyomandibular
	1 parasphenoid

Table 9.8. Fish Bones from a Depth of 30–50 cm on Terrace 11 of Quebrada 6.

Cartilaginous fish	16 ossified vertebral centra (possibly from one dogfish shark or ray)		11 first vertebrae (MNI = 11) 98 other vertebrae
Pacific sardine	7 neurocrania	Chilean jack mackerel	1 maxilla (MNI = 1)
	4 maxillae		4 vertebrae
	4 supramaxillae2	Coco	3 operculars (MNI = 3)
	5 dentaries		1 otolith
	6 articulars		17 vertebrae (MNI = 4, based on size)
	1 dentary/articular/ premaxilla/maxilla/ supramaxilla1 & 2	Ayanque	2 hyomandibulars (MNI = 2)
	2 quadrates		
	18 hyomandibulars	Lorna	1 maxilla
	8 preoperculars		16 vertebrae (MNI = 2, based on size)
	15 operculars (MNI = 10)		
	1 interopercular	Róbalo	1 vertebra (MNI = 1)
	2 suboperculars	Mojarrilla	1 dentary (MNI = 1)
	1 ethmoid	Small-to-medium Sciaenids	2 premaxillae
	5 prootics		1 articular
	2 ceratohyals		3 otoliths (MNI = 2)
	1 posttemporal		16 vertebrae
	7 cleithra	Pacific mackerel	5 vertebrae (MNI = 3, based on size)
	3 basioccipitals		
	2 tripuses	Bonito	3 vertebrae (MNI = 2, based on size)
	3 complete tails		
	278 vertebrae	Left-eye flounder	1 premaxilla (MNI = 1)
Engraulidae, cf. *E. ringens*	2 complete heads	Unidentified fish	2 lenses from eyes
	3 maxillae		1 pharyngeal area
	3 dentaries		76 vertebrae
	2 articulars		406 unspecified elements
	8 dentary + articular joined		8.24 g of scales
	1 quadrate		
	1 articular + quadrate	**Totals:**	
	1 dentary/articular/quadrate	**NISP = 614**	
	1 articular/ceratohyal/ epihyal/maxilla	**Unidentified = 485**	
	2 ceratohyals		
	1 ceratohyal + epihyal		
	1 ceratohyal/epihyal/ hyomandibular		
	3 hyomandibulars		
	1 ethmoid		
	2 prootics		
	3 cleithra		
	1 complete tail		

	1 complete tail
	6 vertebrae
	35 unspecified elements
Coco	2 vertebrae
Róbalo	1 very large vertebra
Bonito	1 vertebra
Unidentified fish	1 ceratohyal
	1 vertebra
	4 unspecified elements

The uppermost part of Terrace 11—from the surface to a depth of 10 cm—had been impregnated with salt "until it resembled cement" (Marcus 2008:297). Bone preservation was poor in this area of salitre, but improved at a depth of 10–20 cm. Marcus' crew fine-screened the midden at this depth, recovering the following fish bones:

Cartilaginous fish	1 ossified vertebral centrum (diameter 26.8 mm, possibly a large shark)
Pacific sardine	5 articulated vertebrae
Engraulidae, cf. *E. ringens*	2 complete heads and 3 neurocrania (MNI = 5)
	2 maxillae
	3 dentaries
	2 articulars
	2 quadrates
	2 frontals
	2 hyomandibulars
	1 hyomandibular + preopercular joined
	1 preopercular
	1 subopercular
	4 ceratohyals
	1 parasphenoid
	2 posttemporals
	1 otolith
	2 vertebrae
	170 fragments too delicate to separate and count
Coco	1 frontal (MNI = 1)
	1 preopercular
	1 opercular
	3 vertebrae (one with exostosis)
	3 unspecified elements
Small-to-medium Sciaenids	1 vertebra
Unidentified fish	1 scale

Quebrada 6, Terrace 12

Terrace 12 of Quebrada 6 turned out to be quite different from Terrace 11. Instead of having been used as a place to dump midden debris, it had been used as the venue for a burial cist, a storage cist, and an earth oven, all set in a matrix of beach gravel. The only refuse found on Terrace 12 appeared to be material that had eroded downslope from Terrace 11, which lay immediately uphill from Terrace 12.

Marcus' crew screened this superficial layer of brown-to-beige midden, whose contents proved to be similar to those of the Terrace 11 midden. The fish bones recovered were as follows:

Cartilaginous fish, cf. *Carcharhinus*	1 ossified vertebral centrum, possibly requiem shark
Pacific sardine	1 preopercular
	5 articulated vertebrae
Engraulidae, cf. *E. ringens*	1 complete head and 3 maxillae (MNI = 4)
	5 dentary + articular joined
	1 dentary/articular/quadrate/ceratohyal/epihyal
	2 ceratohyals
	3 hyomandibulars
	2 operculars
	2 frontals
	1 ethmoid + parasphenoid
	4 prootics
	1 cleithrum
	1 basioccipital
	2 otoliths
	4 complete tails (MNI = 4)
	34 vertebrae
Chilean jack mackerel	1 dentary (larger than average)
Coco	1 vertebra (with exostosis)
Bonito	1 complete tail
	1 vertebra
Unidentified fish	1 vertebra

The fish bone from Terrace 12 is predictably similar to that of Terrace 11, from which it is ultimately derived. Terrace 12 also provided us with a possible second specimen of requiem shark.

Summary and Conclusions

It is likely that the inhabitants of Cerro Azul caught hundreds of thousands of fish—millions in the case of sardines and anchovetas—over the course of the Late Intermediate period. The list of species is long, but not nearly as long as the roster of fish caught in the 1980s with modern watercraft and nylon nets (Chapter 6).

Judging by the remains we analyzed, there were at least three different Late Intermediate patterns. Some households—like those whose refuse ended up in Zones A and B of Terrace 9, Quebrada 5a—do not seem to have been involved in the drying and storing of small fish for export. Those households subsisted mainly on mullet, medium-sized drums like the lorna and coco, grunts, mackerel, jack mackerel, and bonito. They ate cartilaginous fish as well, but mainly dogfish sharks and eagle rays rather than larger species.

Groups of individuals dedicated to the "industrial level" processing of sardines and anchovetas displayed two different patterns. Middens associated exclusively with commoner overseers showed greater consumption of anchovetas and sea catfish than was the case with noble families. Commoners involved in the export of small fish also seem to have eaten fewer mullet, rock bass, jack mackerel, and large drums like the róbalo and corvina. They did, however, get their share of cartilaginous fish, grunt, scaleless blenny, and small-to-medium drums like the mismis, lorna, and mojarrilla.

Elite residential compounds—which housed both noble families and their commoner-class servants—consumed the richest and most varied seafood. Their refuse (which could still be recognized even after it had been carried to a quebrada) indicates that they ate three times as many sardines as anchovetas. Their favorite medium-sized fish was the lorna, but they also ate mullet, rock bass, jack mackerel, grunt, corvina, pintadilla, scaled and scaleless blenny, bonito, clingfish, left-eye flounder, and eight species of drum. Elite compounds also consumed their share of cartilaginous fish, ranging in size from the eagle ray to the requiem shark.

In addition to the three refuse patterns discussed above, there were individual rooms (like Room 8 of Structure D) that were dedicated to anchoveta storage. The percentage of anchoveta bones in the sand filling these rooms was so high that such deposits could be recognized even when the sand had been "borrowed" for use elsewhere (see, for example, the sand used to level Floor 2 of Room 3, Structure 11).

We cannot rule out the possibility that additional patterns of fish use remain to be discovered at Cerro Azul. Nine more elite residential compounds, numerous smaller structures, and many hectares of quebrada terraces remain to be excavated. As work at Cerro Azul proceeds in the future (see, for example, Marcone Flores and Areche Espinola 2015) it will be interesting to see how many of the patterns we detected are reinforced, and how many new patterns come to light. In particular, we hope that the commoner households responsible for the midden debris in Zones A and B of Terrace 9, Quebrada 5a will one day be discovered.

Finally, let us attempt to answer the questions posed at the end of Chapter 6. We will begin by saying that the Late Intermediate ocean seems to have been rich indeed, but the fishing technology of that era limited the species that could have been taken. Late Intermediate fishermen had no trouble hauling in sardines, anchovetas, grunts, drums, blennies, jack mackerels, flounders, and any other fish that could be caught within 100 m of the shore. Deep-sea fish like the Pacific hake, however, were evidently beyond their reach.

Second, there is no clear evidence in Cerro Azul's fish remains for El Niño years, or for the 15–20 year cycles of sardine/anchoveta dominance recorded today. It may be that Features 6 and 20 (our two largest fish samples) accumulated during a period too brief to include an ENSO event. It is also the case that those middens likely accumulated during A.D. 1469–1470, a time when (because of the "Little Ice Age") neither sardines nor anchovetas are believed to have been significantly abundant (Gutiérrez et al. 2009).

We found at Cerro Azul no remains of pompano, dolphinfish, arched blue crab, or shrimp—four species that visit Cañete in El Niño years. (Our inability to find any remains of silversides is just as likely the result of poor preservation as evidence for ENSO conditions.) Thus, while we did find variations in fish remains, most seemed to reflect the difference in diet between nobles and commoners, rather than the kinds of changes brought about by El Niño.

10

The Hunting of Birds and Mammals

Joyce Marcus

Fish were not the only wild source of meat available to Cerro Azul; its inhabitants also hunted birds and marine mammals. In this chapter I look at the Late Intermediate hunting technology of the Cañete Valley.

In the Andean highlands, both deer and camelids were hunted with the *atlatl* or spearthrower. We need not discuss this type of hunting, since our excavations at Cerro Azul did not produce a single atlatl point, nor can we point to a single bone of deer or wild camelid. That still leaves three other types of hunting, to wit (1) the use of the sling, (2) the use of the bolas, and (3) the use of nets. Cerro Azul produced numerous slings and sling fragments, numerous whole or fragmentary bolas, and scores of net fragments. Unfortunately, we found no easy way to distinguish the nets used for birds from the nets used for fish.

Slings

The sling is one of the oldest hunting devices known; it is found throughout the Old World (York and York 2011) and likely accompanied the first immigrants across the Bering Straits. As long ago as 1200 B.C., a six-year-old boy was buried in Lovelock Cave, Nevada with a small sling wrapped around his neck (Loud and Harrington 1929:105, Plates 10, 53a; Heizer and Johnson 1952).

This child had presumably begun practicing with his own sling, and those burying him believed that he would continue to need it in the afterlife.

The idea that a boy of six years would already be practicing with his sling is not surprising, since it takes time to become skilled. It is known that novices sling stones less accurately than those who began practicing in childhood (Brown Vega and Craig 2009:1264; Griffiths and Carrick 1994:8). Experienced adult slingers, on the other hand, were once considered a match for an archer. Korfmann (1973) tells us that in ancient Mesopotamia, Greece, and Rome, it was not unusual for skilled slingers to make throws of more than 200 m. A stone launched accurately at shorter distances could drop an animal or an enemy warrior in his tracks.

Andean Slings

The sling (*warak'a* in Quechua) was one of the Andes' oldest weapons, and slings are still used by Peruvian herders to keep predators at bay. Andean slings are usually made by braiding wool, cotton, grass, bast fibers, or a variety of plant stems (Means 1919). Cahlander (1980:29) notes that the yarns used for slings in the highlands were spun from llama or alpaca fibers. Still other slings employed fibers from bromeliads such as *Puya* and *Tillandsia*, or members of the agave family such as *Agave*, *Furcraea*, or yucca.

The three main landmarks of a sling are (1) a finger loop (at one end of the retention cord) that stays on the slinger's hand after his throw; (2) a centrally placed cradle (the pouch in which the missile is placed); and (3) a tassel (at one end of the release cord) that travels away from the slinger during his throw.

We found two styles of cradles on our Cerro Azul slings. One style, called a webbed cradle, is shown on the sling in Fig. 10.1. This intact sling was found in Structure 5, a looted burial cist on Terrace 9 of Quebrada 5a. It has a well-preserved finger loop on its retention cord and a decorative tassel of tough grass at the end of its release cord. There is an area of thick, decorative wool between the webbed cradle and the retention cord.

The second style of sling at Cerro Azul bears what is called a slit cradle, consisting of two parallel braided bands (Fig. 10.2). The specimen shown in Fig. 10.2 has a thick finger loop on the retention cord; unfortunately, its tassel is too worn to be analyzed.

I cannot speak to the relative advantages of a webbed cradle vs. a slit cradle; I assume that each was appropriate for a particular type of missile. Nor do I know why some tassels were made from bromeliad fiber, others from agave fiber or grass. In the case of the tassel, it would seem that stylistic preference is a more likely cause than function.

The History of the Andean Sling

Slings dating to 2500 B.C. have been found at the coastal site of Asia (Engel 1963: Fig. 140). These preceramic slings were made of reed stems, and the cradles formed one solid pouch rather than being slit or webbed. In Grave 23 at Asia, one of these slings was found tied around a man's waist; this may have been the way he carried it while hunting. At other coastal sites (e.g. Bird and Bellinger 1954, Engel 1963, Kroeber and Wallace 1954:140), male burials have been found with slings wound around their foreheads (see the pottery vessel illustrated in Cahlander 1980:7, Fig. 1.5). Still other slings were found at early sites such as Río Seco and Otuma (Engel 1963:57).

Slings in the Cañete Valley

In Tomb A-16 at the Middle Horizon site of Cerro del Oro, only five km from Cerro Azul, Kroeber (1937:259) found the sling shown in Fig. 10.3. This sling, which measures 1.87 m in length, has three-strand braids and a long oval cradle. At the end of the sling is a huge tassel made by looping untwisted agave strands through the last twists of the braid, then tying the braid ends about them. The lower strands of the tassel have been drawn together, divided into two sections, and woven in figure-8 fashion. A finger loop finishes off the retention cord (see O'Neale 1937:268, 271).

Virtually all the slings we recovered at Cerro Azul came from looted burial cists or looters' debris on Terrace 9 of Quebrada 5a. We suspect that all our slings accompanied male burials; certainly, none occurred with the female burials we excavated (Marcus 2015, 2016). Our suspicions are strengthened by the observations of Thomas Hutchinson (1874:312), who investigated numerous graves on the Peruvian coast. Hutchinson said, "with the men I generally found slings; and with the women almost invariably needles and buttons, frequently some woollen thread and a distaff."

Fig. 10.4 shows two webbed cradles from looters' backdirt on Terrace 9 of Quebrada 5a. Cradle *a* comes from Stratigraphic Zone A, while Cradle *b* is from Structure 5—a burial cist—and was made by braiding eight strands of an unknown plant fiber.

Fig. 10.5 illustrates three slit cradles from Terrace 9. Cradle *a* was found in Structure 5, not far from the webbed cradle shown in Fig. 10.4 *b*. Both Cradle *b* and Cradle *c* were found in Stratigraphic Zone A, not far from the webbed cradle shown in Fig. 10.4 *a*. Unfortunately, the looting on Terrace 9 precludes our knowing whether (1) the webbed and slit cradle slings were found with different men, or (2) some men were buried with both types of sling.

Note that Cradle *c* is decorated in the space between the slit cradle and the retention cord. This decoration consists of diamond-within-diamond motifs done in red, purple, and white camelid wool over cotton; the cotton was in turn wrapped over a grass fiber core. We wonder whether this decorated area served to identify the sling's owner (see Marcus 2015, 2016).

Figure 10.6 illustrates other components of slings. At *a* is a finger loop found in Structure 5. At *b* we see the final 32 cm of a release cord, ending in a bushy tassel of unidentified grass. The cord itself is a four-strand circular braid, found in looters' debris on Terrace 9.

Finally, Fig. 10.7 shows an unusual double tassel from Structure 5. The tassel itself looks to be made from *Tillandsia* or *Puya*; its wrapping has a z-s spin and its wool yarn is yellow, red, and brown. The braid yarns were s-spun, 0.8 mm tight, and medium brown in color. This unusual tassel is also attached to twin cords, making me wonder whether it might come from a larger-than-average sling designed for ritual. We know that some Andean cultures had a number of rituals (including mock battles) in which slings played a role (Arkush and Stanish 2005, Morris et al. 2011).

Years ago, Junius Bird and Louisa Bellinger (1954:45) detected a possible link between the complex headbands of Paracas mummies and the wrapping of ritual slings around the heads of certain male burials. "It is probable," Bird said, "that the idea for this complex [Paracas] headband was based on the custom of carrying or wearing slings wrapped around the head. There is abundant evidence for the practice in other Peruvian periods and areas in the form of modeled and painted pottery and bodies with slings so placed. Usually they are plain, functional slings like those found with Paracas bodies. Others, from late period graves in the Nasca-Ica region, are elaborately designed and prepared, sometimes found as joined pairs, and as such are completely nonfunctional except as headbands."

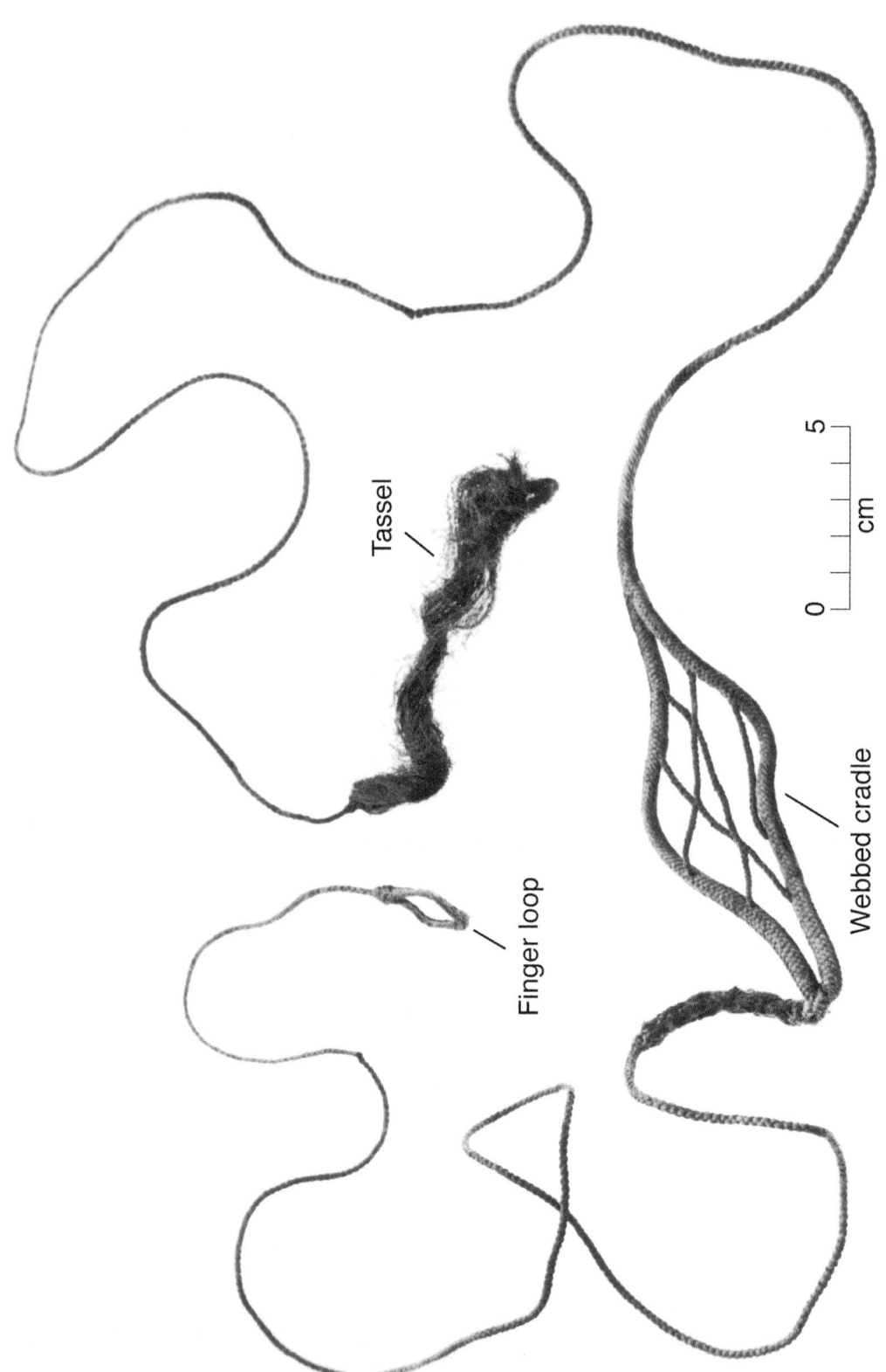

Figure 10.1. Complete sling with webbed cradle from Structure 5 on Terrace 9, Quebrada 5a.

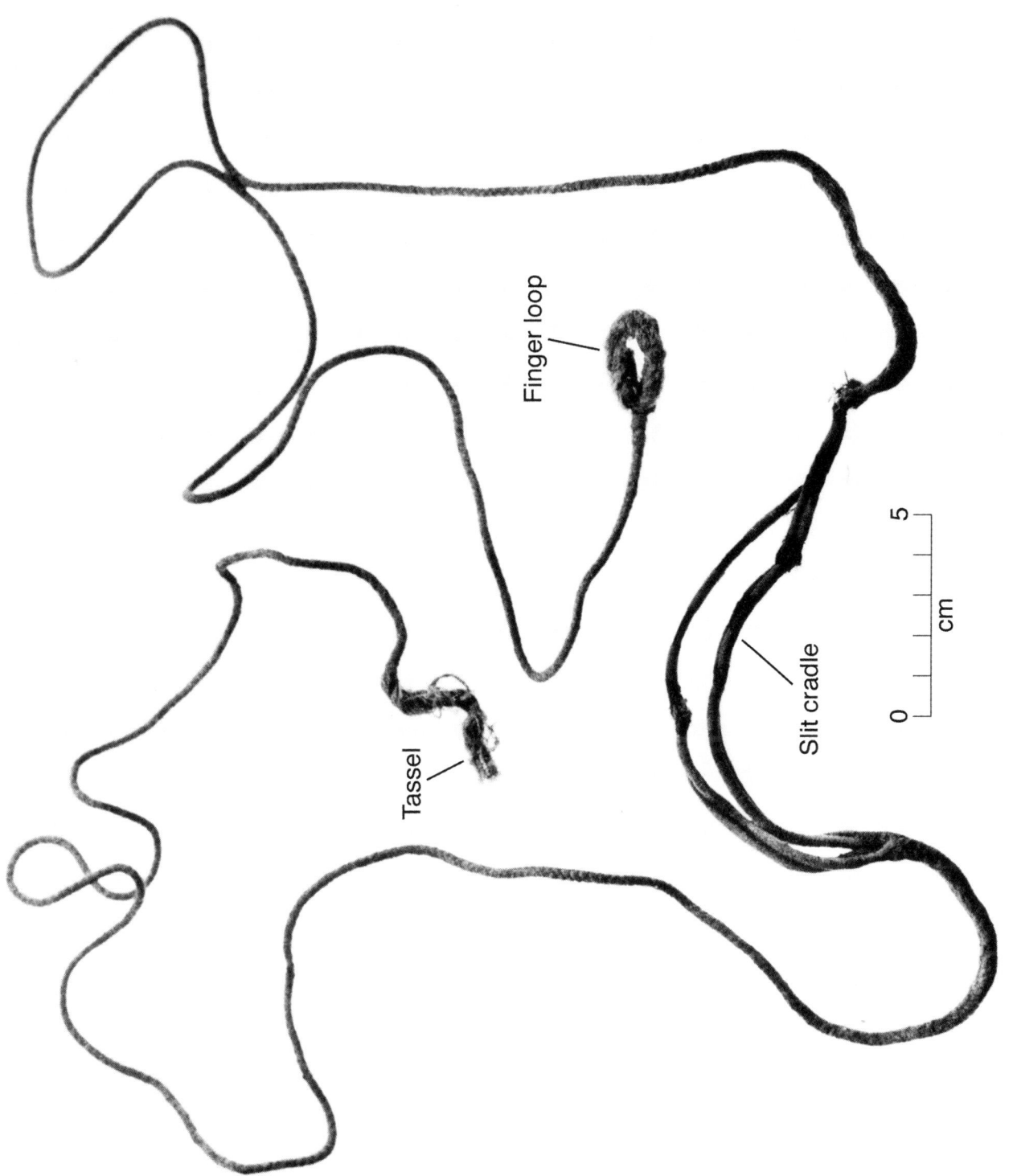

Figure 10.2. Complete sling with slit cradle from Zone A of Terrace 9, Quebrada 5a.

Figure 10.3 (above). Agave fiber sling from Tomb A-16 at Cerro del Oro, Cañete. (Redrawn by Kay Clahassey from Kroeber 1937: Plate XC).

Figure 10.4 (below). Webbed cradles from Late Intermediate slings. *a* is from Zone A, Terrace 9, Quebrada 5a. *b* is from Structure 5, Terrace 9, Quebrada 5a.

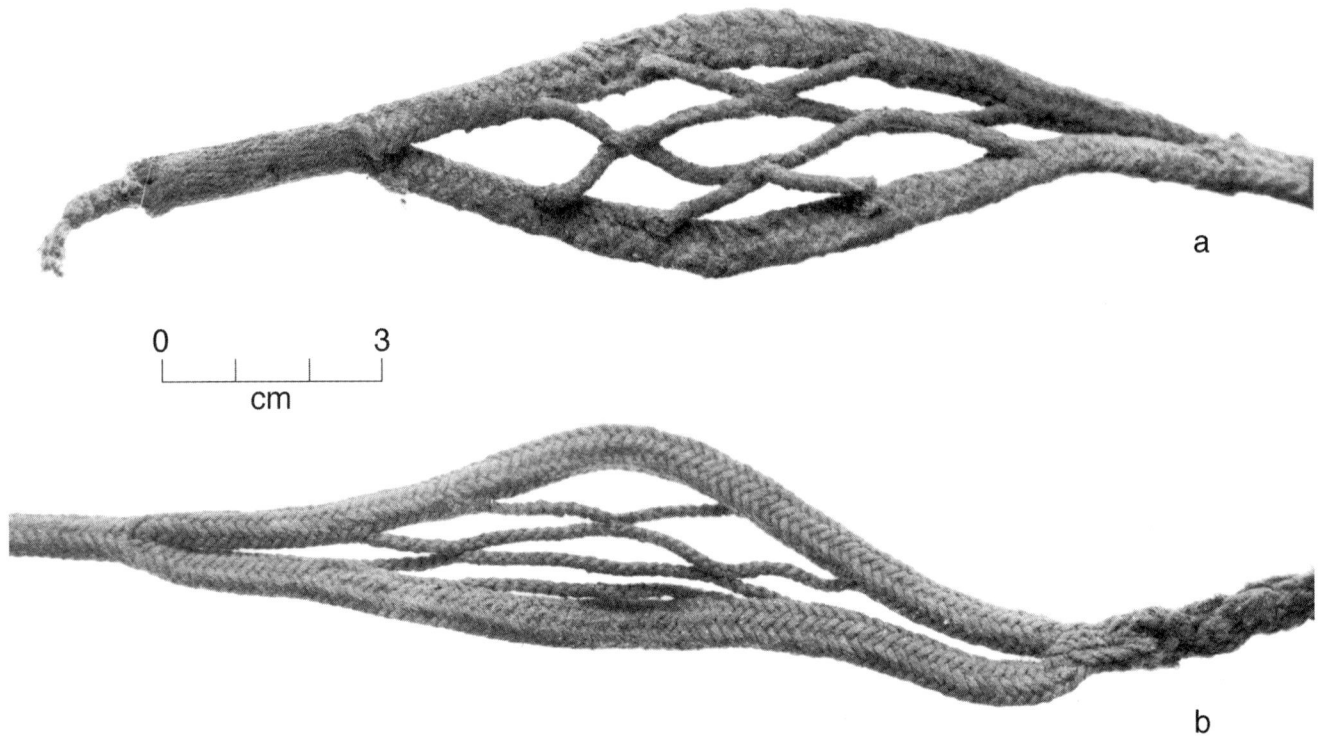

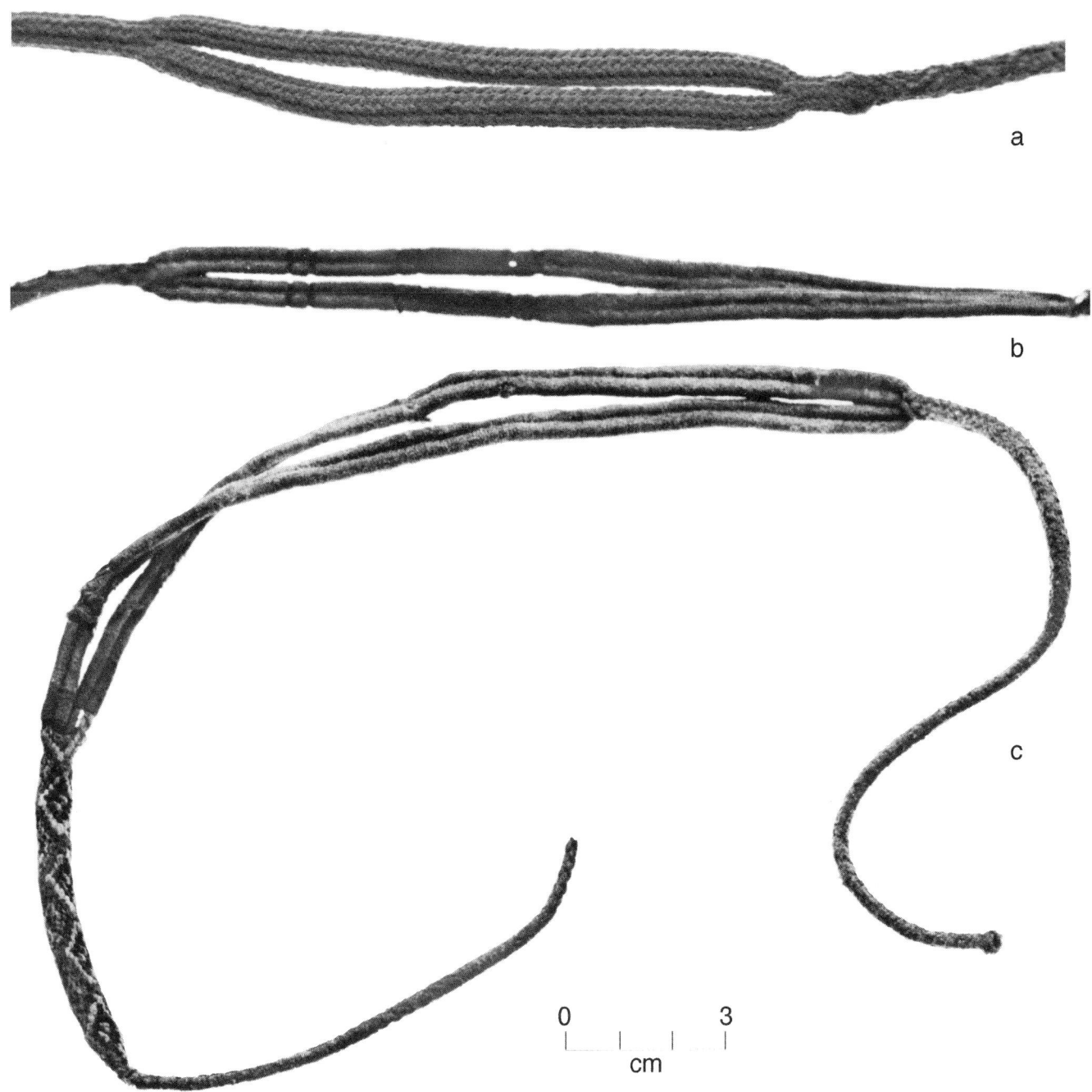

Figure 10.5. Slit cradles from Late Intermediate slings. *a* is from Structure 5, Terrace 9, Quebrada 5a. *b* and *c* are from Zone A of Terrace 9, Quebrada 5a.

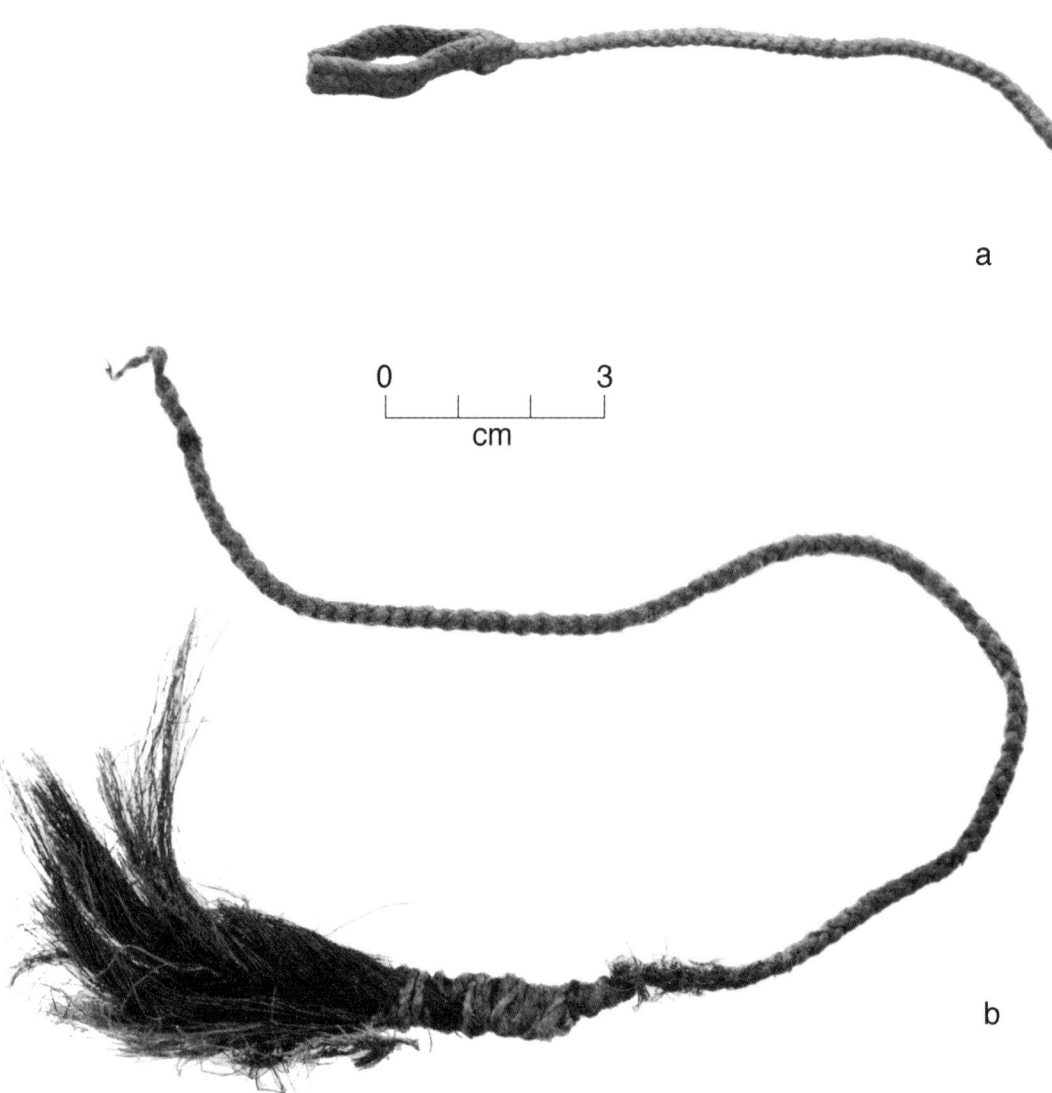

Figure 10.6. Components of Late Intermediate slings. *a*, finger loop from Structure 5, Terrace 9, Quebrada 5a. *b*, decorative tassel of vegetal fiber from the end of a sling (found in looters' debris in Quebrada 5a).

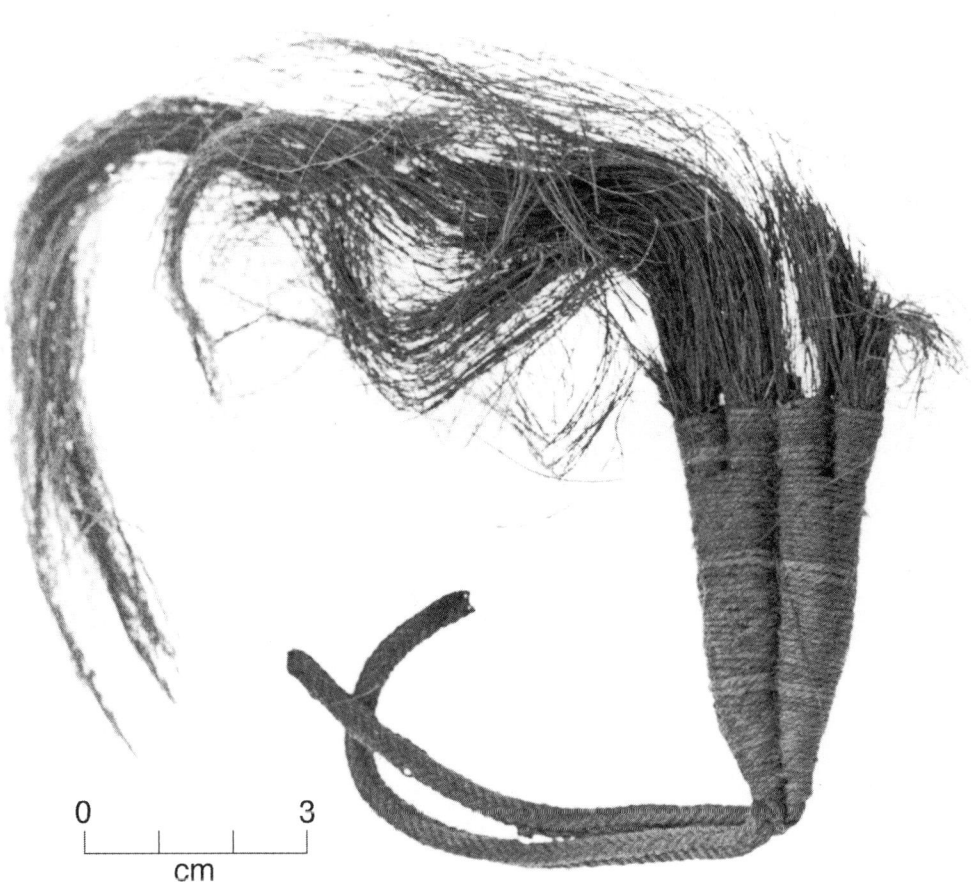

Figure 10.7. Unusual double tassel of vegetal fiber, possibly part of a ritual sling, from Structure 5, Terrace 9, Quebrada 5a.

Experiments with Slings

Brown Vega and Craig (2009) conducted sling experiments in the Department of Puno in June, 2008. The participants used slings made of braided camelid wool, 1.8 to 1.9 m in length, and similar to known archaeological examples from the coast. The best adult slinger managed to cast a stone 130 m.

Brown Vega and Craig found that the mean distance of a novice male's throw (56 m) was only slightly above that for experienced female slingers, and more than 20 m shy of the mean distance for experienced male slingers. All adult slingers could throw stones farther than younger participants, with experience and strength being the deciding factors.

Bolas

Another hunting device that may well have been brought to the New World by Pleistocene foragers was the bolas (*liwi* in Quechua). This device is known both in the Far East and throughout the Americas. Groups as widely separated as the Chukchi of Siberia, the Inuit of Alaska, and the Tehuelche and Mapuche of Patagonia have been observed using the bolas to hunt birds and mammals.

The bolas consists of multiple braided cords, radiating out from a central point where the proximal ends of all cords are connected. The distal ends of most cords end in a closed loop, into which a stone or wooden sphere can be placed. Depending on the region, the ethnic group, and the purpose to which the device is to be put, bolas can have as few as two loops for stones or as many as eight.

When the stones have been removed, the bolas becomes a light and easy-to-carry instrument. Once game has been located, the hunter can then add stones as needed.

Ethnographic Descriptions

We have eyewitness descriptions of bolas-aided hunting from both sides of the Bering Straits. In the late 19th century, Edward Nelson (1899) observed the Chukchi of Siberia hunting ducks and geese and provided the following description:

> When in search of game, the bolas is worn around the hunter's head with the balls resting on his brow. When a flock of ducks, geese, or other wild fowl pass overhead, at an altitude not exceeding 40 or 50 yards, the hunter by a quick motion untwists the sling. Holding the united ends of the cords in his right hand, he seizes the balls with the left and draws the cords so tight that they lie parallel to each other; then, as the birds come within throwing distance, he swings the balls around his head once or twice and casts them, aiming a little in front of the flock. When the balls leave the hand they are close together, the cords trail behind, and they travel so swiftly that it is difficult to follow their flight with the eye. As they begin to lose their impetus they acquire a gyrating motion, and spread apart until at their highest point they stand out to the full extent of the cords in a circle four or five feet in diameter; they seem to hang thus for a moment, then, if nothing has been encountered, turn and drop to the earth.... if a bird is struck it is enwrapped by the cords and its wings so hampered that it falls helpless (Nelson 1899:134–135).

In Alaska, Fitzhugh and Kaplan (1982) noted that the Inuit use bolas to hunt birds on both land and sea. The bolas used on land employed stones, whose weight gave the weapon greater range. When hunting waterfowl from their kayaks, on the other hand, the Inuit used round wooden balls that would float on the water until retrieved. Given the similarity of the caballito de totora to a kayak, this strategy might have worked well on the Peruvian coast.

North of the Yukon, Eskimos have been observed using the bolas to capture low-flying ducks and geese as they fly between the mainland and a series of offshore islands. The birds fly too fast to be speared or shot with arrows, but their flypath often takes them within range of bolas hunters concealed behind rocks or in boats.

Native groups of Patagonia and the Argentine pampas used the bolas to capture guanacos and sea lions on land, and sea birds over both land and water. Both two-stone and three-stone bolas were used, with stones varying from the size of an egg to the size of a fist. In the case of birds, the bolas immobilized their wings and/or legs; in the case of the guanaco, the animal's legs became entangled. Guaman Poma de Ayala (1936:204, 206; 1980) shows Inca boys hunting birds both with two-stone bolas and with nets (Rowe 1946: Fig. 26).

Three-stone bolas usually had one long cord carrying a lightweight stone and two shorter cords carrying heavier stones. The heavier stones would fly out first, parallel to each other; each heavy stone struck on either side of the animal's legs. Following behind was the lighter stone on the longer cord, which wrapped around the legs of the animal before it could flee.

We suspect that the bolas was used by the occupants of Cerro Azul to hunt sea lions. Given the size and weight of these animals, a three-stone bolas might have been required. Sea lions were hunted not only for their meat but also for their hides, which could be inflated to create a watercraft. A 16th-century chronicler, José de Acosta ([1590] 1954:181–182) tells us that

> otro indios de los valles de Ica solían ir a pescar en unos cueros o pellejos de lobo marino, hinchados, y de tiempo a tiempo los soplaban como a pelotas de viento para que no se hundiesen.

At Cerro Azul, female sea lions were the most frequently hunted wild mammal. In Quebrada 5 we discovered Feature 22, a basin-shaped hearth where female sea lions had been roasted over heated stones (Chapter 12).

Bolas Found at Cerro Azul

At Cerro Azul we recovered complete or fragmentary bolas in two archaeological contexts: (1) looted burial cists and (2)

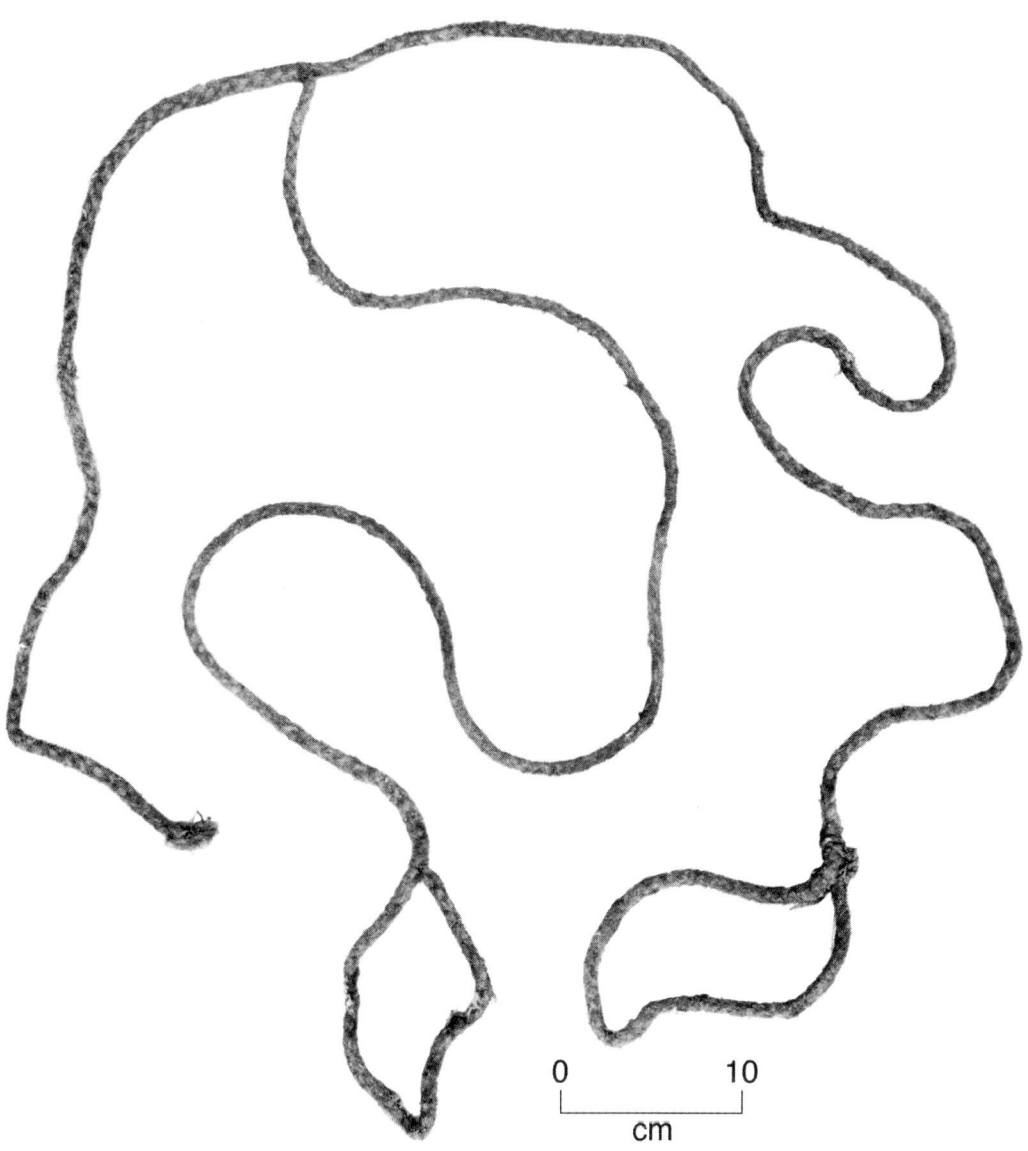

Figure 10.8. Two-stone bolas from Structure 6, Terrace 9, Quebrada 5a.

Figure 10.9. Close-up of the bolas from Figure 10.8, showing the way coarse fiber was used to make a repair to one of the stone-carrying loops.

storage bins. The most common type of bolas we found was the two-stone variety, which we suspect was used for hunting waterfowl. Hendrick Ottsen (1617) saw similar two-stone bolas being used on the Río de la Plata (Argentina) in 1603.

Fig. 10.8 shows a complete bolas from Structure 6 of Cerro Azul. This structure, a burial cist on Terrace 9 of Quebrada 5a, had been looted prior to our arrival. Despite this looting, our team was able to recover the remains of 8 adults and one child from the cist. A number of bolas were found in Structure 6, but owing to the aforementioned looting it was not possible to associate any bolas with a specific skeleton.

The bolas shown in Fig. 10.8 was made by braiding four strands into cords that were then connected to each other. Its two longest cords end in loops for large stones or wooden spheres. The loop on the right in the photo had been repaired with a z-twisted plant fiber; a closeup of that repair is shown in Fig. 10.9. This weapon resembles the two-stone bolas seen on the Río de la Plata by Ottsen in 1603.

A second bolas from Structure 6 is illustrated in Fig. 10.10. This hunting device consists of cords made by braiding four strands of tough grass stems. It resembles the previous bolas by having two cords ending in loops; it differs in that its longest cord is the one with no loop. (When we first examined this bolas, we thought that the conformation of the longest cord might reflect a third loop that had been pulled out of place by looters [Marcus 1987a: Fig. 49]. Further analysis, however, suggests that the partial loop on the long cord more likely resulted from the act of tying the bolas around a burial.)

For a close-up of the way the cords of a bolas were connected, the reader can turn to Fig. 10.11. What that photo shows is the point where the more slender central (upper) cord is attached to the slightly thicker cord that leads left and right to a pair of loops. The specimen in Fig. 10.11 was also found in Structure 6.

Fig. 10.12 shows the way the stone loop of a bolas was created by interbraiding the cords at the point of intersection. This bolas featured a three-strand braided cord of tough grass fiber. It was found in Collca 1 in Structure D, a storage bin set against the south wall of the building's Northeast Canchón.

Conclusions

The occupants of Cerro Azul braided slings and bolas for hunting birds and sea lions. The slings, made from camelid wool or coarse plant fiber, had either webbed cradles or slit cradles; their decorative tassels were made of fibers from bromeliads, tough grasses, or members of the agave family. The majority of the bolas at Cerro Azul were made of coarse plant fiber and featured loops for two stones or wooden spheres.

Although our sample of burial cists in Quebrada 5a was not large, it is of some interest that slings predominated in Structure 5, while Structure 6 tended to produce bolas. This difference could signal a certain degree of specialization in hunting, but

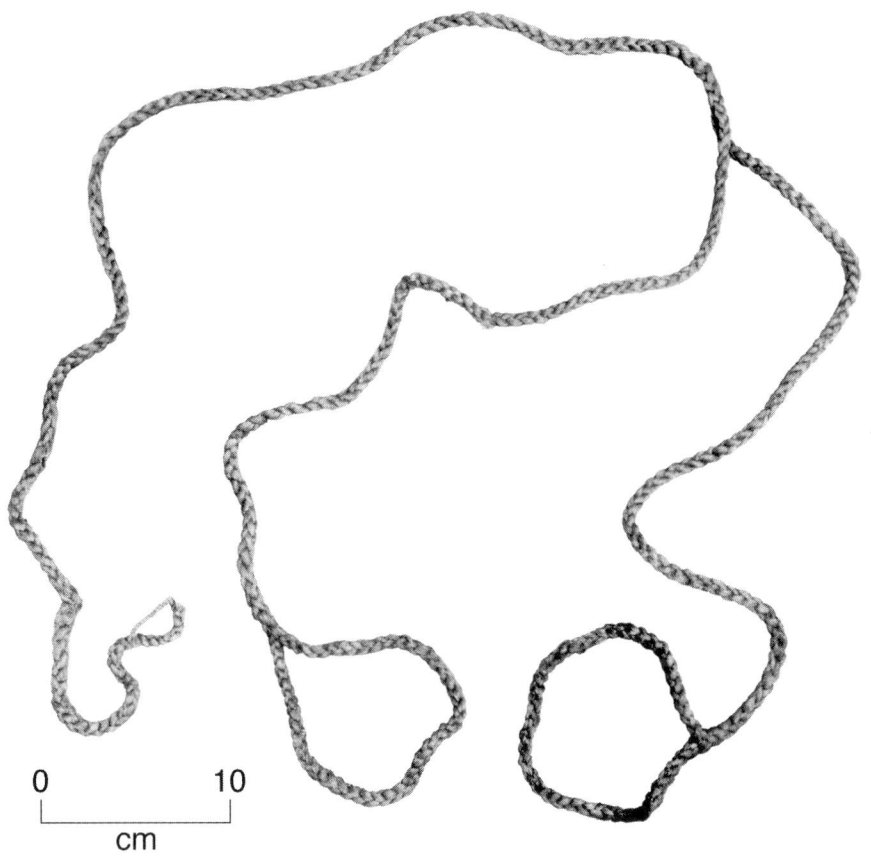

Figure 10.10. Two-stone bolas from Structure 6, Terrace 9, Quebrada 5a.

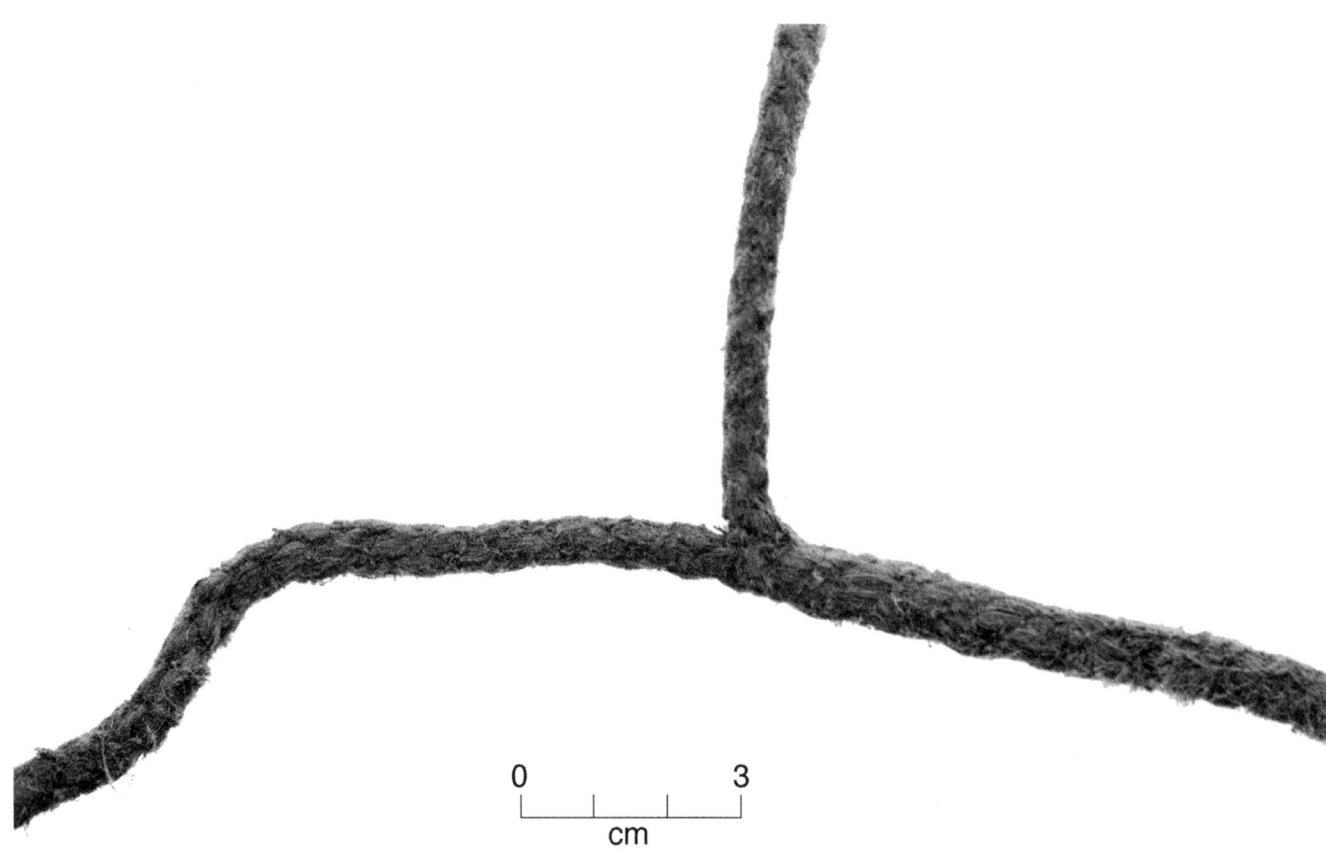

Figure 10.11. Close-up of bolas from Structure 6, showing how a central (upper) cord was attached to a thicker cord carrying the right and left stone-carrying loops.

given the degree of looting in Quebrada 5a I am reluctant to push the data too far.

Chapters 11 and 12 reveal some of the birds and mammals hunted at Cerro Azul. The most frequently captured birds were cormorants and pelicans, both of which could be hunted with bolas from caballitos de totora. Pelicans in particular, because they fly low over Cerro Azul Bay in linear formations, would have been susceptible to bolas hunting like that described for the Chukchi.

As for sea lions, they could have been attacked with slings or bolas when they left the ocean. Wounded or disabled animals might have been finished off with clubs or spears. As discussed in Chapter 12, it appears that Cerro Azul hunters concentrated on female sea lions, declining to provoke the larger and more aggressive males.

Figure 10.12. Close-up of a bolas from Collca 1, Structure D, showing how the stone-carrying loop was attached to the main cord.

11

The Bird Life of Cañete and the Avifauna of Cerro Azul

Joyce Marcus and Christopher P. Glew

The bird life of the lower Cañete Valley is extremely rich, including as it does both marine species and the birds of the inland plain (Koepcke 1964; Schulenberg et al. 2007). Unfortunately, when the University of Michigan project arrived in Cañete many species were suffering from food shortages brought on by the 1982–83 El Niño (Glynn 1988). Of the two authors of this chapter, it fell to Marcus to skeletonize those birds that died of starvation near the archaeological site, and to observe the rebound of bird populations during 1984–86. Using Marcus' bird skeletons, as well as the collections of the University of Michigan's Museum of Zoology, Glew identified the hundreds of bird bones recovered from Late Intermediate Cerro Azul (Table 11.1). Other project members, including Kent Flannery and Ramiro Matos, also contributed to our study of bird life in Cañete.

Behavior and Availability of the Birds Recovered at Cerro Azul

Many species of Peruvian seabirds compete for the same fish; this is especially true for birds relying heavily on the anchoveta. This competition may be partially reduced by the fact that different species of birds use different foraging techniques. While pelicans and gulls use surface seizing, cormorants use surface diving; boobies plunge vertically from a great height; and penguins use underwater pursuit (Duffy 1983: Table 8).

Certain species tend to arrive first when schools of anchovetas are present, and their behavior may be used as a guide by other species. Cormorants and pelicans, for example, are likely to notice the mass "dive-bombing" of Peruvian boobies and move to the action. When several species are feeding on the same school of fish, larger birds such as pelicans may engage in piracy, stealing fish from the smaller birds.

Humboldt Penguins

The Humboldt penguin (*Spheniscus humboldti*) uses underwater pursuit to capture an average of 560 g of small fish per bird per day. For half the year penguins may be well out to sea, beyond the range of the indigenous caballito de totora. During the Andean summer they are usually found closer to shore, laying their eggs in sea caves or nesting on rocky guano islands (Marcus 1987a).

Once there would have been hundreds of thousands of these penguins on the Peruvian coast, but their numbers have been greatly reduced by the commercial mining of guano from the islands on which they nest. Paredes et al. (2003) estimate that the 1982–83 El Niño reduced the Humboldt penguins of Peru and Chile from 20,000 to 6,300. Penguins are most vulnerable to human predation when molting (January to March) or guarding their eggs (anywhere from March to December).

Marcus observed small numbers of penguins swimming among sea otters and sea lions off the Cerro Azul cliffs, and she was able to collect one that died of starvation (Fig. 11.1). Cerro Azul does not have as many suitable nesting areas for penguins as locales like Chincha and Pachacamac (Paredes et al. 2003). It is clear, however, that these birds were occasionally captured at Cerro Azul, since Glew identified a penguin phalanx from Structure 9, and a patch of penguin skin from the Feature 6 midden in Structure D (Fig. 11.2).

Nonbreeding Visitors

While penguins are resident at Cerro Azul, several other bird genera occupy the role of nonbreeding visitors. Various species of albatross (*Diomedea* sp.) visit the waters of the Humboldt Current, but none actually nest in Peru; some albatrosses may come from areas as distant as New Zealand and the Galápagos Islands (Brooke and Cox 2004). At least one Late Intermediate albatross did leave his bones at Cerro Azul.

Table 11.1 Birds Identified from Archaeological Contexts at Cerro Azul.

Humboldt penguin	*Spheniscus humboldti*
Albatross	*Diomedea* sp.
Shearwater	*Puffinus* sp.
Guanay cormorant	*Phalacrocorax bougainvillii*
Red-legged cormorant	*Phalacrocorax gaimardi*
Peruvian booby	*Sula variegata*
Peruvian pelican	*Pelecanus thagus*
Duck family	Anatidae
Duck (unspecified)	*Anas* sp.
Rail family	Rallidae
Plumbeous rail	*Pardirallus sanguinolentus*
Common gallinule	*Gallinula chloropus*
Andean coot	*Fulica ardesiaca*
Sandpiper family	Scolopacidae
Lesser yellowlegs?	cf. *Tringa flavipes*
Sandpiper (unspecified)	cf. *Calidris* sp.
Gull-tern family	Laridae
Inca tern	*Larosterna inca*
Grey gull	*Larus modestus*
Band-tailed gull	*Larus belcheri*
Dove-pigeon family	Columbidae
Pigeon (unspecified)	*Columba* sp.
West Peruvian dove	*Zenaida meloda*
Eared dove	*Zenaida auriculata*
Parrot family	Psittacidae
Thrush family?	cf. Turdidae

Figure 11.1. This Humboldt penguin (*Spheniscus humboldti*) died during the El Niño of 1982–83.

In addition, the domestic chicken (*Gallus gallus domesticus*) was identified from a Spanish Colonial refuse deposit.

Figure 11.2. Patch of skin from a Humboldt penguin. Found in Feature 6, Structure D.

Figure 11.3. A paloma del cabo (*Daption capensis*) rests on Structure D of Cerro Azul between migrations (August 12, 1984).

Other non-nesting visitors include shearwaters and petrels (Ayala et al. 2013). For example, a petrel known as the paloma del cabo (*Daption capensis*) visited our excavation in 1984 before continuing one of its long migrations (Fig. 11.3). In addition, at some time in the Late Intermediate, an unidentified shearwater (*Puffinus* sp.) was captured at Cerro Azul.

Cormorants

While the occupants of the site captured an occasional nonbreeding migrant, they of course relied most heavily on birds that nest locally. The most frequently eaten birds seem to have been cormorants, two species of which are resident in the area. Both species were considered delicacies by the older workmen at Cerro Azul.

The guanay cormorant (*Phalacrocorax bougainvillii*) lives in extensive colonies on offshore guano islands and protected inshore promontories, where predators cannot reach it. Even in the aftermath of the 1982–83 El Niño, Marcus sometimes observed as many as 100 guanay cormorants resting on rocky islands 40–50 m from the Cerro Azul lighthouse. To be sure, some of their number were dying of starvation at that time (Fig. 11.4). For every 100 guanay cormorants, Marcus observed only a dozen or so chuitas or red-legged cormorants (*Phalacrocorax gaimardi*), who sometimes roosted on the rocky ledges of Cerro del Fraile (Fig. 11.5).

When two species of cormorants live in the same area, they usually reduce competition by eating different fish or having one species dive deeper than the other (Lack 1945). Marcus observed both species floating loon-like in the water for a time, then diving and pursuing fish underwater. Guanay cormorants are known to dive to depths of 30 m, although most of their feeding is done within 8 m of the surface (Austin 1961:43–44). Although faster fish usually escape them, they are successful with the smaller and slower species. Löfgren (1984:217) estimates that an individual guanay cormorant can consume half a kilo of anchovetas per day.

Both guanay cormorants and chuitas were eaten at Cerro Azul, although decidedly more of the former (Figs. 11.6–11.8). This probably reflects the fact that while *P. bougainvillii* lives in colonies of hundreds (if not thousands), *P. gaimardi* tends to live in much smaller colonies, or even in pairs.

Boobies

Another sea bird that lives in large rookeries at Cerro Azul is the piquero or Peruvian booby, *Sula variegata* (Fig. 11.9).

Figure 11.4. A starving guanay cormorant (*Phalacrocorax bougainvillii*) rests on a cobble beach at Cerro Azul during the aftermath of the 1982–1983 El Niño.

Figure 11.5. A chuita or red-legged cormorant (*Phalacrocorax gaimardi*) sits on the rocky ledge of a sea cliff at Cerro Azul.

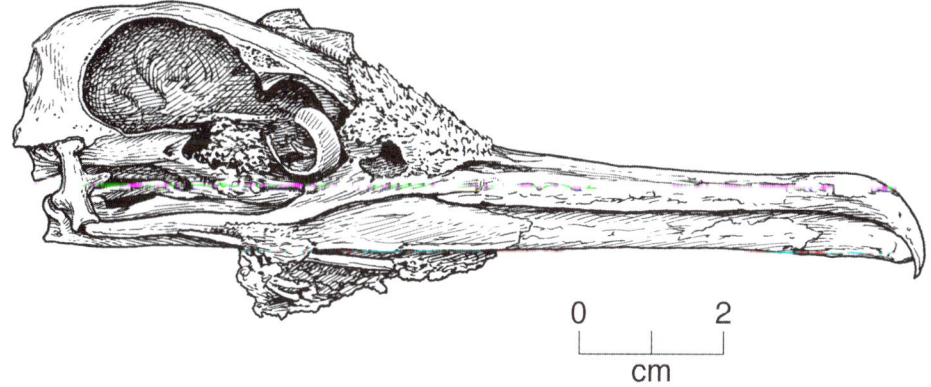

Figure 11.6. Cranium of guanay cormorant from a refuse deposit in Terrace 11 of Quebrada 6.

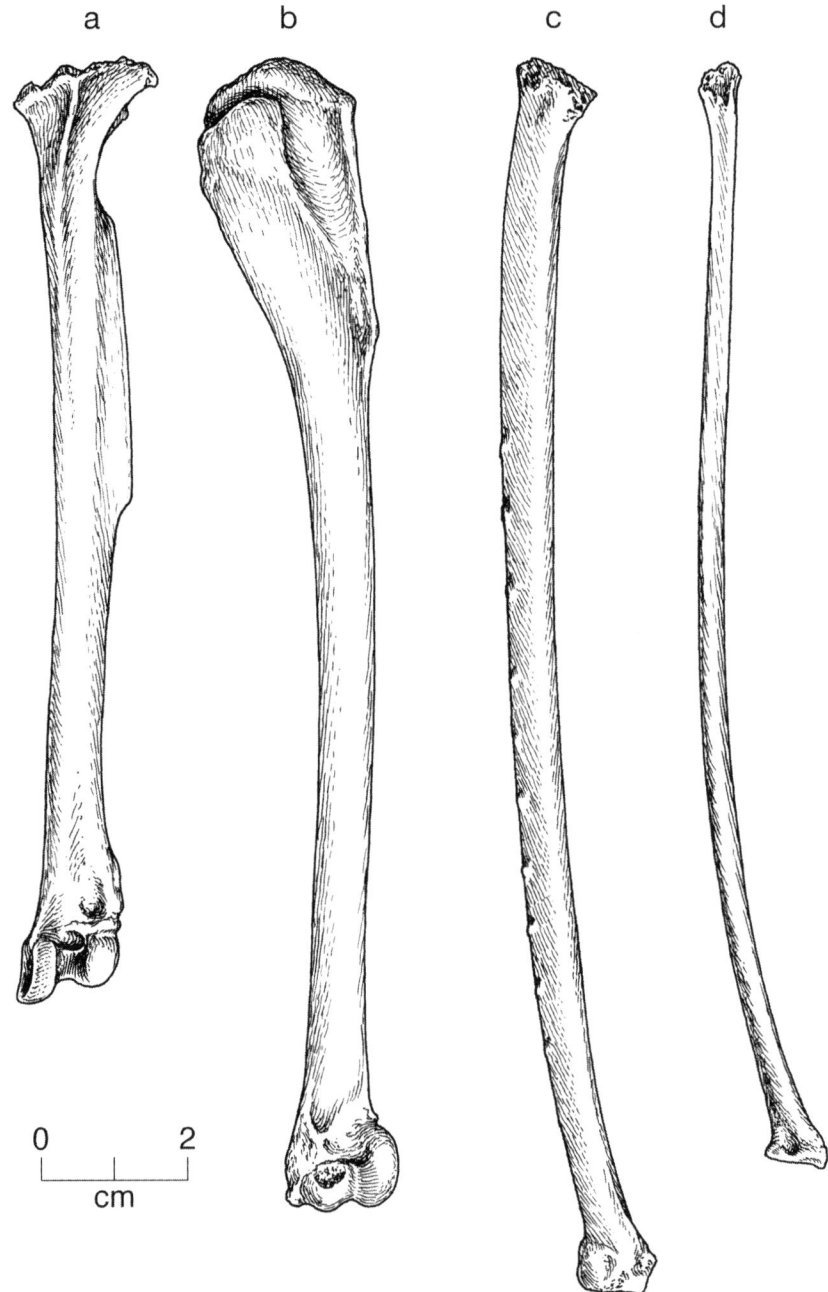

Figure 11.7. Bones of guanay cormorant from Feature 6, Structure D. *a*, tibiotarsus. *b*, humerus. *c*, ulna. *d*, radius.

In contrast to the cormorants, boobies are plunge divers who locate a school of fish and dive upon them precipitously, often reaching underwater depths of 7–9 m. Boobies usually dive in large groups in order to enhance their success, creating such a spectacular display that other bird species may be attracted to the same school of fish. Marcus observed hundreds of piqueros on the sea cliffs at Cerro Azul and found their periodic episodes of mass dive-bombing to be truly entertaining. Glew identified their bones in the Cerro Azul refuse.

A second species, the blue-footed booby (*Sula nebouxi*), has been reported from the Cerro Azul area, but Marcus never observed a living example nor came upon a starved one on the beach.

Pelicans

The Peruvian pelican (*Pelecanus thagus*) often flew low over the ocean at Cerro Azul, using its three-gallon gular pouch as a dip net (Fig. 11.10). Pelicans dive on their prey in an ungainly manner and splash hard, trying to trap the fish in their pouch without actually going under water. Austin (1961:40–42) estimates that a pelican the size of *P. thagus* can capture about two kilos of "trash fish" per day—a good thing, since it may need as many as 70 kilos of fish to raise its chicks until they are large enough to fly. Because of this high food need, pelicans occur in smaller numbers than boobies or cormorants. They were, however, captured and eaten at Late Intermediate Cerro Azul (Fig. 11.11).

Ducks, Rails, Coots, and Gallinules

Glew identified ducks in the Cerro Azul refuse, but none of the remains were complete enough to identify to species. He also succeeded in identifying the plumbeous rail, *Pardirallus sanguinolentus*, whose favorite habitats are bulrush marshes and riverside vegetation. Another resident of bulrush marshes—the common gallinule (*Gallinula chloropus*)—was present in the debris from Cerro Azul. So too was the huayno or Andean coot, *Fulica ardesiaca*.

This cluster of ducks, rails, coots, and gallinules reinforced all the stories Marcus had been told about Huaca Chola, a water-filled, bulrush-bordered depression that once lay just inland from Cerro Azul. Huaca Chola would have been the perfect habitat for the birds listed above, but had been filled in by local hacienda owners during the "golden age" of cotton production.

Sandpipers

Numerous members of the sandpiper family live in the Cerro Azul area, but their fragmentary archaeological remains proved difficult to identify to species. Glew found evidence for both the lesser yellowlegs (*Tringa flavipes*) and a member of the genus

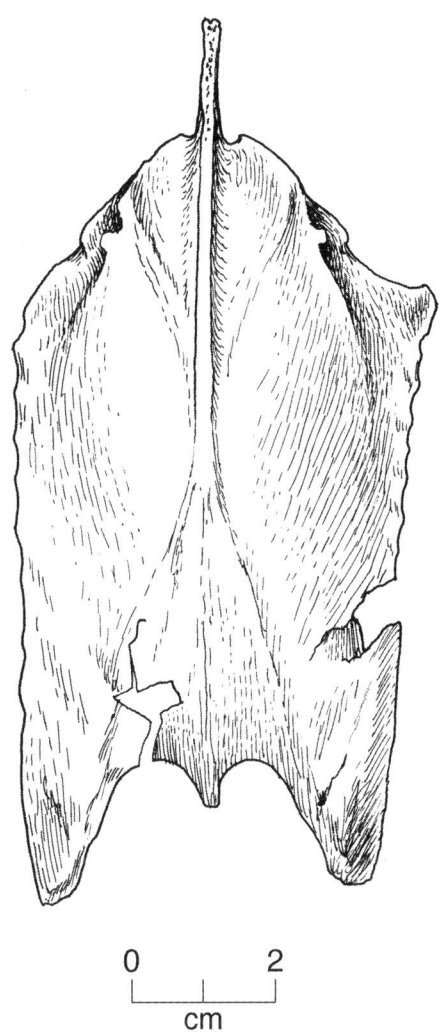

Figure 11.8. Sternum of red-legged cormorant from Feature 6, Structure D.

Figure 11.9. A group of piqueros or Peruvian boobies (*Sula variegata*) resting on the cliffs of Cerro del Fraile between dives.

Figure 11.10. Peruvian pelicans (*Pelecanus thagus*) flying low over Cerro Azul Bay.

Calidris. Both birds would be considered summer visitors to the beaches of the Cañete region.

Terns and Gulls

Members of the Family Laridae (terns and gulls) were numerous at Cerro Azul, and at least three species could be identified in our Late Intermediate refuse. The attractive zarcillo or Inca tern (*Larosterna inca*) roosted on the sea cliffs at Cerro Azul (Fig. 11.12) and could occasionally be seen plunge-diving into the water in pursuit of fish. Glew identified numerous bones of Inca tern from Late Intermediate refuse (Fig. 11.13), and Marcus was lucky enough to find the mummified remains of a complete specimen from post-Conquest deposits in Structure 1 (Fig. 11.14). This mummified tern lay in the same stratigraphic level as a weathered copy of the *Panama Star and Herald*, dating to the 1850s.

The *garuma* or grey gull (*Larus modestus*) was particularly common on the sandy beaches of Cerro Azul Bay (Fig. 11.15). There it could be seen running along the edge of the surf in search of its favorite food, the *muy muy* or mole crab (*Emerita analoga*), which was abundant in certain seasons (see Chapter 4). Bones of grey gull were present in the midden deposits of the quebradas at Cerro Azul.

We also recovered remains of the simeón or band-tailed gull (*Larus belcheri*), a Larid strongly attracted by the fish resources of the Humboldt Current. This gull nests on rocky sea cliffs and offshore guano islands.

Doves and Pigeons

The Family Columbidae (doves and pigeons) is well represented in the Cañete Valley; most species, however, frequent the irrigated croplands inland from Cerro Azul. Glew identified both the cuculí, or West Peruvian dove (*Zenaida meloda*), and the rabiblanca, or eared dove (*Zenaida auriculata*) in refuse deposits from the site. Both doves would today be found either in irrigated fields or around the brushy margins of the Río Cañete and its tributaries (Schulenberg et al. 2007).

Owls

One of the most interesting birds observed by Marcus was the diminutive burrowing owl (*Speotyto cunicularia*). Among the hilltop ruins of the Fortress of Ungará these owls could be found living in burrows they had dug themselves (Marcus 2008:9–11). Although their main prey seemed to be grasshoppers and crickets, the owl pellets near their burrows revealed that these birds also ate rice rats (*Oryzomys xanthaeolus*). Glew recovered no owl bones in the Late Intermediate refuse.

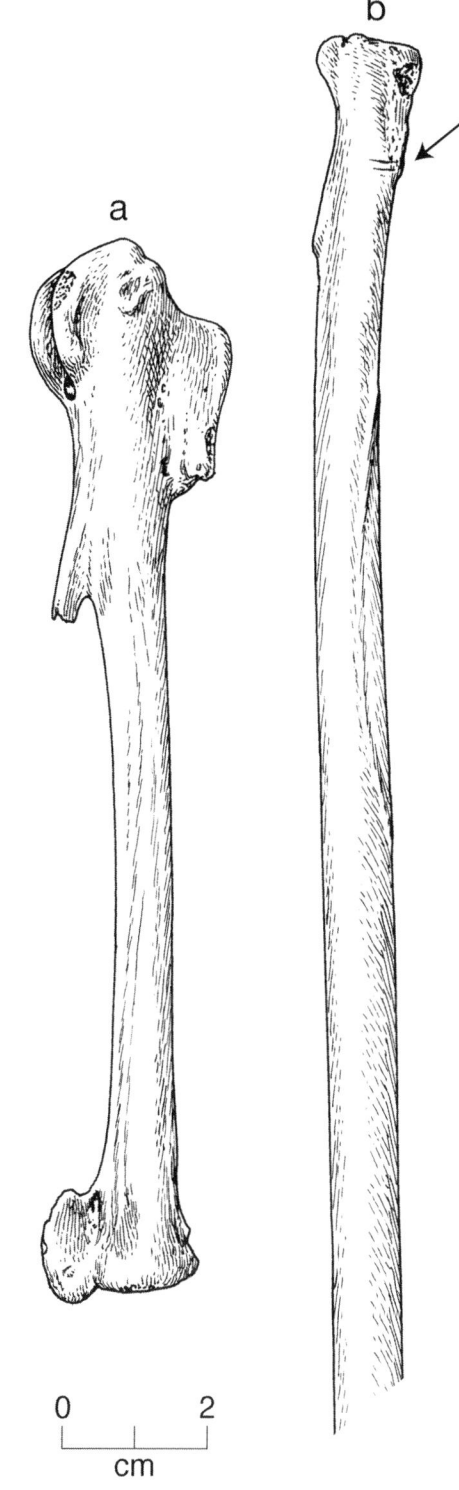

Figure 11.11. Bones of the Peruvian pelican from archaeological refuse. *a*, carpometacarpus from Feature 6, Structure D. *b*, radius with cut marks near the bicipital tubercle (arrow) from Feature 20 of Structure 9.

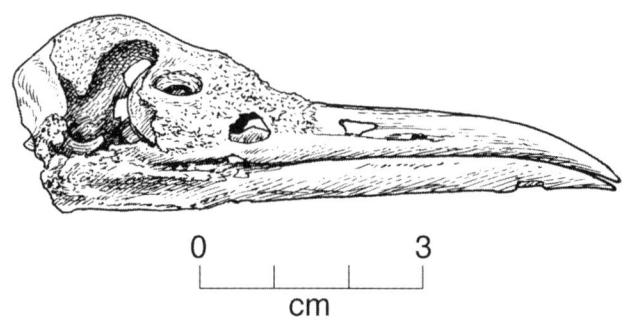
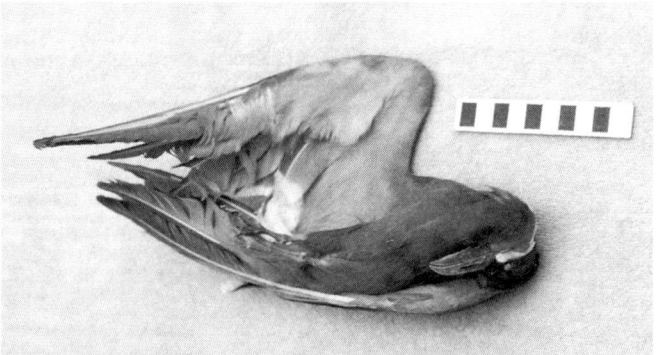

Clockwise from upper left:
Figure 11.12. The zarcilllo or Inca tern (*Larosterna inca*) frequently rests on the sea cliffs at Cerro Azul between dives.
Figure 11.13. The cranium of an Inca tern from the fill of the Northeast Canchón, Structure D.
Figure 11.14. This Inca tern died in the ruins of Structure 1 and became completely mummified. It was found in the same stratum as an 1850s copy of the *Panama Star and Herald*.
Figure 11.15. Gulls on the beach at Cerro Azul Bay. In the foreground are two simeones or band-tailed gulls (*Larus belcheri*). In the background we see a line of four garumas or grey gulls (*Larus modestus*), searching for mole crabs (*Emerita analoga*) along the tide line.

Exotic Birds

Finally, let us turn to two exotic species found in the ruins of Cerro Azul. The humerus of a parrot (Family Psittacidae) was found near Burials 6 and 7 in Quebrada 5a. Parrots occur above 1000 m elevation in the Cañete region, but do not live close to Cerro Azul. Our guess is that this particular bird might have been sought by the Cerro Azul elite for its feathers, but without knowing the species we cannot guess the parrot's place of origin.

During the course of excavating Structure 1, a Late Horizon building on the summit of Cerro del Fraile, Marcus found evidence that Spanish Colonial squatters had occupied the building during the 16th century. From their refuse, Glew identified the bones of a domestic chicken (*Gallus gallus domesticus*).

The Bird Remains from Structure D

Let us look now at the archaeological bird remains from Cerro Azul, beginning with the Southwest Canchón of Structure D. The largest sample of fauna from this part of the building came from **Feature 6**, a midden thought to represent debris swept up from the canchón floor. This midden, which was screened through both 1.5 mm and 0.6 mm mesh, produced 115 specimens of birds (Table 11.2). Among the 22 minimum individuals represented were a penguin; guanay and red-legged cormorants; a pelican; a gallinule; a coot; several sandpipers; a tern and a gull; and several doves and pigeons. While cleaning the floor of the Southwest Canchón not far from Feature 6, Marcus' crew recovered an additional pelican vertebra.

Table 11.2. Bird Bones from Feature 6, Structure D.

Humboldt penguin (MNI = 1)	Patch of skin with feathers	Sandpiper (MNI = 2)	3 coracoids 1 humerus
Guanay cormorant (MNI =2)	3 fragments of cranium 1 sternum 1 furculum 3 fragments of coracoid 2 fragments of scapula 4 fragments of humerus 1 ulna 1 radius 1 ulnar carpal 2 fragments of carpometacarpus 1 tibiotarsus	Gull or tern (MNI = 1)	1 femur
		Inca tern (MNI = 1)	1 coracoid 1 humerus
		cf. Band-tailed gull (MNI = 1)	1 tibiotarsus
		Dove or pigeon (?) (MNI = 1)	1 coracoid 1 ulna
		Unspecified pigeon (MNI =1)	1 ulna
		cf. West Peruvian dove (MNI = 1)	1 radius
Red-legged cormorant (MNI = 1)	1 sternum 1 tibiotarsus	cf. Eared dove (MNI = 1)	1 humerus
Unspecified cormorants (MNI =2)	2 fragments of coracoid 3 quadrates 2 fragments of femur 28 vertebrae 11 ribs	Unidentified birds (MNI = 3)	1 premaxilla 1 dentary 3 vertebrae 6 ribs 2 synsacra 1 humerus 2 fragments of tibiotarsus 1 femur 3 other bone fragments 4 feathers (from 3 different birds) 2 patches of skin with feathers
Peruvian pelican (MNI = 1)	1 carpometacarpus		
Common gallinule (MNI = 1)	1 nasal + premaxillary joined		
Andean coot (MNI =1)	1 humerus 1 tibiotarsus		
Unspecified Scolopacid (MNI =1)	1 synsacrum 1 ulna 1 tarsometatarsus	**Totals:** **NISP = 89** **Unidentified = 26**	

The North Platform, Southwest Canchón

Additional bird bones were recovered in and around the North Platform of the Southwest Canchón. Included were one vertebra of albatross; one coracoid, radius, quadrate, and scapula of pelican; one coracoid of Peruvian booby; two coracoids and a vertebra of guanay cormorant; one humerus of plumbeous rail; one humerus of common gallinule; and the tibiotarsus of an unidentified member of the rail family. The pelican scapula displays cut marks.

The South Corridor

Lying on the floor of the South Corridor, which led from the Southwest Canchón to the interior of Structure D, were one tibiotarsus of guanay cormorant; one cormorant vertebra, unidentifiable to species; one humerus of Peruvian booby; and one carpometacarpus of pelican.

The Northeast Canchón

In the fill of the Northeast Canchón lay one cranium of Inca tern (Fig. 11.13); one carpometacarpus and first phalanx of Inca tern; and one ulna of pelican. An ash-filled hollow in the floor near the northwest corner of the canchón produced the pelvis of a pelican.

The North Central Canchón

The fill of the North Central Canchón produced a synsacrum and femur of guanay cormorant; the ulna of an unidentified shearwater; and the sternum of an unidentified member of the sandpiper family.

Room 1

Room 1 began its history as part of an elite residential unit. No bird bones were associated with this stage of the room's use.

At some point, Room 1 suffered earthquake damage and was evacuated. It was then filled with sand and used for fish storage. Still later in the room's history, bits of trash were thrown in on top of the sand. Included in this trash was the sternum of a dove or pigeon.

Room 3

Room 3 was originally part of the same residential unit as Room 1. Associated with the room's residential stage were the humerus and femur of a guanay cormorant, as well as one tarsometatarsus of an unidentified bird.

Like Room 1, Room 3 eventually suffered earthquake damage and was converted to fish storage. Still later, domestic trash was tossed into the room and came to rest on the sand layer. Included in this trash were the sternum, furculum, coracoid, and rib of a pelican; one tibiotarsus of Andean coot; one humerus of West Peruvian dove; and the humerus of an unidentified member of the thrush family.

Room 4

Room 4 had a complex history. It was originally created as a storage room, with its floor lying 3.73 m below the arbitrary datum point established for Structure D. At a later date, fill was added to the room and a new floor laid at a depth of 2.23 m below datum.

Resting on the original floor (depth 3.73 m) were the tibiotarsus, radius, and femur of a guanay cormorant; the carpometacarpus and ulna of a red-legged cormorant; one cormorant vertebra, not identified to species; and one tibiotarsus of pelican.

In the fill just below the second floor (depth 2.23 m), Marcus' crew found the synsacrum, coracoid, humerus, femur, and tibiotarsus of a guanay cormorant; one tibiotarsus of a red-legged cormorant; one rib of cormorant, not identified to species; one tibiotarsus of a Peruvian booby; and one humerus of an unidentified duck.

Once her crew had finished excavating Room 4, Marcus decided to make a deep sounding below the room's original floor. Her goal was to search for earlier stages of Structure D. This sounding found only tapia fill to a depth of 4.73 m below datum. This fill included one tibiotarsus of plumbeous rail.

Room 6

Room 6 was originally a storage room with its floor lying 1.15 m below the arbitrary datum point for Structure D. Later in its history it was converted to a residential unit, and a new floor was created at a depth of 85 cm below datum. In the fill between the two floors was one coracoid of guanay cormorant.

Room 7

Room 7 had originally been connected to the East Platform of the Southwest Canchón by a narrow corridor. After the room had suffered earthquake damage, the corridor was deliberately blocked and Room 7 became a convenient place to dump refuse. Included in this refuse were the premaxilla of a red-legged cormorant and the dentary of a pelican.

Room 8

Room 8 was a sand-filled fish storage unit. After the abandonment of Structure D, wind erosion caused debris from the Southwest Canchón to drift downward into Room 8. Included in this debris was one coracoid of a guanay cormorant.

Rooms 9 and 10

Rooms 9 and 10 were dedicated to the raising of guinea pigs. No bird bones were found on the floor of either room. After the abandonment of Structure D, however, wind erosion caused debris from the North Platform area of the Southwest Canchón to drift downslope into Rooms 9 and 10. Included in this debris was the tibiotarsus of a guanay cormorant.

A Late Horizon Squatters' House

Upon excavation, "Room 2" of Structure D turned out to be a small interior patio. After Structure D ceased to function as an elite residential compound, Late Horizon squatters built a kincha house in "Room 2." A domestic midden, designated **Feature 4**, accumulated next to this house. Included in the midden was one vertebra from a Peruvian booby.

The Bird Remains from Structure 9

Structure 9, a fish storage facility, lay only 22 m from Structure D. Our most productive faunal collection from Structure 9 was provided by **Feature 20**, a midden presumed to contain the domestic refuse of the commoner-class overseer of the building.

Like Feature 6 of Structure D, the Feature 20 midden was screened through both 1.5 mm and 0.6 mm mesh. This screening produced 28 bones belonging to birds (Table 11.3). Among the 10 minimum individuals recorded by Glew were two guanay cormorants; one red-legged cormorant; a fourth cormorant, not identified to species; a pelican; an unspecified rail; an unspecified sandpiper; an unspecified pigeon; and two birds that could not even be identified to family, owing to the fragmentary nature of their bones. One pelican radius displayed cut marks near its bicipital tubercle (Fig. 11.11).

Table 11.3. Bird Bones from Feature 20, Structure 9.

Guanay cormorant (MNI =2)	1 cranium
	1 scapula
	2 coracoids
	1 ulna
	1 carpometacarpus
	1 first phalanx
Red-legged cormorant (MNI = 1)	1 tibiotarsus
Unidentified cormorant (MNI = 1)	1 ulna
	1 femur
	1 phalanx
	1 rib
	5 vertebrae
Peruvian pelican (MNI = 1)	1 radius (with cut marks near the proximal epiphysis)
Unspecified rail (MNI = 1)	1 ulna
	1 radius
Unspecified Scolopacid (MNI = 1)	1 humerus
Unspecified pigeon (MNI = 1)	1 radius
Unidentified birds	2 humeri
	1 ulna
	1 tibiotarsus
	1 vertebra
	1 other bone

Totals:
 NISP = 22
 Unidentified = 6

The South Entryway

The main entryway into Structure 9 was from the south, through a passageway between storage rooms. The South Entryway had two superimposed floors, the earliest of which lay at a depth of 1.77 m below the arbitrary datum established for Structure 9. In the fill just above this early floor, Marcus' crew found the first phalanx of a penguin; one humerus from a guanay cormorant; and one femur from a Peruvian booby.

Room 5

Rooms 5 and 6 had once been a single long, narrow room. Later they were divided into two smaller rooms by the addition of two tapia blocks. Room 5 was then filled with sand and used for fish storage. After its use for fish storage had ended, the room became a convenient place to dump trash. Included in this trash were the following bird bones:

1 ulna and 2 tibiotarsi of guanay cormorant
1 first phalanx of pelican
1 humerus of common gallinule
1 vertebra and 2 ulnae of Peruvian booby
1 radius of duck
1 humerus of plumbeous rail
1 vertebra, gull or tern
1 rib, 1 femur, 1 humerus, and 1 tibiotarsus of unidentified birds

Room 6

Room 6, immediately to the south of Room 5 and once connected to it (see above), was not immediately used for fish storage after its creation; only later were sand and fish added to the room. Finally, after its period of use for fish storage was over, Room 6 became a convenient place to throw trash. Included in this trash were one vertebra from an unspecified cormorant and the synsacrum, femur, tibiotarsus, and phalanx from an unidentified bird.

Bird Remains from the Quebradas of Cerro Camacho

The bird remains from the quebrada terraces were all in tertiary context, and few were as well preserved as the bones from Structures D and 9.

Quebrada 5, Terrace 16

Terrace 16 of Quebrada 5 was 2.2 m deep. It had served as a place to dump domestic refuse from one or more of the nearby residential compounds, and its nine strata reflected two phases of activity. From perhaps A.D. 1000 to 1300 (Stratigraphic Zones G to D1), ashy midden was repeatedly dumped on Terrace 16. Then came a period of disuse, during which sterile deposits accumulated (Stratigraphic Zone C). Next, from perhaps A.D. 1300 to 1470, refuse dumping resumed (Stratigraphic Zones B-A1). Bird bones were found in three levels, as follows:

Zone F (the oldest refuse level)
One femur of pelican, with cut marks on the distal portion
Zone D1 (the last refuse layer before the Zone C hiatus)
One tarsometatarsus of guanay cormorant
Zone A1 (the uppermost refuse layer)
One tarsometatarsus of guanay cormorant

Quebrada 5a, Terrace 9

Terrace 9 of Quebrada 5a was 2.6 m deep, and its strata reflected three phases of activity. Stratigraphic Zone C, the oldest, produced no bird bones.

Stratigraphic Zone B was an ashy Late Intermediate midden whose most common mollusc was the coquina clam *Donax obesulus*. This midden had to be sampled with care, because burial cists (and individual burials) from Zone A had penetrated it in places. Table 11.4 lists the bird bones from three samples taken in the upper half of the Zone B midden: (1) an area undisturbed by burials; (2) a sample taken near Burials 6 and 7; and (3) a sample taken near Burial 9. Note that our lone specimen of parrot came from near Burials 6 and 7, and might have been associated with one of them.

Table 11.4. Bird Bones Recovered from Upper Zone B of Terrace 9, Quebrada 5a.

Undisturbed Zone B
Guanay cormorant: 1 tarsometatarsus
Unspecified cormorant: 1 vertebra
Unspecified duck: 1 humerus

Near Burials 6 and 7
Guanay cormorant: 1 cranium, 1 synsacrum, 1 ulna, 1 carpometacarpus, 1 first phalanx
Red-legged cormorant: 1 coracoid
Unspecified cormorant: 1 phalanx
cf. Peruvian booby: 1 phalanx
cf. Lesser yellowlegs: 1 tarsometatarsus

cf. Inca tern: 1 coracoid
Unspecified parrot: 1 humerus

Near Burial 9
Peruvian pelican: 1 tarsometatarsus
cf. Andean coot: 1 tibiotarsus

Totals:
NISP = 16
Unidentified = 0

Structure 5, one of the stone masonry burial cists originating in Zone A of Terrace 9, penetrated Zone B as well. Below the floor of this cist, however, Zone B was undisturbed and could be fully screened. In Table 11.5 we present the bird bones from lower Zone B by depth increments: (1) 1.9 m below the surface of the terrace, (2) 1.9–2.2 m below, and (3) 2.2–2.4 m below.

Table 11.5. Bird Bones Recovered from Zone B below Structure 5 on Terrace 9, Quebrada 5a.

Depth 1.9 m below surface
Guanay cormorant: 1 coracoid; articulated humerus, radius, and ulna
Unspecified duck: 1 tarsometatarsus
cf. Andean coot: 1 ulna
Unspecified Scolopacid: 1 synsacrum
Inca tern: 1 humerus
Grey gull: 1 first phalanx
Band-tailed gull: 1 first phalanx
Unidentified birds: 1 radius, 1 tarsometatarsus, 1 phalanx

Depth 1.9–2.2 m below surface
Peruvian booby: 1 vertebra, 1 coracoid

Depth 2.2–2.4 m below surface
Peruvian booby: 1 first phalanx, 1 tibiotarsus

Totals:
NISP = 14
Unidentified = 3

Quebrada 6, Terrace 11

Marcus' attention was drawn to Terrace 11 of Quebrada 6 because of its distinctive coal-black color. She eventually concluded that the black color had resulted from the actions of salt fog on rich organic debris, including fish oil.

The bird remains from Terrace 11 were as follows:

Guanay cormorant: one cranium (Fig. 11.6), one scapula, one tibiotarsus
Unspecified cormorant: 3 vertebrae
Peruvian booby: one cranium, one humerus, one ulna
Pelican: 2 phalanges
Grey gull: one humerus
Unidentified bird: one tibiotarsus

Quebrada 6, Terrace 12

Terrace 12 lay immediately downhill from Terrace 11, and some of the midden debris from the higher terrace had slid down onto Terrace 12. Included in the debris were one coracoid of guanay cormorant; one vertebra and one tibiotarsus from an unspecified cormorant; and the radius and ulna of an Inca tern.

Spanish Colonial Bird Remains

After their conquest of the Kingdom of Huarco in A.D. 1470, the Inca built several structures on the sea cliffs of Cerros Centinela and El Fraile (Marcus et al. 1985). One of these was Structure 1, an adobe building with trapezoidal niches in Late Horizon style.

During the 16th century, several rooms in Structure 1 were occupied by Spanish Colonial squatters. Among the refuse left by those squatters were a number of Old World domestic plants (Chapter 13). They also left the tarsometatarsus of a domestic chicken on the lower part of the building's south stairway.

Summary and Conclusions

Of all the birds recovered from Late Intermediate refuse at Cerro Azul, the guanay cormorant seems to have been the one most often eaten. Whether one is talking about elite residences, commoner residences, or the refuse on quebrada terraces, cormorants are the birds whose bones were most frequently found. Cut marks on the bones of pelicans indicated that they were butchered as well, but they seem to have been a distant second in popularity. Both cormorants and pelicans could have been hunted with slings or bolas (see Chapter 10).

Among the smaller birds boobies, terns, gulls, doves, and pigeons were common enough in the refuse to suggest that they may have been eaten. The fact that one Inca tern seems to have died of natural causes in Structure 1, however, raises the possibility that not all of our bird remains were the product of human activity.

It is interesting that our only parrot bone was found near two burials in Quebrada 5a, hinting that parrot feathers may have been sumptuary goods. Other than that, there is little evidence to show that the Cerro Azul elite had greater access to any particular species of bird, relative to commoner families.

12

The Hunting of Marine Mammals

Christopher P. Glew and Kent V. Flannery

Among the wild animals eaten at Cerro Azul were a series of marine mammals, all of which would be considered Trophic Level 4 predators. These creatures were so large that they outweighed Andean land mammals like the guanaco. Included were sea lions—which may have been hunted with slings, bolas, clubs, and spears—and dolphins, which may have been surrounded and killed when they accidentally beached themselves.

Sea Lions and Fur Seals

Cerro Azul lies well within the range of the South American sea lion *Otaria byronia*, as well as the Southern fur seal *Arctocephalus australis* (Bonner 1981, King 1983, Oliva 1988, Vaz Ferreira 1981). These two members of the Clade Pinnipedia overlap in range, but generally do not live together; sea lions like sandy beach nesting areas, while fur seals prefer rockier environments (King 1983). Male *Otaria* also tend to drive *Arctocephalus* away by attacking the females and eating their young.

The 1982–83 El Niño reduced the food supply for both species of pinnipeds, yet Marcus and Flannery found 10–15 sea lions still occupying offshore islands near the Cerro Azul lighthouse (Fig. 12.1). These protected localities were used whenever the sea lions wanted to leave the water, or "haul out," as sea lion biologists call it (Fig. 12.2).

At no point during the 1980s did Marcus and Flannery observe fur seals living at Cerro Azul, nor was there much evidence of seals in the animal bones from the archaeological site. Only two of the 81 sea mammal fragments from Cerro Azul were considered to be possible fur seal bones (see below).

In 1979, three years before the 1982–83 ENSO, Tovar Serpa (1979) conducted a census of pinnipeds on the coast of Peru. Of the country's estimated 25,055 sea lions, 34 were living on islands near Asia; 213 were living at Chincha Sur; and 1,384 had colonized Zárate Island near the Paracas Peninsula. None of these localities had substantial colonies of fur seals, although 1,877 *Arctocephalus* were occupying Punta los Mártires on the Paracas Peninsula.

South American sea lions establish colonies in well-protected areas, which at Cerro Azul would include the sea caves at the base of Cerros Centinela and El Fraile. There they live in groups whose moderately polygynous males are outnumbered three or four to one by females.

Sexual dimorphism in sea lions is pronounced; males can reach a length of 2.24–2.38 m and weigh 196–300 kg, while females are generally 1.8–1.9 m long and weigh only 107–144 kg (King 1983, Vaz Ferreira 1981). This sexual dimorphism

Figure 12.1. This rocky island near the Cerro Azul lighthouse is a favorite "haul out" locality for the local sea lions (arrow).

is readily apparent in sea lion bones found on the surface of the Cerro Azul site, as can be seen in Fig. 12.3. It also affects hunting behavior; Marcus' workmen said that they had captured juvenile and female sea lions in the past, but would hesitate to take on an angry adult male. Apparently the ancient fishermen exhibited the same concern, because all our sea lion bones from Late Intermediate levels were in the size range of females.

The fact that female sea lions overlap in size with male fur seals poses a problem for faunal analysis (see discussion later in this chapter). Male fur seals measure 1.8–1.9 m and weigh 150–160 kg; females measure only 1.40 1.43 m and weigh 45–50 kg (Bonner 1981, King 1983).

The sea lions at Cerro Azul are known to eat sardines, anchovies, and medium-sized schooling fish; they also pursue squid, octopus, and penguins. Sea lions can even show an interest in large shellfish such as *Concholepas*, leaving the smaller molluscs to the local sea otters (*Lutra felina*).

South American sea lions give birth during the Andean summer (mid-December to early February). Females are protective of their pups except in ENSO years, when food stress may compel them to abandon their young. This happened during the 1982–83 El Niño at Cerro Azul, and resulted in a number of sea lion pups dying of starvation (Fig. 12.4).

Figure 12.2. A young sea lion (*Otaria byronia*) resting on a beach boulder at Cerro Azul.

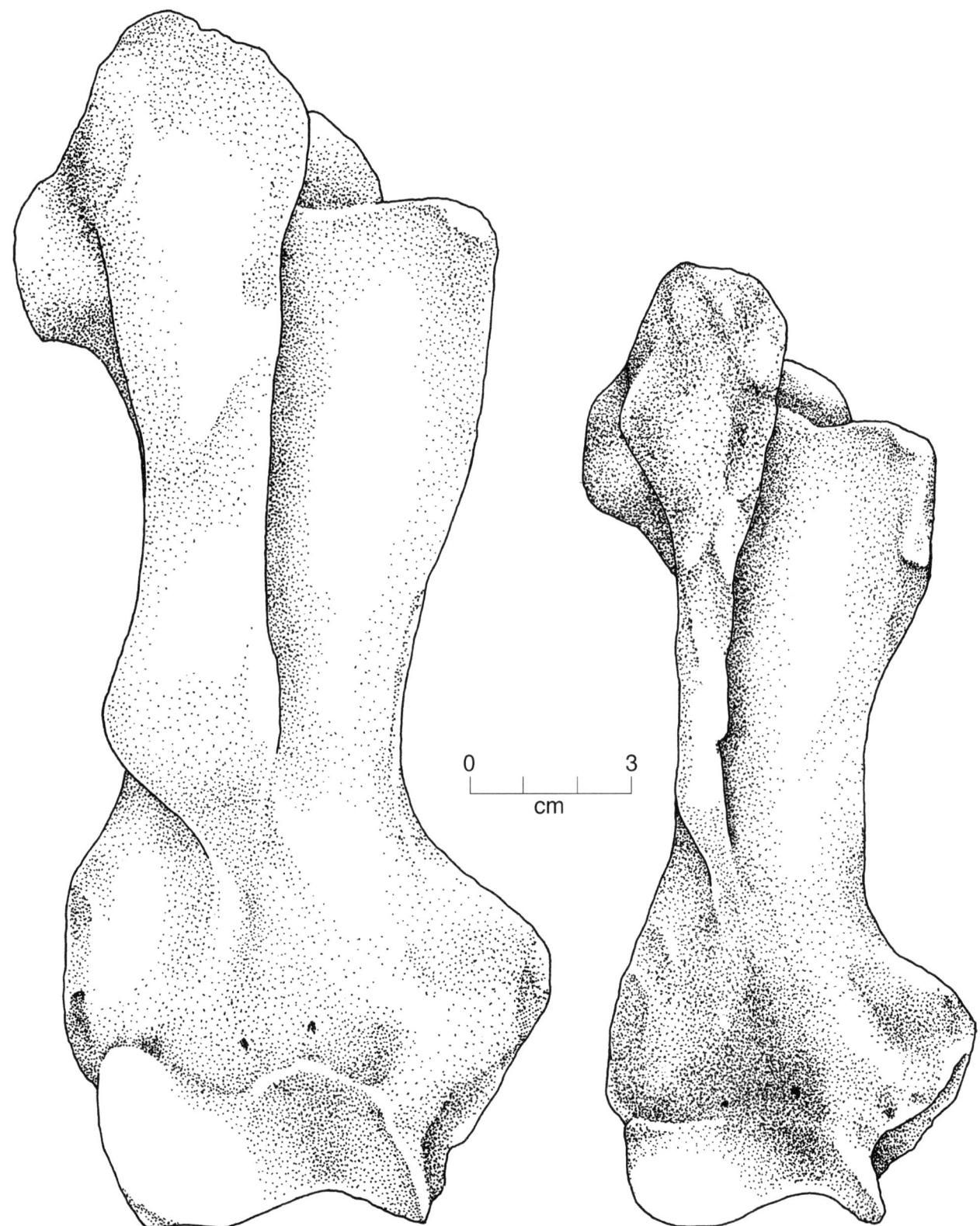

Figure 12.3. Two humeri from sea lions (*Otaria byronia*), illustrating sexual dimorphism. The humerus on the left is from a male, the one on the right from a female. These bones were found on the surface of the Cerro Azul site.

Figure 12.4. During the El Niño of 1982–83, a number of female sea lions responded to food stress by abandoning their young. Many abandoned juveniles, including the one shown here, died of starvation.

Late Intermediate Use of Sea Lions

The Late Intermediate refuse from Cerro Azul revealed two contexts for sea lion bones. In Structure 9 and the quebradas of Cerro Camacho, we found evidence for the butchering, roasting, and eating of sea lions. In one roasting pit (or earth oven) in Quebrada 5 Marcus found evidence for a *pachamanca*, or cookout, involving two female *Otaria*. Bones from two of the sea lions involved in this cookout are shown in Figures 12.5 and 12.6. The mandible in Figure 12.5 was found in the roasting pit itself, while the femur in Figure 12.6 came from an associated layer of refuse. The cut marks shown on the femur were probably made when the lower limb was separated from the pelvis.

Bleached sea lion bones found on the surface of the site (e.g., Fig. 12.7) suggest that such large communal sea lion feasts were not an uncommon occurrence, and that this cooking was usually done outdoors.

In addition to these outdoor cookouts, which produced mandibles and limb bones of sea lions, we also found evidence that sea lions were valued for their hides. In Structure D and on Terrace 11 of Quebrada 6, we found only the kinds of carpals, tarsals, and phalanges that would have been left behind when the flippers were trimmed from a sea lion hide. We know that coastal people used pinniped hides to make inflatable watercraft, and we suspect that the trimming off of the flippers was one of the preliminary steps in manufacture.

The Capture of Dolphins

During the El Niño of 1982–83, one of the ways that artisanal fishermen responded to the reduction of their fish supply was by capturing dolphins and selling the meat (Fig. 12.8). Flannery's examination of the butchered remains left on the beach suggested that at least two genera were involved: the bottlenose dolphin *Tursiops* sp., and the Pacific white-sided dolphin *Lagenorhynchus* sp. (Fig. 12.9). Since the remains of butchered dolphins were mutilated and incomplete, he did not attempt to identify them beyond genus (Fig. 12.10).

Bottlenose dolphins can measure between 1.7 m and 4.0 m in length and weigh 90–650 kg. In the Cerro Azul region they tend to eat rays, bony fish, octopus, and squid in normal years, and add shrimp to their diet in El Niño years. White-sided dolphins are somewhat smaller, measuring 1.5–3.0 m in length and weighing between 75 and 90 kg. In the Cerro Azul region they tend to eat Pacific hake, mackerel, anchovetas, squid, crustaceans, and even some shellfish.

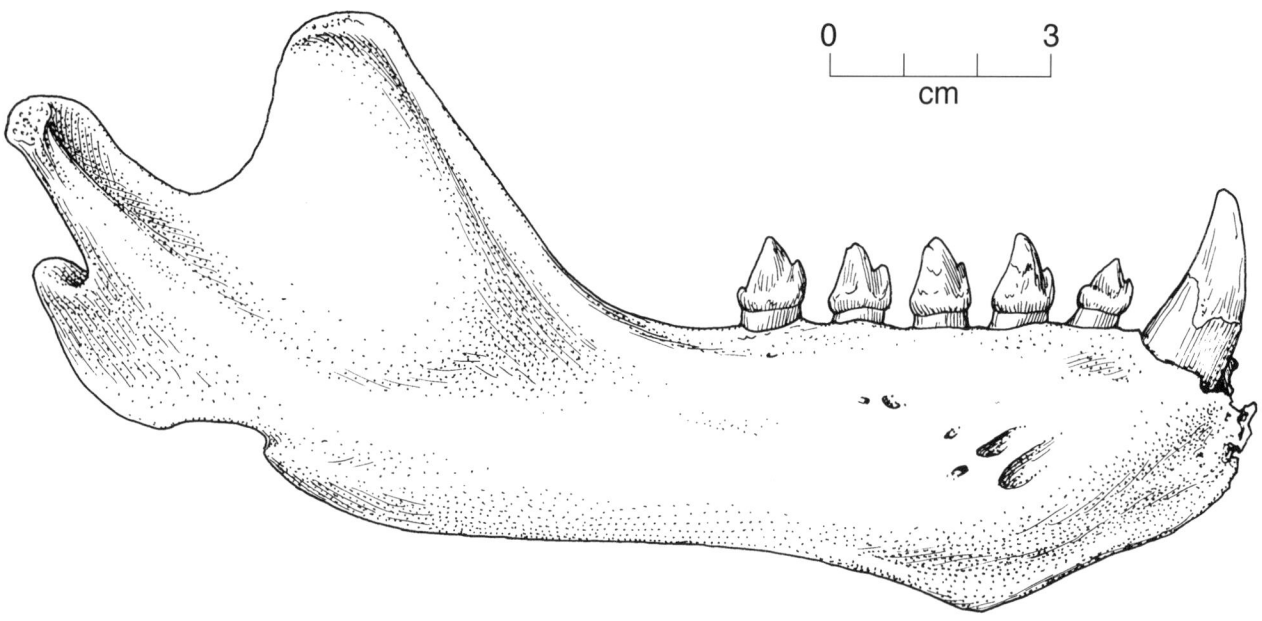

Figure 12.5. Right mandible of sea lion (*Otaria byronia*) from Feature 22, Terrace 16, Quebrada 5.

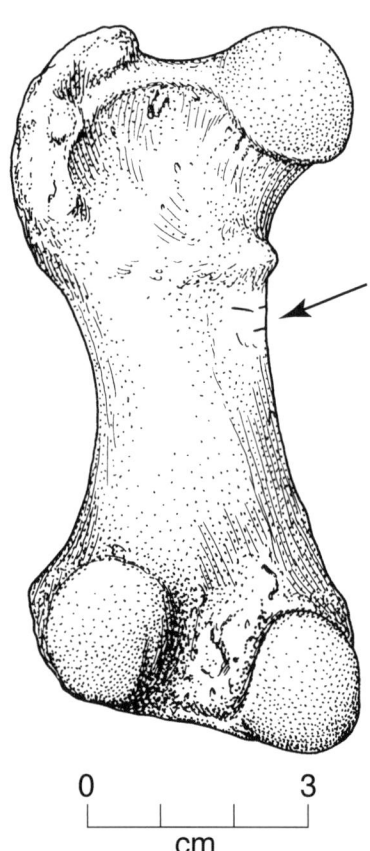

Figure 12.6. Left femur of sea lion from Stratigraphic Zone A, Terrace 16, Quebrada 5. The arrow points to a series of cut marks.

We are confident that the occupants of Late Intermediate Cerro Azul were willing to eat dolphins. The fact that they did not eat them often probably has something to do with the problems of taking on a 500 kg dolphin while in a caballito de totora. The small numbers of dolphin bones found at Cerro Azul lead us to suspect that the occupants of the site probably took advantage of individuals who had beached themselves.

Figure 12.11 presents two of the dolphin vertebrae we recovered. One, belonging to a white-sided dolphin, was found in a pit used to roast two female sea lions (see below). The other, belonging to a bottlenose dolphin, was found in an area of disturbed Late Intermediate burials. Both finds were made in the quebradas of Cerro Camacho.

The Archaeological Remains of Marine Mammals

Marcus' crew recovered some 81 fragments of marine mammals from archaeological contexts at Cerro Azul. So far as can be determined, the bulk of those fragments came from female sea lions of the species *Otaria byronia*. Glew found only two bones that could possibly have been from the Southern fur seal, *Arctocephalus australis*. He discovered, in addition, two vertebrae belonging to dolphins. One was from the bottlenose dolphin *Tursiops*, the other from the white-sided dolphin *Lagenorhynchus*.

We should begin by mentioning some of the difficulties inherent in distinguishing sea lions from fur seals. As we have seen, both *Otaria* and *Arctocephalus* are sexually dimorphic. Male *Otaria* and female *Arctocephalus* are easily told apart, because the former are so much larger than the latter. Unfortunately, female *Otaria* and male *Arctocephalus* overlap in size, making them trickier to separate.

Glew took the Cerro Azul pinnipeds to the Smithsonian Institution's Division of Mammals for analysis, since that comparative collection features multiple skeletons of males and females. There he was able to distinguish between female *Otaria* and male *Arctocephalus*, the former being more gracile and the latter more robust. These distinctions, however, were most pronounced in the cranium and least pronounced in the distal limbs (metapodials to phalanges). As a result, Glew relied more heavily on the crania and proximal limbs for species identifications.

The fact that Glew found virtually no fur seals in our Cerro Azul sample fits with Flannery's lack of sightings of fur seals at Cerro Azul, and King's (1983) observation that *Otaria* and *Arctocephalus* tend not to live together.

We were struck by the fact that Marcus' quebrada excavations produced many more sea lion specimens than either Structure D or Structure 9. This leads us to suspect that most butchering and cooking of marine mammals took place outdoors.

Structure D

The Southwest Canchón of Structure D produced a modest number of pinniped bones, mostly from flippers. Two right first phalanges and one second phalanx from a forelimb were found in the Feature 6 midden. One right astragalus was found on the canchón floor, near the blocked doorway leading to Room 11. Finally, a layer of relatively late fill above the North Platform of the canchón produced one first and one third phalanx from a hind flipper.

In the northwest corner of the **Northeast Canchón**, Marcus found a number of ash-filled hollows in the floor. The relatively late debris in one of these hollows produced the left scapula of a female sea lion.

Collca 1 was a storage unit set against the south wall of the Northeast Canchón. It contained one left metacarpal from a pinniped.

Room 1 had been part of an elite residential unit before it suffered earthquake damage. Just above its original floor lay one left second phalanx from a pinniped forelimb.

Room 3 had also been part of an elite residential unit before suffering earthquake damage; afterwards, it was converted to a fish storage room. From its earlier (residential) stage we recovered one left radial carpal and one second phalanx from the forelimb of a female *Otaria*. From its later (storage) stage we recovered one right first phalanx from a pinniped forelimb.

Room 4 had a complex stratigraphic history. Its original floor lay 3.73 m below the datum point established for Structure D. A deep sounding below this floor produced one right second phalanx from the forelimb of a pinniped.

Later in the history of Room 4, a new floor was created at a depth of 2.23 m below datum. Just below this floor we found

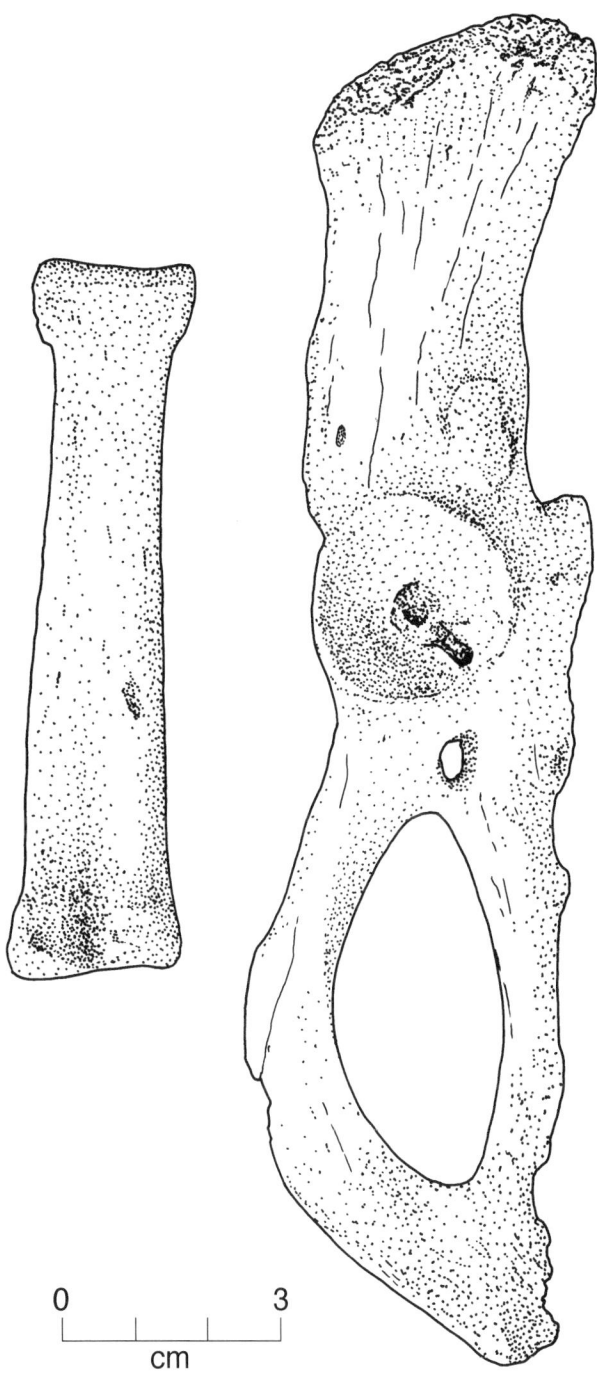

Figure 12.7. Eroded phalanx and innominate of female sea lion, recovered from the surface of the Cerro Azul site.

Figure 12.8. The carcass of a butchered dolphin, discarded on the beach at Cerro Azul during the 1982–83 El Niño.

Figure 12.9. The two dolphins most frequently taken by Cerro Azul fishermen during the 1980s were the bottlenose dolphin (*Tursiops* sp.) (*a*) and the Pacific white-sided dolphin (*Lagenorhynchus* sp.) (*b*).

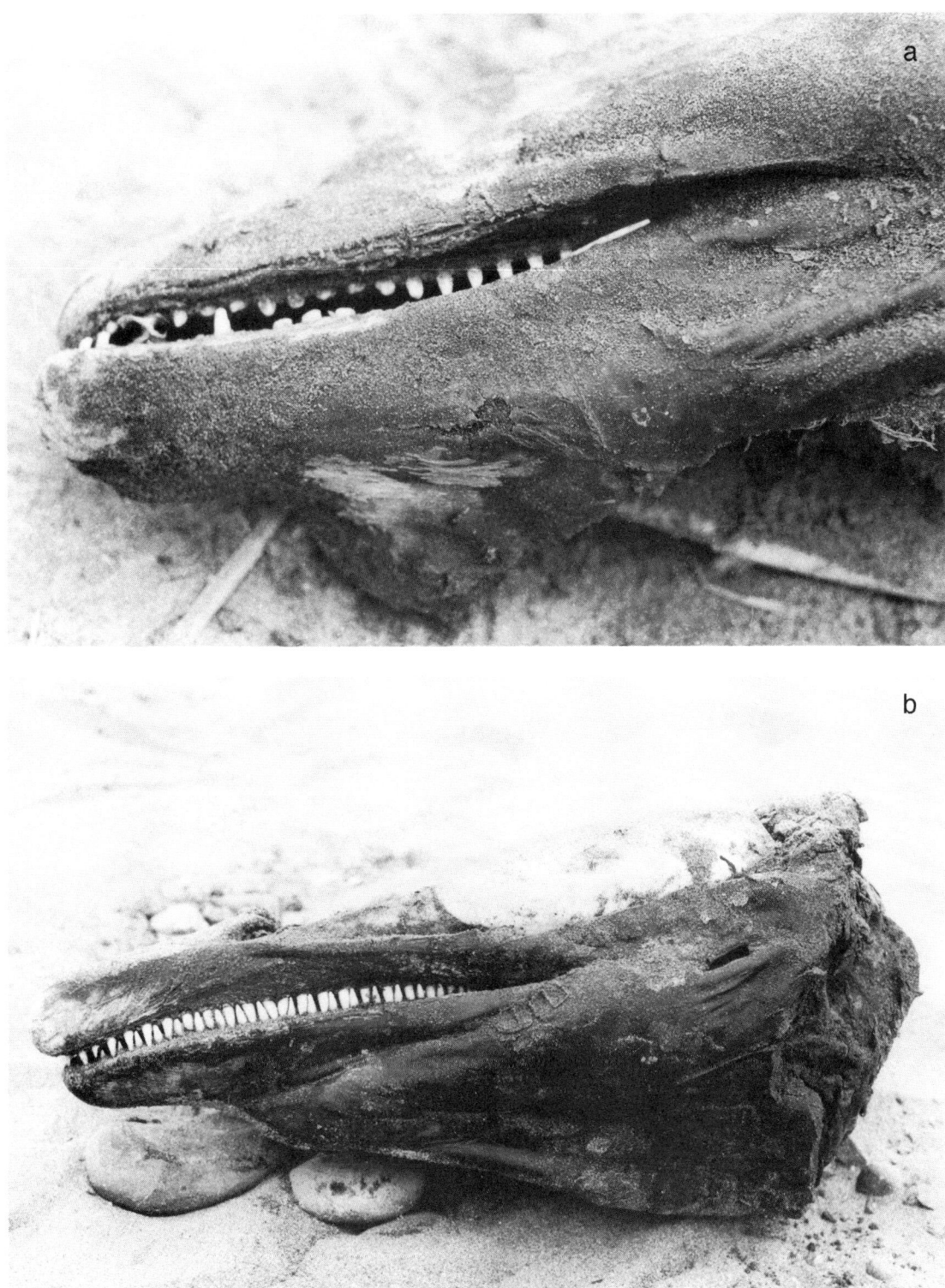

Figure 12.10. Heads of butchered dolphins, discarded on the beach at Cerro Azul. *a*, *Tursiops* sp. *b*, *Lagenorhynchus* sp.

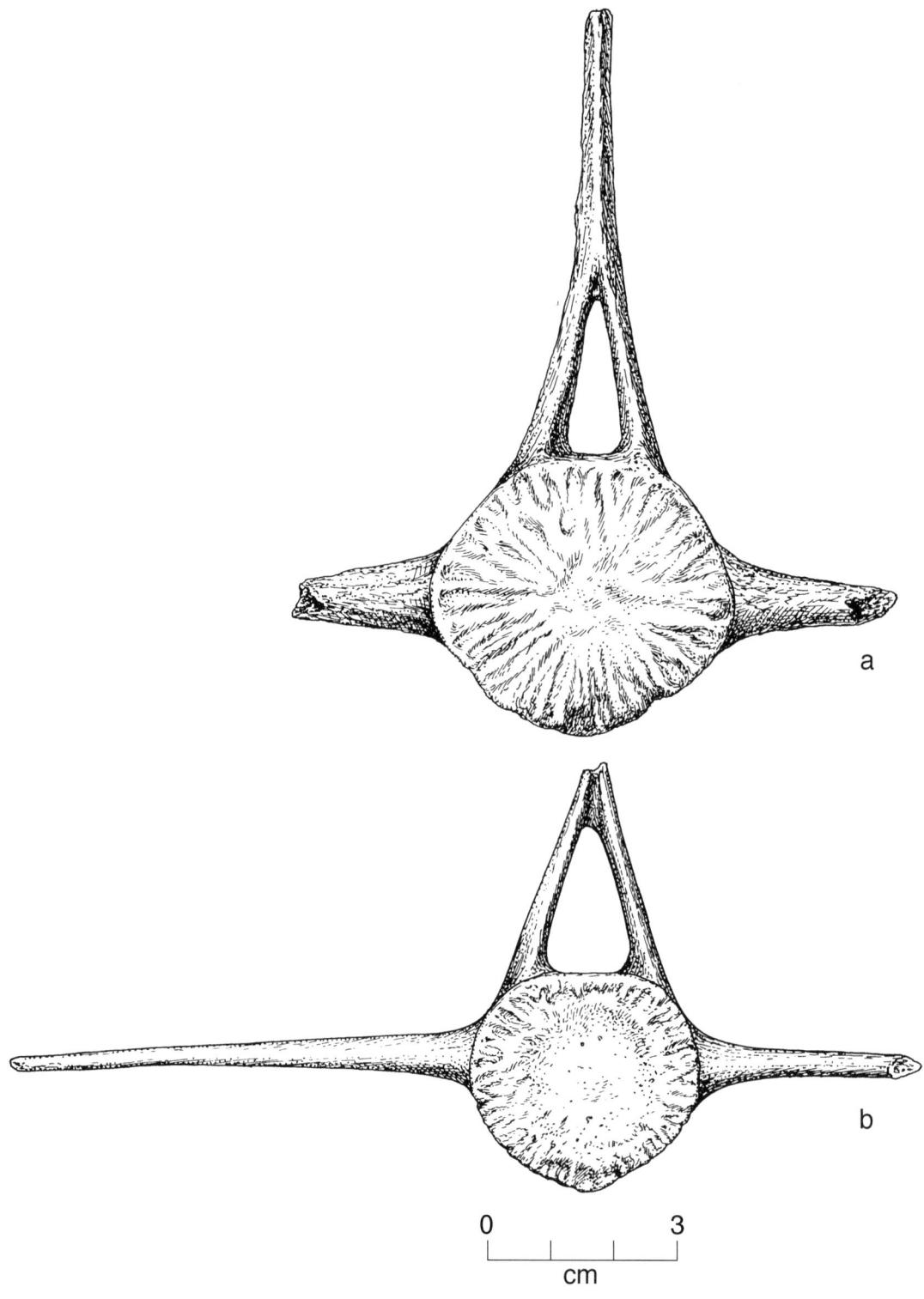

Figure 12.11. Vertebrae from two genera of dolphins eaten at Late Intermediate Cerro Azul. *a*, *Tursiops* sp. from Stratigraphic Zone A, Terrace 9, Quebrada 5a. *b*, *Lagenorhynchus* sp. from Feature 22, Terrace 16, Quebrada 5.

two pinniped phalanges, one of which was a left first phalanx from a forelimb. One and a half meters below this—in the fill between the room's earlier and later floors—we recovered first and second phalanges from the right hind limb of a pinniped.

Finally, we come to **Feature 4**. This feature was the ashy midden associated with a Late Horizon kincha house, built in the "Room 2" patio of Structure D by squatters after the building's main period of occupation was over. This midden produced two pinniped bones: (1) one right metacarpal from a forelimb and (2) one right third phalanx from a hind limb.

We are struck by the fact that virtually all of the pinniped specimens from Structure D were small bones from trimmed-off flippers. This suggests the preparation of sea lion skins for some activity—perhaps the making of inflatable watercraft. Structure D produced no evidence for the butchering or eating of sea lions, only the use of their skins in craft activity.

Structure 9

Marcus and her crew screened **Feature 20**—a midden resting against the east side of Structure 9—through both 1.5 mm and 0.6 mm mesh. In spite of this screening, the midden produced only one left third carpal from a female *Otaria*.

The bulk of the sea lion remains from Structure 9 came from **Room 5**, a small storage unit on the building's west side. The fill of this room included greenish gray sand with small fish remains. This fill produced the left maxilla and left patella of a female sea lion, as well as one left second metacarpal of a pinniped.

After its main use as a storage room had ended, Room 5 was used as a convenient place to dump trash. Since this later trash included large items that did not appear to have come from far away, Marcus (2008:234) suggested that it might be domestic refuse from the nearby wattle-and-daub house of the building's commoner-class overseer. The pinniped remains from this refuse consisted of 18 fragments, at least three of which were clearly from a female sea lion. The collection is as follows:

Cranial elements
    R. petrous bone: 1
    L. petrous + fragment of zygomatic arch: 1
    Fragment of cranium: 1
Vertebral elements
    Atlas: 1
    Fragments of other vertebrae: 8
Limb elements
    L. ulna, epiphyses unfused: 1
    R. third metacarpal: 1
    R. tibia (shaft only): 1
    L. fifth metatarsal: 1
Other fragments: 2

Total:
    NISP = 18
    MNI = one young female *Otaria*?

Room 7 was a long, narrow fish storage unit in the southwest corner of Structure 9. Late in the room's history, the right patella of a female sea lion was discarded there.

Room 8 was a small unit in the northeast quadrant of the building. Its contents included large sherds from a series of restorable storage and cooking vessels, perhaps discarded not far from their place of use. Included among these sherds was one right first phalanx from a pinniped.

We are struck by the fact that Structure 9, a fish storage facility with a resident commoner-class administrator, produced more evidence of sea lion butchering than did Structure D. We assume that this butchering (and subsequent eating) was done by the family of the building's overseer.

Marine Mammal Remains from the Quebradas of Cerro Camacho

Quebrada 5 produced our largest single collection of sea lion remains, and gave us our best look at how these creatures were cooked. It was on Terrace 16 of this quebrada that Marcus' crew found Feature 22, a stone-lined, basin-shaped hearth or earth oven.

Feature 22 was found in Stratigraphic Zone A2 of Terrace 16. It was oval in outline (90 x 100 cm) and had been excavated 22 cm into Zone B, an underlying layer of hard, sterile red earth. This hearth had been used to roast two female sea lions over heated stones, using a version of the cooking technique called *pachamanca* by Quechua speakers.

Marcus' crew found bones of *Otaria byronia* both in the hearth itself and beside it, in Stratigraphic Zone A2. They also recovered 26 false abalones from the hearth and 8 more from Zone A1, suggesting that the sea lion main course might have been preceded by ceviche. The sea lion bones are listed in Tables 12.1 and 12.2.

Table 12.1. Marine Mammal Bones from Feature 22, Terrace 16.

Definitely South American sea lion
    Complete R. mandible: 1 (see Fig. 12.5)
    Complete L. mandibles: 2 (MNI =2)
    L. ulna (distal end unfused): 1
    L. tibia (proximal end unfused): 1
    R. tibia (shaft only): 1

Probably South American sea lion
    Cranial fragment: 1
    R. auditory bulla: 1
    Loose cheek teeth: 4
    Fragment of sternum: 1
    Fragment of vertebra: 1
    Ribs: 2
    First phalanx: 1

Male fur seal or female sea lion?
    L. maxilla + zygomatic arch: 1

Pacific white-sided dolphin
    Vertebra: 1 (see Fig. 12.11, *b*)

Totals:
    NISP = 19
    MNI = 3?

Table 12.2. Marine Mammal Bones from Stratigraphic Zone A2, Terrace 16.

Definitely South American sea lion
    R. mandible: 1 (MNI = 2 if combined with R. mandible from Feature 22)
    R. humerus (epiphyses fused): 1
    R. scapula: 1
    L. scapula (epiphysis unfused): 1
    R. radius (epiphyses unfused): 1
    L. femur: 1 (shows butchering marks; see Fig. 12.6)

Probably South American sea lion
    L. radial carpal: 1
    Rib: 1

Totals:
    NISP = 8
    MNI = 2 (see above)

Overall, it appears that Feature 22 was used to roast two female sea lions, at least one of which was young enough to have unfused limb bones. Whether the dolphin vertebra was an addition to the feast or simply a stray bone that came to be redeposited in the feature cannot be determined. The 34 false abalones from Feature 22 and Zone A1 were too numerous to be accidental inclusions (see Fig. 3.6).

Quebrada 5a, Terrace 9

Terrace 9 had a long and complex history of use. No marine mammal bones were recovered from Stratigraphic Zone C, the oldest layer of refuse on the terrace. Our most ancient sea mammal specimen came from lower Stratigraphic Zone B, a layer of Late Intermediate midden that Marcus' crew screened through two sizes of mesh. The remains consisted of the upper left molar of a pinniped.

Our second oldest specimen came from the upper part of Zone B, not far from two intrusive Late Intermediate burial cists (Structures 4 and 5). It consisted of one pinniped left ulna, robust enough to be that of a male fur seal (although it overlaps in size with the ulna of a female sea lion).

Our youngest sea mammal specimen from Terrace 9 was found in an area of looted Late Intermediate burials in Stratigraphic Zone A. It consisted of one vertebra from a young bottlenose dolphin (see Fig. 12.11, *a*). Given the amount of looting in this area, we cannot determine whether dolphin *ch'arki* had been left as food for the afterlife, or if this vertebra had simply come from the midden into which the burial cists were excavated.

Quebrada 6, Terrace 11

Terrace 11 of Quebrada 6 provided a good sample of highly organic Late Intermediate midden. A screened sample of this midden, taken at a depth of 30–50 cm, produced two pinniped bones: one left fifth metacarpal and one left second phalanx from a forelimb. Two additional specimens from a pinniped forelimb were found at a depth of 30 cm; included were one left second metacarpal and another left second phalanx.

Higher up in the midden, at a depth of 0–20 cm, Marcus' crew found three more pinniped bones: two second phalanges and one right second tarsal from the hind limb of a Pacific sea lion.

The fact that all the pinniped bones from the Terrace 11 midden were elements from flippers makes the Quebrada 6 refuse reminiscent of that from Structure D. Instead of documenting the butchering and cooking of sea lions, it seems to reflect only the trimming of flippers from sea lion skins. Perhaps, therefore, the midden on Terrace 11 represents domestic refuse from one of Cerro Azul's elite residential compounds. Both the fish remains (Chapter 9) and the plant remains (Chapter 13) from Terrace 11 lead us to the same conclusion.

Structure 1, Cerro del Fraile

After their conquest of the Kingdom of Huarco in A.D. 1470, the Inca built a number of administrative and ritual buildings on the sea cliffs at Cerro Azul (Marcus 1987a; Marcus et al. 1985). One of these buildings was Structure 1, a 12-room adobe building on the summit of Cerro del Fraile.

On the lower part of this building's main stairway, Marcus' crew recovered one first phalanx from the right forelimb of a pinniped. This discovery suggests that the working of marine mammal skins continued after the Inca conquest.

Conclusions

The Late Intermediate occupants of Cerro Azul made use of sea lions in at least two ways. On those occasions when they killed a sea lion, they seem to have shared their 100–140 kilogram windfall with others by cooking it outdoors, *pachamanca* style. They also seem to have utilized the skins of sea lions, possibly converting them into inflatable watercraft. As evidenced by the small bones from trimmed-off flippers, much of this skin preparation was done indoors.

Dolphins were also eaten at Cerro Azul, but this activity was so infrequent as to suggest that it occurred only when one of those creatures beached itself.

As for the other marine mammals of the central coast, they seem barely to have been utilized. Fur seals, despite their valuable coats, were represented by only a few possible fragments. Sea otters—which Marcus and Flannery saw often between 1983 and 1986—did not show up in the archaeological refuse, and we found no remains of beached whales.

Part III

The Use of Plants at Cerro Azul

13

Edible, Ritual, and Medicinal Plants

C. Earle Smith, Jr. and Joyce Marcus

Editor's note: Dr. Smith's untimely death in 1987 precluded his writing this chapter. Therefore, using his field notes and plant counts, Marcus has done her best to create a document of which he would approve. It goes without saying that the late Dr. Smith is not responsible for any botanical errors.

Preservation of plant remains at Cerro Azul was excellent. In all, more than 3600 specimens of edible, ritual, and medicinal plants were recovered by the University of Michigan excavations (Table 13.1). So varied were the remains of maize that they are discussed further in Chapter 14. In this chapter we present all the edible, ritual, and medicinal plants in phylogenetic order, using Simpson (2010) as our guide.

Table 13.1. Edible, Ritual, and Medicinal Plants Recovered from Late Intermediate Contexts at Cerro Azul.

Cherimoya	*Annona cherimolia*
Achira	*Canna edulis*
Maize	*Zea mays*
Euphorbia	*Euphorbia* sp.
Coca	*Erythroxylum coca*
Ciruela del fraile	*Bunchosia armeniaca*
Common beans	*Phaseolus vulgaris*
Lima beans	*Phaseolus lunatus*
Canavalia beans	*Canavalia* sp.
Erythrina	*Erythrina* sp.
Peanuts	*Arachis hypogaea*
Pacay	*Inga feuillei*
Unspecified cucurbit	either *Cucurbita* or *Apodanthera*
Butternut squash	*Cucurbita moschata*
Yuca de monte	*Apodanthera* sp.
Achiote	*Bixa orellana*
Guava	*Psidium guajava*
Lúcuma	*Pouteria lucuma*
Sweet potato	*Ipomoea batatas*
White potato	*Solanum tuberosum*
Unspecified chile pepper	*Capsicum* sp.
Kellu uchu	*Capsicum baccatum*

Plants Recovered from Cerro Azul

Cherimoya

The cherimoya or custard apple (*Annona cherimolia*) was one of the fruits eaten at Cerro Azul (Fig. 13.1). Whether it was actually grown in the Cañete Valley or imported from elsewhere is unknown.

Weberbauer (1945) found wild cherimoya trees growing in Peru's "warm and dry northern valleys," and botanists believe that the cultivar of this species originated in the Andes at elevations of 700–2400 m (van Zonneveld et al. 2012). Bonavia et al. (2004) reported seeds of cherimoya from preceramic levels at Los Gavilanes in the Huarmey Valley, as well as the nearby Middle Horizon site of Tuquillo. A mature cherimoya tree can produce 34–45 kilos of fruit per year, making it a good investment wherever environmental conditions permit.

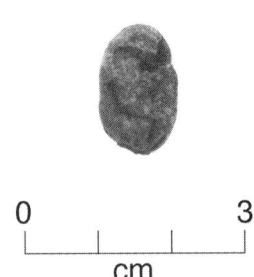

Figure 13.1. A seed from the cherimoya fruit (*Annona cherimolia*), found in Feature 20 of Structure 9.

Achira

Achira, or Queensland arrowroot (*Canna edulis*), was still being grown in the Cañete Valley in the 1980s (Fig. 13.2). This plant grows to be 1–2 m high and is cultivated for its edible rhizomes, which are a useful starch source (Gade 1966, Cisneros et al. 2009). A single rhizome clump grows to be more than half a meter long, and can be rendered edible by baking it in hot ashes or an earth oven.

While achira is likely native to the eastern slopes of the Andes, it reached the Pacific coast in preceramic times and was being grown at Nasca and Pachacamac during later periods (Ugent et al. 1984). We found leaves of achira at Cerro Azul (Fig. 13.3), suggesting that this plant may have been under cultivation not far away.

Maize

Maize (*Zea mays*) was used extensively by the occupants of Cerro Azul, especially in the production of chicha or maize beer (Marcus 2009). Many varieties of maize were delivered to Cerro Azul, presumably from different localities. These maize varieties are discussed in detail in Chapter 14.

Euphorbia

Euphorbia is a spurge whose milky white, latex-like sap is widely believed to have medicinal properties. It does, in fact, have antibacterial and antioxidant effects, and can prevent the spread of infection. *Euphorbia* can also be used as a sedative or purgative.

Figure 13.2. Achira (*Canna edulis*) is still grown in the Cañete Valley. The dark bars on the scale are each 10 cm long.

The Late Intermediate occupants of Cerro Azul clearly understood *Euphorbia*'s medicinal value. Figure 13.4 illustrates the contents of a "medicine bundle" found in Structure 12, a burial cist in Quebrada 5-south (Marcus 2008:279–281). This collection of plants—found wrapped in a cloth bundle—includes locks of brown cotton and white cotton, a specimen of *Euphorbia*, and an achira leaf tied up in a twisted maize leaf. Although the looting of Structure 12 prevents us from associating this bundle with a specific burial, we suspect that these medicinal plants were assembled for an individual who was extremely ill, and then left with his or her body after death.

Coca

Coca (*Erythroxylum coca*) was perhaps the most important ritual/medicinal plant in ancient Peru. The coca tree is not native to the coast, but can be grown by irrigation in certain parts of the *chaupi yunga* or coastal piedmont (Rostworowski de Diez Canseco 1988; Marcus and Silva 1988).

Figure 13.5 shows a mass of coca leaves that we found in the bottom of an amphora on Terrace 9, Quebrada 5a. The area where this vessel came to light was one of looted Late Intermediate burials. The amphora itself is shown in Figure 3.8 of Marcus (2008), and was almost certainly a burial offering. It is likely that the looters left it behind because it belonged to a utilitarian ware of little commercial value.

Figure 13.3. Disintegrating leaf, probably achira. From the fill of Room 4, Structure D, at a depth of 30–50 cm below Feature 5.

Figure 13.4. The contents of a "medicine bundle" from Structure 12, a looted burial cist in Quebrada 5-south. *a*, brown cotton. *b*, white cotton. *c*, *Euphorbia* sp. *d*, an achira leaf tied with a twisted maize leaf.

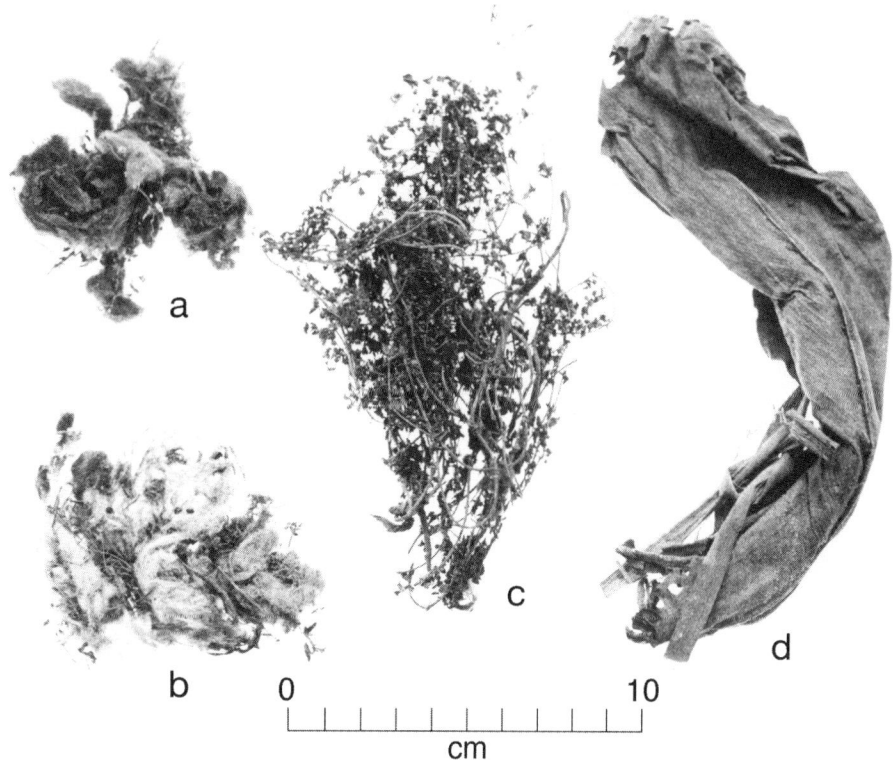

Our assumption is that this amphora contained coca tea—a medicinal beverage—prepared for a terminally ill individual and eventually included in his or her burial. As the tea evaporated, the coca leaves eventually settled to the bottom of the vessel.

Because of the clumped and deteriorated condition of the leaves, we attempted no identification of the variety of coca. The most likely variety is *truxillense*, which will grow in hot, arid regions of the *chaupi yunga* if irrigated (Bohm et al. 1982). Hastorf (1987:295) confirms that *truxillense* "grows along dry coastal areas of northern Peru from 200 to 1200 m." We do not know precisely where the coca used in the Kingdom of Huarco was grown.

Ciruela del Fraile

Ciruela del fraile (*Bunchosia armeniaca*), sometimes called the "peanut butter fruit," is produced by a tree native to northwestern South America. It is found at elevations of 100 to 2600 m, and can grow to a height of 20 m. It may have been brought to Cerro Azul from regions farther up the Río Cañete. Figure 13.6 shows two seeds of ciruela del fraile from a refuse deposit on Terrace 11 of Quebrada 6.

Beans of the Genus Phaseolus

Beans were an important part of the Late Intermediate diet at Cerro Azul, and would have ranked high among local plant protein sources. Multiple genera, species, and varieties of beans were eaten at the site; included were common beans and lima beans (both *Phaseolus*) as well as *Canavalia*.

Based on recent DNA evidence, *Phaseolus vulgaris* (common beans) and *Phaseolus lunatus* (lima beans) belong to different subclades within the genus *Phaseolus* (Delgado-Salinas 1999:441). *Vulgaris* beans are characterized by wide bracteoles bearing three nerves, while *lunatus* is distinguished by seeds with fine lines radiating from the hilum along the surface of the testa. *P. vulgaris* occurs in the wild state in the Basin of Mexico, Puebla, and Jalisco in Mexico, and in Peru, Ecuador, Colombia, and Argentina in South America; *P. lunatus* occurs in the wild state in Jalisco and Veracruz (Mexico), and Colombia and Peru in South America (Delgado-Salinas 1999: Appendix A).

Domestic varieties of both common beans and lima beans were being grown in Peru's Callejón de Huaylas (Ancash) between 8500 and 5500 B.C. (Kaplan et al. 1973). The common beans were dark red and red-brown, while the lima beans appeared reddish in color.

Common beans went on to appear in 50 or more varieties in Peru, where they are known as *nuña* in Quechua (Tohme et al. 1995). The senior author of this chapter recognized nearly a dozen varieties of *Phaseolus* beans in the Late Intermediate deposits at Cerro Azul. He loaned samples of these to Lawrence Kaplan, a leading expert on the genus, for further study. Kaplan's analysis of these samples—which included both common and lima beans—appears as Chapter 15.

Figure 13.7 presents a sample of *Phaseolus* pods and pod valves from the floor of Structure D's South Corridor. Figure 13.8 shows specimens of *P. vulgaris* from a looted burial cist in Quebrada 5-south. Finally, Fig 13.9 illustrates a seed of *P. lunatus* from Feature 6, a midden in Structure D's Southwest Canchón.

Beans of the Genus Canavalia

Beans of the genus *Canavalia* were eaten in large numbers at Cerro Azul. This pantropical genus is represented by both wild and domestic species. The most common domesticate in Mexico is *C. ensiformis*, known commonly as the jack-bean (Sauer and Kaplan 1969). The most common domesticate in Peru is *C. plagiosperma*, which has no common name in English. It has been cultivated on the coast of Peru for at least 4000 years.

According to Sauer and Kaplan (1969:417), the wild species of *Canavalia* "are mostly large seashore and riverbank vines, bearing conspicuous pods full of attractive, dark brown seeds." As a rule, the wild races are "poisonous unless carefully processed." Cultivated *C. plagiosperma*, known archaeologically from the Valleys of Pisco and Nasca, is normally chestnut brown in color.

Figure 13.10 illustrates three seeds of *Canavalia* from Structure D of Cerro Azul. All could be described as chestnut brown.

Erythrina

Erythrina is a flowering deciduous tree, growing 10–20 m tall under the right conditions. It is native to the humid slopes and borders of streams and rivers in Peru, Bolivia, and northern Argentina (Etcheverry and Trucco Alemán 2005). This tree requires 1300 mm of seasonal rain, making it unlikely that it grew in or near the Kingdom of Huarco.

Erythrina trees are not hugely productive. They bear brown pods with 1–7 seeds per pod, more useful as medicine than as food. In those cases where *Erythrina* is cultivated, it is often as an ornamental tree.

Figure 13.11 shows an *Erythrina* pod from Structure 5, a looted burial cist on Terrace 9 of Quebrada 5a. Given *Erythrina*'s alleged healing properties, we suspect that the pod we found was medicine procured for a gravely ill individual. When the patient died, any leftover *Erythrina* was probably placed in his or her grave.

Peanuts

The peanut (*Arachis hypogaea*), known in Peru as *maní*, was regularly eaten by both commoners and elite at Cerro Azul. Figure

Edible, Ritual, and Medicinal Plants

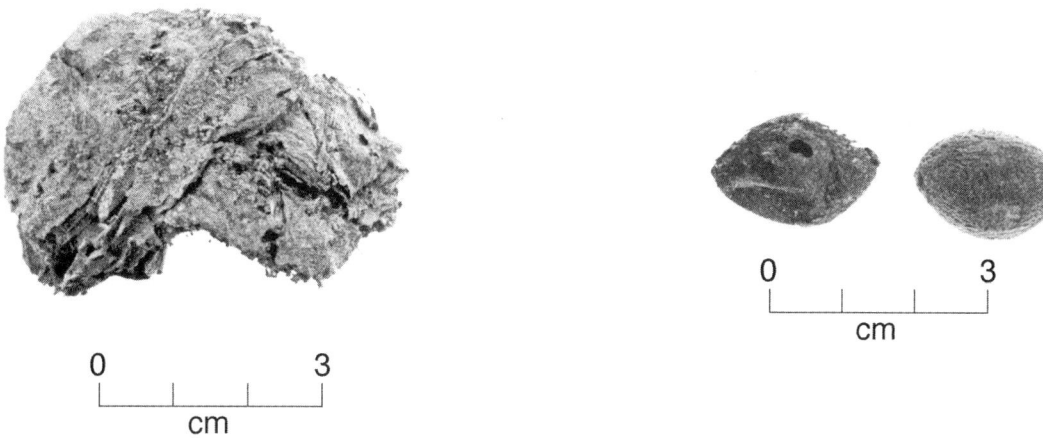

Figure 13.5. A mass of coca leaves (*Erythroxylum coca*) from the interior of a Camacho Reddish Brown amphora from Stratigraphic Zone A of Terrace 9, Quebrada 5a.

Figure 13.6. Two seeds of ciruela del fraile (*Bunchosia armeniaca*). Found in a refuse deposit in Quebrada 6 (Terrace 11, at a depth of 30–50 cm).

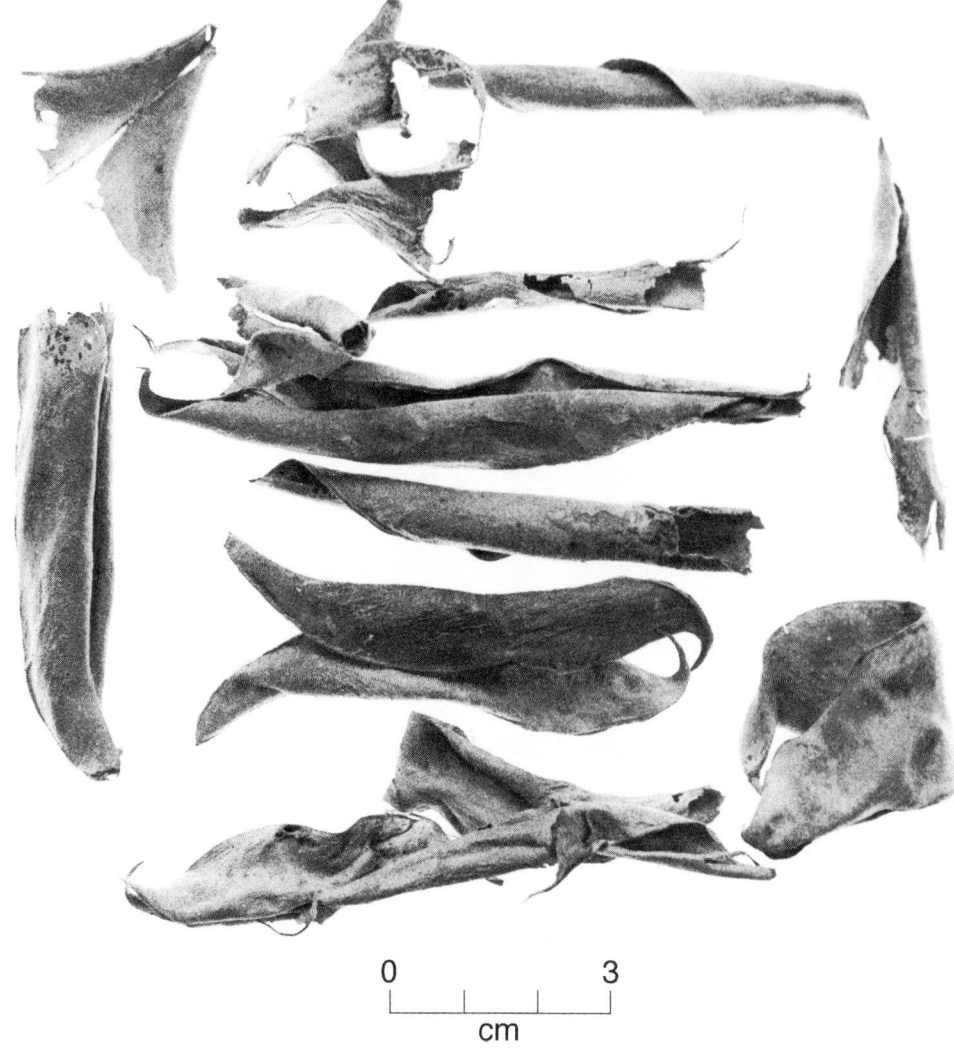

Figure 13.7. A sample of *Phaseolus* pods and valves from the floor of the South Corridor, Structure D.

Figure 13.8. Pod and seed of common bean (*Phaseolus vulgaris*) from Structure 12, a looted burial cist in Quebrada 5-south.

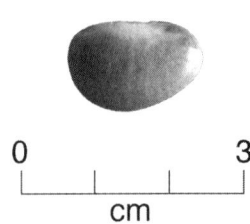

Figure 13.9. Seed of lima bean (*Phaseolus lunatus*) from Feature 6, Structure D.

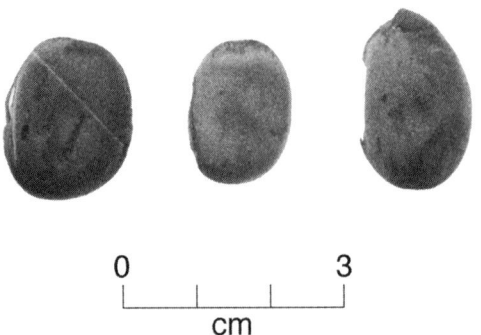

Figure 13.10. Three seeds of *Canavalia* beans, found in the fill of an earthen ramp on the west side of "Room 2," Structure D.

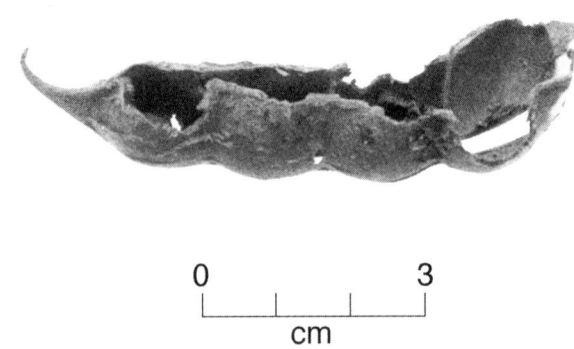

Figure 13.11. *Erythrina* pod from Structure 5, a looted burial cist in Quebrada 5a.

13.12 displays peanut hulls from middens in both Structure D and Structure 9.

Although the peanut was well established on the Peruvian coast by Early Horizon times, it was not native to Peru. Genetic analyses (Kochert et al. 1996) suggest that the domestic peanut—which is a tetraploid—was created through the hybridization of two diploid wild species, *Arachis duranensis* and *Arachis ipaensis*.

The center of diversity for the wild peanut lies in western Brazil, Bolivia, Paraguay, and northern Argentina. *A. duranensis* is common in northwest Argentina, while *A. ipaensis* is known only from southern Bolivia. The hybridization is believed to have taken place in one of those two regions, with *A. duranensis* as the female parent. Once the domestic tetraploid arose, however, the Ucayali region of eastern Peru became the region where the greatest diversity of cultivated varieties developed (Rimachi et al. 2012).

Pacay

The pacay or "ice cream bean" (*Inga feuillei*) comes from a leguminous tree that can reach 15–30 m in height. While not native to the arid coast of Peru, the tree can be grown there with irrigation (National Research Council 1989, Lim 2012:720).

Pacay trees produce flattened, four-sided pods some 30–50 cm long, filled with green seeds embedded in a sweet, white, cottony pulp. This pulp is prized as a snack today, and was evidently popular in the Late Intermediate period as well. We found empty pods in both elite and commoner-class debris.

The leaves of the pacay tree were also widely used at Cerro Azul; at least one burial cist, for example, had been lined with them. Figure 13.13 shows a pacay leaf from one of the quebrada middens. The ready availability of such leaves suggests that the trees themselves were probably grown not far away.

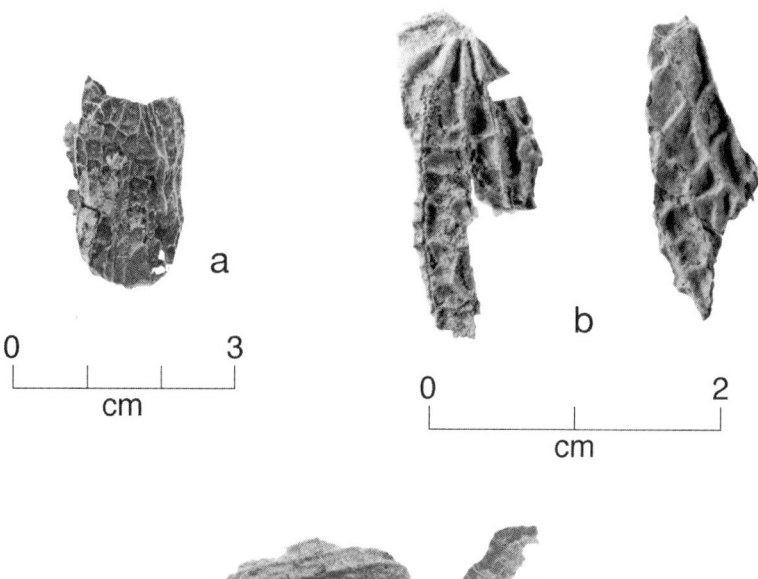
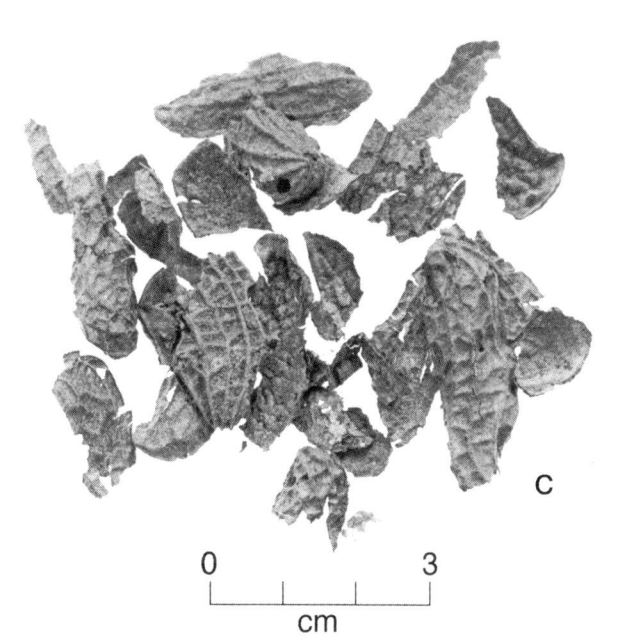

Figure 13.12. Both elite families and commoners at Cerro Azul enjoyed peanuts (*Arachis hypogaea*). *a*, a hull from the fill just above the floor of Room 1, Structure D. *b*, two hulls from Feature 6, Structure D. *c*, hulls from Feature 20, Structure 9.

Butternut Squash

Squashes of the genus *Cucurbita* are widespread in the tropical and warm-temperate New World; at least 12–14 species are known from the region between the United States and Argentina (Nee 1990). Archaeologists working in Peru commonly find the remains of *C. moschata* or *C. maxima*.

While not all the squash remains from Cerro Azul could be assigned to a species, those that could be identified appeared to be from the crookneck or butternut squash, *Cucurbita moschata* (Fig. 13.14). While this squash is cultivated mainly for its pulp, each fruit also contains 30–150 g of seeds (Ortiz Grisales et al. 2014). The eating of these seeds has known medicinal benefits; many Native American societies, for example, found that the seeds of the butternut squash helped purge tapeworms (Lim 2012:277). While we found intact *Cucurbit* seeds in the archaeological debris at Cerro Azul (Fig. 13.15), we did not find sufficient evidence for the roasting or grinding of seeds to prove that they were eaten.

Yuca de Monte

Some wild cucurbits are still eaten in Peru today. Clark et al. (2012) report that *Apodanthera biflora*, a wild cucurbit from the dry forests of northwestern Peru, is still collected for its edible root and seeds.

Apodanthera biflora, known as yuca de monte, is a perennial herbaceous vine that grows from a starchy, tuberous root. While its fruit is not edible, the latter contains 20 or so oily seeds that can be toasted. In Tumbes and Lambayeque the fruits mature in March, and the plant is harvested for its edible root in May and June.

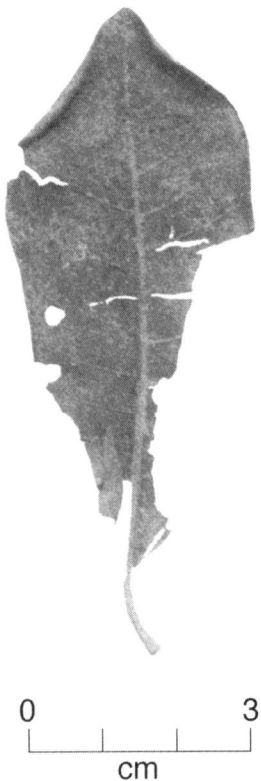

Figure 13.13 (left). A leaf from the pacay tree (*Inga feuillei*). Found in a refuse deposit in Terrace 11 of Quebrada 6 (depth 30–50 cm).

Figure 13.14 (below). Three stems of butternut squash (*Cucurbita moschata*). *a* was found in Structure 10, a storage cist on Terrace 12 of Quebrada 6. *b* was found in a refuse deposit on Terrace 11 of Quebrada 6. *c* was found in a midden between Structures 11 and 12 of Quebrada 5-south.

Figure 13.16 illustrates a desiccated fruit found in Collca 1, a storage cell in Structure D. This wild cucurbit appears to be *Apodanthera*, and while we lack confirmation, it could possibly be *A. biflora* or one of its relatives. We do not know, of course, from how far away it came.

Achiote

Achiote (*Bixa orellana*) is a tropical fruit whose seeds produce red pigment. It qualifies both as a food plant and a ritual plant, since its pigment can be used either as food coloring or body paint. At Cerro Azul one of its possible uses was as ritual face paint; many Late Intermediate jar necks and figurines depict individuals with zones of red facial coloring (Marcus 1987a: Fig. 21, Marcus 2008: Plate XVI).

Moreira et al. (2015) consider it likely that *Bixa orellana* was first domesticated in the Amazon River drainage, where many of its wild relatives live. Its most likely progenitor was *Bixa urucurana*, a wild shrub occupying riverside bluffs in Amazonia. Achiote's domestication took place at least 2400 years ago.

Among the changes following domestication were (1) an increase in pigment production, and (2) a transformation of the pod from indehiscent (non-rupturing) to dehiscent (rupturing at maturity to expose the red seeds). Figure 13.17 shows a dehiscent achiote pod from Structure D.

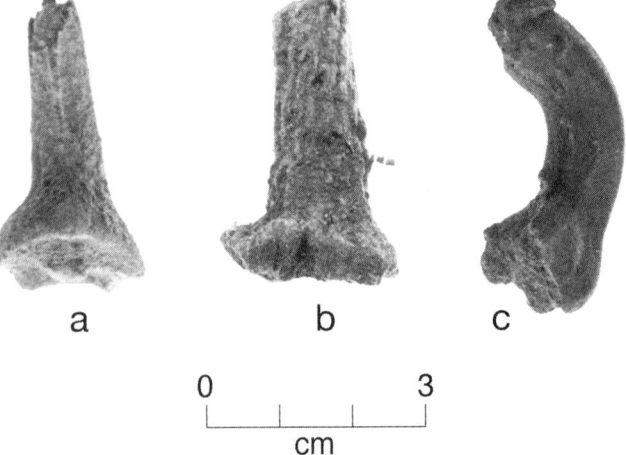

Guava

Another of the tropical fruits consumed at Cerro Azul was the guava (*Psidium guajava*). While its exact region of domestication is still unknown, guava is believed to have originated in southern Mexico or Central America. By the Late Intermediate period it had spread to South America.

Guava is yet another plant that can be considered both edible and medicinal (Lim 2012:684–727). Its fruit can have a strong odor, but is very nutritious; tea made from guava leaves has antioxidant, antimicrobial, and antidiarrheal properties. In addition, the bark of the guava tree is a source of tannin.

Lúcuma

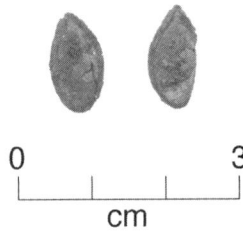

Figure 13.15. Two squash seeds from the Southwest Canchón, Structure D. These seeds were found in the earth above the North Platform of the canchón.

The lúcuma (*Pouteria lucuma*) is a widespread fruit tree of South America's tropical lowlands. Brazil alone has more than 120 species of *Pouteria* (Alves-Araújo et al. 2014). We do not know where *P. lucuma* was first domesticated, but it was being grown on the Peruvian coast 3500 years ago, presumably by irrigation (Quilter et al. 1991).

Both fruits and seeds of lúcuma were found in the room fill of Structure D (Fig. 13.18), indicating that the Late Intermediate elite at Cerro Azul had access to this tropical delicacy.

Sweet Potatoes

The camote or sweet potato (*Ipomoea batatas*) has a long history of cultivation in Peru. While no one is sure where this important root crop was first domesticated, Austin (1988) has suggested that its homeland may lie somewhere between Mexico's Yucatán Peninsula and the mouth of Venezuela's Orinoco River. In that region, he hypothesizes, *I. trifida* hybridized with *I. triloba* to produce the hexaploid *I. batatas*.

By 2500 B.C. the domestic sweet potato was widely dispersed in Central and South America. Peru and Ecuador have a large number of cultivars, differing in their DNA profile from the sweet potato races of Central America (Zhang et al. 2000).

Figure 13.19 shows a desiccated sweet potato from Structure D's Southwest Canchón. Perhaps because it was discarded raw, rather than having been softened by cooking, this tuber survived the ravages of time.

White Potatoes

The papa or common potato (*Solanum tuberosum*) was also eaten at Cerro Azul. Figure 13.20 illustrates a large white potato from looters' backdirt near Structure 5, a burial cist on Terrace 9 of Quebrada 5a. It is likely that this potato was food for the afterlife, left with one of the Late Intermediate burials that the looters disturbed.

Based on a lifetime of eating Andean potatoes, our colleague Ramiro Matos assigned this specimen to the cultivar *tomasa* (*Solanum tuberosum* subspecies *andigena*, variety *tomasa*). He also described this coastal cultivar as "insipid," compared to potatoes grown in the highlands of his native Huancavelica.

In their study of potatoes grown in the coastal desert of Ica, Schafleitner et al. (2007: Table 4) describe *tomasa* as a late-maturing variety, harvested in early September. Under irrigation it can produce 1,347±52 grams of potatoes, with individual tubers averaging 72 grams. Unfortunately, our Late Intermediate specimen was so desiccated that its weight could not be used for comparison.

Recent genotypic evidence (Spooner et al. 2005) suggests that the potato was domesticated only once, and that its likely wild ancestor was a member of the *Solanum brevicaule* complex, which is distributed from central Peru to northern Argentina. Spooner et al. lean toward Peru as the potato's place of origin.

Ugent et al. (1987) suggest that the higher tuber yield of the domestic potato is due to a doubling of the chromosome number (from diploid to tetraploid), a change that took place after domestication.

Small cultivated white potatoes were being grown on the desert coast of Ancash by Late Archaic times (2000 B.C.), and larger white potatoes have been found at the later coastal site of Pachacamac. We do not know how near to Cerro Azul the white potato was being grown in Late Intermediate times. In the 1980s,

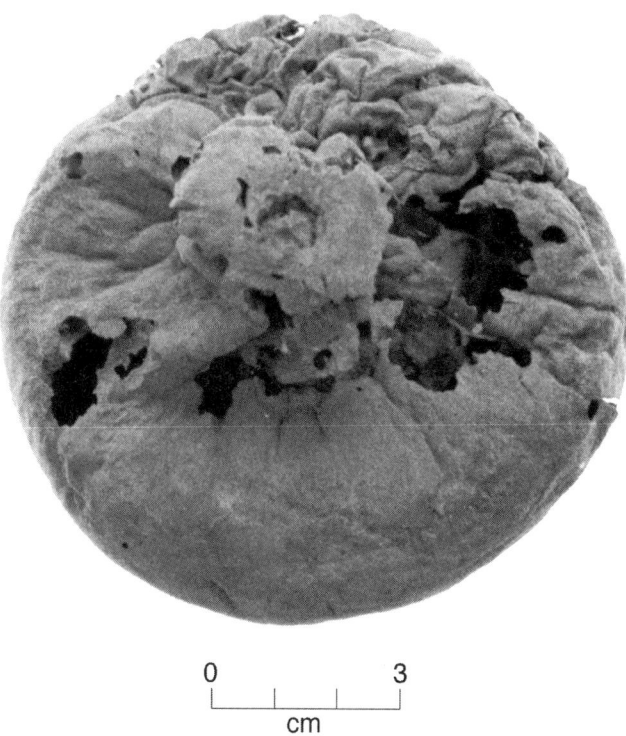

Figure 13.16. Desiccated fruit of yuca de monte (*Apodanthera* sp.) from Collca 1, a storage cell in Structure D.

Figure 13.17. Ruptured pod of achiote (*Bixa orellana*) from the fill of Room 4, Structure D (1.5 m below Feature 5).

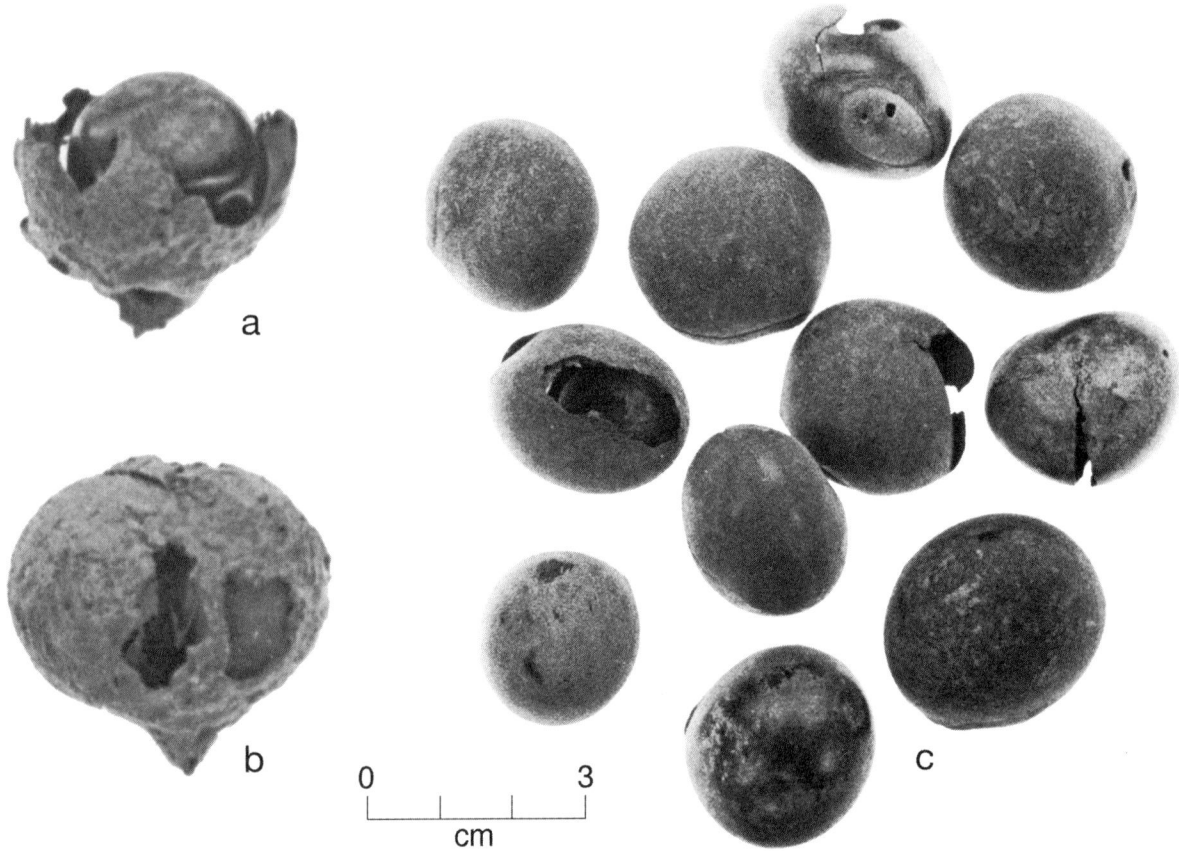

Figure 13.18. Lúcuma (*Pouteria lucuma*) from the fill of Room 4, Structure D. *a, b*, unopened fruits with seeds still inside. *c*, eleven seeds found 1.5 m below Feature 5.

the Cañete market featured both irrigated coastal varieties and rainfall-grown highland varieties.

Chile Peppers

Two species of *uchu* or chile pepper, both native to South America, were recovered at Cerro Azul. All the Late Intermediate chiles that Smith identified to species were *Capsicum baccatum*, believed by Pickersgill (1969) to be the first pepper domesticated in the Andes. Recent DNA evidence suggests that most cultivars of *C. baccatum*, known in Quechua as *kellu uchu*, are closely related to wild *C. baccatum* from Peru and Bolivia (Albrecht et al. 2012:528). This species was being grown on the Peruvian coast during the Archaic (Chiou et al. 2014).

A second chile pepper, *Capsicum chinense*, was found only in Spanish Colonial refuse at Cerro Azul. The wild ancestor of this cultivar, known in Quechua as *chinchi uchu*, is believed to be native to the Amazon Basin (Pickersgill 1969, Chiou et al. 2014). Domestic *C. chinense* had reached the Peruvian coast by the Initial Period. Today its varieties include the habanero pepper.

Smith relied on the morphology of the calyx (and the presence or absence of calyx teeth) to identify the Cerro Azul chiles. He also compared the seeds to those of known chile specimens from local markets. When specimens were damaged or their landmarks unclear, he recorded them simply as *Capsicum* sp. Owing to Smith's untimely death in 1987, he was unable to take advantage of Chiou and Hastorf's recent suggestion that 27 attributes of the seeds can be used to identify chiles to species (Chiou and Hastorf 2014).

Figure 13.21 shows some of the stems and seeds of chile peppers from Feature 6, a midden in the Southwest Canchón of Structure D. *Capsicum baccatum* was the only pepper identified from this building.

Wasp Galls

Wasps of the Family Cynipidae oviposit in the stems of certain woody plants. The plant reacts by producing a

Edible, Ritual, and Medicinal Plants

Figure 13.19 (left). A sweet potato (*Ipomoea batatas*) from the northeast corner of the Southwest Canchón, Structure D. Desiccated and riddled with insect holes, this root crop lay near the tapia wall added to create Room 4 (see Marcus 2008: Fig. 4.35).

Figure 13.20 (below). A white potato (*Solanum tuberosum*) from Late Intermediate debris near Structure 5, a looted burial cist in Quebrada 5a.

hypertrophy or enlargement known as a gall. Not only do these galls protect the eggs, but the wasp larvae also feed on the gall as they develop. Many wasp galls resemble dough balls, and can be eaten by humans as well (Smith 2009:267).

In Room 4 of Structure D we recovered four stems from an unidentified plant, each bearing a wasp gall (Fig. 13.22). The fact that four of these galls had been collected by the residents of Structure D suggests that they were considered edible.

Unidentified Plants

A handful of plants from Cerro Azul—mostly members of the legume family—could not be identified to genus, either because Smith lacked suitable comparative material or because the parts we found were undiagnostic. Examples are given in Figures 13.23 and 13.24.

Plant Remains from Archaeological Contexts

The University of Michigan project recovered in excess of 4000 plant specimens from Cerro Azul. In this chapter we deal only with the roughly 3600 items considered to be edible, ritual, or medicinal in nature. "Industrial" plants—those used as raw material for construction, basketry, weaving, or other tasks—will be discussed in Chapter 16.

Structure D

As was the case with so many archaeological materials, our largest single sample of plant remains came from **Feature 6**, a midden in the Southwest Canchón of Structure D. Table 13.2 gives the complete inventory of edible, ritual, and medicinal plants from that midden, which was screened through both 1.5 mm and 0.6 mm mesh.

Figure 13.21. Stems and seeds of chile pepper (*Capsicum* sp.) from Feature 6 of Structure D.

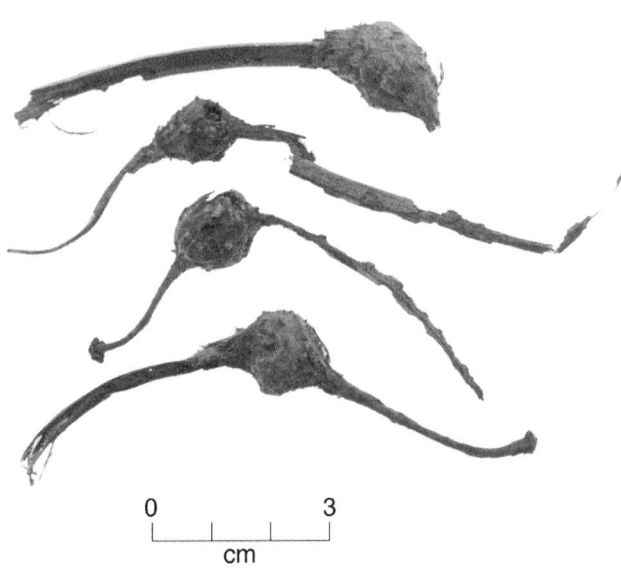

Figure 13.22. Four stems of an unidentified plant, each bearing a wasp gall. Found in the fill of Room 4, Structure D, 1.5 m below Feature 5.

Table 13.2. Edible, Ritual, and Medicinal Plants from Feature 6 of Structure D.

Cherimoya	1 seed
Maize	1328 cobs of 6 varieties (see Chapter 14)
	96 kernels (32 of them imbricated)
	7 stalk fragments
	6 fragments of husks or leaves
Ciruela del fraile	183 seeds
Phaseolus beans	277 pod valves
	42 seeds
	7 cotyledons
Canavalia beans	3 pod valves
	29 seeds
	2 fragments of seed coat
	1 cotyledon
Erythrina	7 seeds
Peanuts	9 pod fragments
Pacay	7 pod fragments
Butternut squash	2 stems (Fig. 13.25)
	3 seeds (Fig. 13.26)
Squash, possibly butternut	3 stem scars
Cucurbit, genus unspecified	1 stem
	1 seed
Guava	2 fruits
Lúcuma	1 whole seed
	28 seed fragments
Chile peppers, cf. *C. baccatum*	12 stems
	1 dried fruit (Fig. 13.27)
	16 seeds

Totals:
 NISP = 2,075

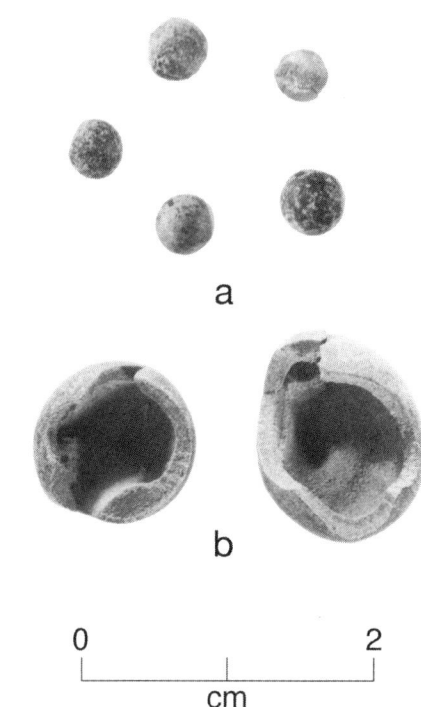

Figure 13.23. Unidentified plants from Feature 6, Structure D. *a*, five seeds. *b*, two nut hulls.

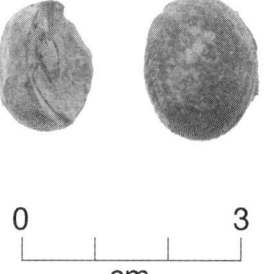

Figure 13.24. Unidentified nuts from the fill of Room 4, Structure D, at a depth of 1.5 m below Feature 5.

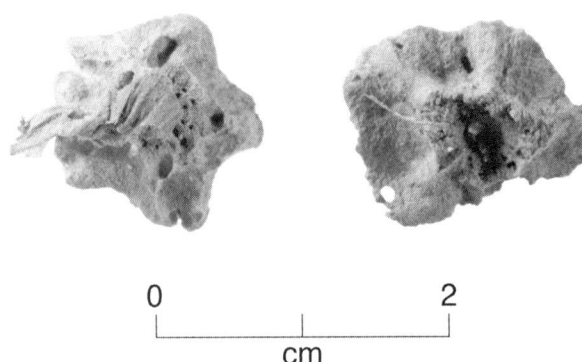

Figure 13.25. Squash stems from Feature 6, Structure D. Shown at twice natural size.

Figure 13.26. Two incomplete squash seeds from Feature 6, Structure D. Shown at twice natural size.

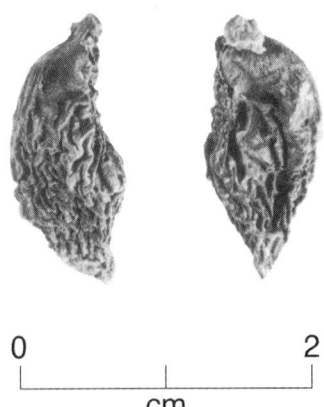

Figure 13.27. Two views of a dried chile pepper from Feature 6, Structure D. Although not every chile pepper could be identified to species, many appeared to be *Capsicum baccatum*.

Most refuse from the residential compounds at Cerro Azul is believed to have been collected in baskets and carried to the quebradas of Cerro Camacho. The fact that this did not happen to the Feature 6 midden suggests that it accumulated late in the occupation of Structure D, and was left behind when the building was abandoned. The contents of Feature 6, therefore, may tell us something about plant use during A.D. 1469–1470, the year leading up to the Inca conquest of Cerro Azul.

We are struck by the fact that two-thirds of the identified specimens from Feature 6 were maize cobs. In our opinion this reflects the making of chicha, or maize beer, in large quantities. There is evidence that the North Central Canchón of Structure D was used as a brewery, one whose storage capacity may have approached 5000 liters of maize beer (Marcus 2008:187).

The maize used in Structure D seems to have come from multiple sources. In Chapter 14, Smith describes at least half a dozen varieties of maize from Feature 6, based on his detailed study of the cobs. It is unlikely that all these varieties of corn were grown in the same locality, since this would have led to hybridization. It is more likely that Cerro Azul received shipments of maize from several different inland agricultural communities, perhaps even from as far away as Lunahuaná.

It appears that beans of various species were also an important part of the food supply in Structure D. Specimens of *Phaseolus* (both common beans and lima beans) amounted to 16 percent of the plants from Feature 6. If one adds all the *Canavalia* specimens to the *Phaseolus* count it brings the total to 18 percent, making beans second in frequency only to maize.

It is worth noting that no leaves, vines, or roots of beans were found in Feature 6, and that only one percent of our maize specimens were stalks, leaves, or husks. This fact strongly suggests that Cerro Azul was obtaining its two most important food plants from elsewhere, rather than growing them itself. Such evidence reinforces the ethnohistoric model of specialized fishermen who got their crop plants from inland agricultural communities (Chapter 21).

The elite occupants of Structure D had access to a number of tropical fruits, not all of which appeared in commoner-class refuse at Cerro Azul. The most abundant of these was ciruela del fraile, followed by lúcuma. Guava and cherimoya were uncommon even in Structure D.

The Southwest Canchón

The Feature 6 midden is believed to represent debris swept up from the floor of Structure D's Southwest Canchón. Elsewhere on the floor of this same canchón we recovered one seed of lúcuma and two seeds of *Canavalia*.

In the northern part of the Southwest Canchón we uncovered a series of platforms, believed to be used by the commoner-class overseer of the canchón. In the earthen fill above these platforms we found the rachis from a ten-row ear of maize; two seeds and one stem of a cucurbit that Smith believed to be butternut squash; and one plant stem with a gall created by a cynipid wasp.

Finally, in the extreme northeast corner of the canchón, we

found the sweet potato shown in Figure 13.19. This tuber had been left on a patch of whitish floor, added at a very late stage in the use of Structure D (Marcus 2008: Fig. 4.32).

The South Corridor

The South Corridor was a single-file passageway that led from the Southwest Canchón to the interior of Structure D. At some point, presumably late in the building's history, someone left a pile of debris on the floor of the corridor. The bulk of the debris was maize cobs stripped of their kernels, presumably so that the latter could be used to make chicha. The total list of food plants from the floor of the corridor was as follows:

Maize	234 cobs of 7 varieties (see Chapter 14)
	3 loose kernels
Phaseolus beans	19 empty pods
Canavalia	3 seeds
Lúcuma	2 seeds

The Northeast Canchón

The Northeast Canchón of Structure D was a large, open-air work space where (among other things) spinning and weaving were carried out. In the northwest corner of the canchón Marcus' crew found a number of ash-filled hollows in the floor. The food plants left in these hollows consisted of 6 maize cobs, 2 seeds of butternut squash, and 5 seeds of lúcuma.

Collca 1

Collca 1 was a storage cell attached to the south wall of the Northeast Canchón. This collca appeared to have been created for the storage of weaving equipment, but at some point it had also become a convenient place to dump the debris from general housecleaning, and that debris included food plants. The list is as follows:

Maize	70 cobs of 11 varieties (see Chapter 14)
	1 husk
Ciruela del fraile	1 seed coat
Phaseolus beans, cf. *vulgaris*	16 empty pods (Fig. 13.28, *a–b*)
Canavalia	11 pods (1 still containing seeds) (Fig. 13.28, *c–d*)
	6 loose seeds
Pacay	1 fragment of tissue

Yuca de monte	1 desiccated fruit (Fig. 13.16)
Lúcuma	2 dried immature fruits
	13 seeds (Fig. 13.29)

The North Central Canchón

The North Central Canchón of Structure D was a kitchen/brewery, with facilities for making and storing chicha. One Camacho Reddish Brown jar from this brewery yielded traces of *jora* (sprouted maize kernels, used as the basis for chicha). Other than that, the only food plants recovered were two *Canavalia* seeds from the floor of the canchón.

Room 1

Room 1 was originally designed as an elite residential unit, but was converted to storage after suffering earthquake damage. Two empty peanut hulls were discarded in the room late in its history.

"Room 2"

"Room 2" turned out, after excavation, to be a small interior patio. On its west side it featured an earthen ramp that led to Rooms 5, 6, and 7. The fill of this ramp yielded 3 seeds of *Canavalia* and 2 seeds of lúcuma.

Room 4

Room 4 had a complex stratigraphic history. Its original floor lay 3.73 m below the datum point established for Structure D. At a later date the room was partially filled in, and a higher floor was created at a point 2.23 m below datum. On this floor the University of Michigan crew discovered Feature 5, a mass of partially restorable vessels crushed by the weight of the overburden. Additional fill was later added, eventually reaching the tops of the walls. Food plant remains were found at two points in the room fill.

On the original floor of Room 4 (3.73 m below datum) we found the following:

A monocotyledonous plant, possibly achira	1 fragment of leaf
Maize	1 ten-row cob
Peanuts	1 empty pod
Achiote	1 pod
	1 peduncle
Lúcuma	11 seeds
Stems with wasp galls	4 (Fig. 13.22)

In the earthen fill above the later floor (< 2.23 m below datum) we found one *Canavalia* seed and 2 desiccated lúcuma fruits.

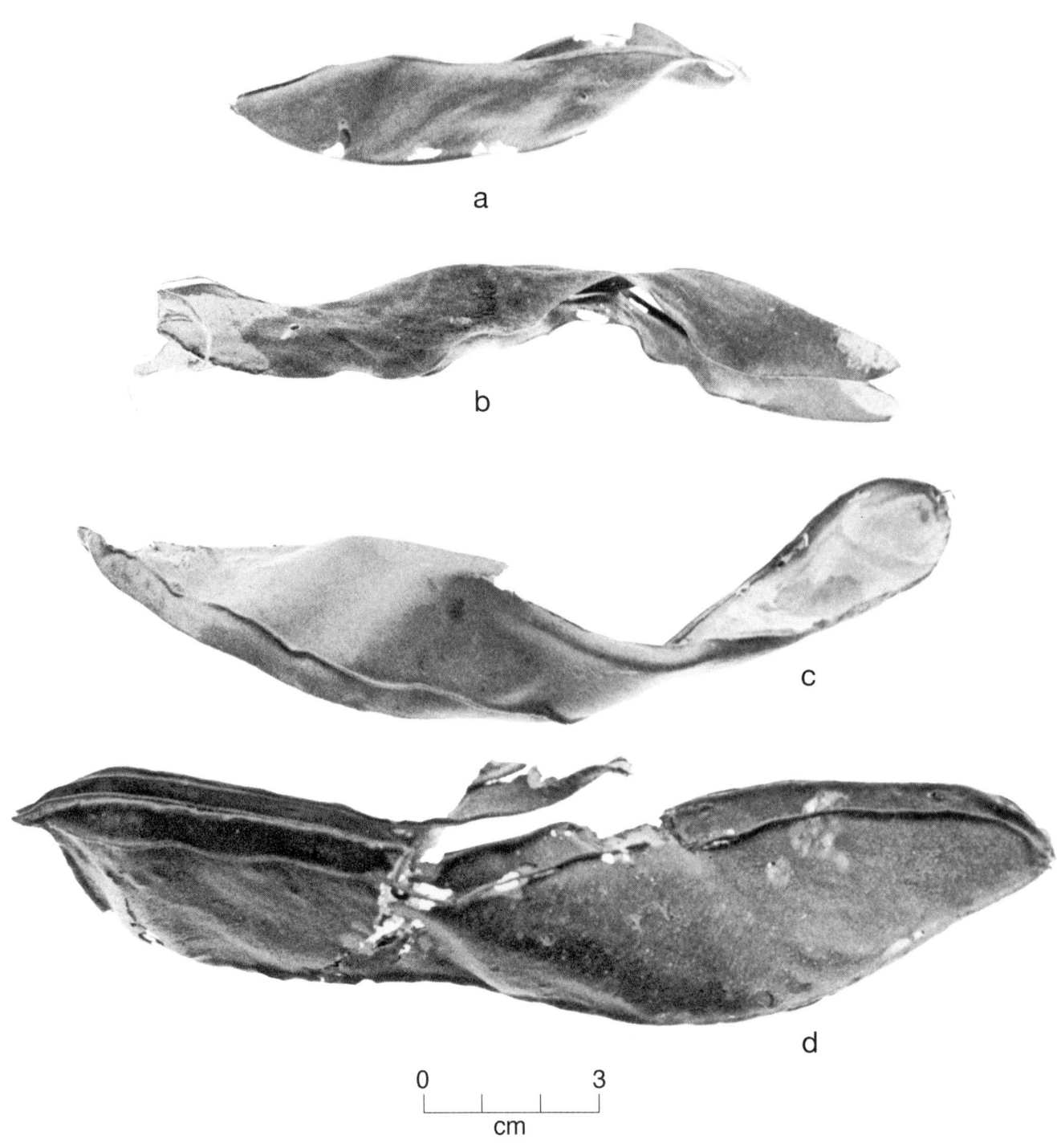

Figure 13.28. A sample of bean pods from Collca 1, Structure D. *a, b, Phaseolus* sp. *c, d, Canavalia* sp.

Room 5

Room 5 also had a complex stratigraphic history. Originally designed as a storage room, it was converted to a residential unit after other rooms in Structure D had suffered earthquake damage. Plant remains were found only in the uppermost fill, associated with the room's period of residential use. Included were 2 ten-row maize cobs, one pod and one seed of *Phaseolus* beans, and a small number of unidentified plant fragments.

Room 7

Room 7 was a small unit whose access doorway was deliberately blocked after the room suffered earthquake damage. After its doorway was blocked, Room 7 became a convenient place to dump refuse. Food plants included in that refuse were as follows:

Maize	4 cobs (2 eight-row, 2 ten-row)
Phaseolus sp.	1 empty pod
Canavalia sp.	11 seeds (Fig. 13.30)
Gourd or squash	1 thick fragment of fruit wall
Lúcuma	3 seeds

Late Horizon Squatters' Debris

At some point after Structure D had ceased to function as an elite residential compound, commoner-class squatters built a wattle-and-daub house in the "Room 2" patio. Pottery from this house suggested that it dated to the Late Horizon.

Feature 4 was a Late Horizon midden associated with the squatters' house. The list of food plants recovered from this midden is as follows:

Maize	83 cobs from several varieties (ten-row; twelve-row; fourteen-row)
Phaseolus sp.	39 empty pods or valves (Figs. 13.31, 13.32)
Canavalia sp.	6 empty pods 8 seeds (Fig. 13.33)
Pacay	5 pod fragments
Lúcuma	8 seeds (Fig. 13.34)

Although admittedly a small sample, the vegetal remains from Feature 4 do not reflect dramatic differences in plant use when compared with samples from Late Intermediate times. In fact, the large number of maize cobs raises the question of whether the squatters were continuing to make chicha in Structure D, using the old Late Intermediate brewing facilities.

Feature 3

Feature 3 was an offering left some time after Structure D had been abandoned; it had been buried in the uppermost layer of postoccupation dust and sand near the center of the building. Included in the offering were four exhausted maize cobs, stripped of their kernels. Three were twelve-row cobs, while one was an

Figure 13.29. Five of the 13 lúcuma seeds from Collca 1, Structure D.

Figure 13.30. Eleven seeds of *Canavalia* sp. from the fill of Room 7, Structure D.

Figure 13.31. A sample of *Phaseolus* pods from Feature 4, the refuse from a Late Horizon squatters' house in "Room 2" of Structure D.

Figure 13.32. Four pods of common bean (*Phaseolus vulgaris*) from Feature 4, the refuse from a Late Horizon squatters' house in "Room 2" of Structure D.

Figure 13.33. Eight seeds of *Canavalia* sp. from Feature 4, the refuse from a Late Horizon squatters' house in "Room 2" of Structure D.

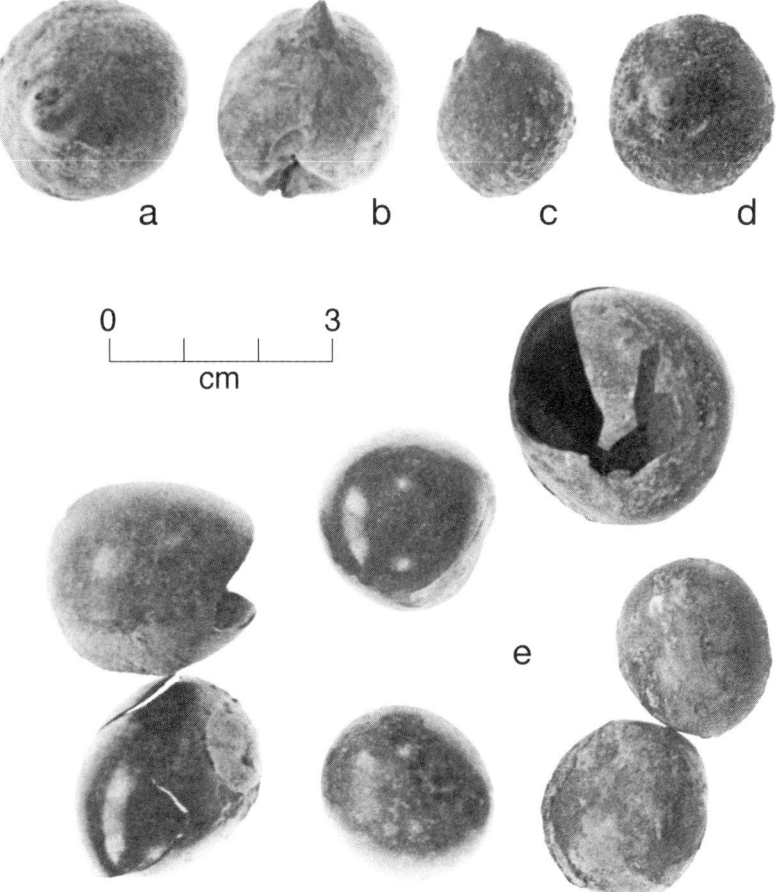

Figure 13.34. Lúcuma from Feature 4, the refuse from a Late Horizon squatters' house in "Room 2" of Structure D.

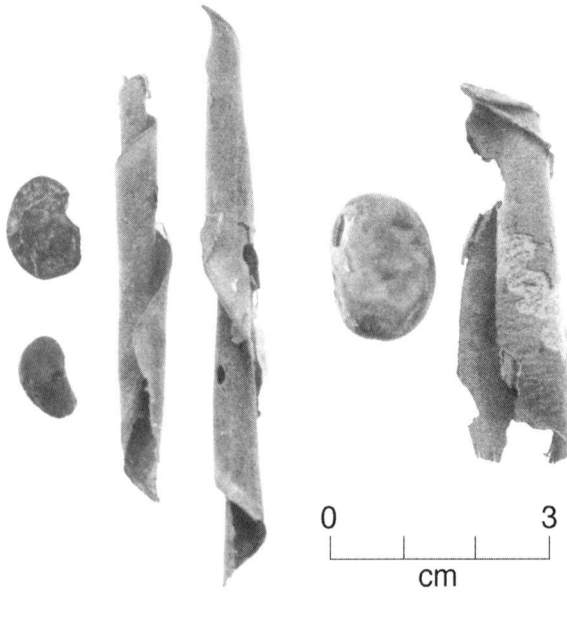

eight-row cob. The offering also included 290 camelid bones and a globular jar.

Structure 9

Our most useful sample of food plants from Structure 9 came from **Feature 20**, a midden banked up against the east wall of the building. This midden is believed to represent refuse from the household of the building's commoner-class overseer, who occupied a kincha house in the center of Structure 9. The list of food plants from Feature 20 is given in Table 13.3.

Table 13.3. Edible, Ritual, and Medicinal Plants from Feature 20 of Structure 9.

Cherimoya	1 seed
Maize	18 cobs (see Chapter 14)
Common beans (*P. vulgaris*)	5 pods and valves (Fig. 13.35, left)
	3 seeds
Canavalia sp.	1 pod (Fig. 13.35, right)
	1 seed
Peanuts	26 hulls
Unidentified legume	2 pod fragments
Butternut squash	15 seeds (Fig. 13.36*b*)
Lúcuma	5 seeds (1 immature) (Fig. 13.37)
	1 dried fruit
Unspecified chile pepper	2 seeds

Totals:
 NISP= 80

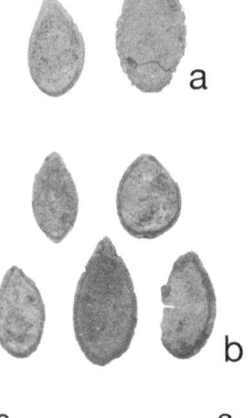

Figure 13.35 (above). Beans from Feature 20 of Structure 9, the midden debris left by a commoner household. On the left we see two seeds and two pod valves of common beans (*Phaseolus vulgaris*). On the right we see one seed and one pod valve of *Canavalia* sp.

Figure 13.36 (left). Seeds of what appears to be butternut squash. *a*, two seeds from an ash-filled hollow in the floor of the Northeast Canchón, Structure D. *b*, five seeds from Feature 20 of Structure 9.

One of the most striking differences between Feature 20 (a commoner-class midden) and Feature 6 (an elite residential midden) is the much smaller number of maize cobs in Feature 20. This presumably reflects the fact that Structure 9 had no brewery; since the commoner-class overseer occupying this small fish-storage structure had no clients on whom to lavish beer, his need for maize was presumably less.

Room 4

Room 4 of Structure 9 was a small storage unit in the extreme northeast corner of the building. At some point it suffered

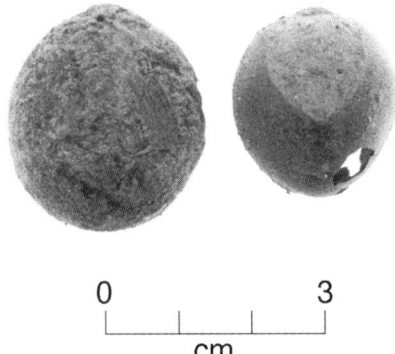

Figure 13.37. Dried fruit and seed of lúcuma from Feature 20, Structure 9.

earthquake damage and its floor buckled badly, making it less useful for storage. After that, it was used as a place to dump trash.

Among the items tossed into Room 4 late in its history were 74 maize cobs with their kernels removed. The source of these cobs is unknown. Room 4 had no roof, and given its location on the very edge of the building, it may have been seen by residents of nearby compounds as a convenient place to toss refuse, thereby avoiding a trip to the quebradas (see Chapter 14).

Room 5

Room 5 was a small fish storage unit on the building's west side. After its main use as a storage unit was over, Room 5 became a convenient place to dump trash. Included in that trash was one seed of *Canavalia*.

Surface Find

While sweeping the surface of Structure 9 prior to excavation, the University of Michigan crew found one possible seed of ciruela del fraile. This surface find was made near the east side of the building.

Food Plants from the Quebradas of Cerro Camacho

Many terraces in the quebradas of Cerro Camacho yielded food plants, discarded as refuse. In other cases, Marcus' crew found plants that had been left in burial cists as food for the afterlife.

Quebrada 5, Terrace 16

Terrace 16 of Quebrada 5a proved to have deep domestic midden deposits. Marcus excavated to a depth of 2.2 m and distinguished 9 "cultural" or "natural" strata. Preservation of plants was generally poor on this salt-encrusted terrace, but Stratigraphic Zone E (third stratum from the bottom) yielded one maize cob and one stem of butternut squash.

Quebrada 5-south

Quebrada 5-south rewarded the University of Michigan crew by producing a multiroom storage facility (**Structure 11**), a looted burial cist (**Structure 12**), and a domestic midden that filled the space between the two structures.

Structure 11. Structure 11 was a tapia building set against the stony slope of the quebrada. Owing to time limitations, Marcus had time to excavate only three of this building's units: two small rooms (Rooms 1 and 2) and a larger unit, Room 3. Marcus' workmen indicated that the quebradas of Cerro Camacho would have been good places to store vegetal foods, because they were farther from the ocean and its destructive salt fog than the residential compounds at Cerro Azul.

Room 1 of Structure 11 produced 14 seeds of *Canavalia* and one seed of lúcuma. Room 2 produced a few *Phaseolus* pods with the beans still inside, one pod and 51 seeds of *Canavalia* (Fig. 13.38), and one stem of butternut squash (Fig. 13.39). Of particular interest were the varied colors of the *Canavalia* seeds. Thirty-eight were small and particolored yellow and tan; 7 were large and particolored tan and reddish brown; 4 were large and particolored yellow and tan; and 2 were small and particolored tan and reddish brown. This variety of seeds raises the possibility that the *Canavalia* stored in Structure 11 had been grown in different locales.

Finally, the ruins of a kincha house in Room 3 produced one maize cob, one stem of bottle gourd, 2 pods from *Phaseolus* beans, and 2 pods and 2 seeds from *Canavalia* beans (Fig. 13.40). Overall, the evidence suggested that Quebrada 5-south was considered a good place to store beans.

Structure 12. Structure 12 was a typical Late Intermediate stone masonry burial cist. It had been looted, but the looters had left behind two gourd bowls containing food for the afterlife.

Gourd 1 had once contained 4 ears of maize, still in their husks. Unfortunately, the local rice rats (*Oryzomys xanthaeolus*) had made their way into the cist and eaten all the kernels; two rats later died there. This same gourd bowl contained 2 seeds of lúcuma.

Gourd 2 of Structure 12 contained 52 *Phaseolus* cotyledons. Possibly because the cist had initially retained moisture, many of the beans had sprouted before becoming desiccated.

Structure 12 was also filled with looters' debris, from which Marcus' crew recovered other plants that may have been left as food for the afterlife. Included were the following:

Maize	1 immature ear, still in its husk (see Chapter 14)
	155 loose kernels
Phaseolus vulgaris	1 pod
	1 loose seed
Canavalia	4 seeds
Butternut squash	1 seed
Lúcuma	12 seeds (Fig. 13.41)
Mummified rice rats that died in the cist	2

Finally, one of the most remarkable discoveries made in Structure 12 was the "medicine bundle" shown in Figure 13.4. Wrapped carefully in a napkin-sized piece of cloth were a lock of *algodón pardo* (brown cotton), a lock of *algodón blanco* (white cotton), a cluster of *Euphorbia* plants, and an achira leaf tied up in a twisted maize leaf. Also present were two splints of caña hueca (*Phragmites australis*). Although caña hueca and cotton are usually treated as "industrial" plants by Andean archaeologists (see Chapter 16), in this context we suspect that medicinal properties were attributed to them.

Figure 13.38. Fifty-one *Canavalia* beans from Room 2 of Structure 11, a storage facility in Quebrada 5-south (see Marcus 2008: Figure 9.16).

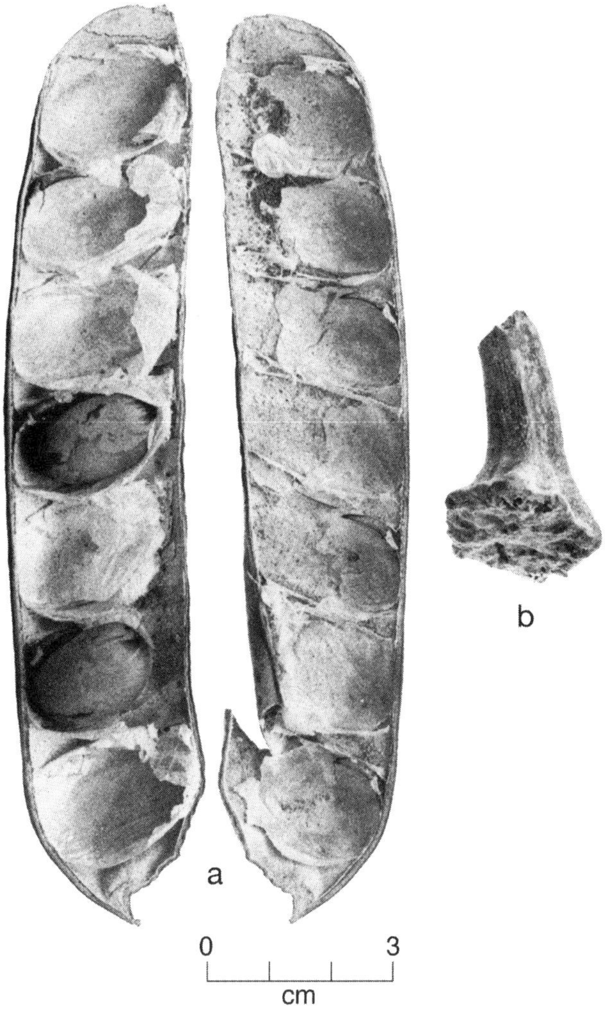

Edible, Ritual, and Medicinal Plants

Figure 13.39. Food plants from Room 2 of Structure 11, a storage facility in Quebrada 5-south. *a*, a pod of *Canavalia* with two seeds still inside. *b*, the stem from a butternut squash (*Cucurbita moschata*).

Figure 13.40. Plants associated with a kincha house built in Room 3 of Structure 11, a storage facility in Quebrada 5-south. *a*, corncob. *b, c*, seeds of *Canavalia* sp. *d*, pod of common bean (*Phaseolus vulgaris*). *e*, stem of bottle gourd (*Lagenaria siceraria*).

Figure 13.41. Three lúcuma seeds from Structure 12, a burial cist in Quebrada 5-south.

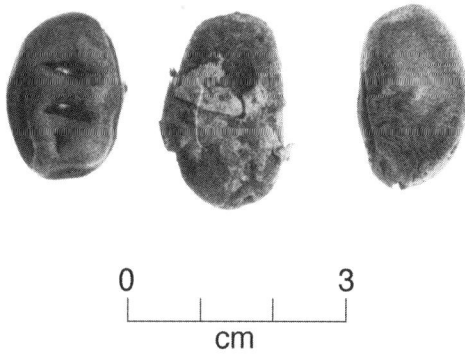

Figure 13.42. *Canavalia* seeds from a refuse deposit in Quebrada 6 (Terrace 11, at a depth of 30–50 cm).

Our guess is that a seriously ill individual was being treated with this bundle of medicinal plants, and that after the patient died, the bundle was simply included with his or her burial for whatever benefits might continue in the afterlife.

The Midden. The small midden between Structures 11 and 12 produced only one stem and one seed of butternut squash.

Quebrada 5a, Terrace 9

Terrace 9 of Quebrada 5a had a complex stratigraphic history. The first use of the terrace dated to a period before the occupation of Structures D and 9, and resulted in a midden designated Zone C.

Stratigraphic Zone B consisted of a Late Intermediate midden in which the shells of small coquina clams were the dominant mollusc remains. During the accumulation of this midden, the ceramics began to resemble those of Structure D.

Eventually, the accumulated midden deposits on Terrace 9 approached a depth of 2.6 m. Such deposits made the terrace a good place to create stone masonry burial cists. The University of Michigan crew discovered four such cists, designated Structures 4–7. These cists—which had been wholly or partially looted—were assigned to Stratigraphic Zone A, although several intruded into Zone B. The ceramics from Zone A equated it chronologically with a late stage in the occupation of Structures D and 9.

The Zone B midden produced the following food plants:

Maize	4 cobs (3 ten-row, 1 eight-row)
	1 kernel
Phaseolus beans	1 empty pod
Pacay	1 pod fragment
	10 leaflets

Structure 5. Structure 5 was a partially looted burial cist. A handful of plants, found in the looters' debris, had probably been left in the cist as food for the afterlife. Included were:

Maize	6 cobs, mostly ten-row (with all the kernels eaten by rice rats)
	2 husks
Erythrina	1 pod
Pacay	58 leaflets (some of which were in the mouths of a few Late Intermediate mummies that had been left behind by the looters)
Lúcuma	2 seed coats

In Looters' Debris Adjacent to Structure 5. One complete white potato (Fig. 13.20), lying among broken burial offerings.

Structure 6. Structure 6 was another looted burial cist, not far from Structure 5. This cist had been lined with pacay leaves (see below). Evidently two of the burials had also been accompanied by "pillows" made from soft willow branches, which are described in Chapter 16. Edible plants found in the looters' debris from Structure 6 were as follows:

Maize	1 ear of the corn variety Confite Puntiagudo, with 12 rows of intact kernels
	6 cobs with their kernels eaten by rice rats
Pacay	>150 of the leaves used to line the cist had been preserved (Fig. 13.43)
	1 pod
	16 stray leaf rachises

In Looters' Debris, Zone A. In looters' debris just to the northeast of the Structure 4 burial cist, Marcus' crew recovered an ear of ten-row Confite Puntiagudo maize with all its kernels present (see Chapter 14). The kernels on this ear were dark reddish brown, suitable for making *chicha morada* in the afterlife.

An Amphora from Zone A. One of the burial offerings left behind by the looters was the Camacho Reddish Brown amphora shown in Figure 3.8 of Marcus (2008). In the bottom of this amphora was the matted cluster of coca leaves shown in Figure 13.5 of this chapter. As mentioned earlier, we suspect that the amphora had once contained coca tea that was prepared for a seriously ill individual and included in his or her burial after death.

Burial 4. Burial 4 included multiple mummy bundles and had been missed by the looters (Marcus 2015: Fig. 11). The following plants were found near Burial 4 (but could not be associated with a specific mummy):

Maize	1 husk
Pacay	4 leaflets

Burial 8. Burial 8—also missed by the looters—was accompanied by one sixteen-row maize cob and two species of cotton (see Chapter 16). Rice rats had eaten all the kernels off the maize cob.

Burial 9. Burial 9 was accompanied by the following food plants:

Maize	1 cob (with the kernels eaten off by rice rats)
	1 tassel (see Chapter 14)
Phaseolus beans	1 empty pod (seeds likely eaten by rice rats)
Pacay	2 pod fragments
Lúcuma	1 seed

Figure 13.43. Leaves of the pacay tree (*Inga feuillei*), used to line the interior of a burial cist (Structure 6, Quebrada 5a).

Quebrada 6, Terrace 11

Terrace 11 of Quebrada 6 was selected for excavation because the coal-black color of its midden deposits piqued Marcus' curiosity. The black color appeared to have resulted from the action of salt fog on richly organic dark brown midden, replete with fish oil. The 162 specimens of food plants from this midden are listed in Table 13.4.

Table 13.4. Food Plants Recovered from Terrace 11 of Quebrada 6.

Maize	73 cobs or fragments thereof (see Chapter 14)
	4 loose kernels
	5 tassel fragments
	5 stalk fragments
	3 stem bases with roots
	2 leaf fragments
	3 husks
Ciruela del fraile	6 seeds
Canavalia beans	5 seeds (Fig. 13.42)
	1 pod fragment
Peanuts	2 empty hulls
Unidentified legume	1 pod fragment
Pacay	5 leaves
	1 pod
Butternut squash	3 stem fragments
Lúcuma	43 seeds

Totals:
   NISP= 162

We have no way of knowing from which residential area the refuse on Terrace 11 came. Based on Table 13.4, however, our guess is that it was an elite compound. For one thing, desirable tropical fruits outnumbered staple foods such as beans. For another, every part of the maize plant was present, from tassels to roots. This diversity of maize parts, coupled with the large number and variety of cobs, suggests a residence well supplied with maize from different sources, probably because it was engaged in chicha production. The presence of llama dung pellets in the Terrace 11 midden indicates that some of these supplies arrived via llama caravans, another feature we associate with elite residential compounds. As we saw in Chapter 9, the fish remains from Terrace 11 led Sommer and Flannery to a similar conclusion.

Quebrada 6, Terrace 12

Terrace 12 lay immediately downslope from Terrace 11 and featured a burial cist, a storage cist, and an ash-filled earth oven. These features were overlain by a shallow brown-to-beige midden, which produced 3 leaflets of pacay.

The storage cist, designated **Structure 10**, had only a handful of food plants inside it, as follows:

Maize	6 cobs (see Chapter 14)
	1 frayed husk
Canavalia beans	1 empty pod
Butternut squash	1 stem

Spanish Colonial Plant Remains from Structure 1

After the Inca had conquered Cerro Azul in A.D. 1470, they built several adobe buildings atop the site's prominent sea cliffs. One of these buildings was Structure 1 on Cerro del Fraile (Fig. 13.44), whose Room 9 featured trapezoidal niches in typical Inca style (Marcus et al. 1985).

After the Spanish Conquest of A.D. 1534, squatters occupied several rooms of this abandoned Inca building. The refuse they left identifies these squatters as Spaniards and dates them to the 16th century. Included among the Spanish olive jars and discarded *alpargatas* (Colonial sandals) was part of a handwritten grocery list or receipt in 16th century calligraphy.

The bulk of the squatters' occupation occurred in Rooms 9 and 10 of Structure 1. In those rooms Marcus' crew recovered a fascinating mix of New World and Old World plants (Table 13.5). They also recovered cobs of Confite Puntiagudo maize that were larger, on average, than those found in Late Intermediate levels. The Colonial cobs had between 12 and 20 kernel rows, instead of the 8–12 rows typical of Late Intermediate cobs.

Table 13.5. Food Plants from 16th-century Colonial Refuse at Cerro Azul.

Cherimoya	*Annona cherimolia*
Maize	*Zea mays*
Barley	*Hordeum* sp.
Common beans	*Phaseolus vulgaris*
Vetch	*Vicia* sp.
Pacay	*Inga feuillei*
Butternut squash	*Cucurbita moschata*
Rue	*Ruta graveolens*
Lúcuma	*Pouteria lucuma*
Basil	*Ocimum basilicum*
Potato	*Solanum tuberosum*
Unspecified chile pepper	*Capsicum* sp.
Habanero pepper	*Capsicum chinense*

Edible, Ritual, and Medicinal Plants

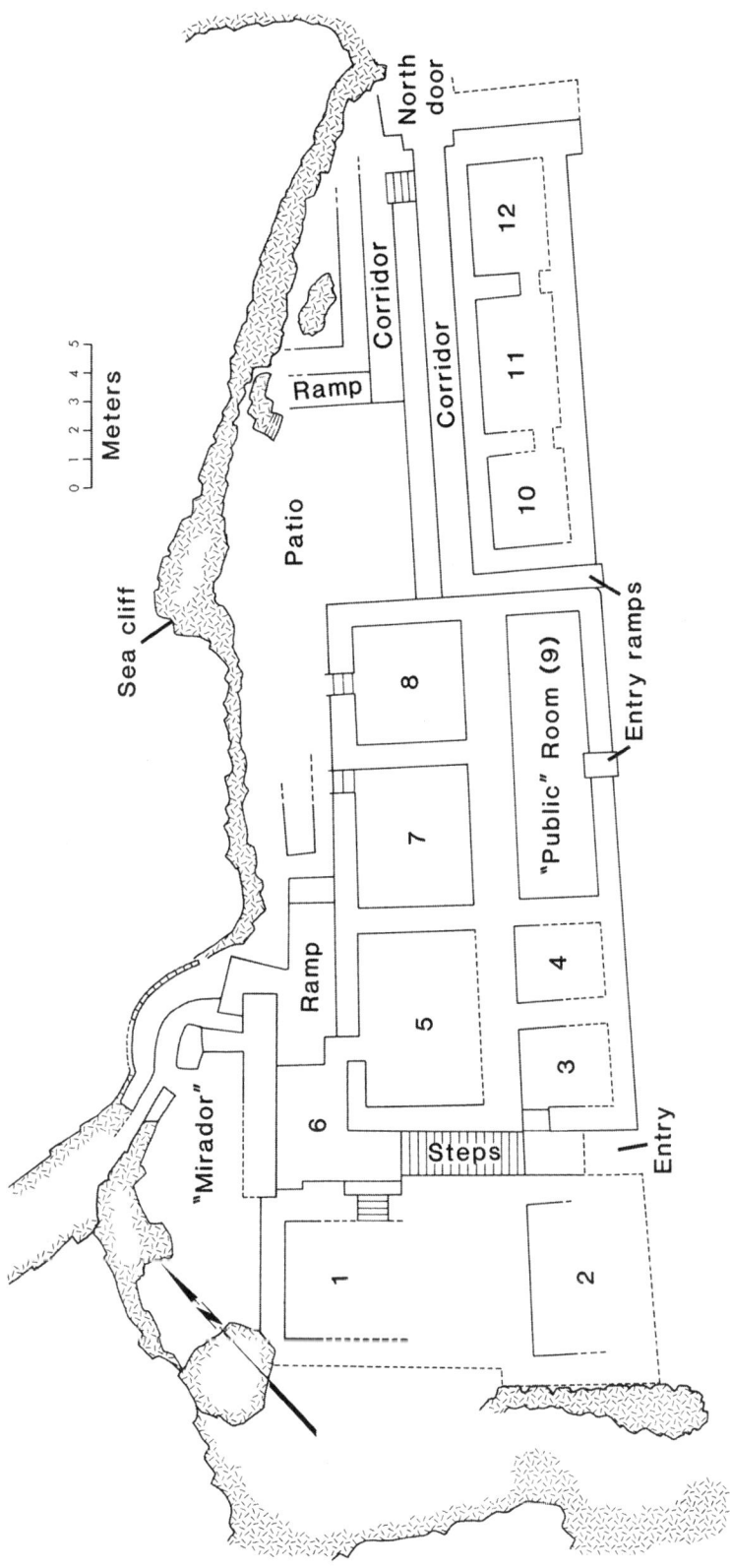

Figure 13.44. Structure 1 of Cerro Azul.

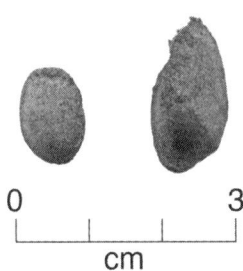

Figure 13.45. Two seeds of cherimoya (*Annona cherimolia*) from 16th-century A.D. refuse in the south end of Room 9, Structure 1.

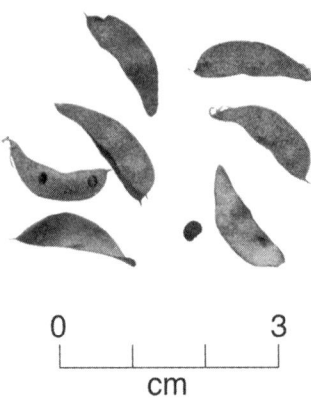

Figure 13.46. Pods, valves, and seeds of vetch (*Vicia* sp.) from 16th-century A.D. refuse in the southwest end of Room 9, Structure 1.

Room 9

Room 9 of Structure 1 was a long, narrow hall decorated with trapezoidal niches. Colonial squatters had left debris in both ends of the room, but kept the entry ramp clean.

In the southwest end of the room Marcus' crew recovered 2 seeds of cherimoya (Fig. 13.45), 1 empty pod from *Phaseolus* beans, 1 fruit and 2 seeds of an unspecified chile pepper, and 3 Old World plants—8 pods and one seed of vetch (*Vicia* sp.) (Fig. 13.46), 3 leaves of rue (*Ruta graveolens*) (Fig. 13.47a), and 7 leaves of basil (*Ocimum basilicum*) (Fig. 13.47b).

In the northeast end of the room, the following plants were recovered:

Maize	3 cobs (fourteen-row and eighteen-row)
Common beans, *Phaseolus vulgaris*	7 tan beans (see Chapter 15) (Fig. 13.48) 9 white-and-tan beans
Lúcuma	6 seeds
Potato	1 roasted tuber (Fig. 13.49a)

Just east of the entry ramp to Room 9, Marcus' crew recovered 2 seeds of butternut squash.

Room 10

Room 10 was a rectangular room to the northeast of Room 9, and separated from the latter by an entry ramp. Squatters had left the following food plants on the floor of this room:

Maize	88 cobs (Confite Puntiagudo, twelve-row to twenty-row) 1 husk
Old World barley	26 heads (Fig. 13.50)
Pacay	1 pod 1 leaflet
Lúcuma	2 seeds
Habanero chile pepper (*C. chinense*)	1 desiccated fruit (Fig. 13.49b)

Edible, Ritual, and Medicinal Plants

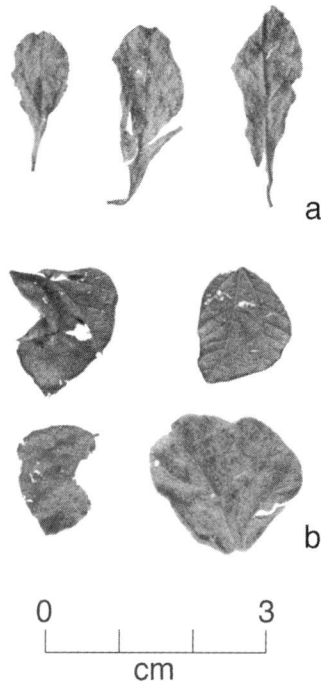

Figure 13.47. Herbs used as seasonings by 16th-century A.D. squatters in Room 9, Structure 1. *a*, three leaves of rue (*Ruta* cf. *graveolens*). *b*, four leaves of basil (*Ocimum basilicum*).

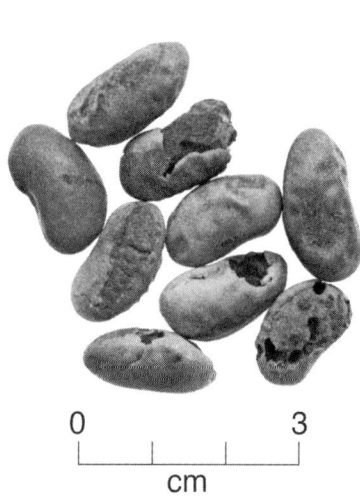

Figure 13.48. Nine seeds of common beans (*Phaseolus vulgaris*) from 16th-century A.D. refuse in the northeast end of Room 9, Structure 1.

Figure 13.49. Crop plants used by 16th-century A.D. squatters in Structure 1. *a*, a roasted potato (*Solanum tuberosum*). *b*, a fruit of habanero pepper (*Capsicum chinense*).

Figure 13.50. Barley (*Hordeum* sp.) from 16th-century refuse in Room 10 of Structure 1.

Other Plants in Structure 1 (date uncertain)

Other proveniences in Structure 1 produced plant remains, but the dating of these plants is uncertain. They could be either Late Horizon or Colonial.

For example, the floor of the corridor immediately to the southwest of Rooms 10, 11, and 12 yielded 3 leaves of achira, 25 cobs of maize, and an empty peanut hull. A niche in the stairway landing just below Room 6 produced one maize cob, a cotyledon and two hulls of peanut. Only direct AMS dating could determine the period to which these plants belong.

Summary and Conclusions

During the Late Intermediate period, Cerro Azul had access to a wide variety of edible, ritual, and medicinal plants. As mentioned earlier, however, the limited range of plant parts in the refuse suggests that Cerro Azul was receiving these plants from inland agricultural communities, rather than growing the crops themselves.

Recently, Peruvian archaeologists of the project "Qhapaq Ñan—Sede Nacional" have discovered a series of Late Horizon storage units upvalley from Cerro Azul, near Lunahuaná and Pacarán. These *collcas* contained many of the same edible or ritual plants found at Cerro Azul, including maize, *Phaseolus* beans, *Canavalia* beans, coca, pacay, peanuts, chile peppers, cherimoya, guava, squash, and lúcuma (Díaz Carranza 2015: Tabla 2). One difference is that the squash found in the Late Horizon *collcas* was the *zapallo* (*Cucurbita maxima*), rather than butternut squash.

Many of these same plants were eaten by the 16th-century Spanish settlers at Cerro Azul. The Spaniards, however, brought with them Mediterranean plants such as barley, vetch, rue, and basil. They also ate habanero peppers (*Capsicum chinense*) rather than the *C. baccatum* peppers that were popular during the Late Intermediate period at Cerro Azul. Some of these differences, of course, could reflect changing exchange routes after the fall of the Inca.

Finally, let us call attention to an important food plant that was notably absent, both in the Late Intermediate remains from Cerro Azul and the Late Horizon remains from the *collcas* found by the Qhapaq Ñan project. That plant is manioc (*Manihot esculenta*), which we observed being cultivated near Zúñiga in the middle Cañete Valley during the 1980s. We leave the explanation of its prehistoric absence to future investigators.

14

Comments on the Late Intermediate Maize

C. Earle Smith, Jr. and Joyce Marcus

Editor's note: Dr. Smith took thousands of measurements of the Cerro Azul maize, and recognized many different varieties. Unfortunately, his untimely death in 1987 precluded his describing all these varieties for inclusion in this chapter. Therefore, using his field notes, Marcus has done her best to create a document of which he would approve. It goes without saying that the late Dr. Smith is not responsible for any botanical errors.

Maize was one of the most common plants recovered at Cerro Azul; more than 1700 cobs were found in Structure D alone. Given that the North Central Canchón of that building was a brewery with the capacity to store 5000 liters of chicha (Marcus 2008:187; 2009:315), we suspect that most of the maize that arrived at Structure D was turned into beer (Fig. 14.1). Marcus' crew even discovered the residue of *jora*—the sprouted maize kernels from which chicha is brewed—clinging to a broken jar in the North Central Canchón (Marcus 2008:183).

The presence of a brewery in an elite residential compound is not unexpected. Andean nobles are known to have used liberal quantities of maize beer as an incentive for work crews of commoners (Moore 1989; Morris 1982; Murra 1960, 1980, Rostworowski de Diez Canseco 1977a; Rowe 1946). They also used chicha to entertain foreign leaders or visiting dignitaries (see also Morris 1979, Prieto B. 2011).

Cerro Azul's impressive maize consumption is all the more interesting in light of the fact that there is little or no evidence that maize could be grown close to the site. If Cerro Azul had been producing its own maize, we would expect to see many more specimens of husks, leaves, tassels, stalks and roots in the site's refuse. In fact, such plant parts were outnumbered more than 100 to 1 by exhausted cobs, those from which all the kernels had been removed.

Where did we find exhausted cobs? Everywhere, especially in the tapia compounds. And where did we find ears of corn with the kernels still on them? In or near elite burials, presumably because mourners assumed that nobles would want to continue consuming chicha in the afterlife. Figure 14.2 illustrates four complete maize ears from looted burials on Terrace 9 of Quebrada 5a. The ear at the upper left is from a variety of dent corn. The other three ears have imbricated kernels that are purple in color, suitable for making *chicha morada*. Smith identified the largest ear in Figure 14.2 as belonging to the race Confite Puntiagudo. Note how the rows of kernels spiral around the cob.

Figure 14.3 shows an immature maize ear, still enclosed in its husk. This ear was found in Structure 12, a looted burial cist in Quebrada 5-south. Figure 14.4 displays an empty husk found near Burial 4 in Quebrada 5a. Some maize ears may have been dislodged from their husks by looters; others had had their kernels

Figure 14.1. The remains of a *hatun maccma*, a chicha storage vessel with a capacity of nearly 2000 liters. This vessel was found in the North Central Canchón of Structure D (see Marcus 2008:187–189).

eaten by the rice rats (*Oryzomys xanthaeolus*) that burrowed their way into many cists (Fig. 14.5).

To be sure, a few husks, stalks, and tassels of maize also showed up in middens or burials. Figure 14.6 shows a stalk and husk from Terrace 11 of Quebrada 6; there are reasons to believe that this midden contained refuse from an elite residential compound, one that may well have been involved in chicha production. Figure 14.7 illustrates a maize tassel found with Burial 9.

Where would this maize have been grown? We suspect that it was brought to Cerro Azul from localities farther inland, perhaps fields irrigated by the canals coming off the Río Cañete (Fig. 1.2). Some of Cerro Azul's maize may even have been grown farther upstream, for example in the Kingdom of Lunahuaná.

In addition to the likelihood that the Structure D maize came from fields outside the immediate confines of Cerro Azul, we are convinced that it came from multiple localities. We say this because, rather than showing uniformity in its color and morphological characteristics, the Cerro Azul maize seems to consist of more than a dozen varieties. In order to maintain their distinct features (instead of hybridizing), these varieties would have to have been grown in different places.

Maize Cobs from the South Corridor and Southwest Canchón, Structure D

One of our most convincing examples of maize diversity comes from a very late stage in the occupation of Structure D. At that time, several of the building's original doorways and corridors had been deliberately blocked. The main route by which cobs left over from brewing would have exited the building was via the South Corridor and the Southwest Canchón. From there, the cobs could be loaded into baskets and carried to one of the quebrada middens.

With the abandonment of Structure D, at least 1562 exhausted cobs never made it to the quebradas. Some 234 cobs were simply left on the floor of the South Corridor. Another 1328 cobs, presumably left on the floor of the Southwest Canchón, were swept up and added to the Feature 6 midden.

These 1562 cobs constitute our largest sample of maize from Cerro Azul, and Smith felt that he could recognize at least 7 varieties within it. These varieties could not be confidently matched to known Peruvian races of maize, since all the kernels had been removed from the cobs. Smith, however, undertook to measure the cobs and look for qualitative differences.

Figure 14.2. Maize ears still bearing kernels, found in the vicinity of looted burials on Terrace 9 of Quebrada 5a. The ear at upper left is a variety of dent corn. The other three cobs have imbricated purple kernels. The largest of these purple cobs, displaying spiral kernel rows, belongs to the race Confite Puntiagudo.

Figure 14.3. An immature maize ear still in its husk. Found in Structure 12, a burial cist in Quebrada 5-south.

Figure 14.4. An empty maize husk found near Burial 4, Quebrada 5a. The ear may have been dislodged from this husk by looters.

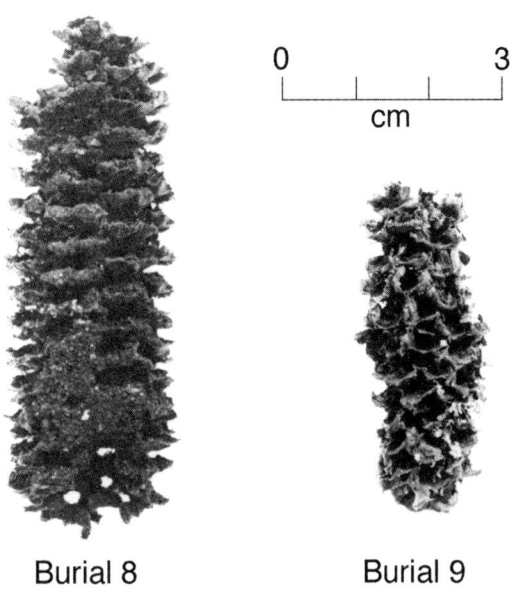

Figure 14.5. Maize cobs found near burials in Quebrada 5a. It appears that local rice rats burrowed into these burials and ate the kernels off both cobs.

Feature 6

Smith began with the whole and partial cobs from Feature 6. He measured all 91 cobs of Type 1; 100 of the 313 Type 2 cobs; 100 of the 272 Type 4 cobs; 100 of the 221 Type 5 cobs; 100 of the 382 Type 6 cobs; and all 49 of the Type 7 cobs. (There were no Type 3 cobs in Feature 6.)

Smith's description of each cob type was as follows:

Type 1 (Fig. 14.8, top; Fig. 14.9, top)

General description	cobs reddish; rachillae prominent
Cob length (based on 10 whole cobs)	mean = 77.4 mm, range = 46.4–111.3 mm
Cob diameter	mean = 16.88 mm, range = 12.5–24.0 mm
Rachis diameter	mean = 8.86 mm, range = 4.8–16.1 mm
Cupule height	mean = 1.63 mm, range = 0.5–2.8 mm
Cupule width	mean = 4.73 mm, range = 2.2–8.0 mm
Row no.	8–14 (mode = 10)

Type 2 (Fig. 14.8, bottom; Fig. 14.9, bottom)

General description	cobs reddish to red; lemmas and paleas often making for a soft surface
Cob length (based on 26 whole cobs)	mean = 77.8 mm, range = 31.9–132.0 mm
Cob diameter	mean = 16.72 mm, range = 12.2–23.2 mm
Rachis diameter	mean = 8.93 mm, range = 5.0–13.6 mm
Cupule height	mean = 1.74 mm, range = 0.6–3.0 mm
Cupule width	mean = 5.15 mm, range = 2.6–7.6 mm
Row no.	8–14 (mode = 10)

Figure 14.7. A maize tassel found near Burial 9, Quebrada 5a.

Figure 14.6. A badly frayed maize stalk and husk from a midden deposit on Terrace 11, Quebrada 6.

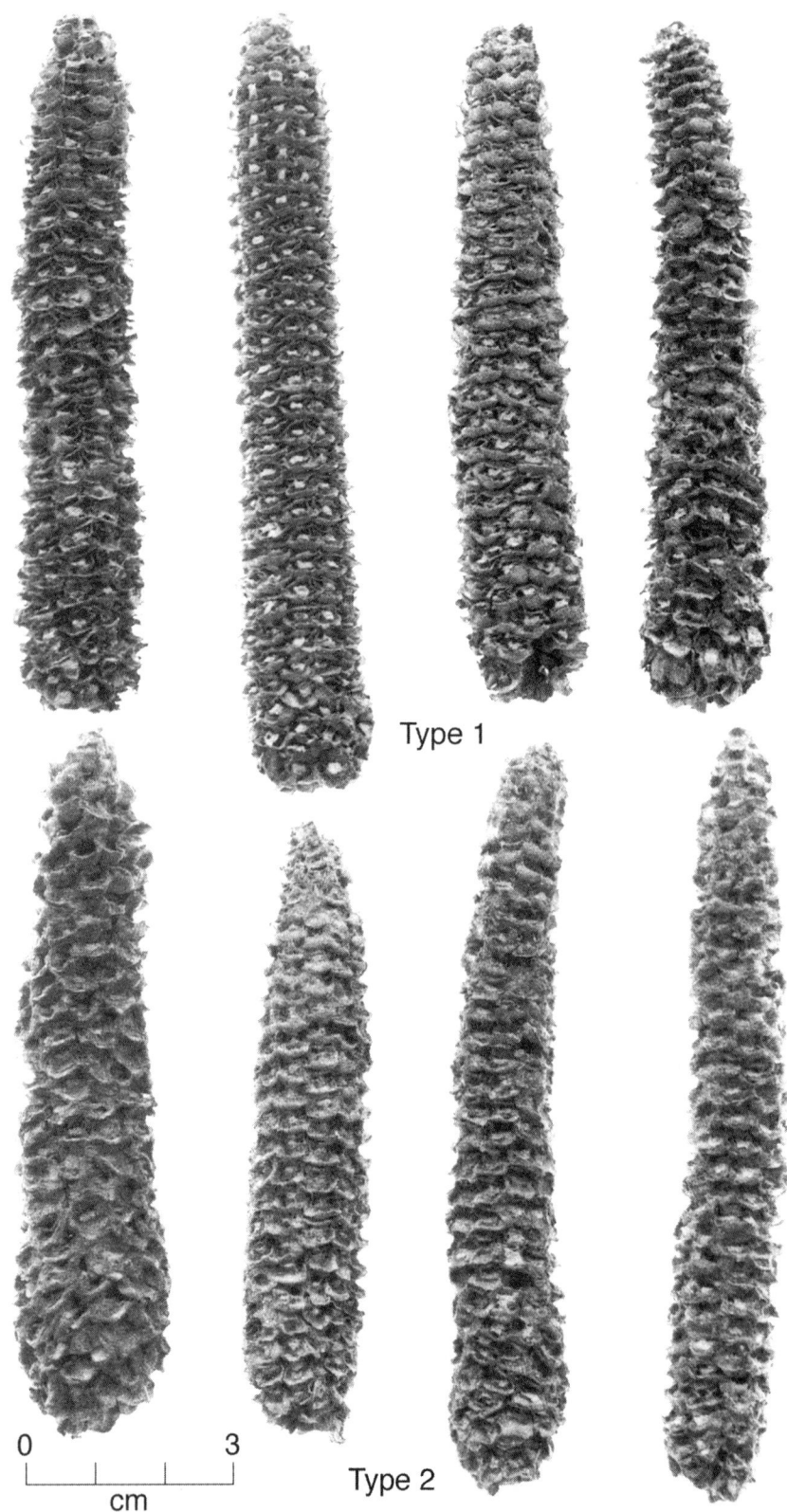

Figure 14.8. Four cobs each of Maize Types 1 and 2 from Feature 6, Structure D.

Figure 14.9. Four cobs each of Maize Types 1 and 2 from the South Corridor, Structure D.

Type 4 (Fig. 14.10, top)

General description	cobs tan; lemmas and paleas often making for a soft surface
Cob length (based on 16 whole cobs)	mean = 80.57 mm, range = 42.4–118.1 mm
Cob diameter	mean = 16.50 mm, range = 12.6–22.1 mm
Rachis diameter	mean = 8.73 mm, range = 5.5–14.5 mm
Cupule height	mean = 1.85 mm, range = 1.2–3.1 mm
Cupule width	mean = 5.01 mm, range = 2.4–7.1 mm
Row no.	8–16 (mode = 10)

Type 5 (Fig. 14.10, bottom)

General description	cobs tan; glumes exserted and very conspicuous
Cob length (based on 14 whole cobs)	mean = 77.08 mm, range = 55.2–120.5 mm
Cob diameter	mean = 16.67 mm, range = 13.0–22.3 mm
Rachis diameter	mean = 8.49 mm, range = 5.6–12.7 mm
Cupule height	mean = 1.83 mm, range = 0.6–2.9 mm
Cupule width	mean = 5.17 mm, range = 3.0–7.7 mm
Row no.	8–14 (mode = 10)

Type 6 (Fig. 14.11, top)

General description	cobs red; glumes exserted and contrasting in color with the rest of the cob surface
Cob length (based on 8 whole cobs)	mean = 65.64 mm, range = 50.5–101.4 mm
Cob diameter	mean = 16.25 mm, range = 11.7–20.9 mm
Rachis diameter	mean = 7.78 mm, range = 5.2–11.4 mm
Cupule height	mean = 1.9 mm, range = 0.9–3.1 mm
Cupule width	mean = 5.07 mm, range = 3.0–7.4 mm
Row no.	8–14 (mode = 10)

Type 7 (Fig. 14.11, bottom)

General description	cobs purplish red to purple, including the lower glumes; lower glumes frequently hard, erect, and conspicuous; rachillae often thickened and prominent
Cob length (based on one whole cob)	mean = 73.5 mm
Cob diameter	mean = 16.74 mm, range = 13.1–22.5 mm
Rachis diameter	mean = 8.50 mm, range = 5.1–12.3 mm
Cupule height	mean = 1.83 mm, range = 0.9–2.7 mm
Cupule width	mean = 5.04 mm, range = 2.6–7.2 mm
Row no.	8–14 (mode = 10)

The South Corridor

After completing his study of Feature 6, Smith turned to the 234 cobs from the South Corridor. The varieties of maize found in this collection overlapped strongly with those in Feature 6. Two differences were that the South Corridor produced no cobs of Type 7, but did have two cobs of Smith's Type 3. Smith also assigned two cobs to Type 2a, a variant of his Type 2.

Type 2a (Fig. 14.12, top left)

Cob length (based on 2 whole cobs)	mean = 72.7 mm, range = 67.0–78.4 mm
Cob diameter	mean = 22 mm, range = 19–25 mm
Rachis diameter	mean = 12.75 mm, range = 12.7–12.8 mm
Cupule height	mean = 1.65 mm, range = 1.6–1.7 mm
Cupule width	mean = 6.95 mm, range = 6.7–7.2 mm
Row no.	10–14

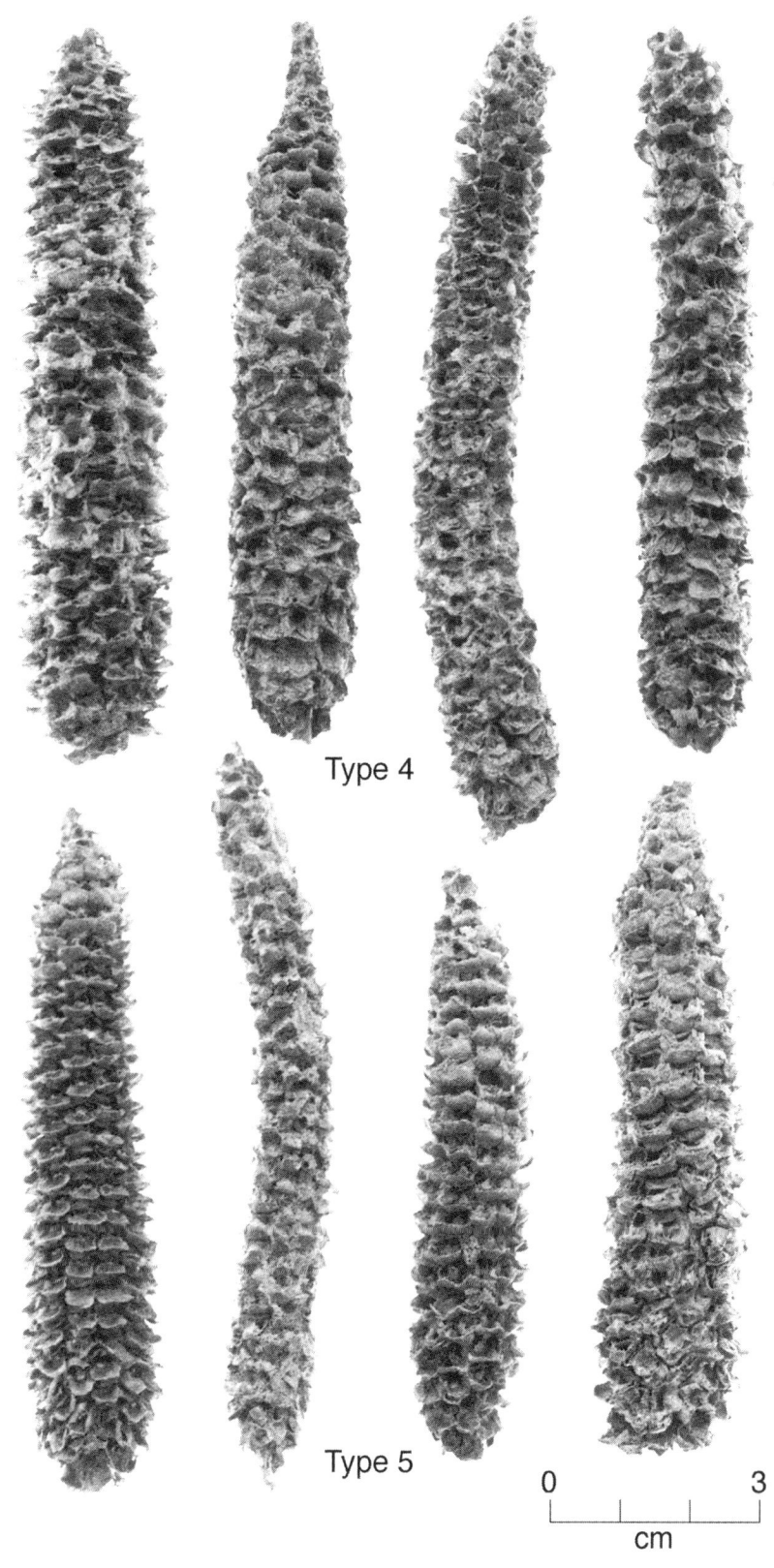

Figure 14.10. Four cobs each of Maize Types 4 and 5 from Feature 6, Structure D.

Type 3 (Fig. 14.12, top right)

Cob length (based on 2 whole cobs)	mean = 62.75 mm, range = 58.2–67.3 mm
Cob diameter	mean = 16.35 mm, range = 15.0–17.7 mm
Rachis diameter	mean = 8.55 mm, range = 7.6–9.5 mm
Cupule height	mean = 2.0 mm, range = 1.7–2.3 mm
Cupule width	mean = 6.1 mm, range = 4.5–7.7 mm
Row no.	10–12

Of the 234 cobs left on the floor of the South Corridor, 34 belonged to Type 1; 59 belonged to Type 2; 2 belonged to Type 2a; 2 belonged to Type 3; 64 belonged to Type 4; 24 belonged to Type 5; 46 belonged to Type 6; and 3 cobs could not be assigned to a type.

While none of the 1562 cobs from these two proveniences had any kernels remaining, Feature 6 also produced some 96 loose kernels. At least 32 of these kernels were imbricated, like those seen on the Confite Puntiagudo ear in Figure 14.2.

As far as can be determined from Smith's study, the maize cobs from Feature 6 and the South Corridor were sufficiently alike to date to the same stage in the history of Structure D. Since the cobs had never been removed from the building, that stage presumably fell late in the building's occupation. At that time Structure D was receiving shipments of maize from at least 7 different localities, each growing its own variety of maize. Some of this maize had red or reddish cobs, while the remainder had tan cobs. Some kernels were imbricated, others not. Kernel rows varied from 8 to 14, with 10 being the most common number.

The Maize from Collca 1 of Structure D

Collca 1 was a storage bin set against the south wall of the Structure D's Northeast Canchón. Its original purpose seems to have been as a place to store loom parts and other craft equipment for the weavers who worked in the canchón (Marcus 2016). At some point, however, Collca 1 became a convenient place to toss unwanted refuse. Included in that refuse were some 70 exhausted maize cobs. Given the proximity of the Northeast Canchón to the North Central Canchón, it is likely that this was a batch of cobs used in chicha making—exhausted cobs that for some reason never made it out of the building.

The Collca 1 cobs did include 13 cobs of Type 2; 11 cobs of Type 4; 3 cobs of Type 5; 7 cobs of Type 6; and 1 cob of Type 7. The remaining 35 cobs, however, were different. They belonged to six new types, to which Smith assigned the letters A through F (Figs. 14.13–14.15).

Unfortunately, Smith's untimely death prevented him from

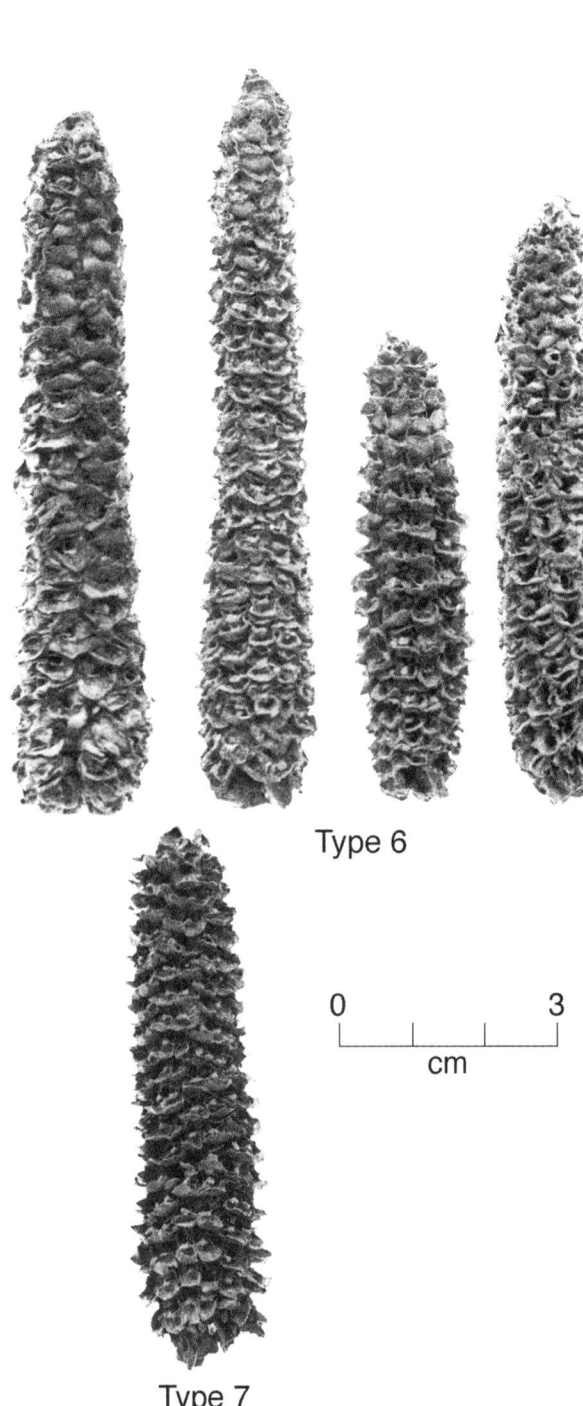

Figure 14.11. Four cobs of Maize Type 6 and one cob of Maize Type 7 from Feature 6, Structure D.

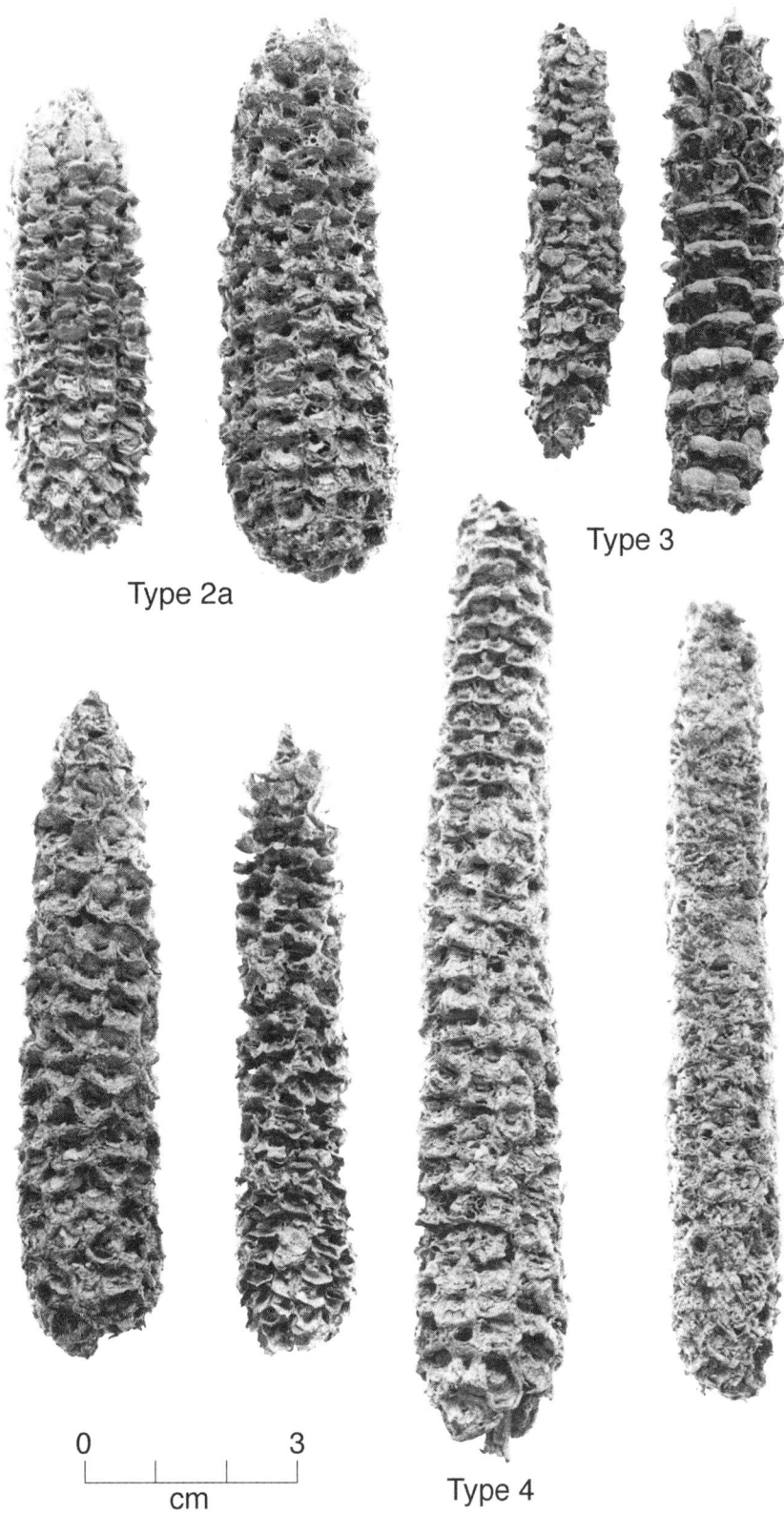

Figure 14.12. Two cobs of Maize Type 2a, two cobs of Maize Type 3, and four cobs of Maize Type 4 from the South Corridor, Structure D.

Figure 14.13. Four cobs each of Maize Types A and B from Collca 1, Structure D.

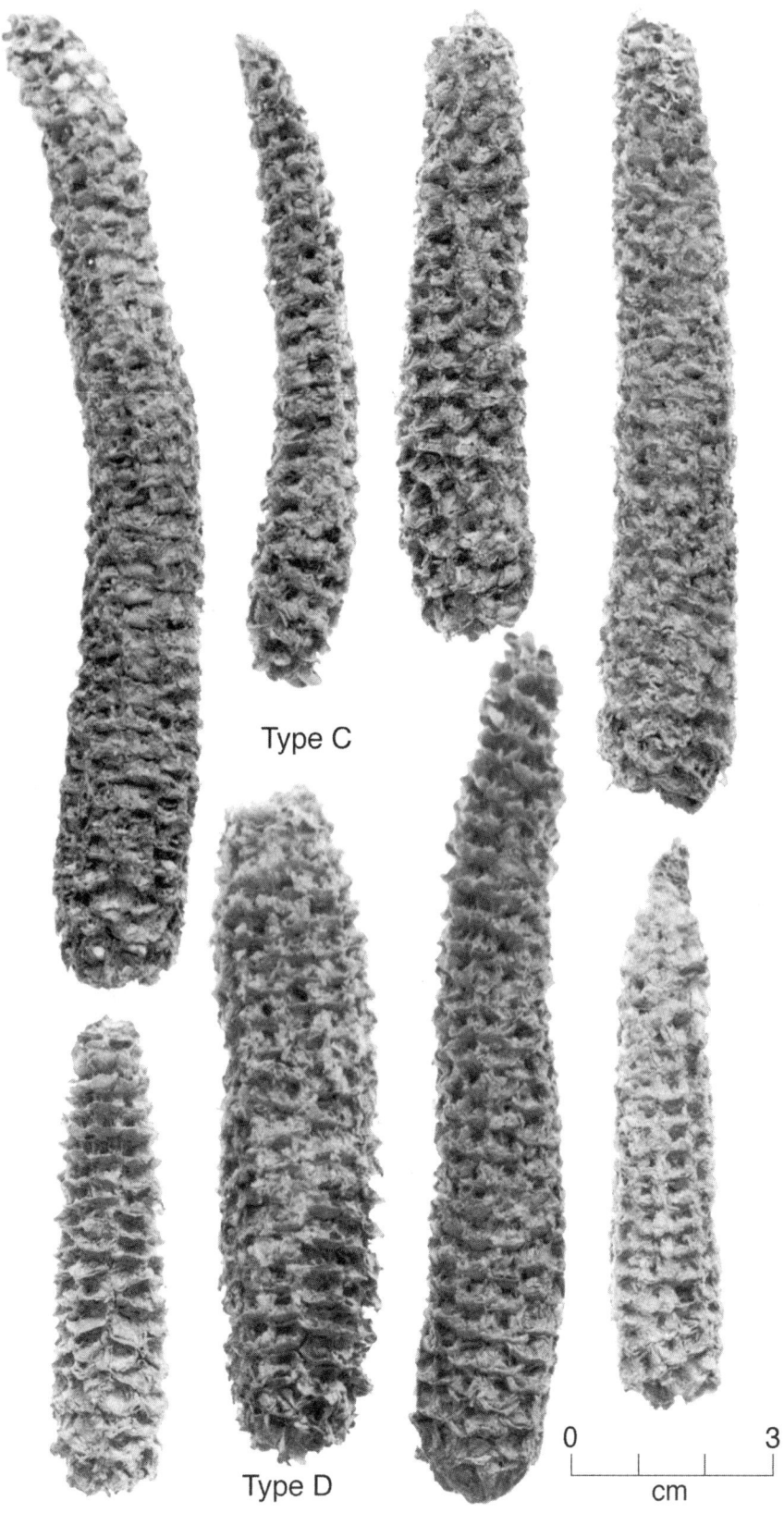

Figure 14.14. Four cobs each of Maize Types C and D from Collca 1, Structure D.

completing his study of Types A-F. We do know that these types differed in color (tan, reddish, red, and maroon) as well as in glume morphology (stiff, erect, and conspicuous vs. small, firm, and inconspicuous).

The 70 cobs from Collca 1 show us that the sources of maize used at Cerro Azul were not only diverse, but varied over time. The maize from Collca 1 almost certainly dates to an earlier stage in the building's history than that from Feature 6. Of the 11 cob types present in Collca 1, only five were still in use at the time Feature 6 was created. By that time, Structure D was using two types not present in Collca 1, and was no longer using Types A-F.

The Maize from Structure 9

Structure 9 was a small fish-storage facility with a commoner-class overseer. Its most informative maize samples came from Feature 20 and Room 4.

Feature 20

Feature 20 was a midden piled up against the east wall of Structure 9. This midden is believed to contain domestic refuse from the household of the building's overseer.

Feature 20 produced 18 exhausted maize cobs, only 14 of which were measurable. Two of the cobs were carbonized, making them even harder to study. This small sample lacked the variety seen in Structure D and consisted mostly of cobs with reddish glumes and pubescent cupules (Fig. 14.16). In the case of one cob, it appears that the kernel rows would have had a spiral configuration like that of Confite Puntiagudo. The measurements of these cobs were as follows:

Cob length (based on 2 whole cobs): 70.3–75.8 mm
Cob diameter: mean = 16.89 mm, range = 11.2–24.3 mm
Rachis diameter: mean = 9.26 mm, range = 7.0–11.3 mm
Cupule height: mean = 1.19 mm, range = 0.5–2.1 mm
Cupule width: mean = 2.99 mm, range = 2.0–5.2 mm
Row no.: 8–14 (70 percent have 12–14 rows)

When we compare Feature 20 of Structure 9 with Feature 6 of Structure D, we are struck by how much less maize was present in Feature 20. Almost certainly this is due to the fact that Structure 9 had no brewery, and would therefore not have had the same need for large quantities of maize.

To be sure, Room 2 of Structure 9 produced a chicha-storage vessel with effigy maize-cob lugs (Marcus 2008:224–227 and Fig. 8.15). This vessel, however, was not set in the floor, nor was there evidence to suggest that any of the chicha stored in this vessel might have been brewed by the building's commoner-class overseer. It seems more likely that the overseer received periodic allotments of chicha from the noble family he served.

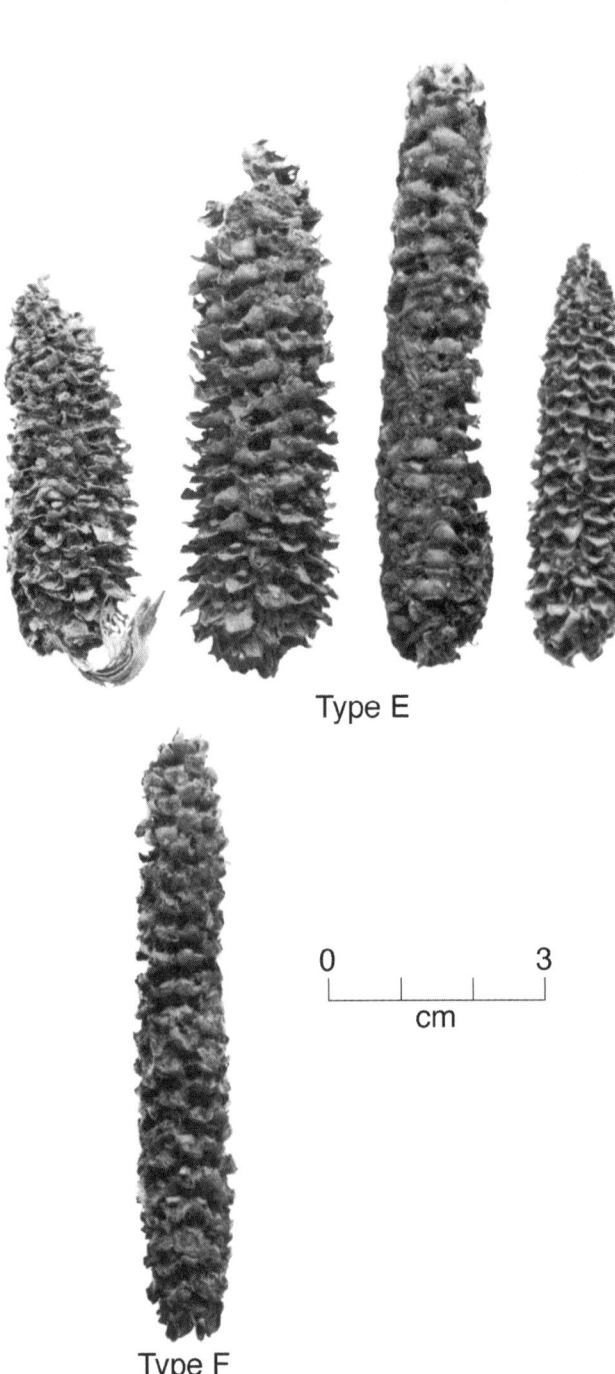

Figure 14.15. Four cobs of Maize Type E and one cob of Maize Type F from Collca 1, Structure D.

Comments on the Late Intermediate Maize

Figure 14.16. Six cobs from Feature 20, Structure 9.

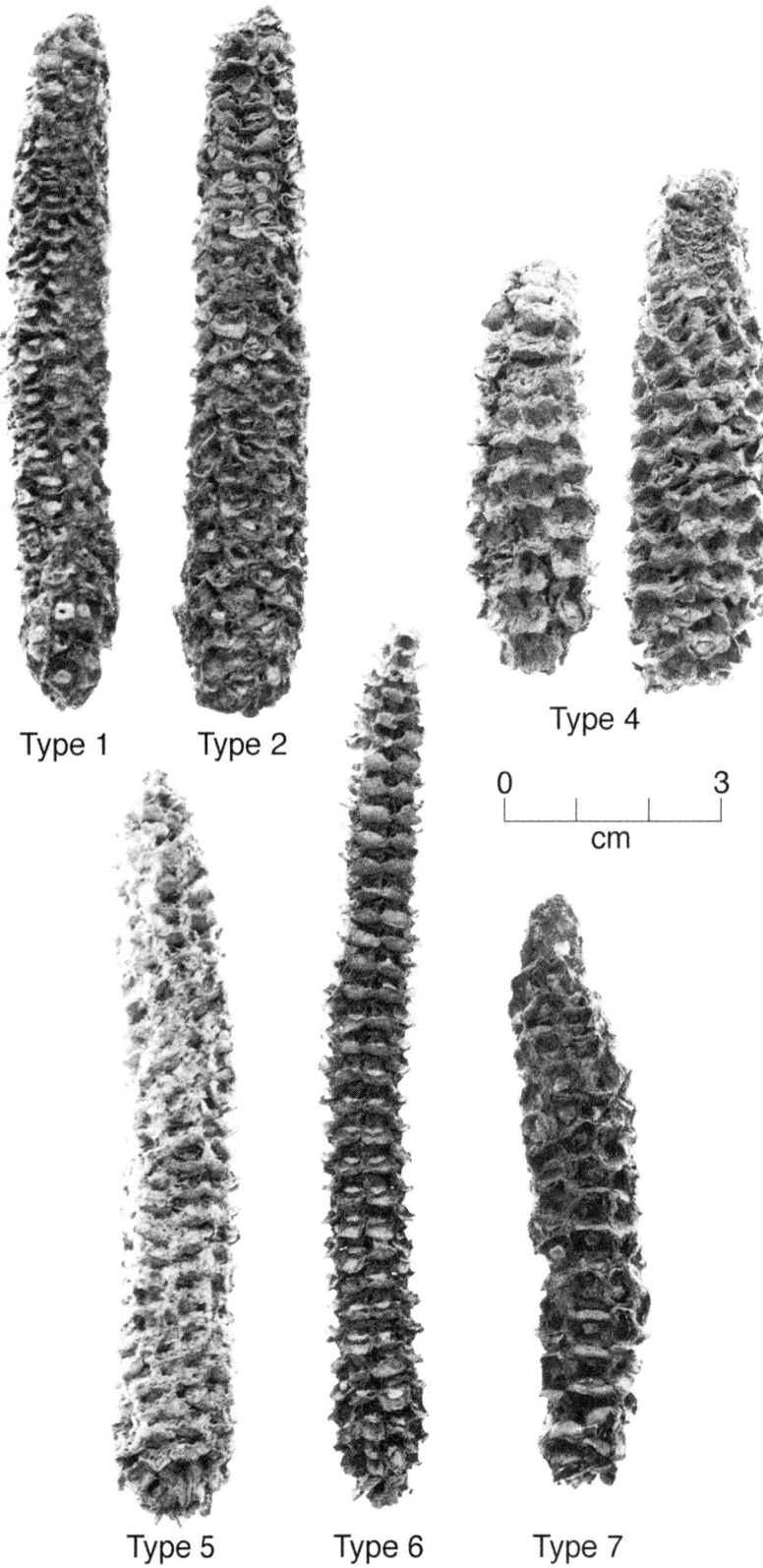

Figure 14.17. Examples of Maize Types 1, 2, 4, 5, 6, and 7 from Room 4 of Structure 9.

Room 4

Room 4 was a small storage unit in the extreme northeast corner of Structure 9. After suffering earthquake damage, Room 4 came to be used as a place to dump refuse. Included in this refuse were 74 exhausted maize cobs.

As mentioned in Chapter 13, we are unsure where all these cobs came from. Room 4 lay on the outer edge of Structure 9, and having no roof, it would have been visible to passersby. Residents of nearby compounds—possibly even including Structure D—may therefore have seen Room 4 as a convenient place to dump refuse, in lieu of carrying it to a quebrada midden.

For all these reasons, it comes as no surprise that the maize in Room 4 of Structure 9 overlaps in variety with the maize from Structure D. Of the 74 cobs from Room 4, 19 belong to varieties seen in Feature 6 of Structure D. There are 7 whole cobs of Type 1; 1 whole cob of Type 2; 6 whole cobs of Type 4; 3 whole cobs of Type 5; 1 whole cob of Type 6; and 1 whole cob of Type 7 (Fig. 14.17).

While the 55 partial cobs from Room 4 are more difficult to classify, they share many attributes of the Structure D maize. While reddish, purplish, and maroon colors are abundant, other cob fragments are tan. Most cobs have stiff and conspicuous glumes, but there are some whose glumes are softer and less conspicuous.

Below we present measurements of the 19 whole cobs from Room 4.

Type 1

Cob length (based on 7 whole cobs)	mean = 74.37 mm, range = 50–102.9 mm
Cob diameter	mean = 17.20 mm, range = 14.0–22.0 mm
Rachis diameter	mean = 9.0 mm, range = 5.4–13.5 mm
Cupule height	mean = 1.73 mm, range = 1.1–2.7 mm
Cupule width	mean = 5.17 mm, range = 4.2–6.4 mm
Row no.	8–14 (86 percent have 8–10)

Type 2

Cob length (1 cob)	44.5 mm
Cob diameter	17.2 mm
Rachis diameter	8.3 mm
Cupule height	0.7 mm
Cupule width	3.6 mm
Row no.	12

Type 4

Cob length (based on 6 whole cobs)	mean = 63.75 mm, range = 52.8–77.7 mm
Cob diameter	mean = 14.79 mm, range = 13.4–20.0 mm
Rachis diameter	mean = 6.9 mm, range = 5.4–10.9 mm
Cupule height	mean = 1.53 mm, range = 0.6–2.1 mm
Cupule width	mean = 6.48 mm, range = 4.7–8.5 mm
Row no.	8–12 (mode = 10)

Type 5

Cob length (based on 3 whole cobs)	mean = 92.87 mm, range = 83.8–102.2 mm
Cob diameter	mean = 17.13 mm, range = 16.5–17.5 mm
Rachis diameter	mean = 8.83 mm, range = 8.5–9.4 mm
Cupule height	mean = 1.67 mm, range = 1.0–2.0 mm
Cupule width	mean = 5.5 mm, range = 4.5–7.0 mm
Row no.	10

Type 6

Cob length (1 cob)	120.0 mm
Cob diameter	13.6 mm
Rachis diameter	5.4 mm
Cupule height	2.7 mm
Cupule width	5.0 mm
Row no.	8

Type 7

Cob length (1 cob)	84.0 mm
Cob diameter	16.0 mm
Rachis diameter	6.5 mm
Cupule height	2.6 mm
Cupule width	5.1 mm
Row no.	8

Figure 14.18. One maize tassel and six cobs from a midden on Terrace 11, Quebrada 6.

Figure 14.19. Six cobs from Structure 10, a storage cist on Terrace 12 of Quebrada 6.

Maize from the Quebradas of Cerro Camacho

As a rule, maize specimens from the quebradas of Cerro Camacho were not as well preserved as those from the tapia compounds; having been moved several times, most quebrada cobs were in tertiary context. An exception to this rule were the ears of maize placed with burials.

Quebrada 6, Terrace 11

Terrace 11 of Quebrada 6 was a coal-black, richly organic midden whose floral and faunal contents suggest that it was refuse from one of the elite residential compounds. This midden produced 73 whole or fragmentary maize cobs, 4 loose kernels, 5 tassels, 5 stalk fragments, 3 root systems, 2 leaves, and 3 husks (see Table 13.4). The diversity of cobs and plant parts in Terrace 11 leads us to suspect that the compound from which this refuse came was involved in chicha production, receiving shipments of maize from multiple sources.

The cobs from Terrace 11 were mostly reddish and bore 8–14 kernel rows. Their glumes varied from stiff and prominent to soft and inconspicuous. Having been moved several times, a number of the cob fragments were too rotten or deteriorated to measure. While it was obvious that several varieties of maize were present, none could be matched to any of the cob types in Structure D. Figure 14.18 illustrates six cobs and a tassel from Terrace 11.

Quebrada 6, Terrace 12

Terrace 12 of Quebrada 6 featured a looted burial cist, a storage cist, and an ash-filled earth oven. The storage cist, **Structure 10**, produced six cob fragments and one husk. The six cob fragments, two of which were too deteriorated to measure, are shown in Figure 14.19. The best preserved cob fragment

Figure 14.20. Three cobs from Structure 5, a looted burial cist in Quebrada 5a.

was reddish in color and had glumes that were prominent and cupped.

Quebrada 5a, Terrace 9

As mentioned earlier, some examples of maize ears with intact kernels were found in looters' debris near burial cists in Quebrada 5a (Fig. 14.2). Not all the maize associated with Late Intermediate burials, however, was intact. Particularly in the case of the stone masonry burial cists in Stratigraphic Zone A of Terrace 9, it appeared that the local rice rats had burrowed into several cists and eaten the kernels off the cob while they were still relatively fresh. Figures 14.20 and 14.21 illustrate cobs from two of these burial cists.

Conclusions

Given the importance of chicha in the social relations between noble patrons and their commoner clients, it is no surprise that our biggest maize samples came from Structure D,

Figure 14.21. Six cobs from Structure 6, a looted burial cist in Quebrada 5a.

an elite residential compound with its own brewery. Cerro Azul's maize was not grown at the site, however, nor did it come from a single locality with morphologically uniform ears. The brewery in Structure D alone seems to have received more than a dozen varieties of maize. These varieties must have been grown on separate fields, or else they would have hybridized.

Intact ears, still bearing kernels, were rare at Cerro Azul. Our small sample of ears includes both dent corn with yellow or tan kernels, and Confite Puntiagudo corn with imbricated purple kernels. While we cannot say how many known Peruvian races were represented at Cerro Azul, our sample of Late Intermediate maize includes tan cobs that might have come from dent ears, and reddish cobs with spiraling kernel rows that might have been Confite Puntiagudo.

15

Phaseolus and *Erythrina* from Cerro Azul

Lawrence Kaplan

The University of Michigan's excavations at Cerro Azul produced literally hundreds of bean specimens. In Feature 6 of Structure D alone there were 367 specimens of *Phaseolus*, *Canavalia*, and *Erythrina*.

In this chapter I look at 11 samples of beans from Cerro Azul, set aside by C. Earle Smith, Jr. because he thought that I might recognize the varieties involved. Those samples are as follows:

Sample 001 (Fig. 15.1). *Phaseolus lunatus* from Feature 6, Structure D.

Sample 002 (Fig. 15.2). *Phaseolus vulgaris* from Feature 6, Structure D.

Sample 003 (Fig. 15.3). *Phaseolus vulgaris* from Feature 6, Structure D.

Sample 004 (Fig. 15.4). *Phaseolus vulgaris* from Feature 6, Structure D.

Sample 005 (Fig. 15.5). *Phaseolus lunatus* from Feature 6, Structure D.

Sample 006 (Fig. 15.6). *Phaseolus vulgaris* from Feature 6, Structure D.

Sample 007 (Fig. 15.7). *Phaseolus lunatus* from Feature 6, Structure D.

Sample 008 (Fig. 15.8). *Phaseolus lunatus* from Feature 6, Structure D.

Sample 009 (Fig. 15.9). *Erythrina* sp. from Feature 6, Structure D.

Sample 010 (Fig. 15.10). *Phaseolus vulgaris* from Structure 12, Quebrada 5-south.

Sample 011 (Fig. 15.11). *Phaseolus vulgaris* from Spanish Colonial refuse in Room 9 of Structure 1, Cerro del Fraile.

Comments on *Phaseolus* Beans

In the 11 samples from Cerro Azul, both lima beans (*Phaseolus lunatus*) and common beans (*P. vulgaris*) were present. Some of the seeds of the two species were not readily distinguishable, even when using characteristics that have proven reliable in other regions. The veining of the seed coat, which in most *lunatus* beans radiates from the hilum, was not visible in all specimens owing to the deterioration of color.

The twin elevations at the micropyle end of the hilum are useful in distinguishing these two species, but can be damaged or difficult to discern in archaeological specimens. In common beans these elevations are more completely joined than they are in most of the Peruvian limas and in the *lunatus* beans of Mexico and the Greater Southwest U.S. The micropyle is a very small pore which, when the seed is opened, is found immediately over

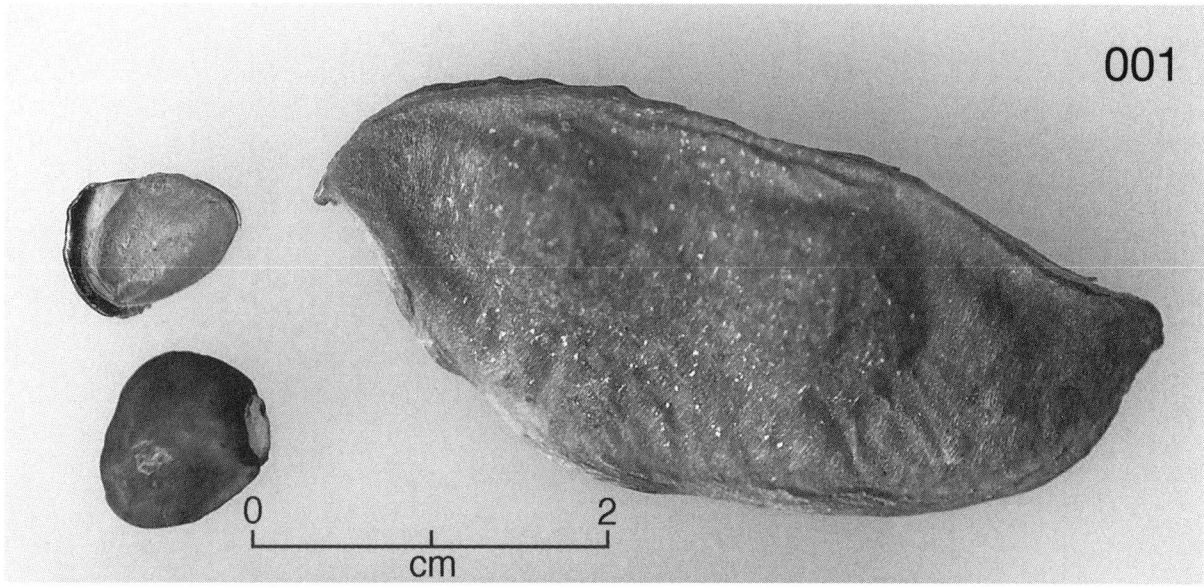

Figure 15.1. Sample 001, a pod of *Phaseolus lunatus* carrying two seeds, found in Feature 6 of Structure D.

Figure 15.2. Sample 002, five of a group of six dark red-brown *Phaseolus vulgaris* seeds found in Feature 6 of Structure D.

Figure 15.3. Sample 003, nine dark red-brown *Phaseolus vulgaris* seeds found in Feature 6 of Structure D. This variety is similar to *P. vulgaris* found at the site of Ancón.

Figure 15.4. Sample 004, eight dark red-brown *Phaseolus vulgaris* seeds found in Feature 6 of Structure D. This variety is smaller and less elongate than *P. vulgaris* from the site of Ancón.

Figure 15.5. Sample 005, two curved and elongate *Phaseolus lunatus* seeds with yellow seed coats and spotted dark areas. These lima beans, found in Feature 6 of Structure D, belong to a variety depicted often on Mochica pottery and resemble specimens from Ancón and Paracas.

Figure 15.6. Sample 006, eleven whole or partial *Phaseolus vulgaris* seeds found in Feature 6 of Structure D. These dark red-brown beans are similar to those in Sample 004.

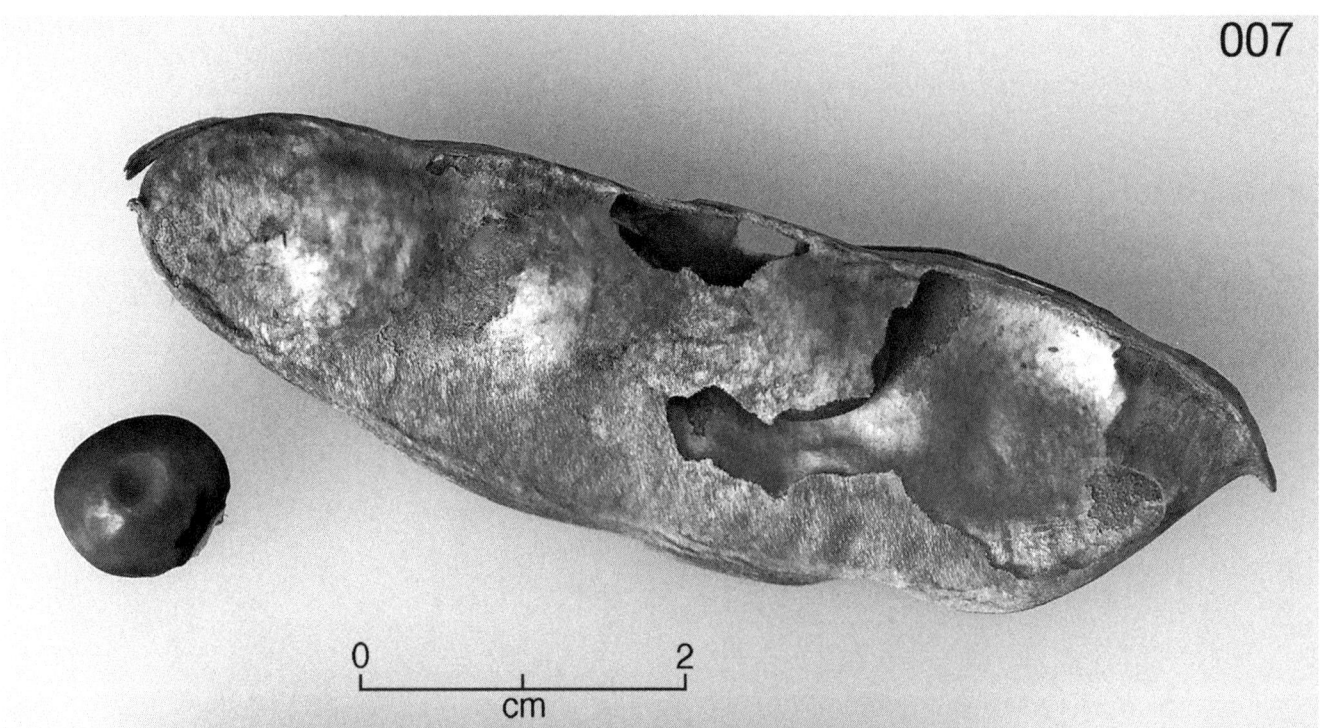

Figure 15.7. Sample 007, a pod of *Phaseolus lunatus* containing four seeds, one of which has been removed for study. Found in Feature 6 of Structure D, this lima bean belongs to the same variety as Sample 001.

Figure 15.8. Sample 008, three *Phaseolus lunatus* seeds found in Feature 6 of Structure D. These lima beans are similar in color and marking to Sample 005, but smaller.

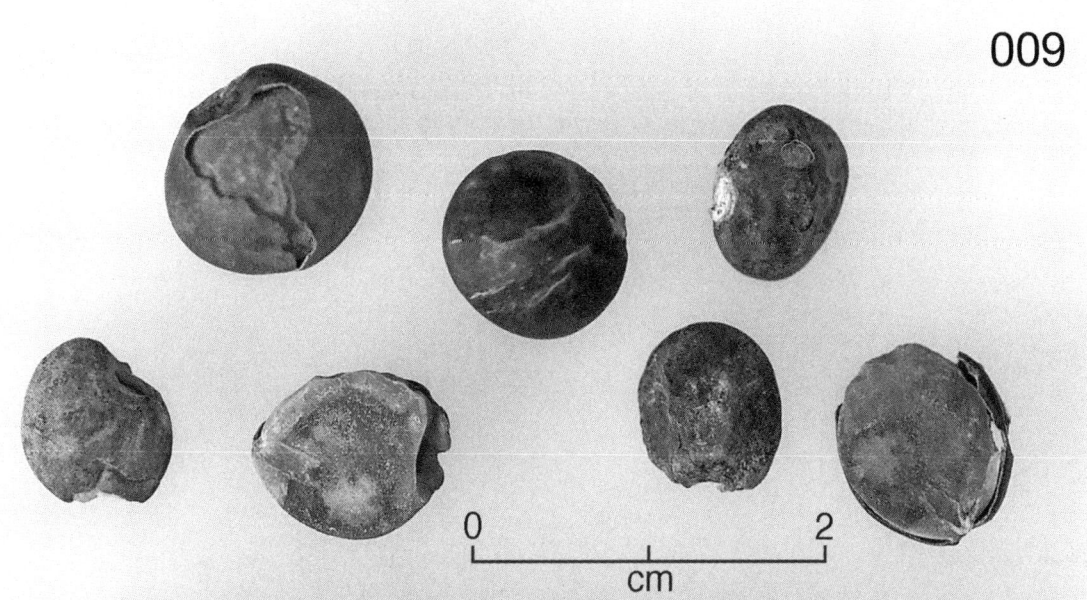

Figure 15.9. Sample 009, seven *Erythrina* seeds found in Feature 6 of Structure D.

the embryonic root tip. Without these distinctive characteristics, the most reliable *lunatus* trait is the pod in which the beans are enclosed. The *lunatus* pod, as exemplified by Cerro Azul Sample 001, is broad, contains few seeds, and always has a tip that is distinctly curved downward, i.e., toward the seam of the pod opposite the one to which the seeds are attached.

Phaseolus vulgaris

Six of the 11 samples from Cerro Azul (002, 003, 004, 006, 010, and 011) belong to the species *P. vulgaris*. Let me begin by discussing the four samples from Feature 6, a Late Intermediate midden in the Southwest Canchón of Structure D.

Sample 002 consists of six seeds, one mature and filled out and five that are immature and shrunken. Their color is dark red-brown. Length 0.95 cm, width 0.83 cm, thickness 0.44 cm.

Sample 003 consists of nine mature seeds, all dark red-brown and not variegated. Lengths range from 1.38 to 1.80 cm, width from 0.9 to 1.1 cm, and thickness from 0.73 to 0.90 cm. These beans are similar to *P. vulgaris* from Ancón in the collections of the Museo Nacional de Arqueología in Lima, Peru.

Sample 004 consists of eight seeds, seven mature and one immature. The color is dark red-brown, though a few appear to have darker longitudinal bands. Lengths range from 1.12 to 1.25 cm, width from 0.75 to 0.90 cm, and thickness from 0.58 to 0.70 cm. These beans are smaller and less elongate than the *P. vulgaris* from Ancón exemplified by Sample 003.

Sample 006 consists of nine seeds (and two fragments) of *P. vulgaris*, similar to Sample 004.

Sample 010 was recovered from Structure 12, a partially looted Late Intermediate burial cist in Quebrada 5-south. It consists of one fragmented pod and one loose seed. The seed has a length of 1.65 cm, a width of 1.10 cm, and a thickness of 0.85 cm. The color of the seed is dark red-brown and not variegated, like the seeds in Sample 003.

Sample 011 was recovered from Room 9 of Structure 1 on Cerro del Fraile. It dates to the 16th century A.D., and was associated with Spanish Colonial squatters who occupied an abandoned Inca building. The sample consists of seven large seeds of *P. vulgaris*. Their color is light cream-tan and not variegated; the seeds are somewhat lunate or curved. Lengths range from 1.5 to 1.75 cm, width from 1.0 to 1.1 cm, and thickness from 0.75 to 0.90 cm. These Spanish Colonial beans are distinct in color from the Late Intermediate *P. vulgaris* at Cerro Azul, and differ from all prehispanic south coast beans in the Museo Nacional de Arqueología that I have examined.

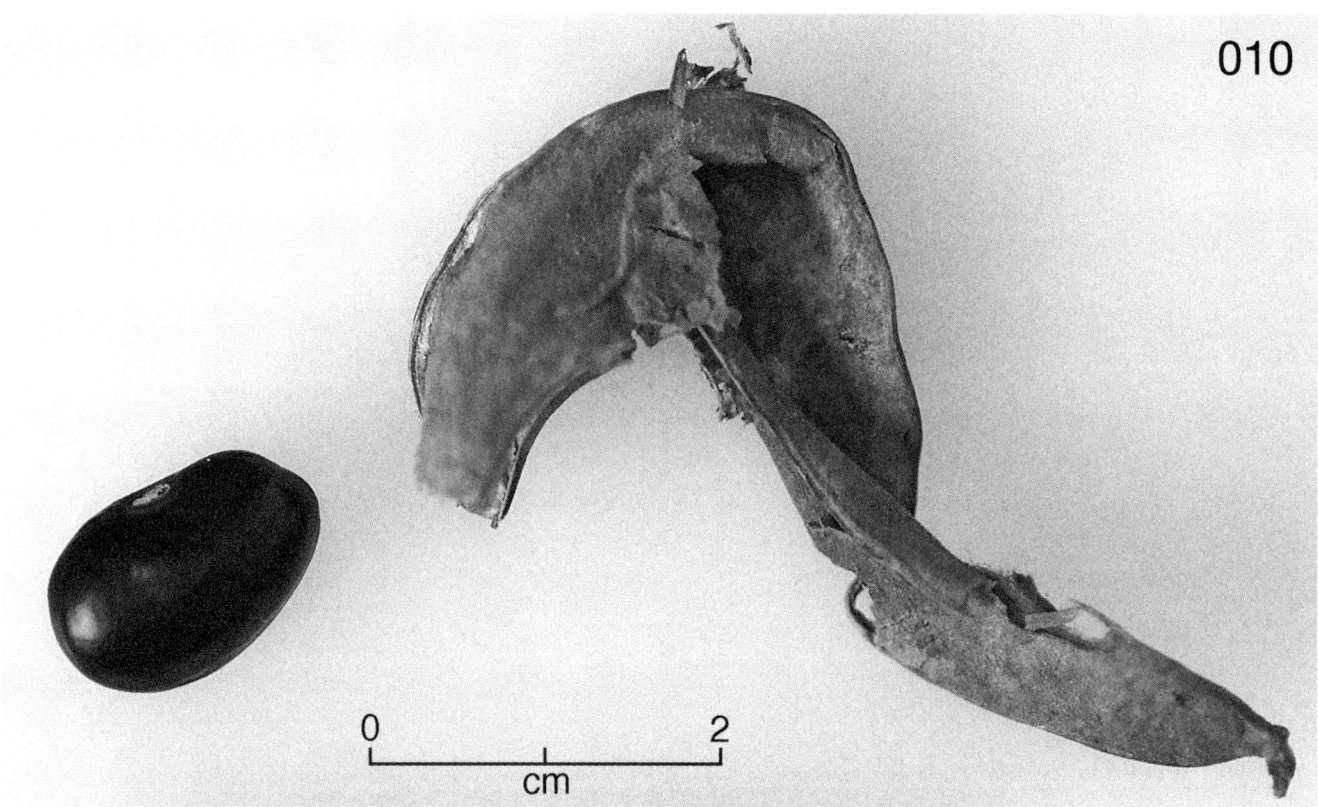

Figure 15.10. Sample 010, one *Phaseolus vulgaris* pod and one dark red-brown seed. Both items were found in Structure 12, a looted burial cist in Quebrada 5-south.

Figure 15.11. Sample 011, seven large *Phaseolus vulgaris* seeds (six intact and one broken in half), found in 16th-century Spanish Colonial refuse in Room 9 of Structure 1. These light cream-tan beans are unlike any of the prehispanic beans from Cerro Azul.

Phaseolus lunatus

Four of the 11 samples from Cerro Azul (001, 005, 007, and 008) are beans of the species *P. lunatus*. All were recovered from Feature 6, a Late Intermediate midden in the Southwest Canchón of Structure D.

Sample 001 consists of one intact pod of *P. lunatus*, containing two well preserved seeds. The pod is brown; the seeds are medium brown and not variegated. The seed length is 1.05 cm, the width (hilum to opposite edge) 0.98 cm, the thickness 0.8 cm, and the hilum length 0.4 cm with a dark eye ring. The length of the pod is 4.86 cm, its width 2.12 cm.

Sample 005 consists of two seeds of the type of large, elongate, and curved lima beans so often depicted on Mochica pottery. The beans have yellow seed coats with solid and spotted dark areas, concentrated at one end but extending over roughly two-thirds of the surface. The most fully intact seed has a length of 2.5 cm, a width of 1.1 cm, and a thickness of 0.6 cm. Similar *P. lunatus* seeds have been found at Ancón and Paracas.

Sample 007 consists of one pod and four seeds, identical to *P. lunatus* in Sample 001.

Sample 008 consists of three *P. lunatus* seeds, similar in color and markings to those in Sample 005, but smaller and with a much lower ratio of length to width. Lengths range from 1.3 to 1.7 cm, width from 0.85 to 1.1 cm, and thickness from 0.5 to 0.7 cm.

Erythrina sp.

One of the 11 samples from Cerro Azul (009) belonged to *Erythrina* sp., a leguminous tree of the tropical highlands of South America. One species of *Erythrina*, *E. edulis*, is grown in the Andes today as a food source for both humans and animals.

Sample 009 consists of seven seeds of *Erythrina* sp. from Feature 6, a Late Intermediate midden in the Southwest Canchón of Structure D. The contexts in which *Erythrina* was found at Cerro Azul suggest that it may have been used both as food and medicine (see Chapter 13).

16

Industrial Plants

C. Earle Smith, Jr. and Joyce Marcus

Editor's note: Dr. Smith's untimely death in 1987 precluded his writing this chapter. Therefore, using his field notes and plant counts, Marcus has done her best to create a document of which he would approve. It goes without saying that the late Dr. Smith is not responsible for any botanical errors.

In Andean archaeology there is a long tradition of separating edible, ritual, and medicinal plants from the so-called "industrial" plants—those used for fiber, cordage, matting, containers, or construction material. In this chapter we continue that tradition. At the same time, we acknowledge that some plants can serve in both edible and industrial capacities. Cattails, for example, can be used to make ropes and baskets, but their rhizomes are also edible. The fruit of the guava tree is a delicacy, but its flexible branches can also be used in the manufacture of crayfish traps (see Chapter 5).

Marcus' crew recovered more than 600 specimens of industrial plants from Late Intermediate contexts at Cerro Azul. As Table 16.1 shows, the list includes cotton for weaving, bottle gourds for containers, *Acacia* and *Prosopis* spines for spindles, prickly pear spines for needles, bulrushes for matting, bromeliads for sling tassels, and soapberries for washing one's hair.

Industrial Plant Remains at Cerro Azul

In this chapter we briefly describe the industrial plants used at Cerro Azul. We then discuss all the specimens recovered from archaeological contexts. We list the industrial plants in phylogenetic order, following the work of Simpson (2010).

Unspecified Palms

Palms do not grow near Cerro Azul, but palm products of some kind may have been imported to the site. Smith was convinced that some of the seeds in his "unidentified plants" category were from palms, but he could not determine the genus. In addition, Linda Perry (personal communication, 2015) detected palm phytoliths in the earthen matrix of the Feature 6 midden (Structure D).

Table 16.1. So-called Industrial Plants Recovered from Late Intermediate Contexts at Cerro Azul.

Unspecified palm	Arecaceae
Chonta palm	*Gulielma ciliata*
Cattail/sacuara	*Typha* cf. *angustifolia*
Bromeliad	cf. *Tillandsia* sp.
Rush	*Juncus* spp.
Bulrush/totora	*Scirpus* cf. *californicus*
Sedge/junco	*Cyperus* spp.
Carrizo/caña hueca	*Phragmites australis*
Caña brava	*Gynerium sagittatum*
Caña de Guayaquil	*Guadua angustifolia*
Unspecified grass	Poaceae
Burr grass	*Cenchrus* sp.
Willow	*Salix* cf. *humboldtiana*
Huarango	*Acacia* cf. *macracantha*
Algarrobo	*Prosopis* cf. *chilensis*
Bottle gourd	*Lagenaria siceraria*
Prickly pear	*Opuntia* spp.
Cotton	*Gossypium barbadense*
Kidney cotton	*Gossypium barbadense* var. *brasiliense*
Achiote	*Bixa orellana*
Unspecified wood	
Guava branches	*Psidium guajava*
Balloon vine	*Cardiospermum* sp.
Soapberry	*Sapindus saponaria*
Almendrillo/ huevo de coto	*Cordia* spp.
Woody vines	Bignoniaceae

Chonta Palm

The chonta palm (*Gulielma ciliata*) is native to eastern Peru. Our workmen were familiar with chonta wood and believed that our Late Intermediate malleros (the rectangular wooden plaques used as templates for spacing the knots during net-making) were made from that material (see Chapter 7).

Cattails

Sacuara or cattails (*Typha* cf. *angustifolia*) grow in marshy areas of the Cañete coast. Their rhizomes can be eaten, and their fibrous stems and leaves can be used for matting, basketry, and ropes. Figure 16.1 displays a cattail stem found in a refuse deposit in Terrace 11 of Quebrada 6. Similar cattail stems were used to make baskets in the Paracas region, south of Cañete (Towle 1961:16).

Figure 16.1. Stem of sacuara or cattail (*Typha* cf. *angustifolia*) from a refuse deposit in Terrace 11 of Quebrada 6.

Figure 16.2. Totora or bulrushes (*Scirpus californicus*).

Figure 16.3. Bulrush and cattail growing intermixed. The dark bars on the scale are each 10 cm long.

Bromeliads

Bromeliads do not grow at Cerro Azul, but *Tillandsia* sp. can be found living on sand dunes and rocky ledges not far inland. *Tillandsia* (a relative of Spanish moss) is known in Peru as *puya*. Its fibers were used as stuffing, pillows, and decoration on the Peruvian coast (Towle 1961:31). At Cerro Azul, *Tillandsia* was also used for the decorative tassels on slings (see Chapter 10).

Rushes

Rushes (*Juncus* spp.) are widespread in the wetlands of the Andes, but tend to prefer cooler habitats than the Pacific coast offers. Rushes were not particularly common at Cerro Azul, perhaps because they had to be imported from higher and cooler elevations. The stems and leaves could be used for cordage.

Bulrushes

The bulrush *Scirpus* cf. *californicus*, known in Peru as totora, grows in swampy depressions from the United States to Argentina. In Peru its fibrous stems and leaves were used to make rope, cordage, mats, baskets, and caballitos de totora (Towle 1961:26). The root-stocks are also edible.

Figure 16.2 shows totora growing near Cerro Azul in the 1980s. Figure 16.3 features a cluster of *Scirpus* growing intermixed with cattails (*Typha*).

At Cerro Azul, totora was sometimes bundled for use as pillows (Fig. 16.4). It might also be braided into rope (Fig. 16.5) or twilled into mats (Fig. 16.6).

Sedges

We found sedges (*Cyperus* sp.) growing near Cerro Azul during the 1980s (Fig. 16.7). Figure 16.8 shows a Late

Figure 16.4. A bundle of totora from Structure 4, a looted burial cist in Quebrada 5a. Such bundles were used as pillows in life, and included in one's mummy bundle.

Intermediate inflorescence of junco, or sedge, from a refuse deposit in Terrace 11 of Quebrada 6. In addition to their usefulness in cordage and basketry, sedges have rhizomes that can be eaten (Towle 1961:26).

Carrizo or Caña Hueca

Phragmites australis, known locally as carrizo or caña hueca, grew in moist locales near Cerro Azul during the 1980s (Figs. 16.9, 16.10). This readily available, sturdy reed was put to multiple uses during the Late Intermediate, showing up in the middens of both elite compounds (Fig. 16.11, *a*) and commoner-class households (Fig. 16.12).

One of the most common uses of *Phragmites* was in the construction of kincha (or wattle-and-daub) houses. Figure 16.13 shows some of the burnt carrizo from the walls of a destroyed kincha house, found resting on Stratigraphic Zone C of Terrace 9, Quebrada 5a (Marcus 2008:286 and Fig. 9.28). Figures 16.14 and 16.15 present examples of burnt clay daub from kincha houses at Cerro Azul, each bearing impressions of the carrizo used in the walls.

Phragmites australis grows to 2–4 meters in height in coastal Peru, often coexisting with caña brava (see below). At Paracas, to the south of Cañete, it was made into reed tubes, combs, spindles, and matting (Towle 1961:19).

Caña Brava

Gynerium sagittatum, known locally as caña brava, is a sturdy cane that forms dense stands up to 10 meters high in swampy areas of the Cañete coast (Fig. 16.16). Its culms can be used to make strong baskets and projectile shafts, and in Paracas its leaves were woven into baskets (Towle 1961:18).

Figure 16.17 shows two poles from a litter that accompanied one of the looted burials in Structure 6 of Cerro Azul; both poles were identified by Smith as *Gynerium*. Figure 16.18 shows a

Figure 16.5. Braided rope from Feature 20, Structure 9; it appears to be made of bulrushes.

deliberately cut and trimmed artifact of caña brava from Room 4 of Structure D.

Caña de Guayaquil

Guadua angustifolia, known locally as caña de Guayaquil, is a species of stout bamboo native to Amazonia. *Guadua* grows 10 meters high and can reach a diameter of 15 cm. It was imported to the coast for heavy duty tasks, including the construction of seagoing rafts (Towle 1961:18).

Unspecified Grasses

Remains of grasses (Family Poaceae) were found in various Late Intermediate contexts at Cerro Azul, but were almost never identifiable to species. Some grasses may have been brought to the site for thatching or weaving. In other cases, we found quids of grass that appeared to have been partially chewed. Such quids suggested that Cerro Azul kept supplies of grass on hand to feed llama caravans when they arrived.

Burr Grass

One of the few grasses identifiable to genus was burr grass (*Cenchrus* sp.). Whether it was used in construction or fed to llamas is unclear.

Willow

Willow trees grow along the Río Cañete to elevations of 3000 meters. None of our willow specimens could be identified to species, but one possibility is *Salix humboldtiana*. Willow branches are flexible enough to be made into osier baskets and other artifacts. At Cerro Azul, willow twigs were often bundled for use as pillows; an example can be seen in Figure 16.19.

Huarango and Algarrobo

Two sturdy, deep-rooted leguminous trees can be found growing on the alluvium of the Río Cañete in the Kingdoms of Huarco and Lunahuaná. One of these trees is the huarango (*Acacia* cf. *macracantha*), whose wood was used in the construction of kincha houses (Marcus 2008:286 and Fig. 9.28). The other is the algarrobo (*Prosopis* cf. *chilensis*), a tree whose pods contain edible gum (Towle 1961:56) and whose wood can be used in construction.

Both *Acacia* and *Prosopis* bear long, sharp spines on their limbs. At Cerro Azul, these spines were abraded into double-

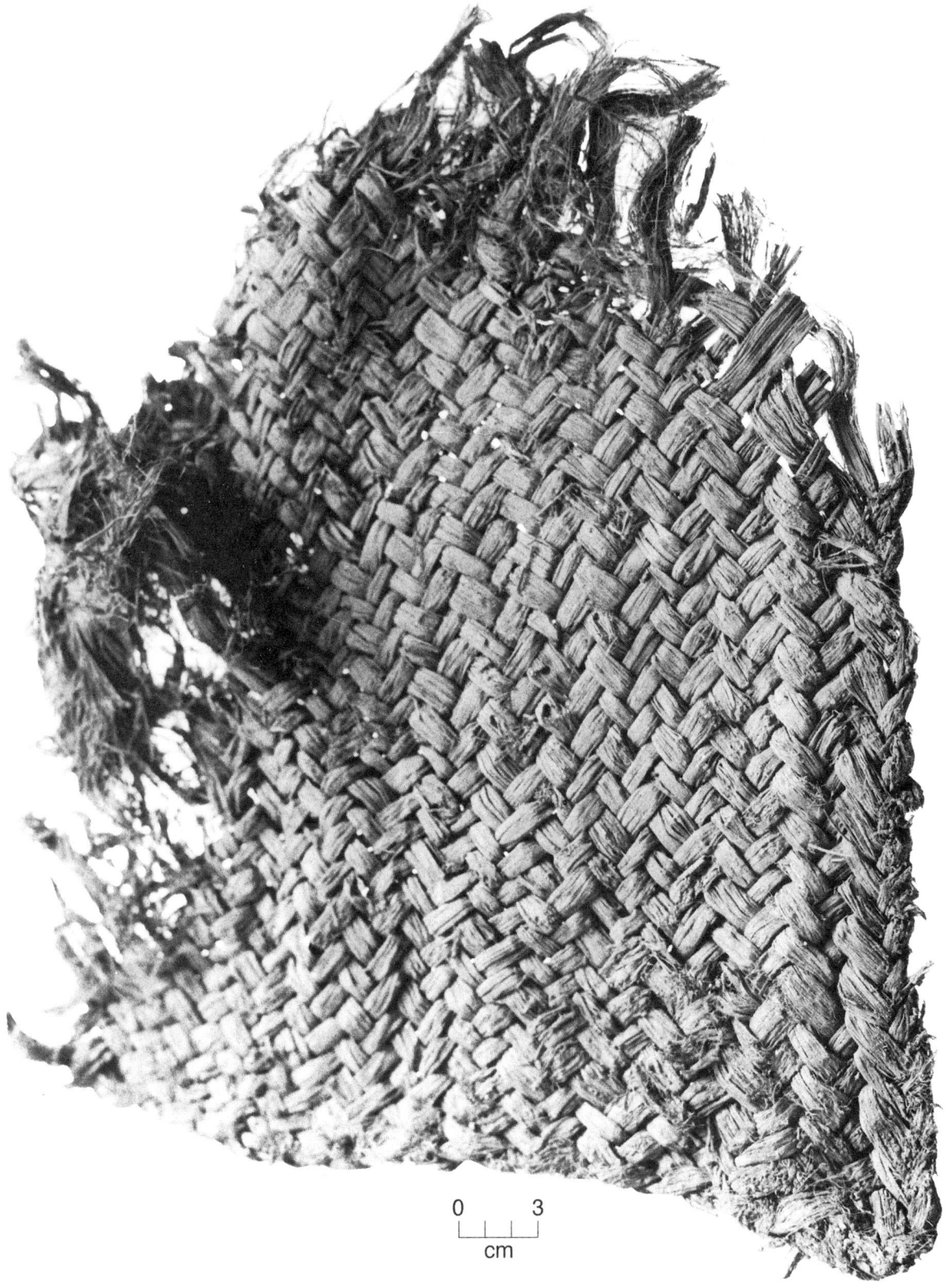

Figure 16.6. Twilled mat from looters' debris, Zone A of Terrace 9, Quebrada 5a. It appears to be made of bulrush.

Figure 16.7. Junco or sedge (*Cyperus* sp.). The total length of the scale is 13 cm.

Figure 16.8. Inflorescence of sedge (*Cyperus* sp.) from a refuse deposit in Terrace 11 of Quebrada 6.

Figure 16.9. Carrizo or caña hueca (*Phragmites australis*). Ramiro Matos serves as scale.

Figure 16.10. Carrizo or caña hueca (*Phragmites australis*). The dark bars on the scale are each 10 cm long.

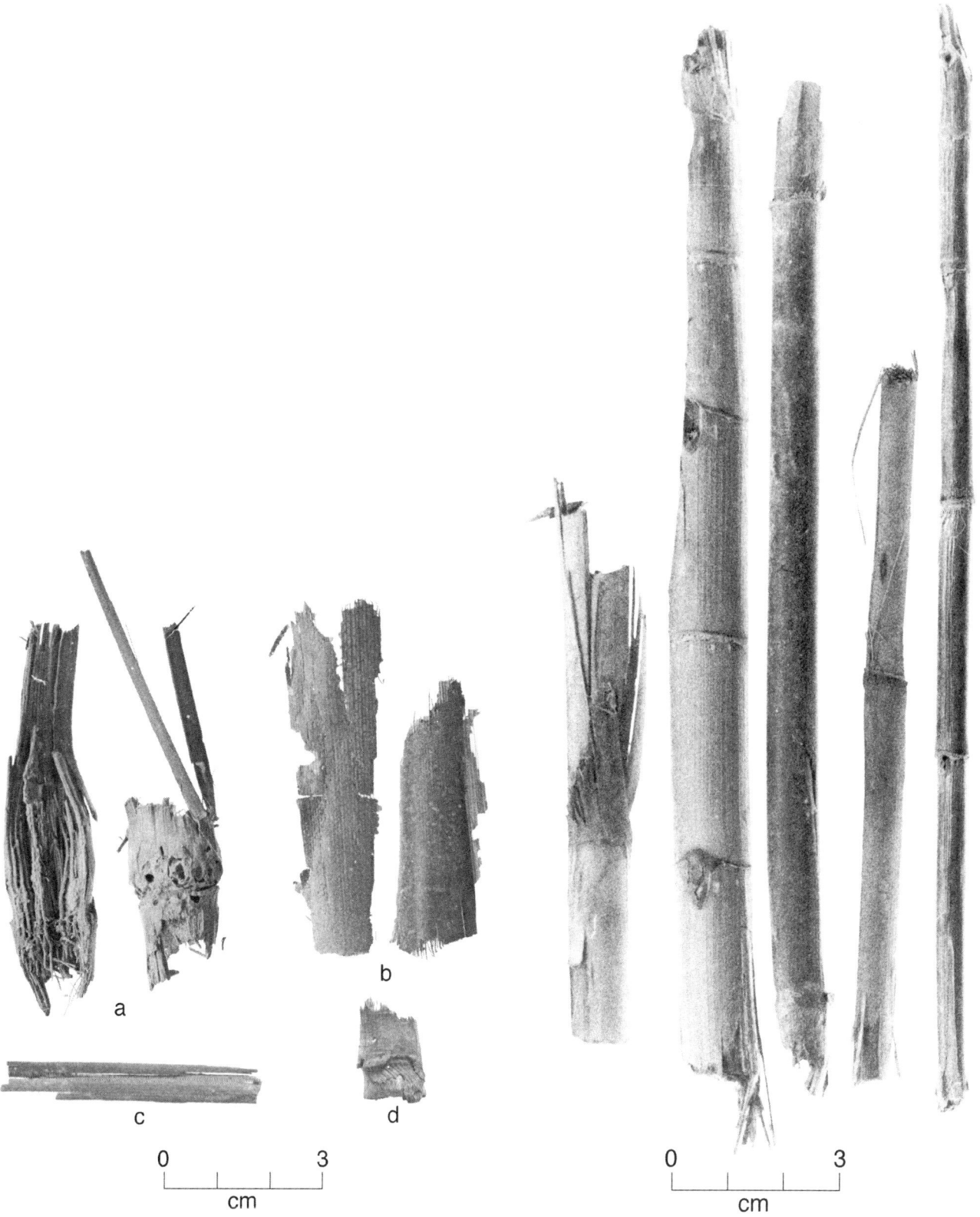

Figure 16.11. A sample of monocotyledonous plants from Feature 6, Structure D. *a*, Phragmites. *b*, Gynerium. *c*, Scirpus. *d*, stalk of *Zea mays*.

Figure 16.12. A sample of reeds and canes from Feature 20, Structure 9. Included are carrizo and caña brava.

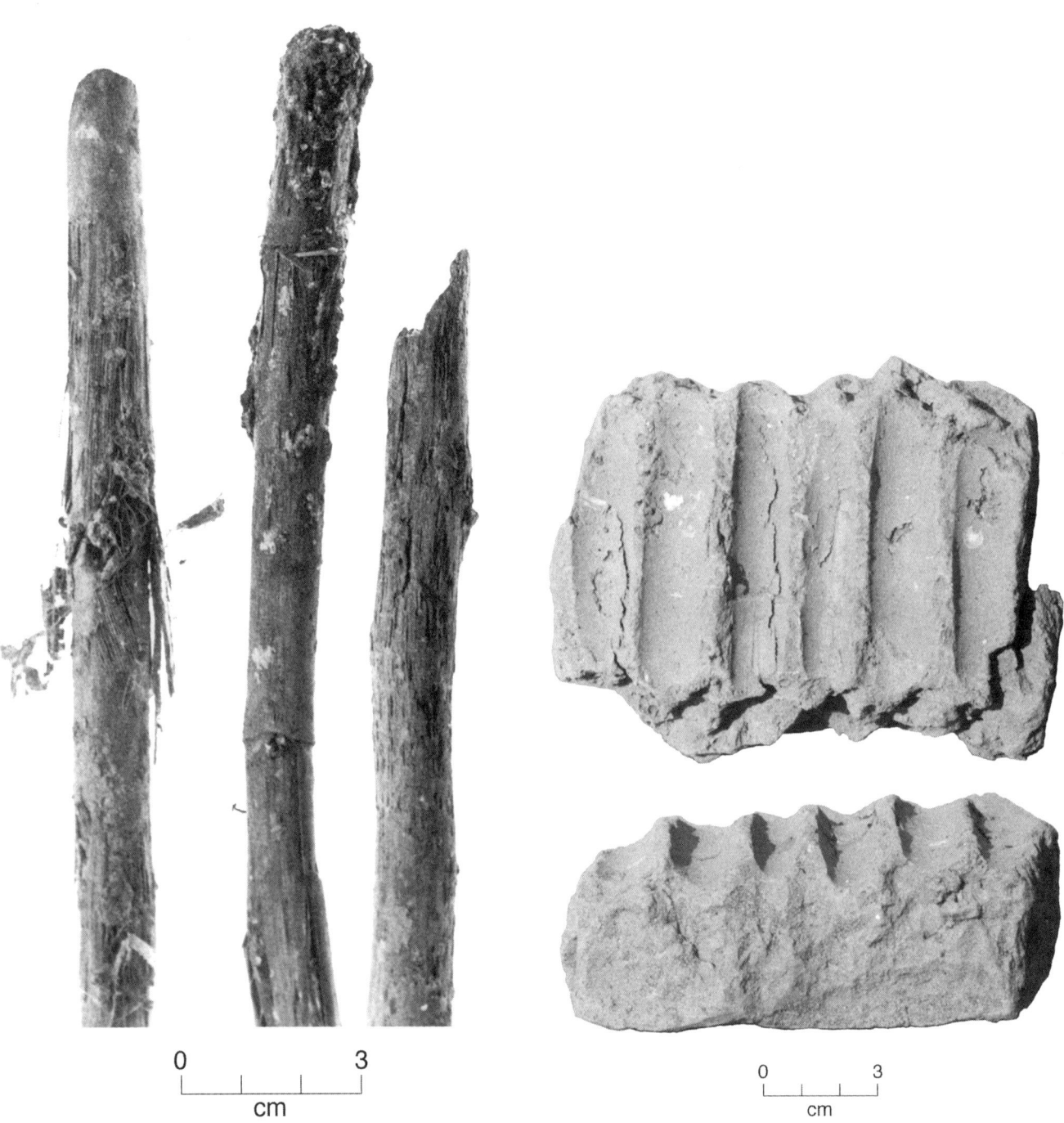

Figure 16.13. Three specimens of caña hueca from the remains of a kincha house. These remains were found atop Zone C of Terrace 9, Quebrada 5a.

Figure 16.14. Two views of a fragment of kincha wall, found on the central platform of Structure 9.

Industrial Plants

Figure 16.15. Three fragments of kincha wall from a squatters' house in "Room 2," Structure D.

pointed spindles for spinning fiber. Examples can be seen in Figure 16.20.

Bottle Gourds

The bottle gourd (*Lagenaria siceraria*), known in Peru as the *mate*, was widely used for containers at Late Intermediate Cerro Azul. Figure 16.21 shows a gourd bowl and a gourd lid, both found in looted burial cists.

Intact burials at Cerro Azul were often accompanied by shallow gourd bowls containing guinea pigs, dried fish, or plant foods. Figure 16.22 shows a gourd bowl found with Burial 2 (Quebrada 5a). Figure 16.23 shows a similar bowl found with Individual 1 of Burial 6 (Quebrada 5a). Burials from this same time period in the neighboring Valley of Chincha were often supplied with food in gourd bowls like ours (Towle 1961:94).

So numerous were bottle gourds at Cerro Azul that we suspect they were grown locally, perhaps by the same farmers who produced butternut squash (see Chapter 13). The biggest unanswered question about the bottle gourd, of course, is its place of origin. No wild bottle gourds have ever been found in the New World, meaning that their origin must be sought in Asia or Africa. Presumably this question will ultimately be answered by DNA analysis, but debates continue at this writing.

Prickly Pear

Prickly pear cactus (*Opuntia* spp.) can be found in the piedmont of the lower Cañete Valley, as well as at higher elevations. This cactus has spines up to 13 cm long, and the Late Intermediate occupants of Cerro Azul turned these spines into sewing needles.

The University of Michigan crew recovered both unmodified and modified *Opuntia* spines. Figure 16.24 shows two unmodified spines found with Individual 2 of Burial 4 (Quebrada 5a). Another large, unmodified spine was found in the fill of an earthen ramp in Structure D (Fig. 16.25). Finally, Figure 16.26 displays both eyed needles and unmodified *Opuntia* spines from an elite woman's weaving workbasket (Burial 4, Quebrada 5a).

Cotton

One of Peru's most important industrial plants was cotton (*Gossypium barbadense*). Its wild form still occurs in the Guayas Basin of coastal Ecuador and in neighboring areas of Peru (Stephens 1975:407). Chaudhary et al. (2008:567) suggest that *G. barbadense* was first domesticated on the western slopes of the northern Peruvian Andes. It was well established during the Archaic at coastal sites like Ancón and Huaca Prieta.

Wild *G. barbadense* was a perennial shrub producing short,

Figure 16.16. Caña brava (*Gynerium sagittatum*). Ramiro Matos serves as scale.

Industrial Plants

0   3
cm

Figure 16.17. Two litter poles made from caña brava. Found in Structure 6, a partially looted burial cist in Quebrada 5a.

0   3
cm

Figure 16.18. Deliberately cut and trimmed section of caña brava, probably an unfinished artifact. Found in the fill of Room 4, Structure D.

Figure 16.19. A bundle of willow twigs and pacay rachises from Structure 6, a partially looted burial cist in Quebrada 5a. Such bundles were used as pillows in life, and included in one's mummy bundle.

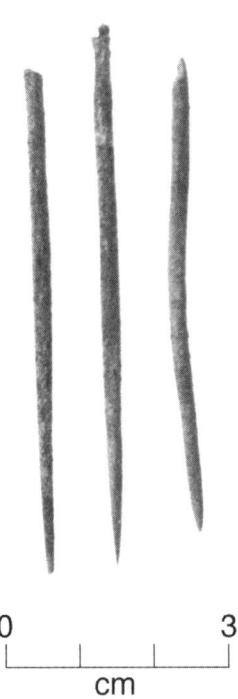

Figure 16.20. Three broken huarango or algarrobo spines from Feature 20, Structure 9.

Industrial Plants 275

Figure 16.21. The bottle gourd (*Lagenaria siceraria*) was used for containers at Cerro Azul. *Above*, the circular lid of a gourd container, found in Structure 12, a looted burial cist in Quebrada 5-south. *Below*, a bowl with restricted orifice from Structure 5, a looted burial cist on Terrace 9 in Quebrada 5a.

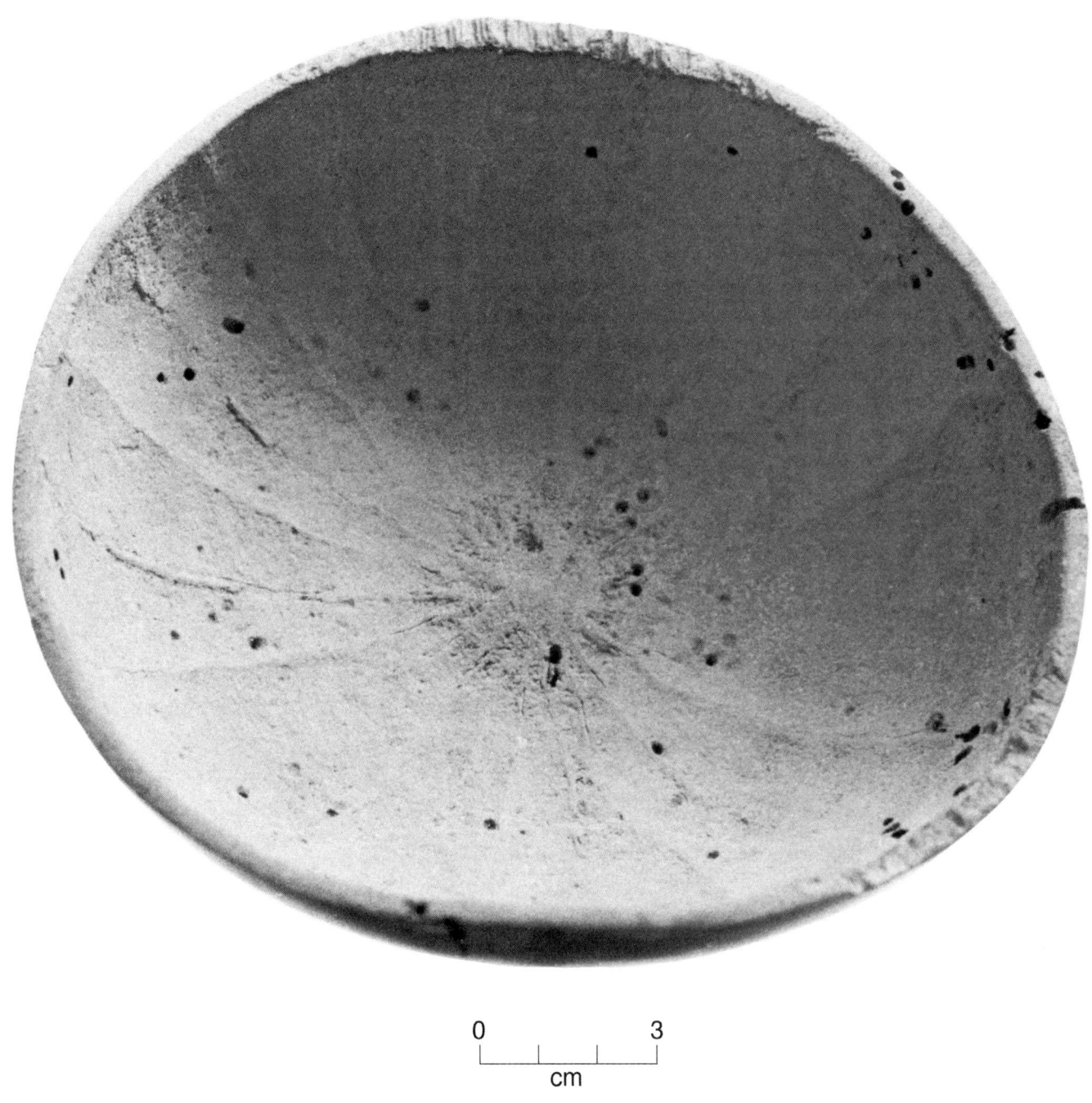

Figure 16.22. A shallow gourd bowl found with Burial 2 (Quebrada 5a, Terrace 9).

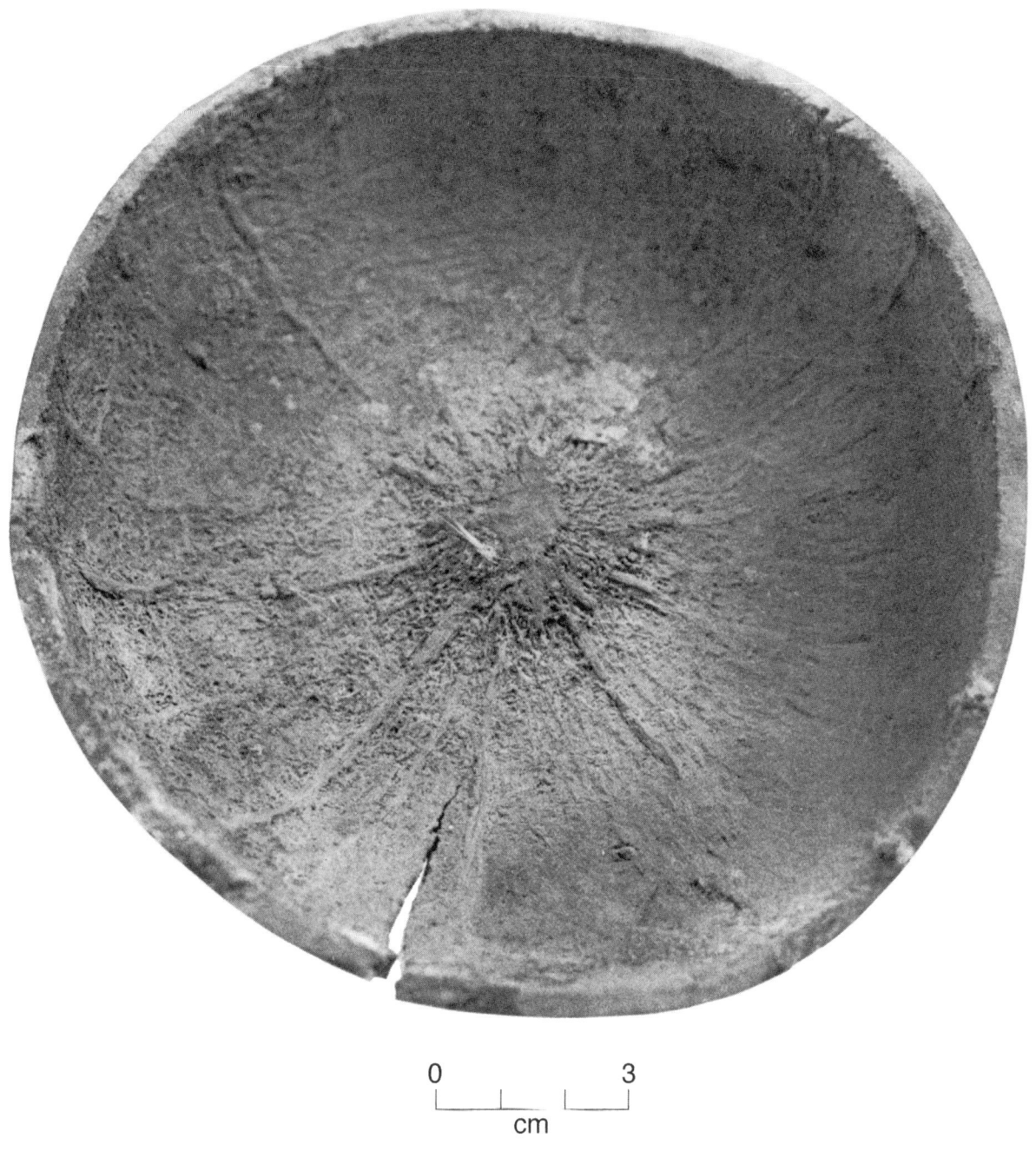

Figure 16.23. A shallow gourd bowl found with Individual 1 of Burial 6 (Quebrada 5a, Terrace 9).

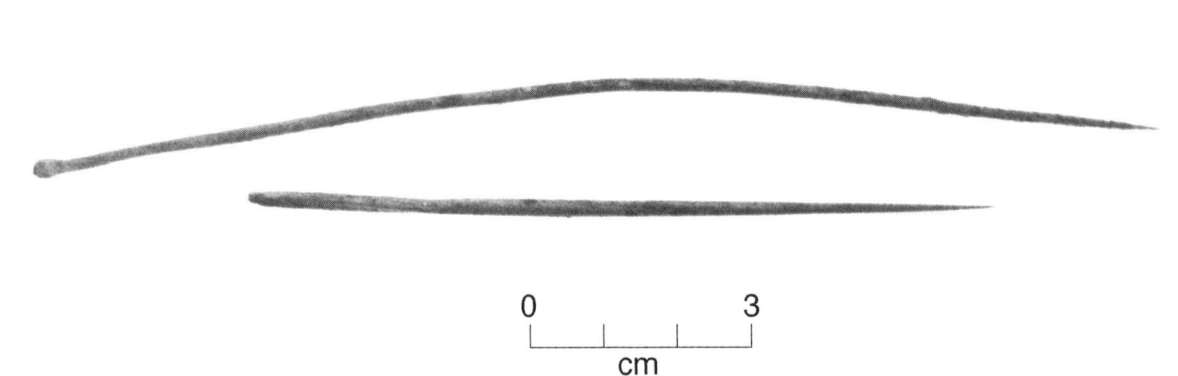

Figure 16.24. Two very large spines of the prickly pear (*Opuntia* sp.) found with Individual 2 of Burial 4 (Quebrada 5a, Terrace 9).

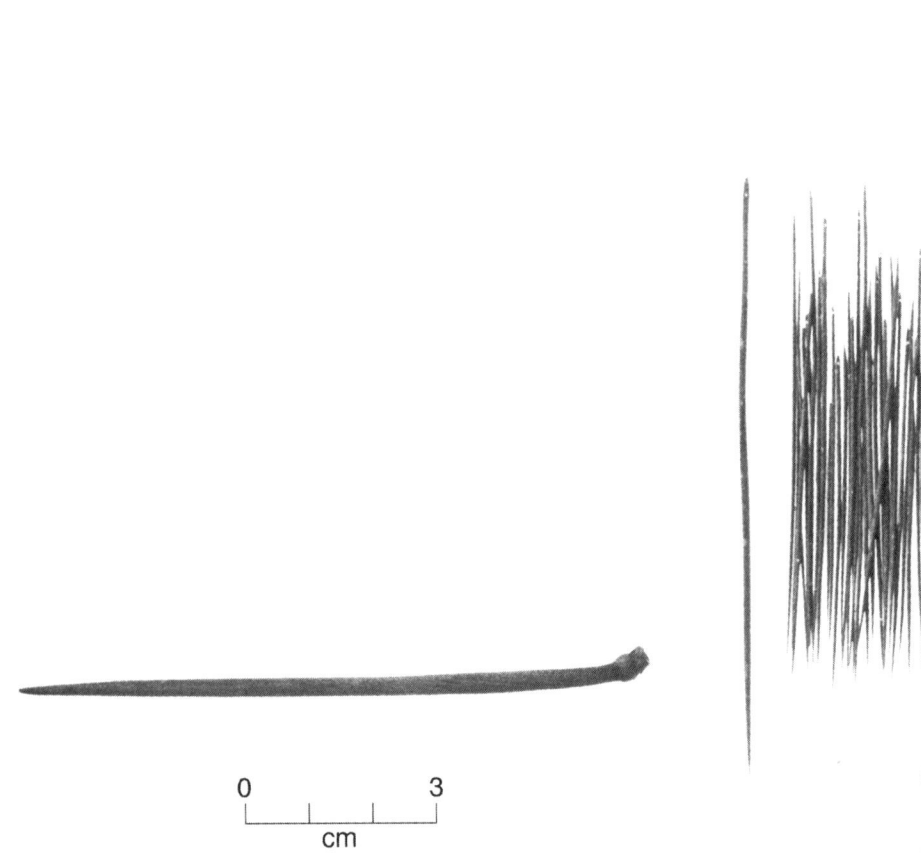

Figure 16.25. Large spine of prickly pear, found in the fill of an earthen ramp on the west side of "Room 2," Structure D. Spines like this were made into needles.

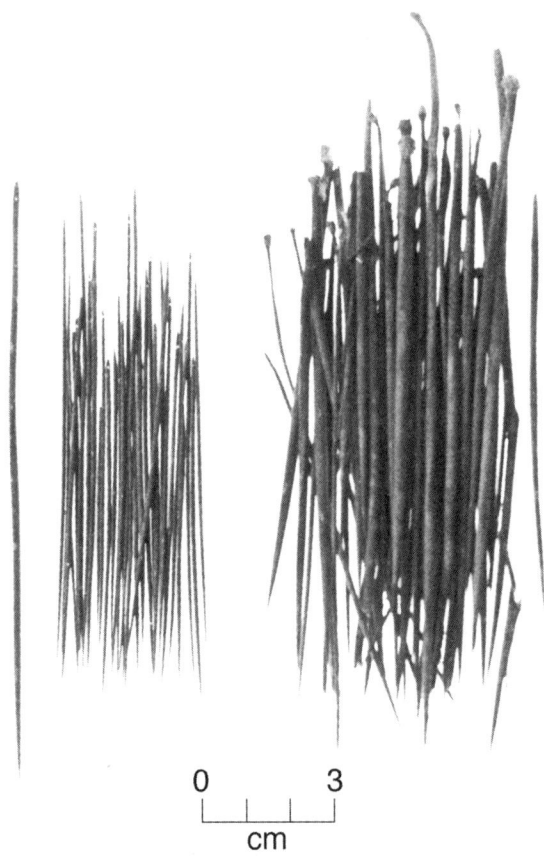

Figure 16.26. Unmodified cactus spines (*right*) and needles with eyes made from cactus spines (*left*). These items came from a workbasket in the vicinity of Individuals 1a and 2 in Burial 4 (Quebrada 5a, Terrace 9).

Figure 16.27. Boll of cotton (*Gossypium barbadense*) from a refuse deposit in Terrace 11 of Quebrada 6.

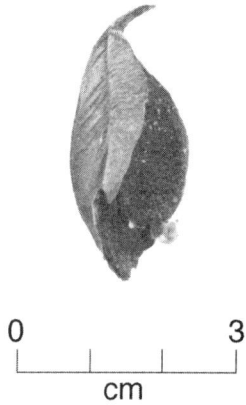

Figure 16.28. Segment of cotton boll (*Gossypium barbadense*) from the floor of the South Corridor, Structure D.

coarse, weak fibers. Following domestication, human selection turned Peruvian cotton into an annual crop plant producing longer, finer, and stronger fibers; a number of genes were involved in this transition (Chaudhary et al. 2008:575).

Wild cotton fiber ranges in color from dingy gray to rusty brown. Under domestication, however, *G. barbadense* developed two contrasting fiber colors: pure white (known in Peru as *algodón blanco*) and chocolate brown (known as *algodón pardo*). Both color variations had been established by Archaic times (Stephens 1975:412–417), and both were present at Cerro Azul during the Late Intermediate (see Fig. 13.4). In fact, some Late Intermediate textiles had alternating stripes of white and chocolate cotton.

Late Intermediate *G. barbadense* produced both three-lock and four-lock bolls (Stephens 1975:407), but few of the specimens found in archaeological refuse were well enough preserved to allow for the counting of locks. Both Figures 16.27 and 16.28 show cotton bolls from Late Intermediate contexts. Both Figures 16.29 and 16.30, on the other hand, illustrate loose cotton seeds.

Kidney Cotton

One variety of Peruvian cotton—*Gossypium barbadense* var. *brasiliense*—is known as "kidney cotton." This shrub gets its name from the fact that it has long, attenuated bolls, bearing large seeds in kidney-shaped masses (Stephens 1975:418). Cotton of the *brasiliense* variety is usually found in the Amazon Basin and the eastern slopes of the Andes. Two locks of kidney cotton, however, were found with Burial 8 in Quebrada 5a (Fig. 16.31). Evidently, in spite of the fact that large amounts of *algodón blanco* and *algodón pardo* were available on the coast, noble families at Cerro Azul occasionally imported kidney cotton from eastern Peru.

Achiote

Achiote (*Bixa orellana*) is one of those plants that can be considered edible, ritual, or industrial (see Chapter 13). Among its industrial uses are as red paint.

Unspecified Wood

In various refuse deposits at Cerro Azul we found pieces of wood or bark, some of which had been burned as fuel. Those specimens that could not be identified to genus or family were simply assigned to the category "unspecified wood."

Figure 16.29. Seeds of cotton (*Gossypium barbadense*) from Feature 6, Structure D. Shown here at twice natural size.

Figure 16.30. Cotton seeds from Feature 20 of Structure 9. Shown here at life size.

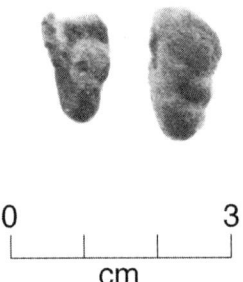

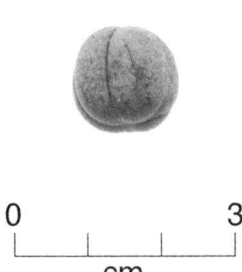

Figure 16.31. Two locks of kidney cotton (*Gossypium barbadense* var. *brasiliense*) found with Burial 8 (Quebrada 5a, Terrace 9).

Figure 16.32. A seed of *suirucu* or soapberry (*Sapindus saponaria*) from the floor of Room 2 in Structure 11, a storage facility in Quebrada 5-south.

Guava Branches

Guava (*Psidium guajava*) has already been discussed in Chapter 13 because of its edible fruit. We found, however, that its flexible branches were also used as industrial raw material. For example, the wooden hoops used in the construction of crayfish traps were made from guava branches (see Chapter 5).

Balloon Vine

Occasional specimens of the balloon vine (*Cardiospermum* sp.) were found in Late Intermediate refuse. The vine itself can be used in construction, and its balloon-like fruits can be used as stoppers.

Soapberry

The *suirucu* or soapberry (*Sapindus saponaria*) is a tree growing 4–10 meters in height. Its fruits contain saponin, which forms a soapy lather in water. Soapberry seems to have been widely used by both nobles and commoner-class families at Cerro Azul (Fig. 16.32), with the washing of one's hair a likely activity. In addition, the fruit of the soapberry contains shiny black seeds that were made into beads for both humans and figurines. Towle (1961:63) reports considerable use of soapberry at Nasca sites, to the south of Cañete.

Almendrillo/Huevo de Coto

Occasional specimens of *Cordia* spp., a shrub or tree known in Peru as almendrillo or huevo de coto, appeared in Late Intermediate refuse. It is not clear to what use this plant was put.

Woody Vines

Woody vines of the family Bignoniaceae were occasionally found in the refuse at Cerro Azul. Smith was unable to identify them to genus, but noted that some of these woody vines would have made useful industrial plants.

Plant Remains from Archaeological Contexts

The University of Michigan project recovered in excess of 600 specimens of industrial plants at Cerro Azul. In the pages that follow, we list all those specimens found in archaeological contexts.

Structure D

Our largest single sample of industrial plants in Structure D came from **Feature 6**, a midden in the building's Southwest Canchón. That midden was screened through both 1.5 mm and 0.6 mm mesh (Marcus 2008:95–96).

Table 16.2. So-called Industrial Plants Recovered from Feature 6 of Structure D.

Tillandsia sp.	3 fragments of fiber
	5 miscellaneous foliage fragments
Bulrush/totora	1 fragment
Cyperus sp.	1 knot tied in plicate leaves
Carrizo/caña hueca	1 fragment
Caña brava	1 fragment
Unspecified grass	1 quid of fibers
Burr grass	1 caryopsis
Prickly pear	1 spine
Cotton	425 loose seeds (Fig. 16.29)
	1 peduncle from a three-lock boll
	3 locks of white cotton with a total of 8 seeds
	2 locks of chocolate cotton with a total of 10 seeds
	1 boll segment
	1 cotton swab
Unspecified wood	3 fragments of wood (1 burned)
	4 fragments of bark
Balloon vine	13 seeds
Soapberry	11 loose seeds (1 immature)
	1 seed in a fleshy layer
	30 fragments
Cordia sp.	2 seeds
Bignoniaceae	1 pod valve

Totals:
  NISP = 513

Let us comment briefly on the list of plants from Feature 6. This midden is believed to represent debris swept up from the floor of the Southwest Canchón, late in the occupation of Structure D. The hundreds of cotton seeds found in Feature 6, therefore, suggest that bolls were opened and the seeds removed from the fiber in this canchón.

The quid of chewed grass suggests that llamas were fed in the canchón, a possibility reinforced by the large number of llama dung pellets found in Feature 6 (see Chapter 17). The large number of soapberry specimens suggests that the seeds were removed from the fruits in the Southwest Canchón, perhaps as a first step in soap making.

Finally, we should add that Linda Perry (personal communication, 2015) detected palm phytoliths in the earthen matrix of Feature 6. We are uncertain whether the source of these phytoliths was debris on the floor of the canchón, or the tapia clay eroding from the surrounding walls. Both materials ended up in Feature 6.

The South Corridor

The South Corridor was a single-file passageway leading from the Southwest Canchón to the interior of Structure D. At some point, presumably late in the building's history, someone left a pile of debris on the floor of the corridor. Included in the debris were three fragments of leaf sheath from caña hueca; one leaf sheath of an unspecified grass; and one boll segment of cotton (Fig. 16.28).

Collca 1

Collca 1 was a storage bin attached to the south wall of the Northeast Canchón. Masses of both white and chocolate cotton seem to have been stored in this bin. This discovery makes sense in light of the fact that spinning and weaving took place in the canchón.

"Room 2"

"Room 2" turned out, after excavation, to be a small interior patio. On its west side it featured an earthen ramp that led to Rooms 5, 6, and 7. At least three huarango posts (two square and one round in cross-section) were part of the construction of this ramp. In the earthen fill of the ramp Marcus' crew found one large, unmodified *Opuntia* spine (Fig. 16.25).

Room 4

Room 4 had a complex stratigraphic history. Its original floor lay at a depth of 3.73 m below the datum point established for Structure D. At a later date the room was partially filled in, and a new floor created at a depth of 2.23 m below datum. Feature 5, a mass of partially restorable ceramic vessels, was found on this higher floor.

In the lower fill of the room, not far above the original floor, Marcus' crew recovered one pod and one peduncle of achiote. In the upper fill of the room, above the second floor, they found one culm section of caña brava trimmed as an artifact (Fig. 16.18).

Room 5

Originally designed to be a storage unit, Room 5 was later converted to residential quarters. Industrial plants were found only in the later (residential) stage of the room; included were an unspecified grass and a fragment of bark.

Rooms 9 and 10

Rooms 9 and 10 were devoted to the raising of guinea pigs. The guinea pigs themselves were kept in Room 9, and their bedding was assigned by Smith to his "unspecified grass" category.

Room 10, in contrast, was where the occupants of Structure D had stored fodder for the rodents. Over the centuries, this fodder had deteriorated to a mushy green carpet that defied identification by macroscopic techniques. We submitted a sample of guinea pig dung to Linda Perry to see if microscopic analysis could shed light on their food. Several clues suggested that their fodder might have included herbs of the chenopod-amaranth group. First, their dung contained the tiny starch grains typical of cheno-ams. Second, no phytoliths were present in the dung, and cheno-ams lack phytoliths. Third, cheno-am greens are rich in nitrogen, and a high-nitrogen food source would account for the bright green color of the vegetal mush on the floor of Room 10. Needless to say, these suggestions are tentative pending further analyses.

Late Horizon Squatters' Debris

At some point after Structure D had ceased to function as an elite residential compound, commoner-class squatters built a kincha house in the "Room 2" patio. Pottery from this house dated it to the Late Horizon.

Feature 4 was a Late Horizon midden associated with the squatters' house. Included in this midden were one prickly pear spine and one small hank of *alogodón blanco*.

Structure 9

The only sample of industrial plants from Structure 9 came from a midden called **Feature 20**. This midden is believed to represent domestic refuse from the kincha house of the building's commoner-class administrator. The industrial plants from Feature 20 were as follows:

Bulrush/totora	1 braided rope (Fig. 16.5)
Carrizo/caña hueca	4 fragments
Caña brava	4 fragments
Unspecified grass	1 hollow stem
Huarango/algarrobo	3 broken spines
Prickly pear	1 spine, modified as a bobbin
Cotton	35 loose seeds (Fig. 16.30) several fragments of fiber
Soapberry	4 seeds (one still inside a fruit)

Industrial Plants from the Quebradas of Cerro Camacho

The University of Michigan crew discovered a number of industrial plants in midden debris on various quebrada terraces. Still other quebrada specimens were found with mummy bundles, or in the debris from looted Late Intermediate burials.

Quebrada 5-south

Excavations in Quebrada 5-south revealed a multi-room storage facility (Structure 11), a looted burial cist (Structure 12), and a small domestic midden. Both of the structures produced industrial plants.

Structure 11. Room 2 of Structure 11 yielded one seed of soapberry (Fig. 16.32). Room 3 produced one stem of bottle gourd.

Structure 12. The looters who broke into Structure 12 left behind two gourd bowls; a gourd lid (Fig. 16.21, *a*); two splints of caña hueca; and several swatches of chocolate-colored cotton.

Quebrada 5a, Terrace 9

The stratigraphy of Terrace 9 in Quebrada 5a was complex. The first use of the terrace (Zone C) was as a place to dump refuse. Atop the refuse were the remains of a kincha house whose industrial plants included huarango (for the upright posts) and caña hueca (for the walls; see Fig. 16.13).

Above Zone C lay Stratigraphic Zone B, a Late Intermediate midden in which the shells of small coquina clams were the dominant mollusc remains. Fibers and threads of cotton were found in this midden.

The uppermost stratigraphic layer of Terrace 9 was Zone A, an ashy midden contemporaneous with the latest occupation of Structures D and 9. Marcus' crew recovered no industrial plants from undisturbed areas of Zone A, but this may be because such undisturbed areas were few. Zone A had been penetrated by four Late Intermediate burial cists (Structures 4–7). There were also non-cist burials in both Zones A and B.

Structure 4. Structure 4, a looted burial cist, produced a pillow made from bulrushes (Fig. 16.4). The cist itself had been lined with reed or bulrush mats.

Structure 5. Structure 5, a looted burial cist, yielded a gourd bowl (Fig. 16.21, *b*), stray cotton yarns, and unspecified grass stems.

Structure 6. Structure 6, a partially looted burial cist, had been lined with pacay leaves. The looters' debris in this cist also provided some details about the industrial plants used in elite burials. Some individuals were evidently placed on a litter made from caña brava (Fig. 16.17) or had been given "pillows" made from bundles of willow twigs and pacay rachises (Fig. 16.19). Structure 6 also yielded a series of ties made from split grass stems; what these strips of grass had been used to tie is unknown.

Burial 2. Burial 2 was accompanied by a gourd bowl (Fig. 16.22).

Burial 4. Burial 4 (Marcus 2015) yielded two bracelets whose beads were made from unidentified black seeds. One bracelet was associated with Individuals 1a and 1b; the other hung around the neck of a figurine found with Individual 3.

Individual 2 was accompanied by two prickly pear spines (Fig. 16.24). There were also cactus spine sewing needles in a weaver's workbasket found near Individuals 1a and 2 (Fig. 16.26).

Burial 6. Individual 1 of Burial 6 was accompanied by a gourd bowl (Fig. 16.23). Individual 2 wore a bracelet of black seeds like those described for Burial 4.

Burial 8. Burial 8 was accompanied by 15 grams of *algodón pardo* and 3 locks of imported kidney cotton (Fig. 16.31).

Burial 9. In the wrapping of Burial 9's mummy bundle there were 15 grams of *algodón pardo*.

Quebrada 6, Terrace 11

Terrace 11 of Quebrada 6 consisted of a coal-black, organically rich deposit of domestic refuse. At a depth of 30–50 cm, in Late Intermediate debris, Marcus' crew recovered the following industrial plants:

Cattail	1 specimen (Fig. 16.1)
Rush	1 fragment of rhizome
Sedge	1 inflorescence (Fig. 16.8)
Caña hueca	7 stem sections
	1 leaf
Unspecified grass	7 fragments of leaves and stems
Bottle gourd	1 fragment
Cotton	1 clump of white fiber (300 g)
	1 boll (Fig. 16.27)
	2 tufted seeds
	scattered locks of chocolate and white fibers
Unspecified wood	2 sticks

At a depth of 0–20 cm on this same terrace, Marcus' crew recovered one stem section of caña brava.

Quebrada 6, Terrace 12

Terrace 12 lay immediately downslope from Terrace 11. A shallow brown-to-beige midden on this terrace produced one fragment of bottle gourd.

Spanish Colonial Plants from Structure 1

After the Inca conquered Huarco in A.D. 1470, they built several buildings on the site's prominent sea cliffs. One of these buildings was Structure 1 on Cerro del Fraile (see Fig. 13.44). After the Spanish Conquest of A.D. 1534, squatters occupied several rooms of this abandoned Inca building.

The 16th-century Spanish squatters continued to use local industrial plants. Colonial refuse in Room 9 of Structure 1 included one inflorescence fragment of what appears to be either bulrush or sedge. Colonial refuse in Room 10 of that same building produced one stem fragment of caña de Guayaquil.

Refuse on the floor of the corridor immediately to the southwest of Rooms 10, 11, and 12 produced a carbonized fragment of unidentified cane and a mass of cotton fiber. Owing to a paucity of ceramics, Marcus was unable to determine whether this refuse was Late Horizon (Inca) or early Colonial.

Summary and Conclusions

The Late Intermediate occupants of Cerro Azul made use of more than two dozen genera of industrial plants. For the most part, our evidence suggests that the bulk of these plants were used both by nobles and commoners. Only a few industrial plants, such as kidney cotton, may have been used mainly or exclusively by the elite, and that presumably reflects the fact that they had to be imported from distant regions.

The majority of Cerro Azul's industrial plants would have been available within the Kingdom of Huarco or the neighboring Kingdom of Lunahuaná. Included were materials for house construction, basketry, weaving, sewing, spinning, and the creation of litters, pillows, beads, bobbins, and a variety of other artifacts.

Plate I. Cañete's *peña*, or rocky coast, supports hundreds of species of molluscs, crustaceans, fish, sea birds, and marine mammals.

Plate II. Cerro Azul's nobles imported the giant Chilean mussel (*Choromytilus chorus*) to use as a cosmetic pigment palette.

Plate III. The embudo, or funnel, from a crayfish trap damaged and repaired at Cañete in 1985.

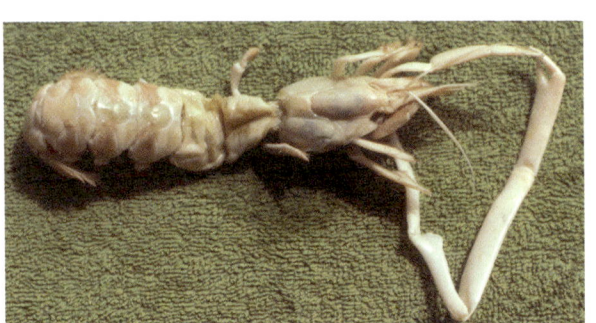

Plate IV. The burrowing of the ghost shrimp (*Callianassa islagrande*) has a profound effect on sandy beach environments at Cerro Azul.

Plate V. The arched blue crab (*Callinectes arcuatus*) expands its range to Cerro Azul during El Niño years.

Plate VI. A fisherman on Cerro Centinela launches his atarraya, or cast net, into the Pacific.

Plate VII. A Late Intermediate cast net, found in a looted burial cist at Cerro Azul.

Plate VIII. The most common member of the grunt family captured at Cerro Azul is the chita (*Anisotremus scapularis*).

Plate IX. The largest member of the drum family captured at Cerro Azul is the róbalo (*Sciaena starksi*).

Plate X. The Peruvian booby (*Sula variegata*) lives in colonies on the sea cliffs at Cerro Azul.

Plate XI. During the El Niño of 1982–83, many Humboldt penguins (*Spheniscus humboldti*) died of starvation.

Plate XII. This mummified Inca tern (*Larosterna inca*) died in the ruins of Cerro Azul during the 1850s.

Plate XIII. Sacuara or cattails (*Typha* cf. *angustifolia*) are among the "industrial" plants harvested near Cerro Azul.

Plate XIV. Some of the elite burials at Cerro Azul were provided with litters combining caña brava (*Gynerium sagittatum*) and totora (*Scirpus* cf. *californicus*).

Plate XV. One ear of dent corn (left) and two ears of imbricated purple corn (center and right) from looted burials at Cerro Azul.

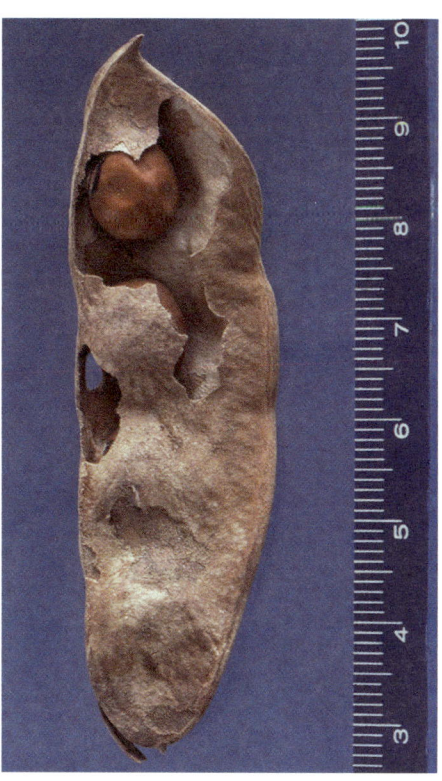

Plate XVI. Many of the lima beans (*Phaseolus lunatus*) at Cerro Azul featured brown seeds.

Plate XVII. A white potato (*Solanum tuberosum*) from a looted burial at Cerro Azul.

Plate XVIII. A sweet potato (*Ipomoea batatas*) found in Structure D at Cerro Azul.

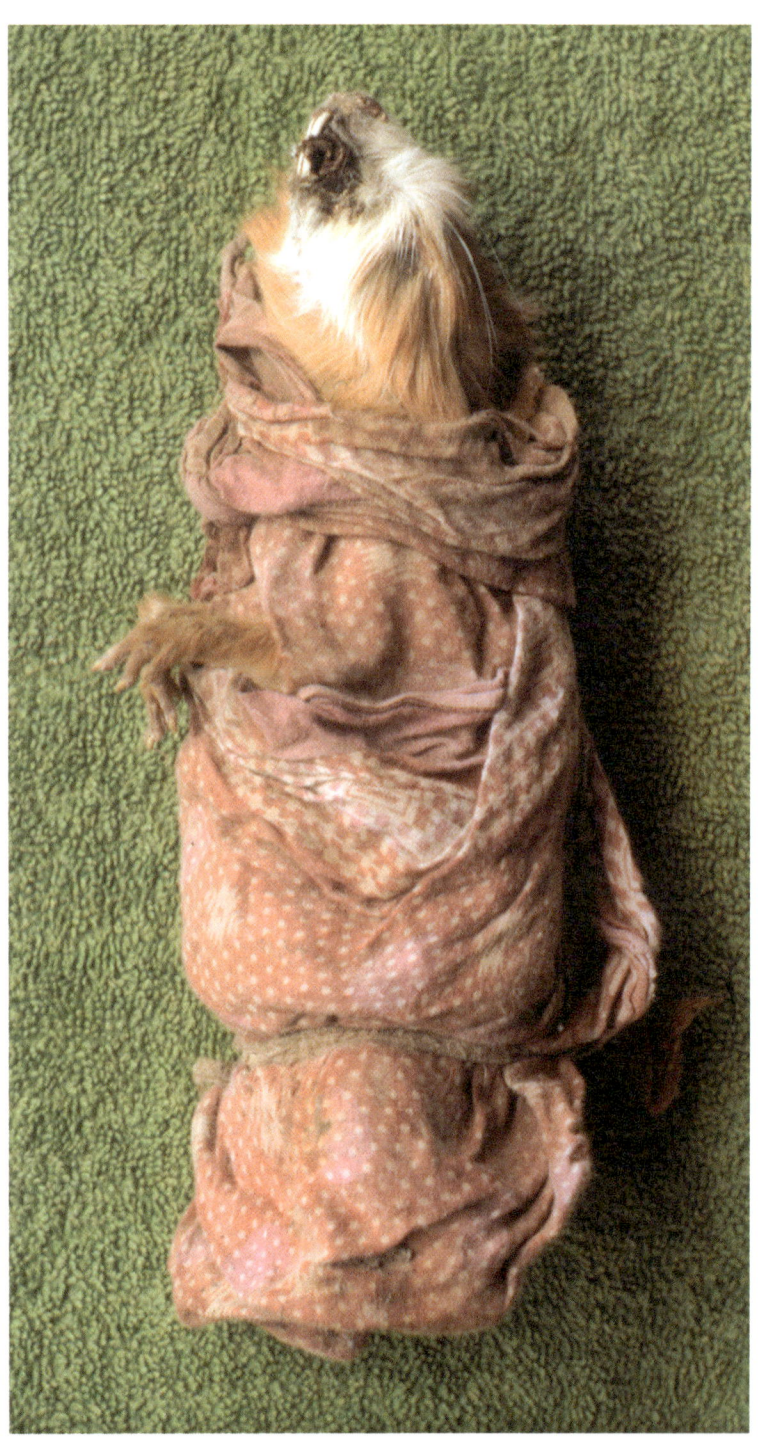

Plate XIX. At some point in the Colonial period, this guinea pig was wrapped in a red polka dot bandana and buried in the ruins of Structure 1 at Cerro Azul. Its mouth was stuffed with coca leaves.

Part IV

The Domestic Animals, Their Skeletal Remains, and Their Byproducts

17

Camelids and *Ch'arki* at Cerro Azul

Christopher P. Glew and Kent V. Flannery

By the time Cerro Azul was founded, there had been domestic camelids in the Andes for thousands of years. The occupants of Cerro Azul made use of camelid meat and camelid byproducts, but we found no evidence that the animals themselves were raised at the site.

To be sure, llama dung from the floor of the Southwest Canchón indicates that camelids spent some time in that part of Structure D (Fig. 17.1). We believe, however, that those llamas were members of the caravans that came to Structure D to deliver agricultural products and carry away dried fish. The area surrounding Cerro Azul has precious little in the way of vegetation on which llamas could feed. A quid of partially chewed grass—found in Feature 6—hints that the occupants of Structure D collected fodder from inland localities to feed the llamas when they arrived (Chapter 16).

Wheeler (1984a, 1984b) and Wheeler et al. (1976) have shown that when communities are involved in breeding domestic camelids, they tend to produce faunal assemblages with high frequencies of neonates, infants, and yearlings. No such assemblage was found at Cerro Azul. The majority of camelid bones we identified were from animals older than three years. This is an age profile consistent with llama transport caravans, which consist mainly of castrated adult males (Flannery et al. 1989).

As we will see, the midden deposits at Cerro Azul display what Miller (1979) calls "the *ch'arki* effect," meaning that much of the camelid meat arrived in the form of small parcels that had been dried and/or salted elsewhere. It appears to us—as stated earlier in Marcus et al. (1999)—that noble families at Cerro Azul had access to both *ch'arki* and whole llamas, while commoners seem to have relied mostly on *ch'arki*. We suspect that Cerro Azul's elite, like most Andean nobles, occasionally sought living llamas for sacrificial events.

As for where the camelid meat consumed at Cerro Azul came from, we may not have to look farther than the neighboring Kingdom of Lunahuaná. As Marcus (2008:1) has pointed out, roadcuts through middens in the Lunahuaná-Inkawasi region expose camelid bones in much greater numbers than do roadcuts in the Huarco region. That is presumably because Inkawasi, 470 m above sea level, lies in a region with enough forage to sustain llama herds. To be sure, some of the *ch'arki* may have been produced at even higher elevations.

In this chapter, we begin by describing the camelid bones from archaeological contexts at Cerro Azul. We then discuss the measurements of the camelids, the signs of arthritis and polydactyly, the evidence for *ch'arki*, and the differences in diet between nobles and commoners.

Structure D

Our largest sample of camelid bones from Structure D came from **Feature 6,** a midden resting against the east wall of the Southwest Canchón. The list of bones is given in Table 17.1.

The bones from Feature 6 represent virtually every major part of the animal, including the head and hooves. Included among the bones are the remains of a minimum of two llamas, with signs of butchering on the atlas vertebra, proximal femur, sacrum, and ribs. In addition, Feature 6 contained hundreds of pellets of llama dung, presumably swept up from the floor of the canchón.

The Floor of the Southwest Canchón

Marcus' crew found additional camelid bones on the floor of the Southwest Canchón. An atlas vertebra with cut marks on its ventral side; a fragment of sacrum; one proximal right metatarsal; and one first phalanx had been left on the floor in the northern portion of the canchón. Still farther to the north, near a blocked door between Room 11 and the Southwest Canchón, lay one cervical vertebra, two thoracic vertebrae, one rib, and a long bone fragment. The canchón floor also produced some additional pellets of llama dung.

The North Platform

The North Platform of the Southwest Canchón is believed to be the "office" from which a commoner-class administrator oversaw the movement of goods into and out of Structure D. Resting on the surface of the platform were one first phalanx, one cervical vertebra, and one rib fragment. In the fill just above the platform were the following camelid bones:

1 R. petrous bone
1 R. proximal radius
1 R. proximal ulna
1 R. proximal tibia (with cut marks)
1 L. unfused proximal tibia
1 L. fused proximal tibia (with cut marks)
1 fragment of sternum
2 vertebrae (with cut marks)
2 ribs (with cut marks)
3 long bone fragments

At a late stage in the occupation of Structure D, the area of the North Platform was built over and covered with fill. This late fill contained a burnt atlas vertebra; one fused distal tibia; one unfused distal tibia; one calcaneum (with cut marks); two cervical and four thoracic vertebrae (some with cut marks); and eight rib fragments.

Table 17.1. Camelid Bones from Feature 6.

Elements of the head and neck	
Fragment of occiput	1
Other cranial fragments	7
Petrous bone	1
Mandible fragments	2
Permanent teeth	Lower R. I3, R. P4, L. I3
Deciduous tooth	1
Hyoid bones	3
Axis vertebra	1 (with cut marks)
Other cervical vertebrae	4
Elements of the forelimb	
Scapulae	2 R., 1 L. (MNI = 2)
Proximal humerus	1 L.
Distal humerus	1 R.
Proximal radii	2 R. (MNI = 2)
Distal radii	1 R., 1 L.
Proximal ulna	1 R.
Accessory carpal	1 R.
Elements of the hind limb	
Proximal femur	1 R. (with cut marks)
Distal femur	1 L.
Patella	1 L.
Distal tibia	1 L.
Calcaneum	1 L. (with cut marks)
Fourth tarsal	1 R.
Os malleolare	1 L.
Elements of the feet	
First phalanges	7, one showing arthritis (MNI = 2)
Second phalanges	3
Third phalanges	2
Elements of the axial skeleton	
Sacrum	1 (with cut marks)
Thoracic vertebrae	8
Lumbar vertebrae	2
Caudal vertebrae	2
Vertebral epiphyses	3
Rib fragments	36 (2 with cut marks)
Costal cartilage fragments	2
Unidentifiable fragments, probably camelid: 39	

Totals
 NISP = 107
 MNI =2
 Unidentified = 39

Figure 17.1. Pellets of camelid dung from Feature 6, Structure D. The presence of this dung on the floor of the Southwest Canchón suggests that llama caravans entered Structure D to load and unload burdens.

The South Corridor

The South Corridor was a single-file passageway leading from the Southwest Canchón into the interior of Structure D. Late in the occupation of the building, a pile of debris was left on the floor of the corridor. While this debris was dominated by maize cobs, it also included portions of the right and left mandibles of a camelid, in addition to one first phalanx and one rib.

The Northeast Canchón

The Northeast Canchón was a large open work area where (among other things) weaving was done. In the northwest corner of the canchón there were a number of ash-filled hollows in the floor, dating to a period late in the building's occupation. The camelid remains from these hollows were as follows:

1 axis vertebra
1 fragment of scapula
1 proximal R. radius and ulna (with cut marks)
1 fragment of innominate (with cut marks)
1 fragment of proximal tibia
1 cervical vertebra
2 thoracic vertebrae
3 rib fragments (1 with cut marks)
4 long bone fragments

Also discovered on the floor of the Northeast Canchón were the mummified remains of a free-tailed bat (*Tadarida* sp.). This animal was presumably a post-abandonment visitor.

Collca 1

Collca 1 was a storage bin set against the south wall of the Northeast Canchón. At some point in the bin's history, a camelid right mandible had been discarded there.

The North Central Canchón

The North Central Canchón was the kitchen/brewery of Structure D. A small number of camelid bones were scattered throughout this large open work space. In the fill just above the floor were one distal humerus and one thoracic vertebra.

In ash deposits on the floor were one proximal tibia, one thoracic vertebra, and one lumbar vertebra. Near Features 15 and 16—two cavities in the floor made by chicha storage vessels—Marcus' crew found two nearly complete innominates from llamas (Figs. 17.2–17.4), as well as one fragment of sternum and one fragment of long bone. In the bottom of Feature 9, an empty storage vessel, someone had discarded a camelid metatarsal.

The Northwest Canchón

Owing to post-occupational wind erosion, not much is known about the Northwest Canchón. A surviving patch of floor did produce one right scapula (Fig. 17.5) and one right astragalus.

Room 1

Room 1 of Structure D was originally designed as part of an elite residential apartment. After suffering earthquake damage, it was converted to a fish storage room. Camelid bones were associated with both stages of the room's use. Just above the room's original floor, Marcus' crew recovered one petrous bone and a thoracic vertebra. In the room's later fill (postdating its conversion to storage) they found one unfused proximal femur; one patella; one first phalanx; one caudal vertebra; and one fragment of rib.

Room 3

Room 3 had a history similar to that of Room 1; originally part of an elite apartment, it was converted to fish storage after suffering earthquake damage. No camelid bones were found in association with the original (residential) stage. After the room's use as a fish storage unit had ended, debris was tossed into it. Included were the following camelid bones:

3 teeth (lower R. I3, M2, and M3)
1 scapula fragment
1 unfused proximal humerus
1 fused proximal radius
1 unfused proximal radius
1 fragment fused innominate
1 fragment unfused innominate
1 distal tibia
1 calcaneum
1 central tarsal
3 first phalanges
1 cervical vertebra
3 thoracic vertebrae
1 lumbar vertebra
1 caudal vertebra

3 rib fragments (1 with cut marks)
7 long bone fragments

Room 4

Room 4 had a complex stratigraphic history. Its original floor lay 3.73 m below the arbitrary datum point established for Structure D. At a later date, fill was added to the room and a new floor created at a depth of 2.23 m. The sacrifice of a guinea pig accompanied the laying of this floor (Chapter 19).

Lying on the upper floor was **Feature 5**, a group of partially restorable pottery vessels that had been crushed by the weight of the overburden. At a later date more fill was added to Room 4, completely covering up Feature 5 and the upper floor.

Marcus decided that Room 4 was an appropriate place to carry out a sub-floor sounding, aimed at discovering whether Structure D had a construction stage that antedated the Late Intermediate. No earlier stage was found, but in tapia fill at a depth of one meter below the room's original floor, Marcus' crew recovered the second phalanx of a camelid. This was the oldest camelid bone from the Room 4 area.

The second oldest sample of camelid bones came from the surface of the room's original floor (3.73 below datum) and represented at least 3 individuals. The list of bones was as follows:

3 fragments of occiput, some with cut marks (MNI = 3)
2 L. proximal humeri (with cut marks)
1 R. distal radius
1 L. proximal ulna
1 L. radial carpal
1 L. intermediate carpal
1 R. third carpal
1 L. proximal metacarpal
1 distal metapodial (with cut marks)
1 R. innominate, not fully fused
1 R. proximal femur
1 R. distal femur
1 R. central tarsal
1 first phalanx
2 second phalanges
1 cervical vertebra
4 thoracic vertebrae
2 lumbar vertebrae
7 long bone fragments

Totals:
 NISP = 26
 MNI = 3
 Unidentified = 7

These bones do not look like the remains of *ch'arki*; they suggest the butchering of complete llamas. Given that a guinea

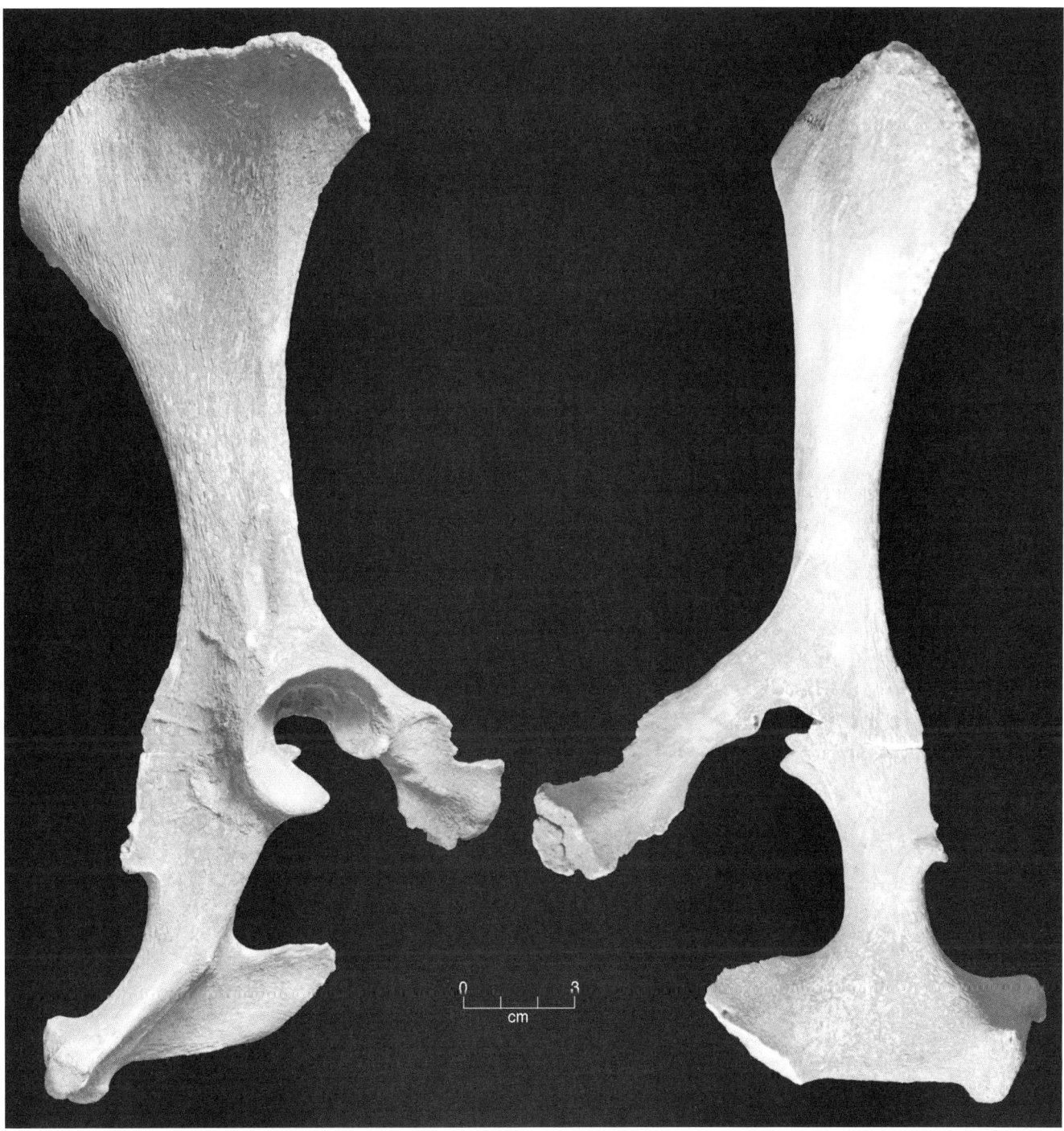

Figure 17.2. Two views of the right innominate of an adult camelid, found in the vicinity of Features 15 and 16 in the North Central Canchón, Structure D.

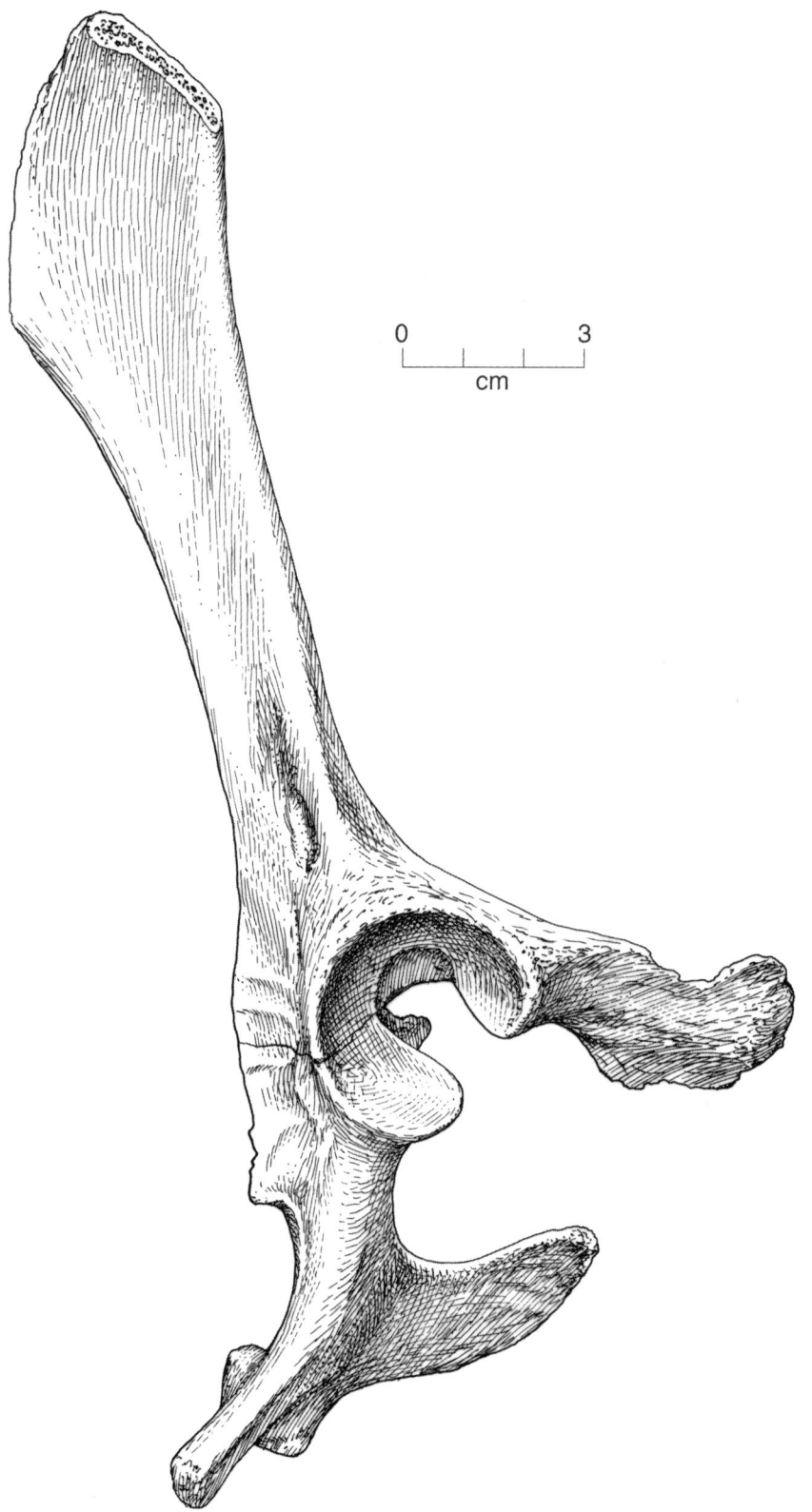

Figure 17.3. Drawing of the same camelid innominate shown in Figure 17.2.

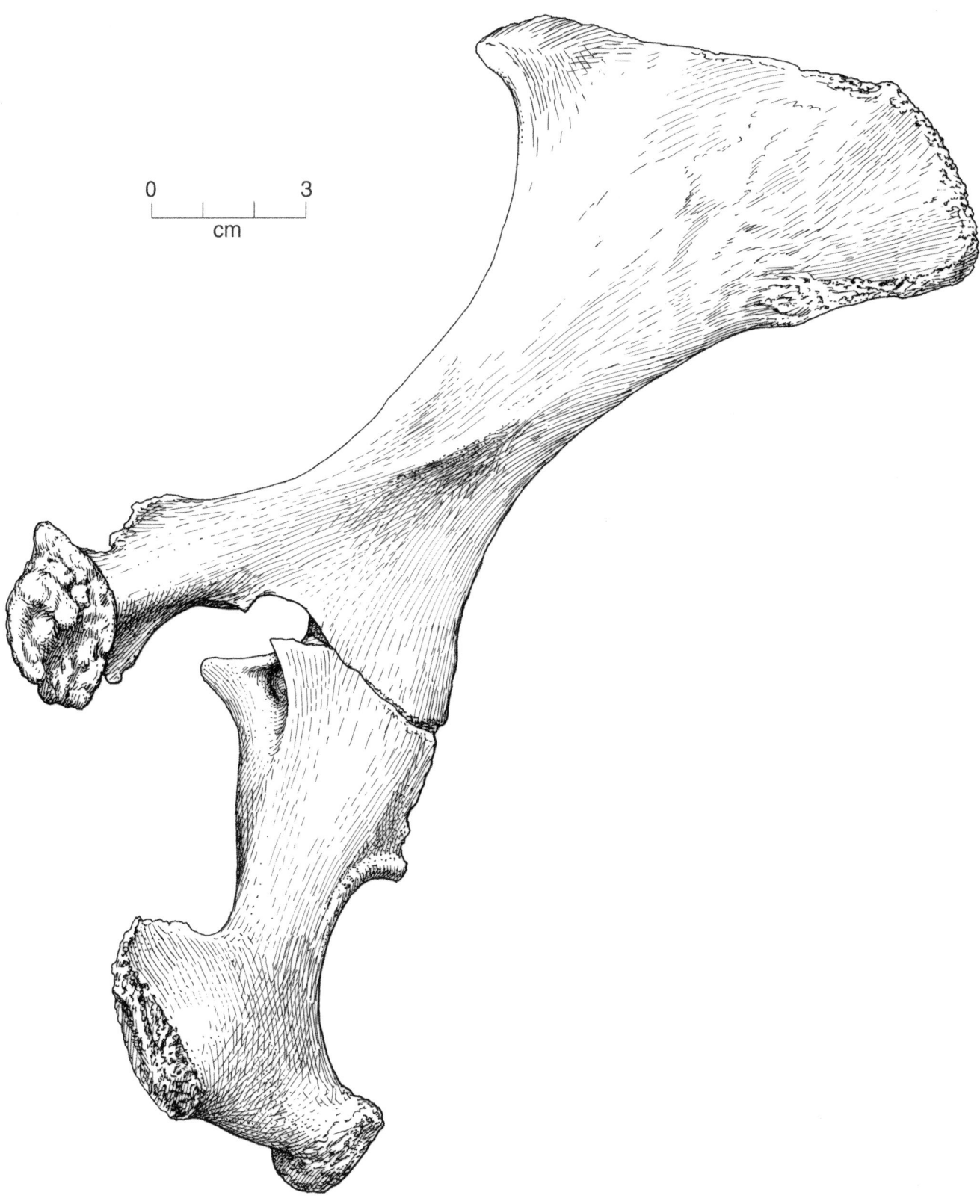

Figure 17.4. Right innominate of a second adult camelid (interior view, showing pubic symphysis). Found near Features 15 and 16 in the North Central Canchón, Structure D.

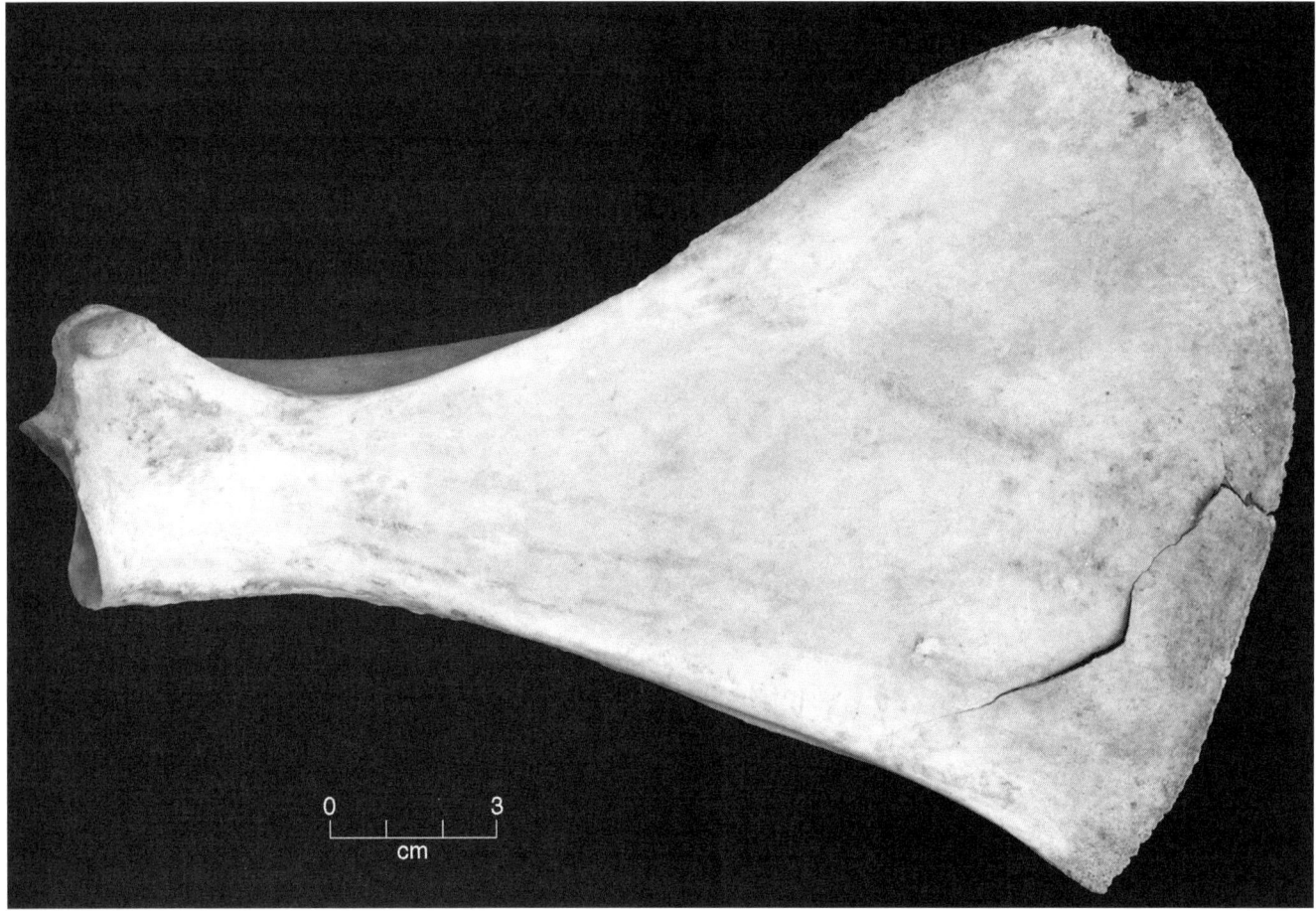

Figure 17.5. Complete right scapula of camelid, found in the fill of Structure D's Northwest Canchón.

pig was sacrificed when the new floor was laid and Room 4 repurposed, we wonder if these camelids might not have been slaughtered as well, shortly before the earthen fill supporting the new floor was added to the room.

The third oldest sample in Room 4 consisted of camelid bones in the fill immediately below the upper floor (2.23 m below datum), and represented at least two more individuals. The list of bones was as follows:

    R. and L. mandibles from the same llama (Fig. 17.6)
    1 scapula fragment
    2 L. proximal humeri, one unfused (MNI = 2)
    1 L. unfused distal humerus
    1 L. intermediate carpal
    1 L. fourth carpal
    1 L. metacarpal (Fig. 17.7)
    1 R. innominate (Fig. 17.8)
    1 L. distal femur (with cut marks)
    1 L. astragalus
    1 R. proximal metatarsal
    1 first phalanx
    10 unspecified fragments

Totals:
  NISP = 13
  MNI = 2
  Unidentified = 10

In all, the remains of perhaps five camelids had been added to the fill between the lower and upper floors of Room 4.

A word should be said about the pair of mandibles in Figure 17.6. Based on the stage of tooth eruption (see Wheeler 1982), this llama should have been between 2 years 9 months and 3 years 8 months old when it died. We suspect that it was a young animal chosen for a feast, rather than a veteran of one of the caravans coming to Structure D.

The fourth oldest sample of camelid bones from Room 4 consisted of eight fragments found among the crushed pottery vessels of Feature 5. The list was as follows:

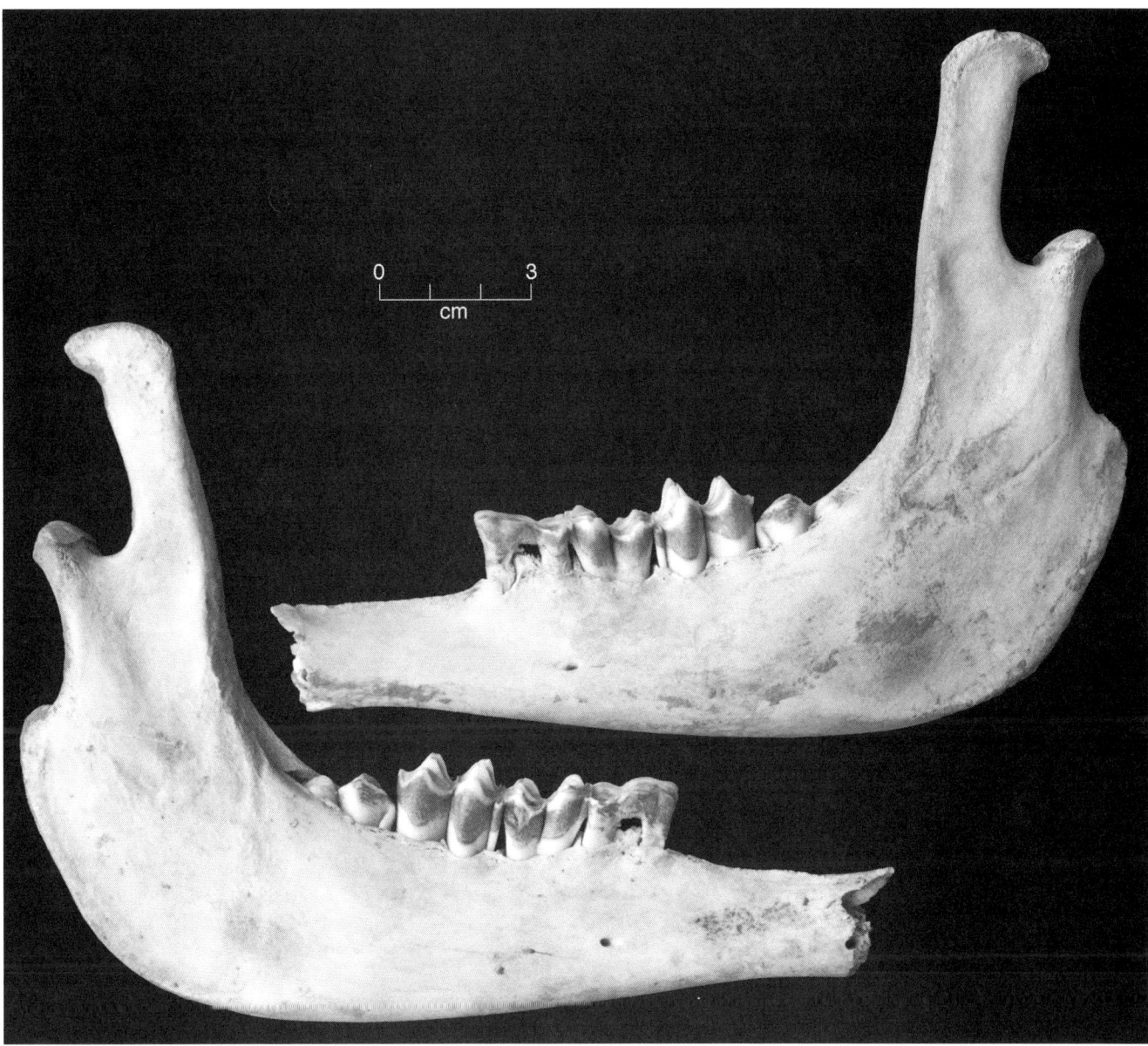

Figure 17.6. Right and left mandibles from the same camelid, found in the fill just below the upper (second) floor, Room 4, Structure D. Its stage of tooth eruption indicates that this animal's age would have been between 2 years 9 months and 3 years 8 months when it died.

1 R. proximal radius (with cut marks)
1 R. proximal ulna (with cut marks)
1 L. proximal femur (with cut marks)
3 ribs (with cut marks)
2 long bone fragments

Room 5

Room 5 had a complex stratigraphic history (Marcus 2008:161–167). Originally designed as a doorless storage room, it was converted to a residential unit after a few other rooms in Structure D had suffered earthquake damage. Camelid bones were found with several stages in the room's history.

During the room's conversion from storage to residence, its earlier floor was covered with fill. Two first phalanges were included in that fill.

One of the residential additions to the room was a "sleeping bench" in the unit's eastern half. One sesamoid bone from the hind leg of a camelid was found on this sleeping bench.

Late in the room's history, its most recent floor was covered up with fill. The camelid bones in this fill were as follows:

1 possible fragment of canine tooth
1 R. distal humerus (with cut marks)
1 L. distal humerus (with cut marks)
1 L. patella (burned)
1 thoracic vertebra
1 caudal vertebra

Room 7

Room 7 was a small unit whose access doorway was deliberately blocked after the room suffered earthquake damage. Late in its history, Room 7 became a convenient place to dump refuse. Included in this refuse were the following camelid bones:

1 L. distal humerus (burned)
1 R. proximal tibia
1 first phalanx
2 second phalanges (1 burned)
1 arthritic cervical vertebra
2 thoracic vertebrae
1 lumbar vertebra
1 long bone fragment

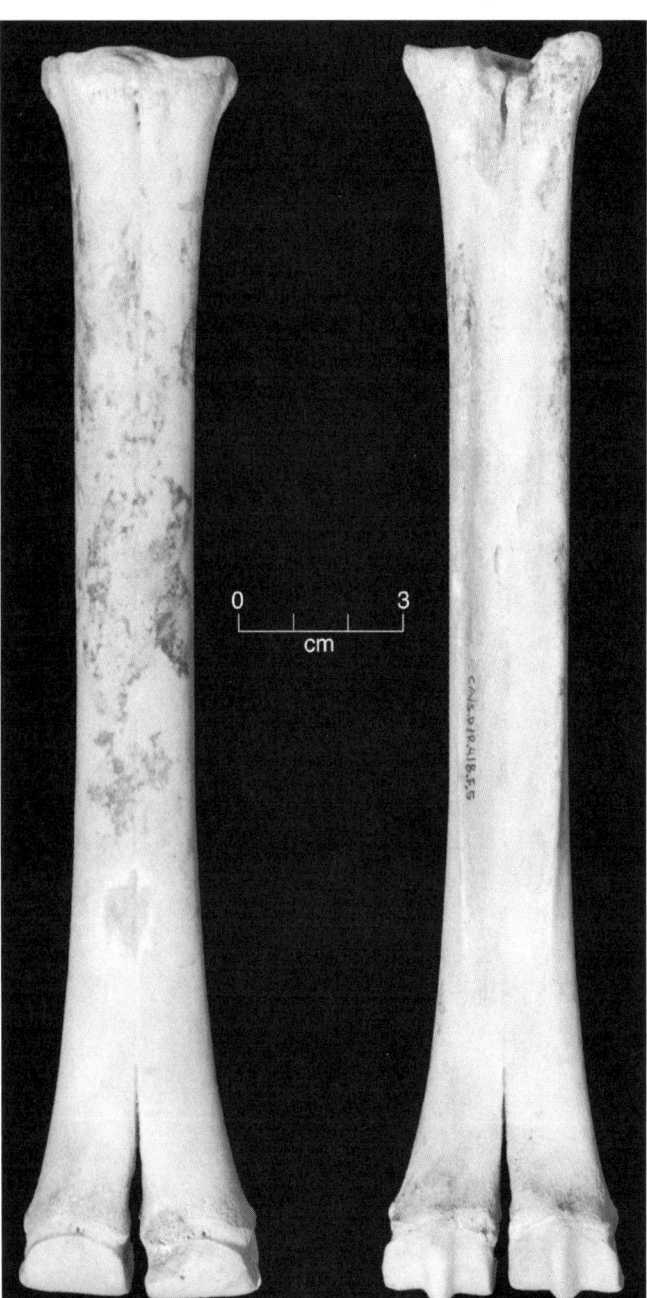

Figure 17.7. Dorsal and plantar views of the complete left metacarpal of a camelid. This bone was found in the fill of Room 4, Structure D, below the level of Feature 5.

Room 8

Room 8 was a sand-filled unit devoted to fish storage. After the abandonment of Structure D, wind erosion caused debris from

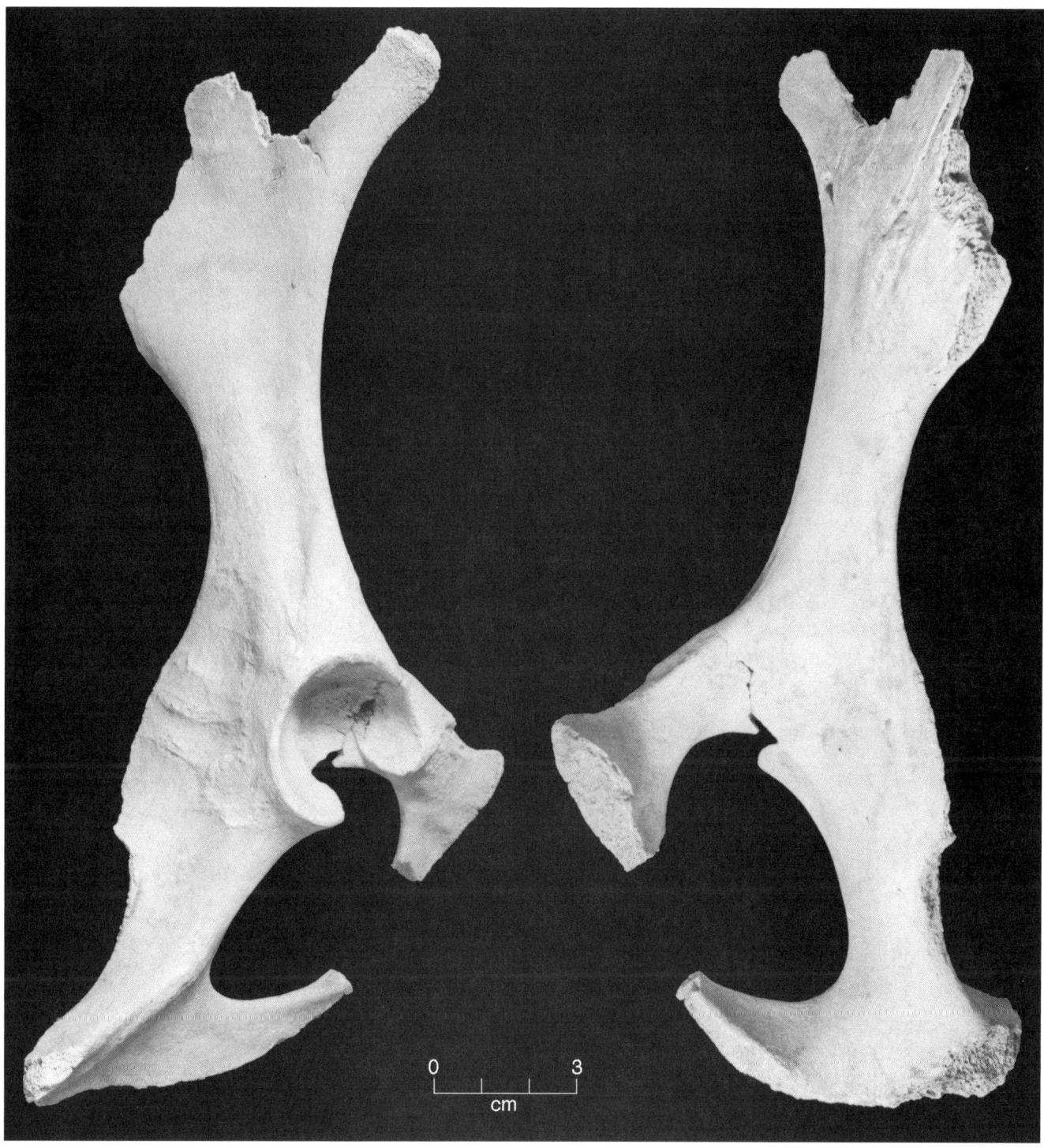

Figure 17.8. Two views of the right innominate of an adult camelid. Found in the fill of Room 4, Structure D, below the level of Feature 5.

the Southwest Canchón to drift downslope into Room 8. Included in this debris was one fragment of camelid rib.

Rooms 9 and 10

Rooms 9 and 10 were devoted to the raising of guinea pigs. No camelid remains were found on the floor of those rooms. During the centuries following the abandonment of Structure D, however, wind erosion caused part of the fill covering the North Platform of the Southwest Canchón to drift downward into Rooms 9 and 10. Included in this fill were the following camelid bones:

1 fragment of L. scapula
1 R. calcaneum (with cut marks)
1 first phalanx
2 rib fragments (1 burned, 1 with cut marks)

Late Horizon Squatters' Debris

At some point after Structure D had ceased to function as an elite residential compound, commoner-class squatters built a kincha house in the "Room 2" patio. Pottery from this house suggested that it dated to the Late Horizon.

Feature 4 was a Late Horizon midden associated with the kincha house. The camelid bones from this midden were as follows:

1 fragment of R. scapula (with cut marks)
1 L. patella
1 fused R. proximal tibia
1 unfused L. proximal tibia
1 L. distal tibia
2 cervical vertebrae (1 with cut marks)
2 thoracic vertebrae (1 with cut marks)

Feature 3

Feature 3 was an isolated feature that could not be associated with a specific room or canchón of Structure D. It appeared to be part of an offering made near the center of the building, sometime after Structure D had been abandoned and was filling up with dust and sand. Remains from the offering consisted of 290 burned bones from a minimum of three llamas; four exhausted maize cobs; and the neck of a Camacho Black jar. The burned llama bones were as follows:

1 fragment of maxilla
1 fragment of mandible
1 petrous bone
2 axis vertebrae (MNI = 2)
2 scapula fragments (MNI = 2)
1 fragment of sacrum
5 innominate fragments (MNI = 3)
28 fragments of burnt vertebrae
249 badly calcined fragments

Totals:
NISP = 290
MNI = 3

Structure 9

The largest sample of camelid bones from Structure 9 came from **Feature 20**, a midden piled up against the east wall of the building. The list of bones is given in Table 17.2.

Table 17.2. Camelid Bones from Feature 20.

Elements of the head and neck	
Fragment of occiput	1
Upper R. M2	1
Elements of the forelimb	
Scapula fragment	1
Distal radius	1 L.
Elements of the hind limb	
Distal tibia	1 L.
Proximal metatarsal	2 L. (MNI = 2)
Distal metapodial	1 L.
Fourth tarsal	1
Os malleolare	1
Elements of the feet	
First phalanges	2
Second phalanges	2
Third phalanges	3
Sesamoids	2
Elements of the axial skeleton	
Thoracic vertebra	1
Lumbar vertebra	1
Caudal vertebra	1
Other fragments of vertebrae	9
Rib fragments	10
Unidentifiable fragments, probably camelid: 22	

Totals
NISP = 41
MNI = 2
Unidentified = 22

Despite the fact that the excavation of Feature 20 involved the fine-screening of 2.5 m^3 of midden, this midden produced only 41 identifiable camelid bones. It also did not contain any pellets of llama dung, such as we found in Feature 6 of Structure D.

The South Entryway

The South Entryway into Structure 9 led directly to a small interior patio, then to the tapia platform on which the building's administrator had built his kincha house. The original floor of the entryway lay 1.77 m below the arbitrary datum point established for Structure 9. Found on this floor were the following camelid bones:

1 fragment of tooth
1 L. radial carpal
1 fused L. distal femur
2 unfused R. distal femora
1 second phalanx
1 cervical vertebra
4 caudal vertebrae
6 fragments of ribs

Later in the history of the South Entryway, a new floor was laid down at a depth of 99 cm below datum. This floor produced no camelid bones.

Room 5

Room 5 was one of a number of sand-filled fish storage units in Structure 9. Late in its history, however, it fell into disuse and became a convenient place to dump trash. Included in the trash were the following camelid bones:

1 atlas vertebra (with cut marks)
1 R. proximal radius (with cut marks)
1 R. proximal ulna (with cut marks)
1 L. ulnar carpal
1 R. third carpal
1 R. patella
1 R. proximal tibia
1 first phalanx (rodent gnawed)
2 cervical vertebrae
2 thoracic vertebrae
1 lumbar vertebra
2 rib fragments (with cut marks)
3 long bone fragments

Room 6

Room 6 lay adjacent to Room 5 and had a similar history. After its use as a sand-filled fish storage unit had ended, a camelid second phalanx and a limb bone fragment were tossed into the room.

Room 7

Room 7, adjacent to the South Entryway, appeared to have been used for fish storage from the outset. Late in its history, the following camelid bones were discarded in it:

1 first phalanx
3 cervical vertebrae (some with cut marks)
1 thoracic vertebra

Room 8

Room 8, although clearly a storage unit, had never been filled with sand and used to store fish. Its contents included broken storage and cooking vessels and other domestic refuse. Marcus' impression was that this refuse had not come from very far away; it may have been domestic trash from the kincha house of the building's overseer.

The camelid bones from Room 8 were as follows:

1 L. mandibular ramus (with cut marks)
1 L. lower I3
1 hyoid
1 L. unfused proximal humerus (with cut marks)
1 L. fused distal humerus (with cut marks)
1 L. accessory carpal (with cut marks)
1 R. patella
1 L. proximal tibia (with cut marks)
1 L. distal tibia
3 rib fragments

Room 12

Room 12 appeared to be another room designed for domestic storage, and never filled with sand and fish. It yielded one right calcaneum with cut marks; one first phalanx; one rib fragment; and two unidentifiable limb bone fragments.

Bones Found While Sweeping the Surface

A few additional camelid bones were found while Marcus' crew was sweeping the surface of Structure 9 prior to excavation.

Included were two astragali (one R., one L.); two first phalanges; one cervical vertebra; four lumbar vertebrae; and one rodent-gnawed rib fragment.

Camelid Bones from the Quebradas of Cerro Camacho

Camelid bones were present in the middens on Cerro Camacho but, owing to the number of times they had been moved, they were neither as numerous nor as well preserved as the bones in Structures D and 9.

Quebrada 5, Terrace 16

The deposits on Terrace 16 of Quebrada 5 were 2.2 m deep. Marcus identified nine stratigraphic zones, with G the oldest and A1 the youngest. Only four stratigraphic zones produced camelid bones.

Zone F, a stratum of soft gray ash, contained only two rib fragments. **Zone D1**, a stratum of ashy midden, was capped by a layer of discarded mussel shells; one camelid calcaneum lay among the mussels.

The most interesting discovery made in **Zone A2** was **Feature 22**, a hearth used to roast sea lions (see Chapter 12). In the midden debris outside the hearth, Marcus found one camelid patella, two thoracic vertebrae, one lumbar vertebra, and three rib fragments. Three more fragments of camelid vertebrae were found in the hearth, but it did not appear that they had been cooked there. Rather, they appeared to be refuse from Zone A2 that had simply spilled over into the hearth in the aftermath of the sea lion cookout.

A few more camelid bones appeared in **Zone A1**, a layer of soft midden that represented the last dumping of refuse on Terrace 16. The list of bones is as follows:

1 R. petrous
1 R. radius and ulna (see Fig. 17.9)
1 R. proximal tibia
1 cervical vertebra
1 thoracic vertebra
2 lumbar vertebrae

Quebrada 5-south

Quebrada 5-south produced a tapia-walled storage unit (Structure 11), a looted burial cist (Structure 12), and a small midden deposit.

Structure 11, Room 3

Room 3 was the largest of the three rooms Marcus excavated in Structure 11. Late in its history, a kincha house had been built in this room. Associated with this house were two camelid rib fragments.

The Midden between Structures 11 and 12

In the midden between Structures 11 and 12, Marcus' crew found the complete metacarpal of a four-toed llama (Fig. 17.10). Such polydactyly represents a genetic anomaly—almost certainly the result of inbreeding—and judged by the amount of arthritis on the metacarpal, it had been deleterious. This was, in fact, not the only example of polydactyly we found at Cerro Azul (see Fig. 17.11).

Quebrada 5a, Terrace 9

The deposits on Terrace 9 of Quebrada 5a were 2.6 m deep and quite complex. Stratigraphic Zone C, the oldest layer, produced no camelid bones.

Stratigraphic Zone B was a thick layer of Late Intermediate midden. Its ceramics indicated that this midden had been laid down at a time when the occupation of Structure D was just beginning. The composition of its mollusc remains, its fish remains, and its edible plants all suggested that this stratum was refuse from commoner households. Zone B was not rich in camelid bones; the list (which follows) amounts to no more than a couple of discarded llama feet.

2 first phalanges
3 second phalanges
1 third phalanx

Structure 6
Structure 6 was a looted burial cist in Stratigraphic Zone A of Terrace 9. Looters' debris in the cist contained only one camelid first phalanx.

Burial 2
Burial 2 had been provided with a cervical vertebra of llama, presumably part of a parcel of *ch'arki*.

Burial 3
Burial 3 was discovered in Zone A, just southwest of Structure 6. The lower leg of a llama seems to have been associated with this burial. The right tibia, right astragalus, right fourth tarsal, and right metatarsal were still articulated, and a patch of llama hide with white hair was included. Also in the fill around the burial were a left astragalus, left calcaneum, left central tarsal, left second + third tarsal, and left fourth tarsal. These bones all appeared to be from a young animal (or animals).

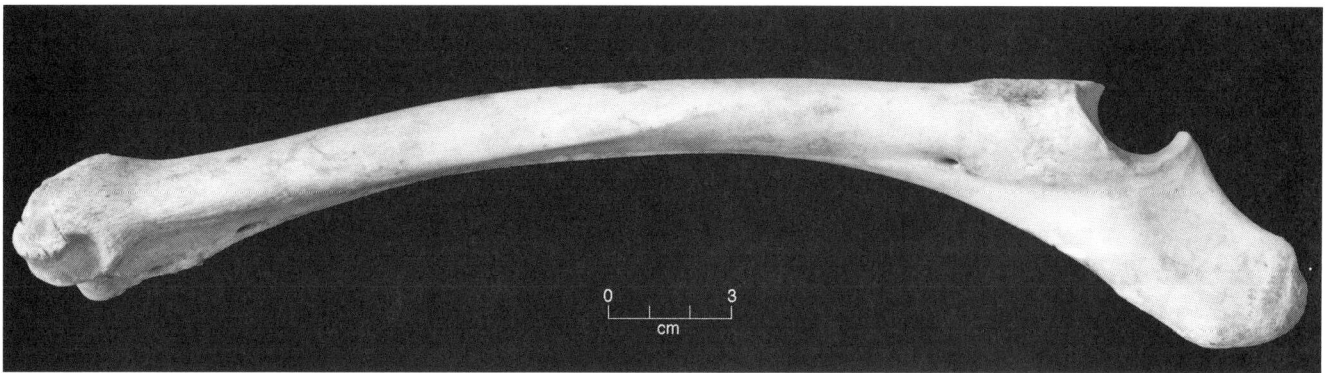

Figure 17.9. Complete right radius and ulna of camelid, found in Stratigraphic Zone A1 of Terrace 16, Quebrada 5.

Quebrada 6, Terrace 11

Terrace 11 of Quebrada 6 was filled with black, heavily organic midden. Its upper 20 cm produced only one fragment of camelid rib. At a depth of 30–50 cm, Marcus' crew found one first phalanx.

Quebrada 6, Terrace 12

Terrace 12 of Quebrada 6 contained mainly midden debris that had drifted downslope from Terrace 11. It produced only one thoracic vertebra of camelid.

The Size of the Cerro Azul Camelids

Having accounted for all the prehistoric camelid bones from Cerro Azul, it is now time to determine whether those bones are from domestic llamas, domestic alpacas, wild guanacos, or wild vicuñas.

We should say at the outset that we encountered no examples of the very distinct incisor teeth of vicuñas at Cerro Azul. The Kingdom of Huarco does not really constitute a suitable environment for the vicuña, which is adapted to the high-altitude *puna* (Koford 1957). In addition, the genetic anomalies we found at Cerro Azul—including three-toed and four-toed animals—strongly suggest that we are dealing with inbred domestic animals, which rules out both guanacos and vicuñas. Our most relevant question, therefore, is whether both llamas and alpacas were being eaten and/or sacrificed at Cerro Azul, or only llamas.

To answer this question, we will use the "decision rules" established by Elizabeth S. Wing (1972 and personal communication, 1973). Wing began by taking a series of measurements on known guanaco, vicuña, llama, and alpaca skeletons (Figs. 17.12, 17.13). She then had those measurements subjected to a stepwise discriminant analysis in order to establish a series of decision rules by which the continuum of sizes could be objectively divided into "large" and "small" camelids (Wing 1972). Her "large" category was made up of guanaco and llama; her "small" category consisted of alpaca and vicuña.

Glew took as many of Wing's measurements as he could, given the fragmentary condition of the Cerro Azul camelid bones. His measurements are given in Table 17.3.

Wing's decision rules make it clear that in the case of virtually every bone, we are dealing with camelids in her "large" category. Among Wing's most reliable measurements are those for the scapula and humerus. Scapula measurements greater than 47.5 mm for c-c, greater than 33.6 mm for b-b, and greater than 29.1 mm for d-d put the animal in the "large" category. For the humerus, measurements greater than 44.2 mm for e-e do the same.

The third most reliable bone is the tibia, where measurements greater than 22.75 mm for d-d, greater than 27.0 mm for c-c, and greater than 58.4 mm for a-a put the animal in the "large" category. Fourth in order of reliability is the metatarsal, where a measurement greater than 210.3 mm for y-z is considered "large."

To be sure, some of the measurements we took on the Cerro Azul camelid bones oscillate around the decision line between "large" and "small." In most cases, however, these borderline measurements occur on bones whose other dimensions place them in the "large" category. In still other cases, the bones with borderline measurements display epiphyses that had only recently fused, suggesting that the bone might not have reached its final size.

We are convinced, in other words, that virtually all the camelid bones at Cerro Azul were from llamas (*Lama glama*). To be sure, we cannot rule out an occasional parcel of *ch'arki* made from alpaca; we simply find the probability low. We also find it unlikely that any of the "large" camelids were guanacos. Not only do we have no evidence for guanaco hunting at Cerro Azul, there are two signs that we are dealing with domestic animals.

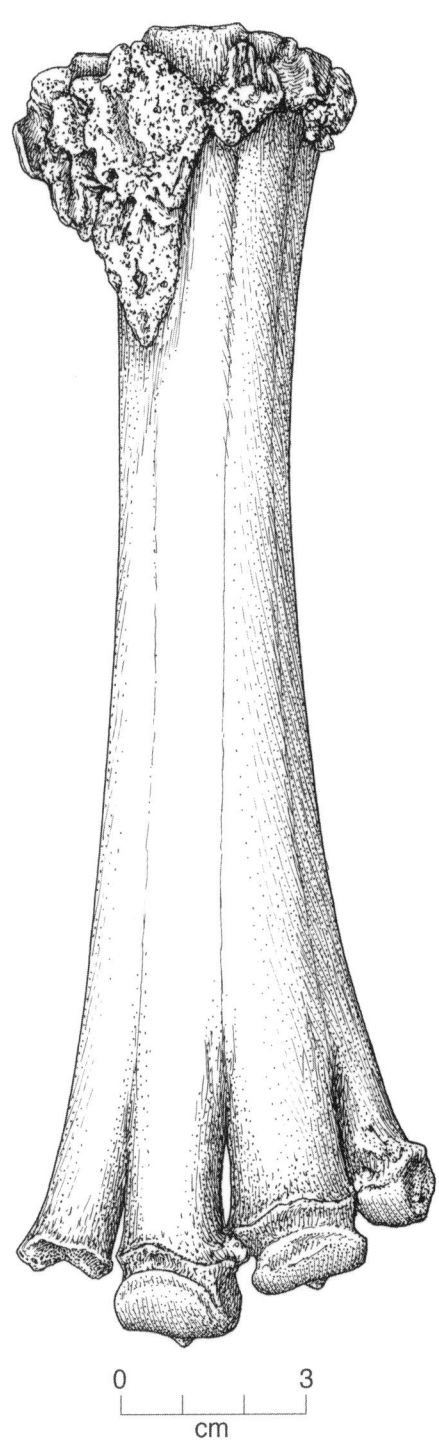
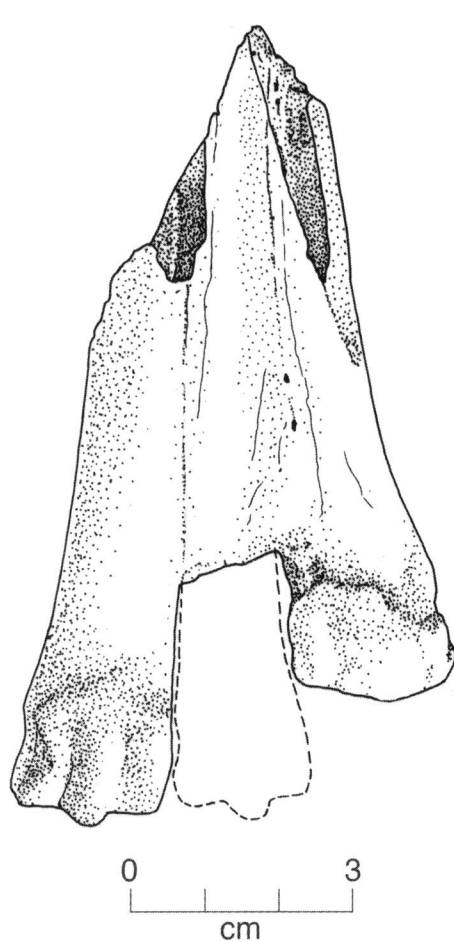

Figure 17.10. Right metacarpal of llama, found in the midden between Structures 11 and 12 in Quebrada 5-south. This young llama would have had four toes on one foot, and shows arthritic calcification of the proximal metacarpal.

Figure 17.11. Distal metapodial of llama, found on the surface of Cerro Azul. This llama would have had three toes on one foot.

One sign is the aforementioned evidence for polydactyly (Figs. 17.10, 17.11). The other likely sign of domestication is arthritis, presumably brought on by a lifetime of carrying burdens (Figs. 17.14, 17.15). Burden carrying, of course, would decrease the likelihood that any of our bones were from alpacas.

The "*Ch'arki* Effect" at Cerro Azul

Let us now consider the question of how much llama meat at Cerro Azul came from whole animals, and how much came from the parcels of dried meat that Quechua speakers refer to as *ch'arki*. *Ch'arki* today (and presumably in the past) is produced by cutting the animal up into relatively small, easily transported packets of meat which are then freeze-dried or salted and dried, depending on the ambient temperature. Most *ch'arki* is produced in the highlands, where it can be dried in the sun during June and July, the coldest months of the Andean year.

In his study of *ch'arki*-making in the sierra northwest of Lake Titicaca, Miller (1979) noted that a key attribute of *ch'arki* is that each packet contains both meat and bones. The grease inside the bones makes the meat more flavorful when it is rehumidified.

Miller observed that the prime *ch'arki*-making parts of a llama include the torso and upper legs. The least favored parts of the animal were the head and lower limbs (metapodials and phalanges); the latter do not have enough meat on them to make them desirable. Based on his observations in the Lake Titicaca region, Miller proposed that something he called "the *ch'arki* effect" should be detectable in those faunal assemblages where packets of dried meat were a major contribution to the food supply.

Specifically, Miller predicted that llama remains from *ch'arki*-producing areas should show lower-than-expected frequencies of bones from the torso and upper limbs; areas receiving *ch'arki*, on the other hand, should show higher-than-expected frequencies of bones from the torso and upper limbs, as well as lower-than-expected frequencies of the skull and lower limb bones.

To test for the *ch'arki* effect at Cerro Azul, we have grouped the llama bones from Structures D and 9 by anatomical region, as follows: (1) head = cranium, mandibles, and teeth; (2) axial skeleton = the vertebral column; (3) forequarter = scapula, humerus, radius, and ulna; (4) hindquarter = innominate, sacrum, femur, and tibia; (5) forefoot = carpals and metacarpals; (6) hind foot = tarsals and metatarsals; and (7) toes = distal metapodials and phalanges. In Figure 17.16, we compare the frequencies of bones from these anatomical regions with the frequencies one would expect to see in a complete llama skeleton.

Before proceeding to our interpretation of Figure 17.16, let us first state our expectations. We expect that Cerro Azul received most of its llama meat in the form of *ch'arki*, but we also expect that its nobles periodically held rituals during which they slaughtered whole animals. We therefore expect to see evidence for the *ch'arki* effect, but not in its purest form.

Now let us turn to Figure 17.16. Both the head and axial skeleton clearly show the *ch'arki* effect; components of the head are strongly underrepresented in Structures D and 9, while the vertebrae are overrepresented. In both cases, Structure 9 shows the *ch'arki* effect more strongly than does Structure D.

Both the forequarters and hind quarters also show the *ch'arki* effect, but in this case Structure D shows the effect more strongly than Structure 9. This may be because cuts of *ch'arki* from those parts of the animal were considered more desirable than those from the vertebral column, making it likely that more of them went to nobles.

Both the forefoot and the hind foot are underrepresented at Cerro Azul. These are considered less desirable parts of the animal, since they provide less meat than the upper limbs. We note that this aspect of the *ch'arki* effect is more strongly expressed in Structure D than in Structure 9. Once again, the difference may reflect the fact that nobles chose preferred cuts of meat and left the least desirable cuts for the commoners.

The only part of Figure 17.16 that does not meet Miller's expectations is the higher-than-anticipated number of phalanges. Miller himself was prepared for such anomalous frequencies, predicting that his expectations might not be met by all communities receiving *ch'arki* (Miller 1979).

Butchering Marks on Bones

One other potential line of evidence for *ch'arki* in a camelid assemblage is the frequency and placement of cut marks on the bones. To be sure, any butchered llama is likely to show cut marks where its limbs were detached from the trunk, where the hooves were cut off and discarded, and so forth. The process of making *ch'arki*, however, can greatly increase the number of cut marks, since virtually the entire vertebral column may be cut up into small parcels of meat and bone. Let us therefore look at some examples of butchering at Cerro Azul.

Figure 17.17 shows two proximal tibiae of adult llamas, both butchered in a way that shows a degree of standardization. Both tibiae were found in Structure D, and appear to have been broken open to expose the marrow.

Figure 17.18 shows the proximal right radius and ulna of an adult llama from Structure D, with a series of cut marks made by someone trying to free the distal condyle of the humerus. We suspect that these cuts were made when a complete llama was being butchered.

Figure 17.19 shows a llama metacarpal from Room 4 of Structure D, with a cut mark on one of the distal condyles. This is the point where the toes would have been removed and discarded.

Figure 17.20 shows a calcaneum from the Feature 6 midden. It bears two cut marks on the processus cochlearis, presumably made by someone trying to separate the hind foot from the hindquarters.

Figures 17.21 and 17.22 present two examples of the atlas vertebra, one from Structure D and one from Structure 9. Both

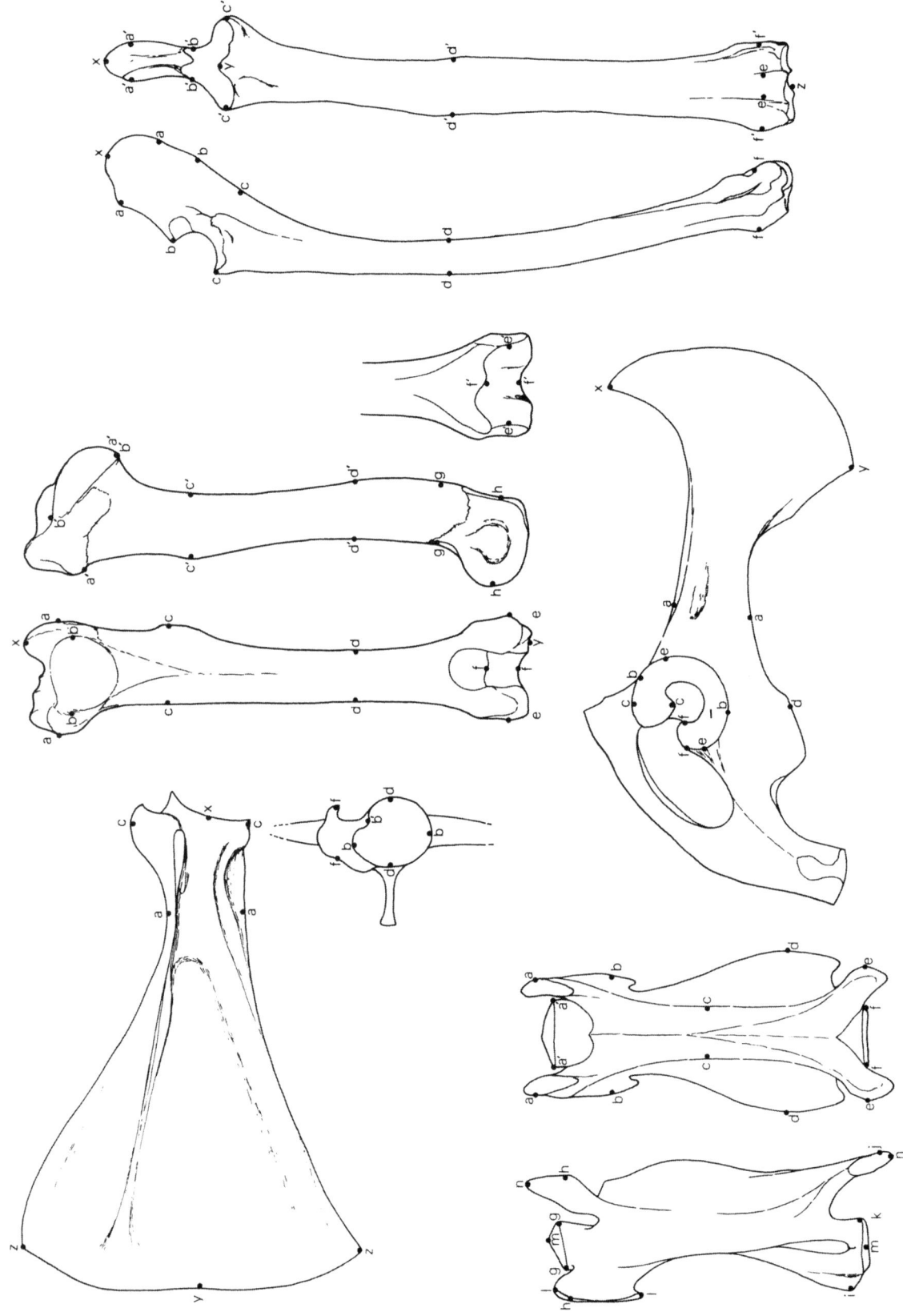

Figure 17.12. Measurements of the scapula, humerus, radius, ulna, cervical vertebra, and innominate used to establish Wing's "decision rule." (Courtesy, Elizabeth S. Wing).

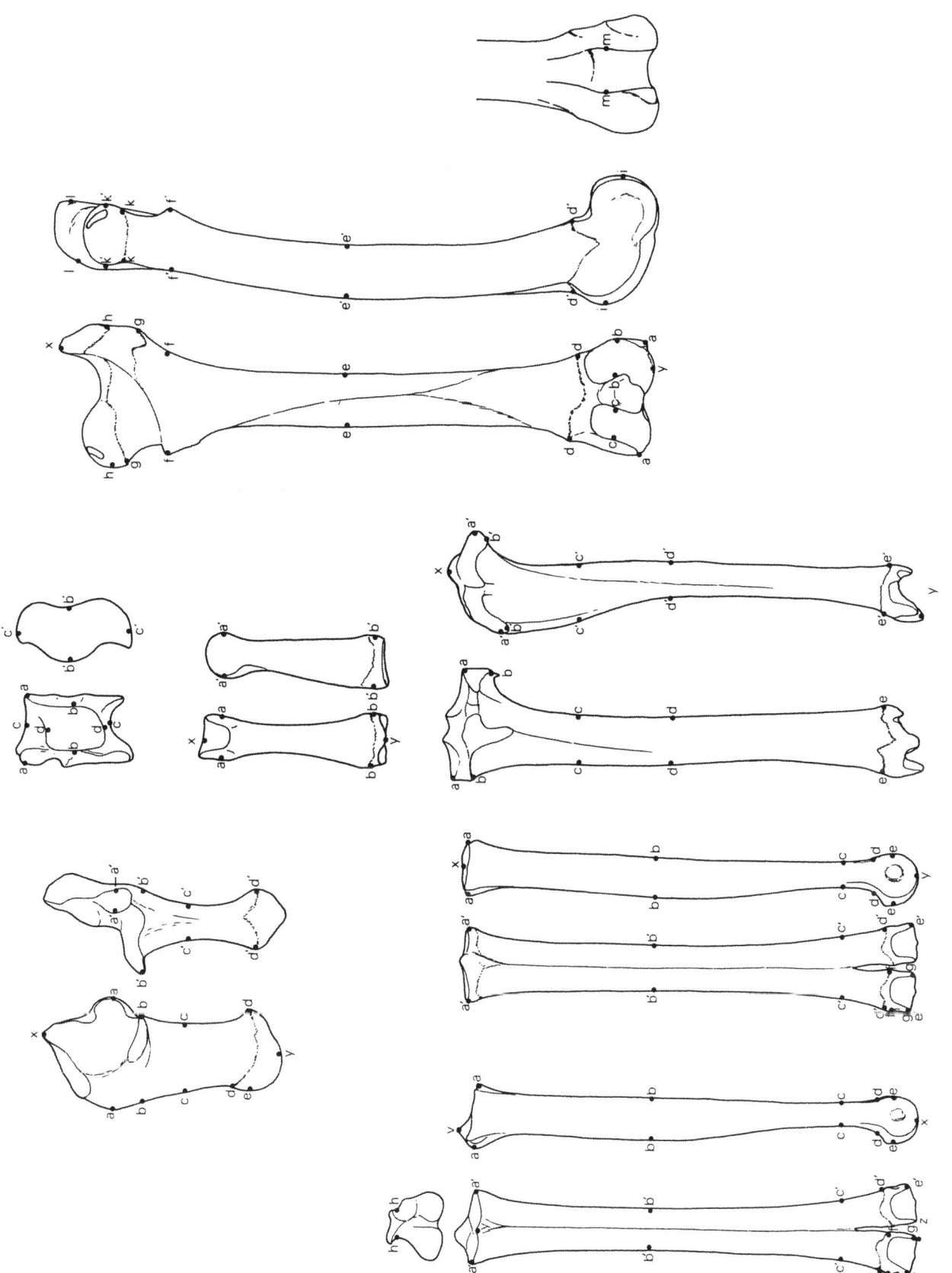

Figure 17.13. Measurements of the calcaneum, astragalus, first phalanx, femur, metapodials, and tibia used to establish Wing's "decision rule." (Courtesy, Elizabeth S. Wing).

Table 17.3. Measurements (in millimeters) of Camelid Bones from Cerro Azul (see Figs. 17.12, 17.13).

Scapula, Structure D

c-c	b-b	a-a	d-d
48.54	34.42	30.62	32.88
50.64	33.75	33.46	30.88

Humerus, Structure D

e-e	f'-f'	h-h	g-g	e'-e'	b'-b'	f-f	b-b
48.53	18.75	42.94	25.35	42.55	—	16.12	—
—	—	—	—	—	37.82	—	37.16
42.02	15.40	31.12	—	35.00	—	14.08	—
47.88	17.86	42.34	23.08	40.44	—	14.68	—

Humerus, Structure 9

e-e	f'-f'	h-h	g-g	e'-e'	f-f
43.62	16.21	38.20	23.14	36.09	13.50

Radius & Ulna, Structure D

f'-f'	f-f	b'-b'	c'-c'	a-a
—	—	16.06	43.56	35.82
40.76	29.32	—	—	—
46.07	30.37	—	—	—
—	—	—	39.20	—
—	—	15.82	40.97	35.67

Radius & Ulna, Structure 9

f'-f'	f-f	b'-b'	c'-c'
45.38	30.76	—	—
—	—	16.29	37.41

Radius & Ulna, Quebrada 5, Terrace 16

f'-f'	f-f	b'-b'	c'-c'	a-a
40.67	28.74	15.74	41.77	32.68

Tibia, Structure D

d-d	c-c	e'-e'	e-e	a-a	b-b	a'-a'	b'-b'
—	—	23.81	40.82	—	—	—	—
—	—	—	—	—	51.50	63.58	55.94
—	32.71	—	—	63.83	59.68	66.73	58.04
25.94	—	23.28	37.98	—	—	—	—
—	—	21.47	37.12	—	—	—	—
—	—	—	—	63.42	59.33	67.72	59.95
—	—	—	—	67.31	63.06	—	—

Tibia, Structure 9

e'-e'	e-e	a-a	b-b	a'-a'	b'-b'
—	—	60.91	53.52	63.92	53.12
23.80	40.44	—	—	—	—
21.01	37.48	—	—	—	—

Tibia, Quebrada 5, Terrace 16

a-a	b-b
55.33	51.34

Table 17.3. Measurements (in millimeters) of Camelid Bones from Cerro Azul (see Figs. 17.12, 17.13).

Astragalus, Structure D

c'-c'	c-c	a-a	b'-b'
37.06	29.74	23.80	17.50
41.20	32.60	28.34	21.34

Astragalus, Structure 9

c'-c'	c-c	a-a	b'-b'
39.28	30.30	26.85	18.74
36.88	28.20	22.67	17.88

Calcaneum, Structure D

x-y	a-a	d'-d'	b'-b'	c-c	b-b	d-d	e-d	c'-c'
78.00	34.93	17.01	21.57	20.88	28.26	22.86	22.24	9.22
73.30	32.98	17.30	22.82	20.14	25.90	22.30	21.84	9.44
79.81	36.96	21.78	25.04	22.04	26.40	27.04	25.90	12.82

Calcaneum, Structure 9

x-y	a-a	d'-d'	b'-b'	c-c	b-b	d-d	e-d	c'-c'
80.75	36.82	21.40	21.81	22.43	26.12	25.82	24.75	13.24

Calcaneum, Quebrada 5, Terrace 16

x-y	a-a	d'-d'	b'-b'	c-c	b-b	d-d	e-d	c'-c'
83.73	36.57	19.32	22.89	23.34	31.29	25.19	24.24	10.44

Metatarsal, Structure D

y-z	a'-a'	h-h	b-b	d'-d'	e'-e'	g-g	e-e	a-a
312.0	31.22	19.72	18.28	38.96	42.59	18.64	20.19	20.79

Metacarpal, Structure D

c-c	e'-e'	e-e	d'-d'	d-d	a-a	a'-a'	g-g
12.23	42.88	21.38	40.05	19.34	25.08	34.65	20.24

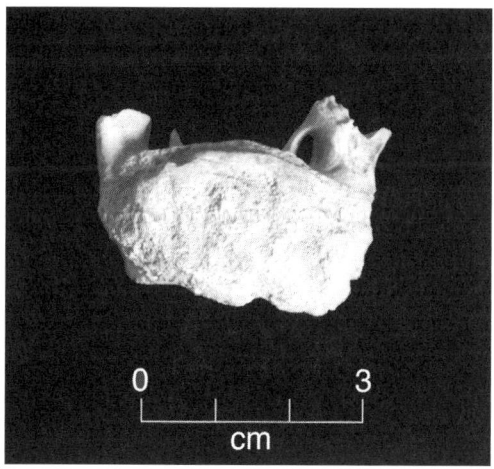

Figure 17.14. Anterior end of llama cervical vertebra, showing signs of arthritis. Found in Room 7 of Structure 9.

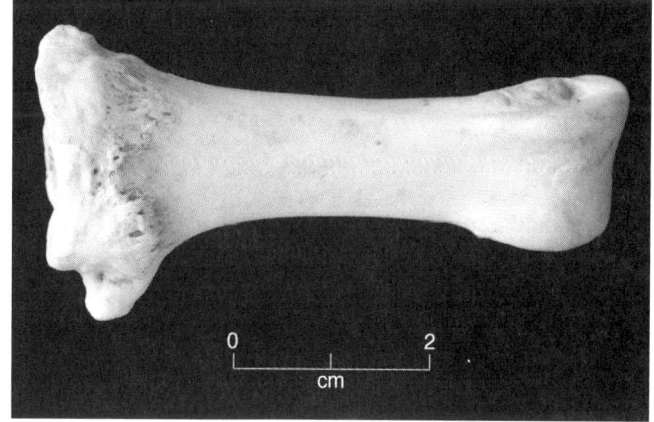

Figure 17.15. First phalanx of a llama from Feature 6 of Structure D, showing arthritic calcification of the proximal end.

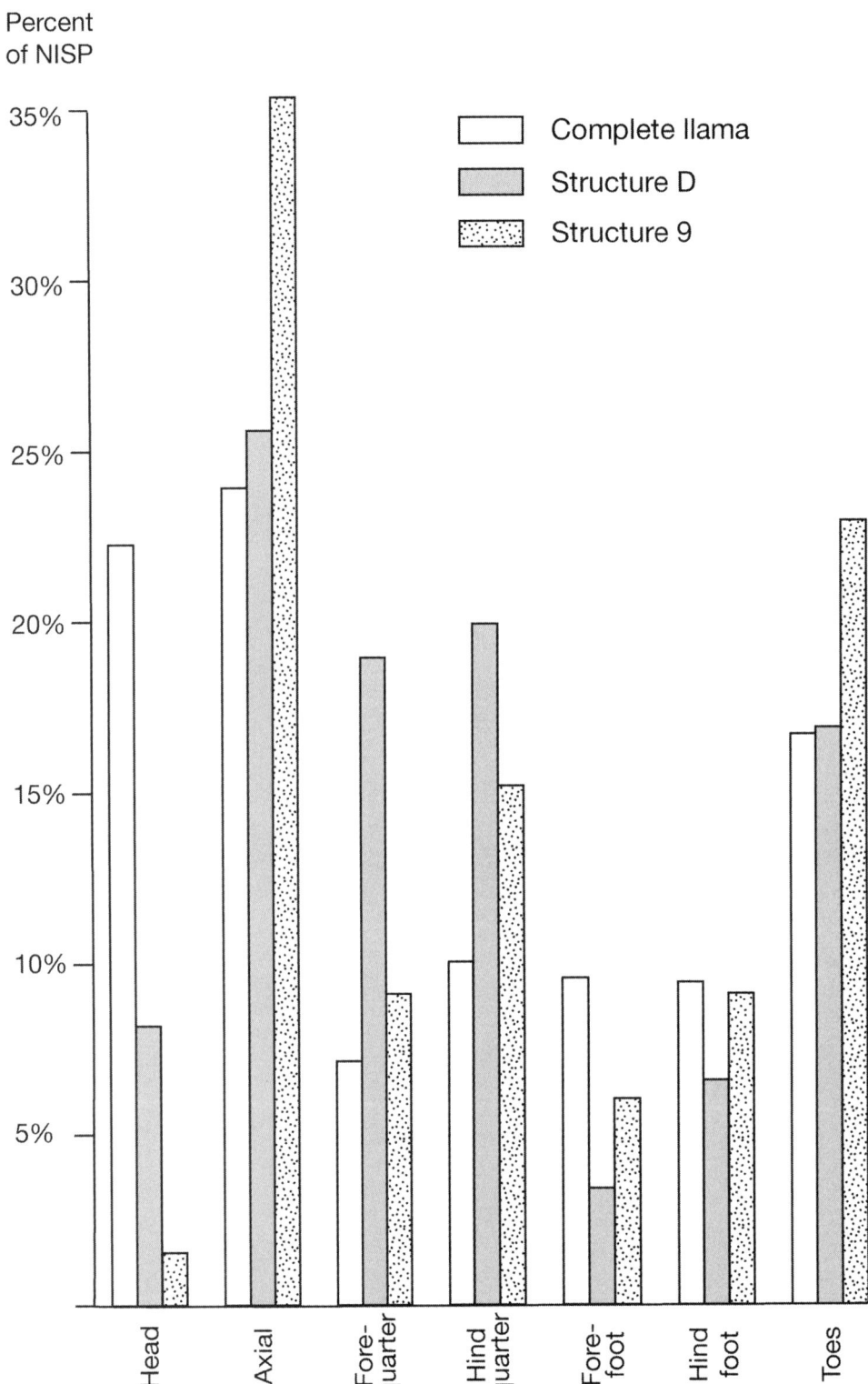

Figure 17.16. The relative abundance of bones from seven anatomical regions in Structure D, Structure 9, and a complete llama skeleton.

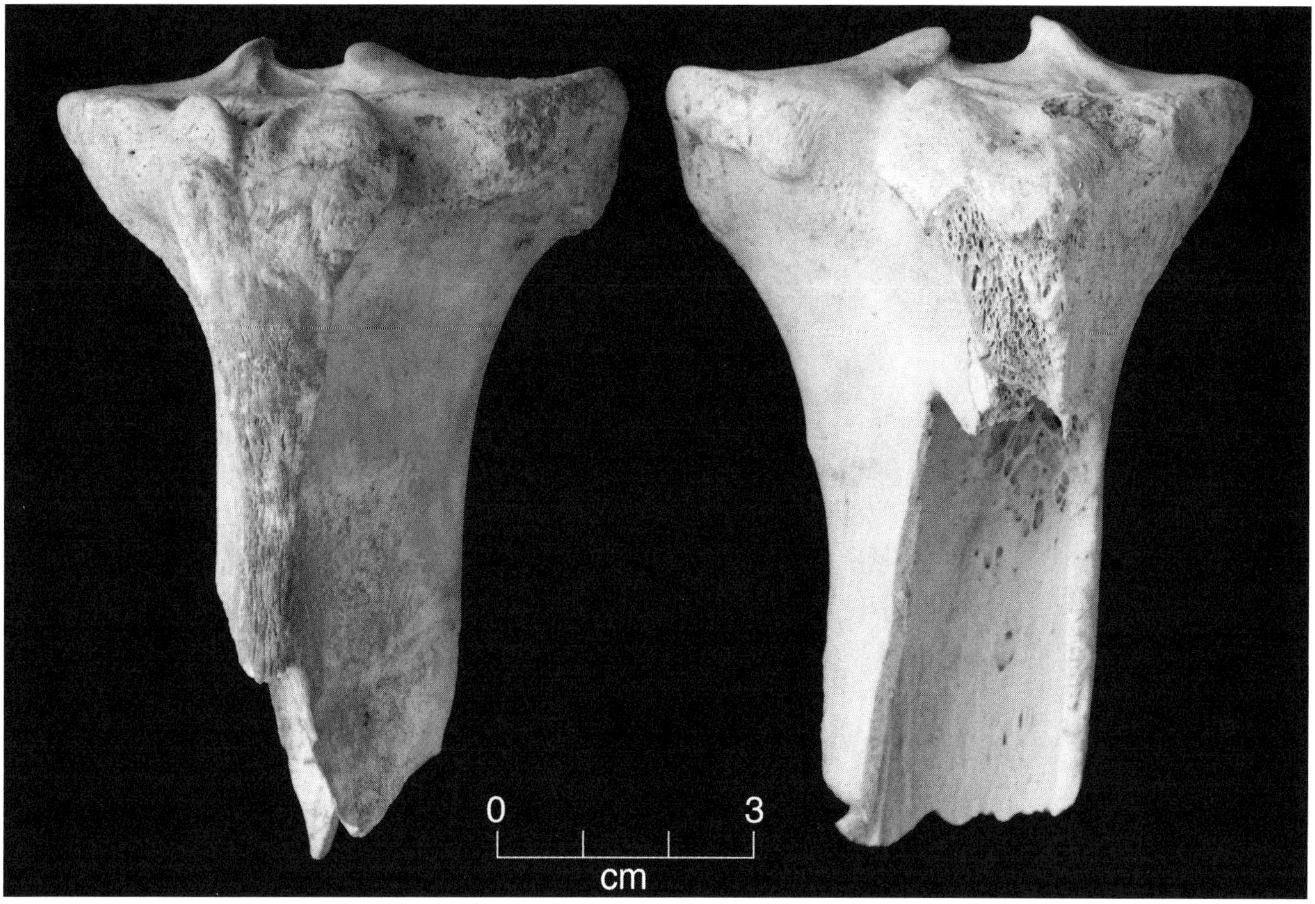

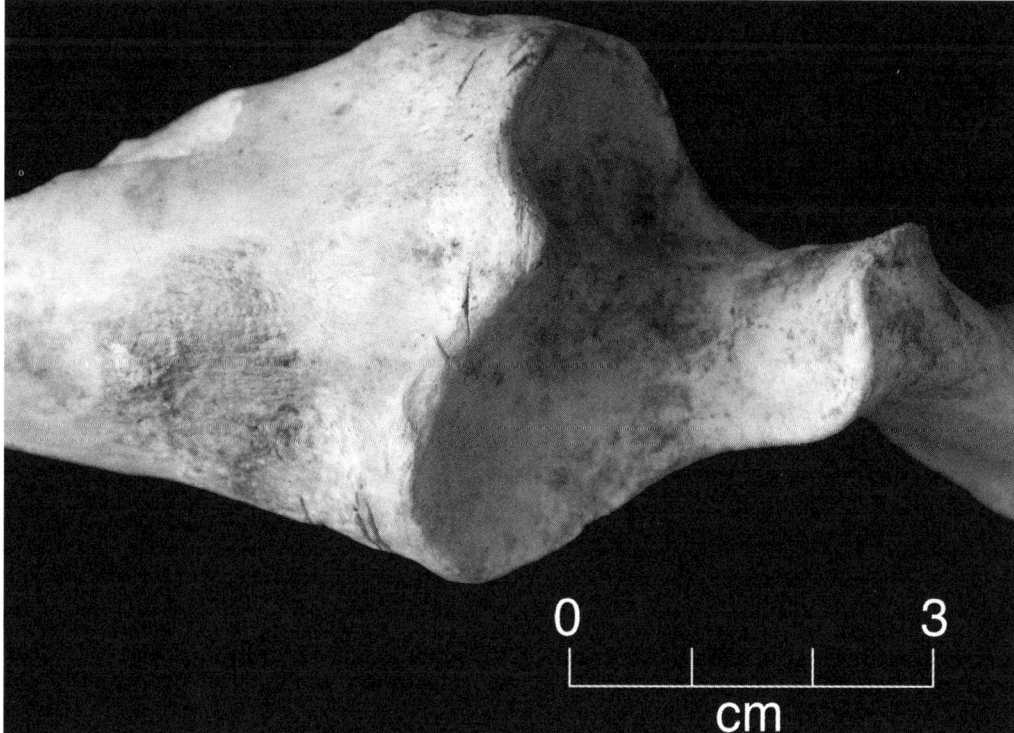

Figure 17.17 (above). Two proximal tibiae of adult llamas, butchered in much the same way. Both are from Structure D. The bone on the right is from the Southwest Canchón (found in the fill above the North Platform). The one on the left is from the Northeast Canchón (found in one of the ash-filled hollows in the floor, northwest corner of the canchón).

Figure 17.18 (left). Fused proximal right radius and ulna of a llama. There are cut marks all along the proximal edge of the radius, presumably made by someone trying to free the distal condyle of the humerus. This bone was found in one of the ash-filled hollows in the floor of Structure D's Northeast Canchón (northwest corner).

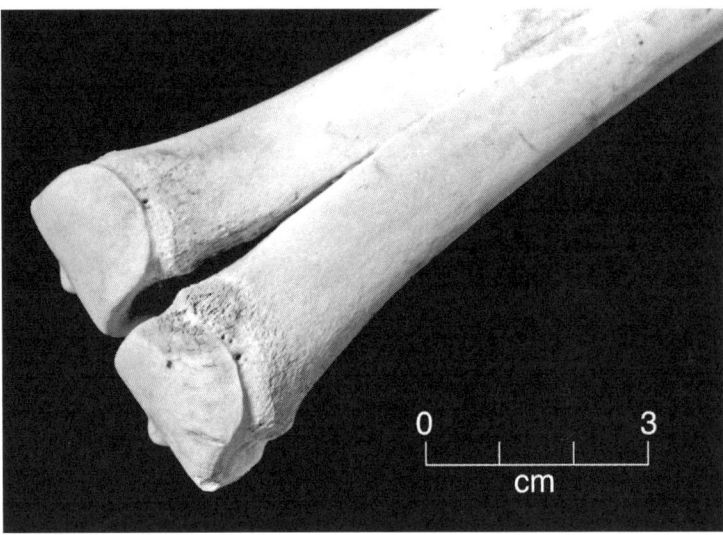

Figure 17.19 (above). Distal end of llama metacarpal, displaying a cut mark on one of the condyles. Found in the fill of Room 4, Structure D, below the level of Feature 5.

Figure 17.20 (below). Left calcaneum of a llama from Feature 6 of Structure D. This bone has two cut marks on the processus cochlearis, probably made by someone trying to separate the lower leg from the distal tibia.

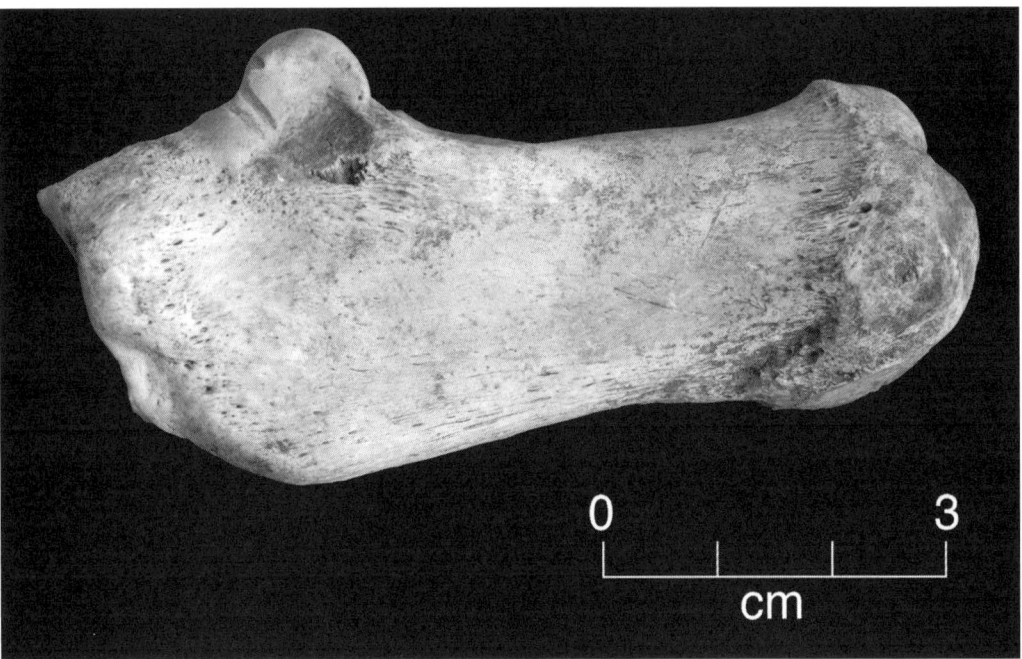

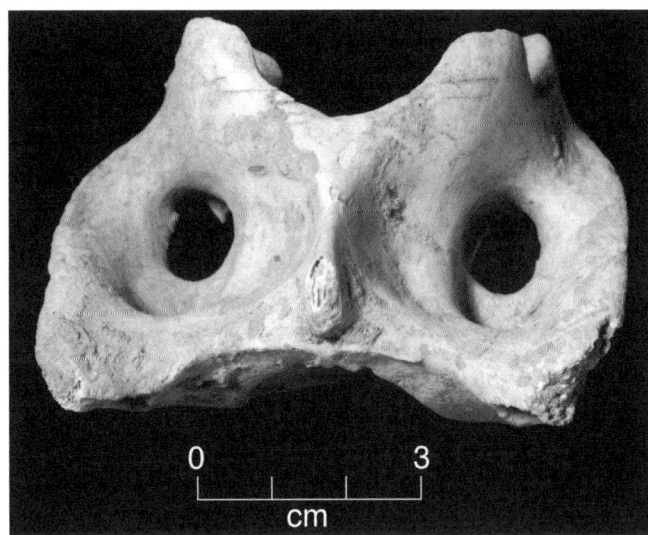

Figure 17.21 (above). The ventral surface of a llama atlas vertebra, displaying numerous cut marks near the anterior articular cavities. Such cuts may have been made by someone freeing the occipital condyles during decapitation. Found in the northern area of the Southwest Canchón, Structure D.

Figure 17.22 (below). The ventral surface of a llama atlas vertebra, displaying numerous cut marks near the anterior articular cavities. Such cuts may have been made by someone freeing the occipital condyles during decapitation. Found in Room 5 of Structure 9.

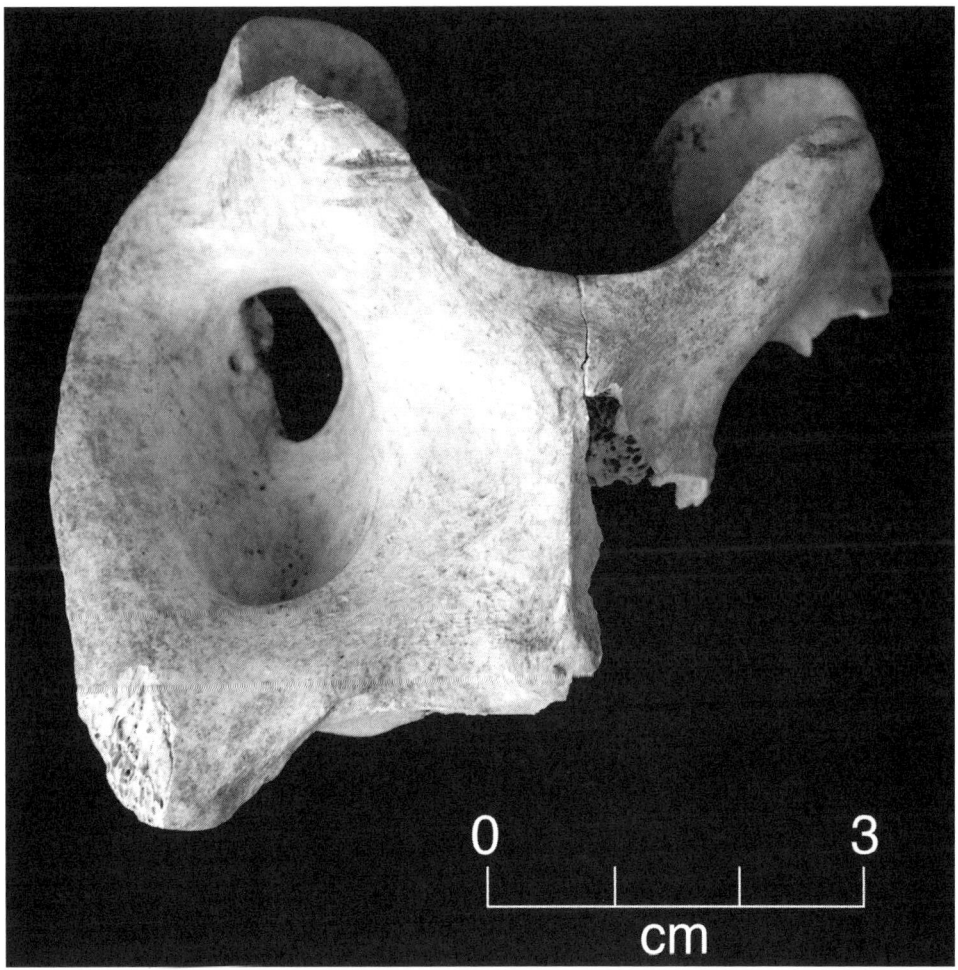

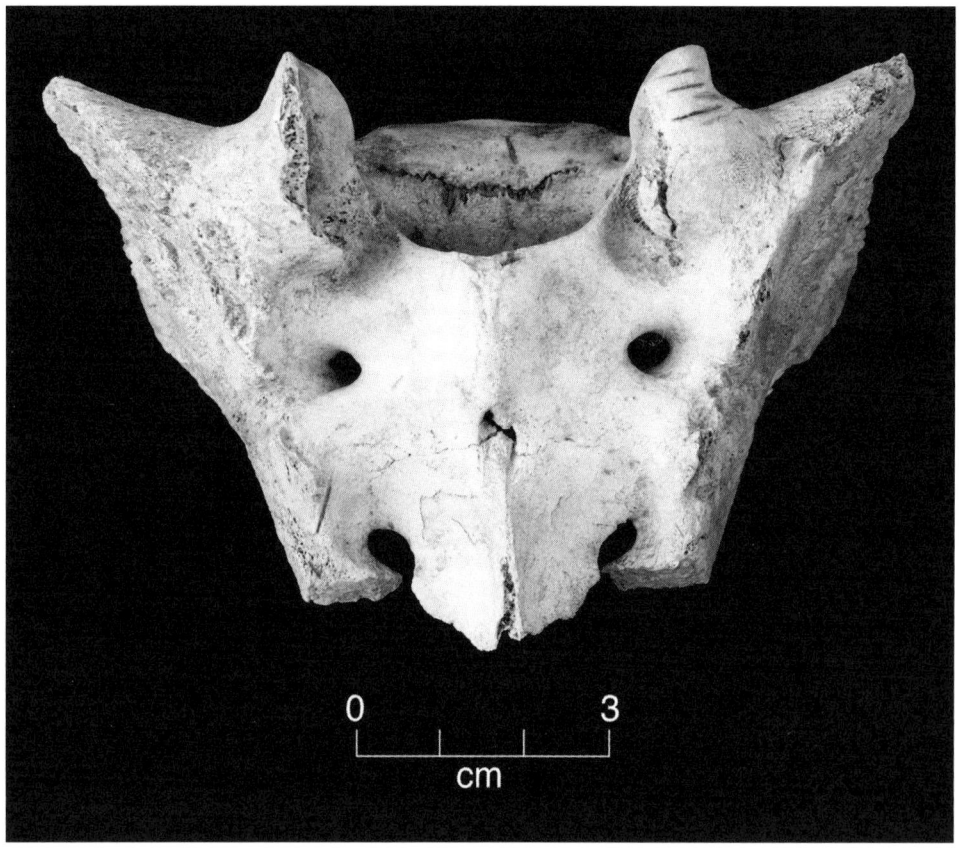

Figure 17.23. Sacrum of llama from Feature 6 of Structure D, showing cut marks on the right anterior articular process. One possible cause for these cuts is an attempt to free the sacrum from the final lumbar vertebra.

atlases show cut marks made during decapitation, when someone was trying to free the occipital condyles of the skull. As in the case of the two tibiae in Figure 17.17, having two atlases with similar cut marks shows a degree of standardization in butchering behavior.

Figure 17.23 shows the sacrum of an adult llama with repetitive cut marks on the right anterior articular process. Here we suspect that we are seeing an attempt to free the sacrum from the final lumbar vertebra.

Now we come to a series of vertebrae that may have been included in parcels of *ch'arki*. The axis vertebra shown in Figure 17.24 comes from Feature 6, and displays repetitive cut marks on one of its posterior articular processes. Since we know from earlier evidence that the skull was usually freed by separating the condyles from the atlas vertebra, Figure 17.24 can be seen as an attempt to separate the individual cervical vertebrae, each with its small parcel of meat.

The bone in Figure 17.25 makes an even more convincing case for *ch'arki*, since it appears to have been placed with Burial 2 as a parcel of food for the afterlife. This thoracic vertebra has cut marks on one of its articular processes, similar to the cuts seen in Figure 17.24.

Next we come to a cervical vertebra from Structure 9, also showing repetitive cut marks on one of its anterior articular

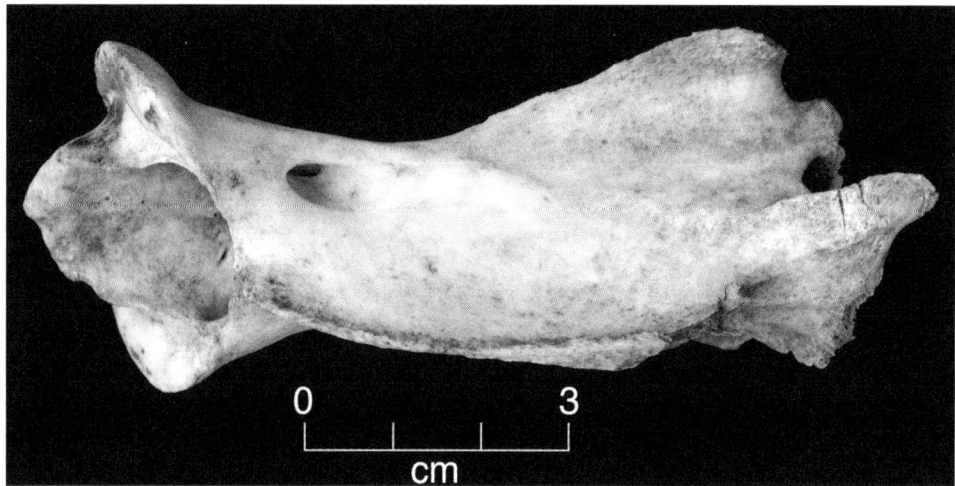

Figure 17.24. Dorsal view of the axis vertebra of a llama from Feature 6 of Structure D, displaying cut marks on one of the posterior articular processes. This vertebra may have been included in a portion of *ch'arki*.

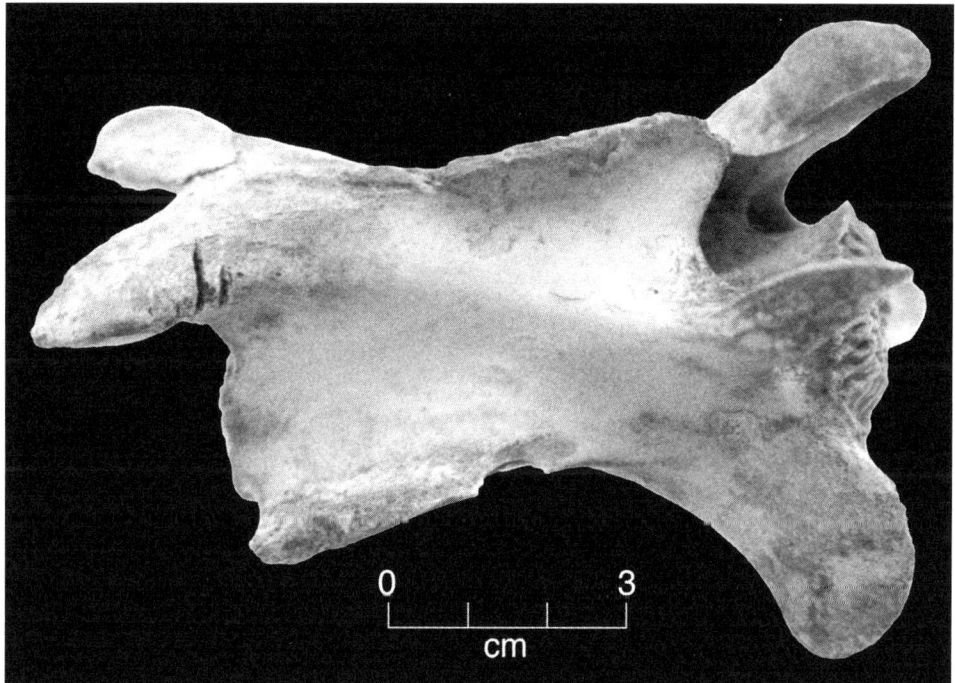

Figure 17.25. Cervical vertebra of llama, displaying cut marks on one of the articular processes. This bone was associated with Burial 2 of Quebrada 5a, and probably was part of a portion of *ch'arki* provided to the deceased for the afterlife.

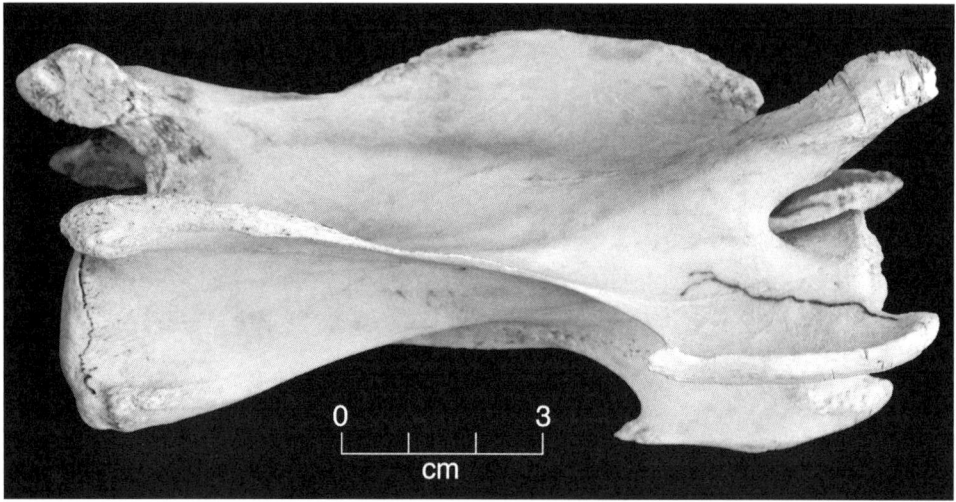

Figure 17.26. Lateral view of a cervical vertebra from a llama, displaying cut marks on one of the anterior articular processes. This vertebra, found in Room 7 of Structure 9, may have been included in a portion of ch'arki.

Figure 17.27. Small pigment pouch made from what appears to be a camelid bladder.

processes (Fig. 17.26). This, too, may represent the bone from a parcel of *ch'arki*.

Camelid Byproducts

Meat was not the only camelid product utilized at Cerro Azul. The Late Intermediate weavers had learned that the long bones of llamas could be cut and polished into *chocchikuna* (or *ruki*), implements used to keep warp and weft separate (Marcus 1987a: Figs. 53 and 54).

This was not the only use of camelid byproducts by the people of Cerro Azul. Figure 17.27 shows a small, tied pouch filled with dry, orangish-red pigment. Such dry pigments were often used in Andean rituals and in many cases had to be imported from distant mineral sources (Villagómez 1649, Siracusano 2011). Keeping the pigment dry required a waterproof container. The pouch in Figure 17.27 appears superficially to be the bladder of a camelid. To be certain, this would have to be confirmed by DNA analysis.

Differences in Diet between Nobles and Commoners

Figure 17.16 raises questions about differences in meat consumption between the nobles occupying Structure D and the commoner-class family occupying Structure 9. Perhaps the best way to examine those differences would be a direct comparison between Feature 6 (a midden from Structure D) and Feature 20 (a midden from Structure 9).

These two middens are compared in Figures 17.28 and 17.29. Each figure displays the skeleton of a llama. In Figure 17.28, we have blackened all those bones present in Feature 6; in Figure 17.29, we have blackened all those bones present in Feature 20. What the two drawings suggest is that both Structures D and 9 received parcels of *ch'arki*; in addition, the occupants of Structure D had access to an occasional complete llama. And even in the case of *ch'arki*, Structure D received more of the desirable parts of the animal than did Structure 9.

We believe that Structure D was occupied by a noble extended family, their servants, and at least a few trusted

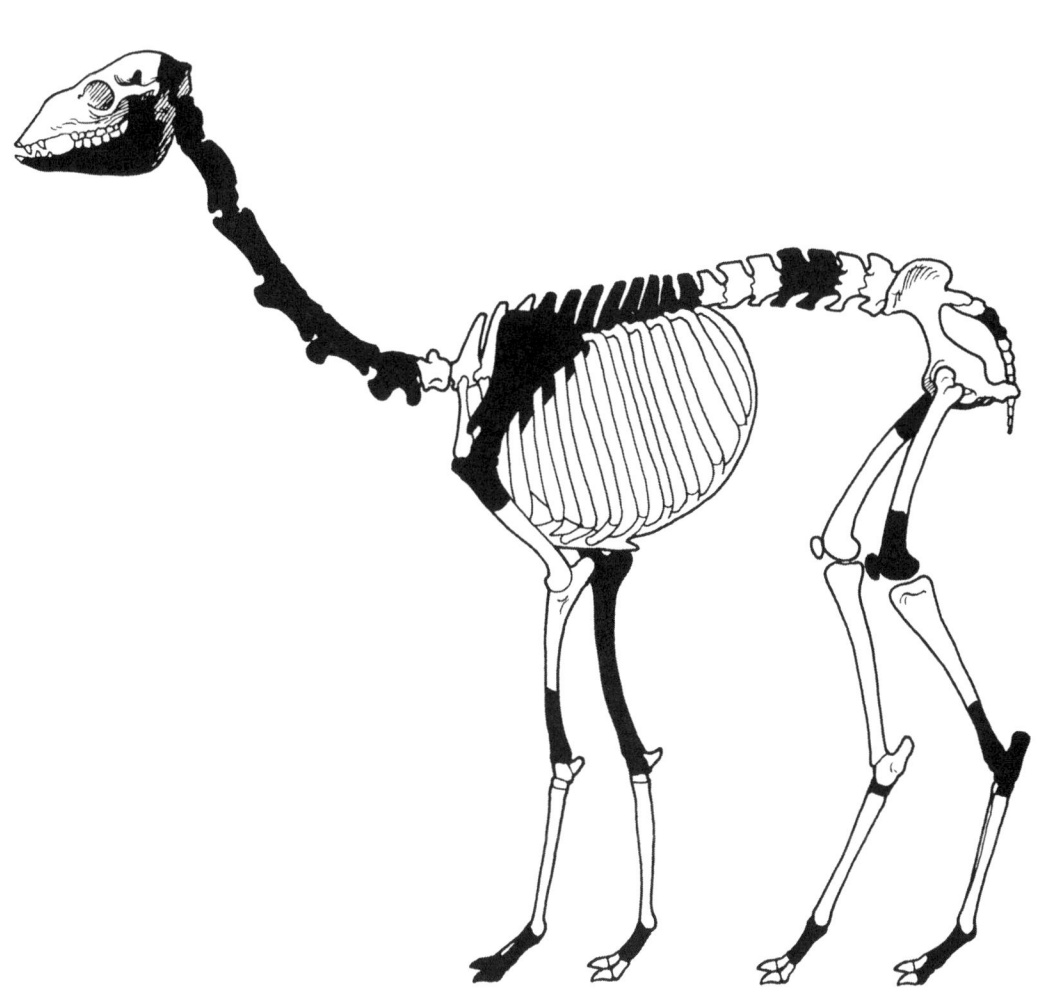

Figure 17.28. On this drawing of a llama skeleton, the bones present in the Feature 6 midden are indicated in black.

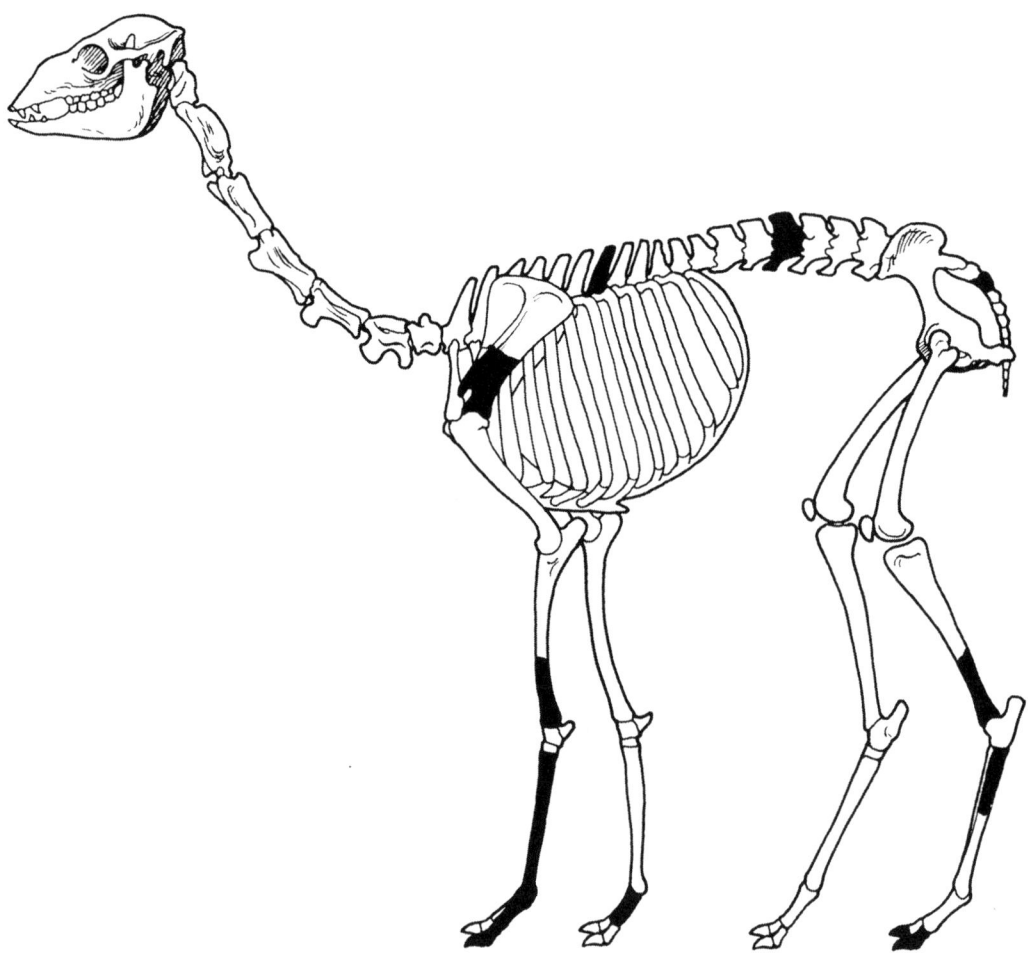

Figure 17.29. On this drawing of a llama skeleton, the bones present in the Feature 20 midden are indicated in black.

commoner-class overseers. Structure 9, on the other hand, was occupied by a commoner-class family whose household head was an overseer. Our comparison of the two middens suggests that, either as the result of feasting or ritual sacrifice, noble families had greater access to complete llamas and/or preferred cuts of *ch'arki*.

Conclusions

Cerro Azul received both dried camelid meat (*ch'arki*) and living llamas from neighboring regions. While *ch'arki* is presumed to have formed part of the diet of both nobles and commoners, it was primarily the nobles who received living llamas. The latter were probably consumed following sacrificial rites or during the course of feasts. It also appears that nobles tended to receive the most desirable parcels of *ch'arki*, i.e., those from the torso, forequarters, and hind quarters.

Both the llamas and the *ch'arki* must have come from at least as far away as the Kingdom of Lunahuaná, and perhaps even from polities higher in the sierra. There is no evidence that camelids were raised at Cerro Azul.

Three lines of evidence suggest that most, if not all, of the camelids eaten at Cerro Azul were llamas (*Lama glama*). First, the measurements of the bones themselves are generally within Wing's (1972) "large" size range. Second, Cerro Azul has evidence of polydactyly, a genetic abnormality that likely resulted from the inbreeding of domestic animals. Third, some weight-bearing bones show signs of arthritis, suggesting that beasts of burden were involved.

18

Domestic Dogs

Kent V. Flannery and Christopher P. Glew

The Late Intermediate occupants of Cerro Azul raised domestic dogs (*Canis familiaris*). We would know little about these animals, however, were it not for the fact that they were sometimes included with upper-class burials. Marcus' excavation of Structure D produced only one dog bone, and Structure 9 produced none at all. On the other hand, at least five dogs were associated with the burials on Terrace 9 of Quebrada 5a.

We do not fully understand why dogs were included with elite burials. Certainly there is no archaeological evidence that dogs were considered food at Cerro Azul, as they were in Mesoamerica. If we could link most of our complete (or nearly complete) dog skeletons to men's burials, we might be able to argue that dogs were hunting companions. This argument is weakened by the fact that one of our dogs was found with Burial 9, a woman accompanied by an infant and a child (see below).

We also cannot picture the dogs from the Quebrada 5a burials as cute little pets—lap dogs for noble women. Their measurements indicate that the dogs found with burials were 75 to 80 percent the size of a golden retriever; they were, in other words, large enough to have been guard dogs. We consider it possible, therefore, that dogs were added to some burial cists to guard nobles during the afterlife.

Dog Bones from the Tapia Compounds

Structure D

One right mandible was found on the floor of the Southwest Canchón, not far from the Feature 6 midden.

Structure 9

No dog bones were found.

Dog Bones from Middens in the Quebradas

Quebrada 5, Terrace 16

The most interesting discovery on Terrace 16 was **Feature 22**, a hearth or earth oven used to cook sea lions (see Chapter 12).

This feature was associated with Stratigraphic Zone A2. Above Zone A2 lay Stratigraphic Zone A1, which was separated from Zone A2 by a layer of salitre or salty crust.

Two badly salt-damaged dog bones were found not far above Feature 22. One was a fragment of right scapula; the other consisted of a still-articulated left maxilla and premaxilla. Because of their level of salt damage, we believe that these dog bones were associated with the salitre level and not the hearth. The latter had good bone preservation, and no evidence for the cooking of dogs.

Quebrada 6, Terrace 11

Terrace 11 featured a coal-black midden with lots of organic refuse. A sample taken at a depth of 30–50 cm below the surface produced the following dog bones:

Articulated left maxilla and premaxilla: 1
Right radius (fully fused): 1
Left distal femur (epiphysis unfused): 1

Totals:
  NISP =3
  MNI=2

It should be noted that the fully fused right radius was from an adult dog—one considerably smaller than the "guard dogs" found with the burials in Quebrada 5a. We can suggest, therefore, that dogs of several sizes were being raised during the Late Intermediate.

Quebrada 6, Terrace 12

Terrace 12 contained three features: a burial cist, a storage cist, and an ash-filled earth oven. Stratigraphically above these features was a layer of brown-to-beige midden that had eroded downslope from Terrace 11 of the same quebrada. This midden layer produced the following dog bones:

Right premaxilla: 1
Right maxilla: 1
Tooth fragments: 2

Totals:
  NISP = 4
  MNI = 3

Dogs Associated with Burials

Our most nearly complete skeletons of Late Intermediate dogs were those included with elite burials on Terrace 9 of Quebrada 5a. Unfortunately, many of those skeletons had been disturbed when the burials were looted.

Burial 1

Burial 1, found in Stratigraphic Zone A of Terrace 9, was accompanied by the dog whose cranium is shown in Figure 18.1. The basioccipital length of this cranium was 15.3 cm. Its maximum width at the zygomatic arches could not be taken, owing to damage from looters.

Structure 6

Structure 6 was a looted burial cist in Stratigraphic Zone A of Terrace 9. At least one dog had been associated with the burials in this cist. The most useful surviving dog bone was a complete right mandible with the third and fourth premolars still in place.

Burial 9

Burial 9 consisted of an adult woman accompanied by an infant and a child (Marcus 2015:8–10). This burial produced the left maxilla and premaxilla of a dog, with all its teeth present except for a few incisors. There was still a great deal of hair clinging to the maxilla, suggesting that the dog had been well preserved before the burial was disturbed by looters.

Other Dogs from Zone A

Near the northeast corner of Marcus' excavation on Terrace 9, her workmen encountered an area of looted burials. These burials had obviously been accompanied by two dogs, but the canine remains were now so badly disturbed that their postcranial skeletons were intermixed. The crania of these two dogs are shown in Figures 18.2 and 18.3. The cranium of Dog 1 had a basioccipital length of 16.3 cm; that of Dog 2 measured 16.0 cm.

Found with these animals was a large piece of skin and fur from a brown-and-tan dog. In the Munsell Soil Color system (Munsell Color Company 1954), the brown areas would be classified as 5YR 3/3 ("dark reddish brown"), while the tan areas would be considered 2.5Y 7/6 ("yellow"). We do not know to which of the two dogs this piece of skin belonged.

Conclusions

Domestic dogs of at least two sizes were raised at Cerro Azul. We have only one bone from the smaller type of dog, which

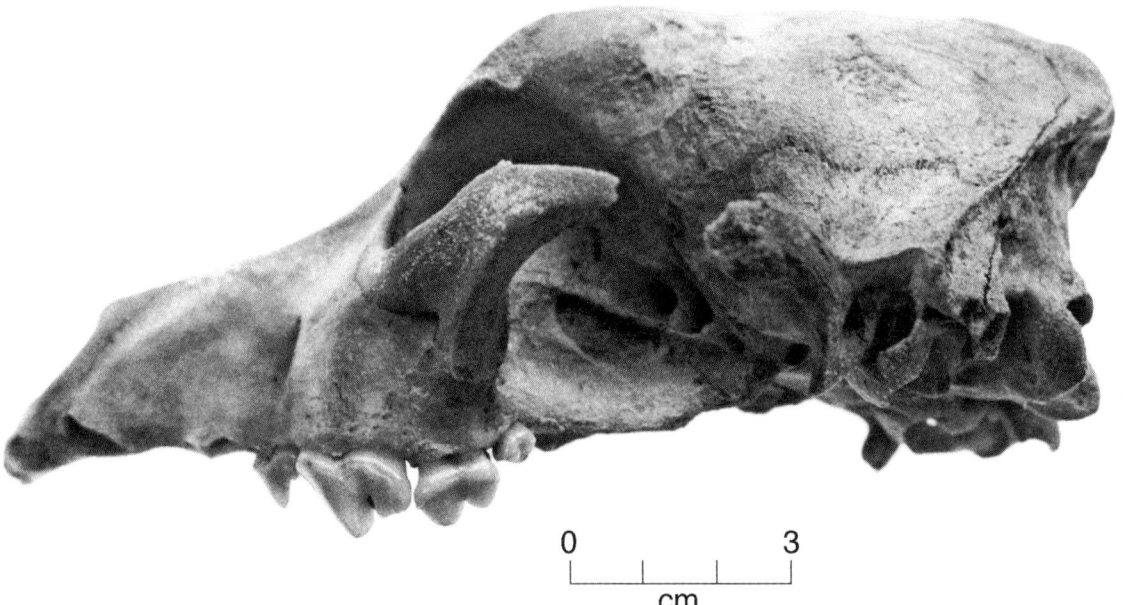

Figure 18.1. The cranium from a dog associated with Burial 1, Terrace 9, Quebrada 5a.

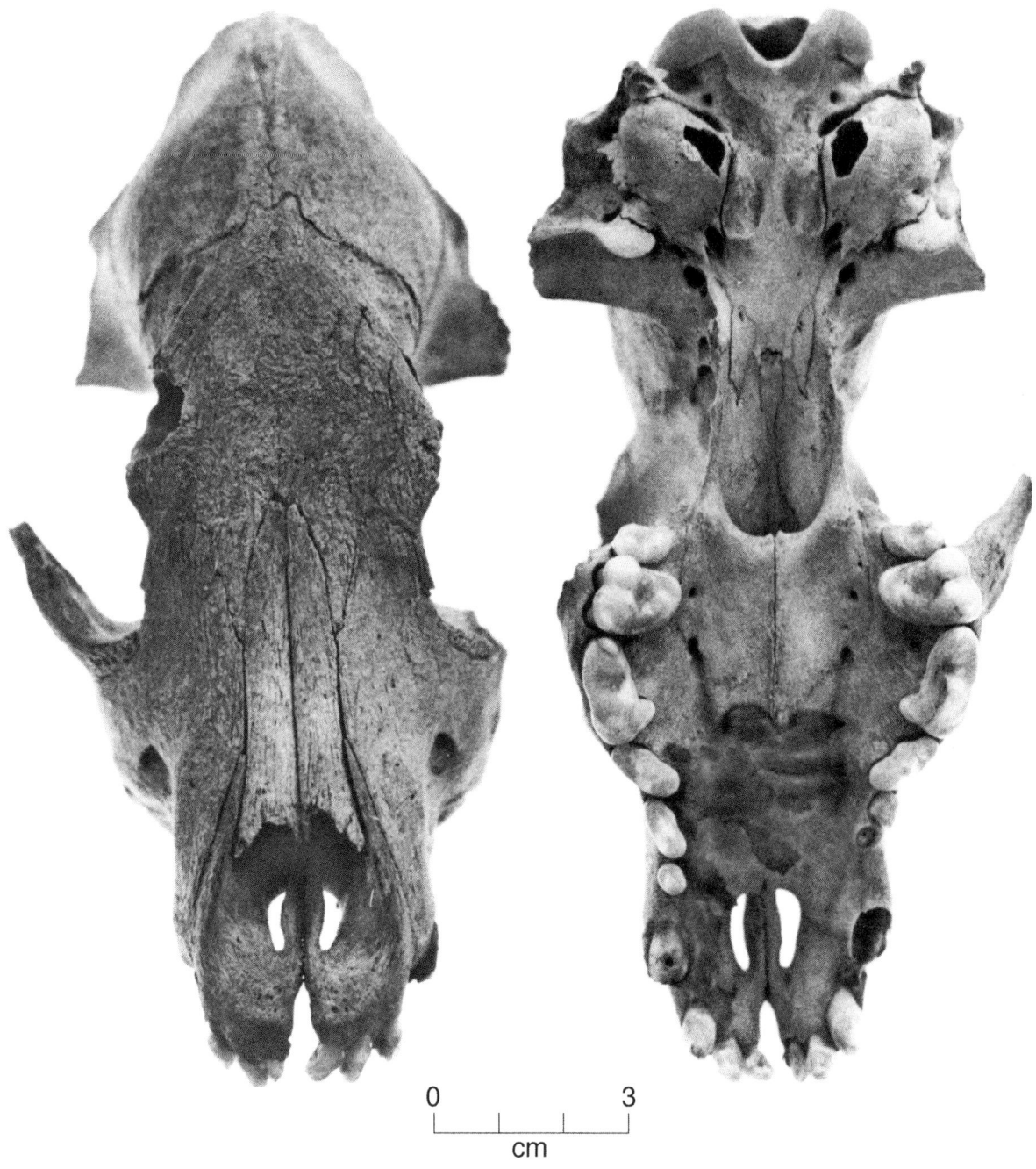

Figure 18.2. Two views of the cranium from the first of two dogs associated with looted burials in Zone A of Terrace 9, Quebrada 5a.

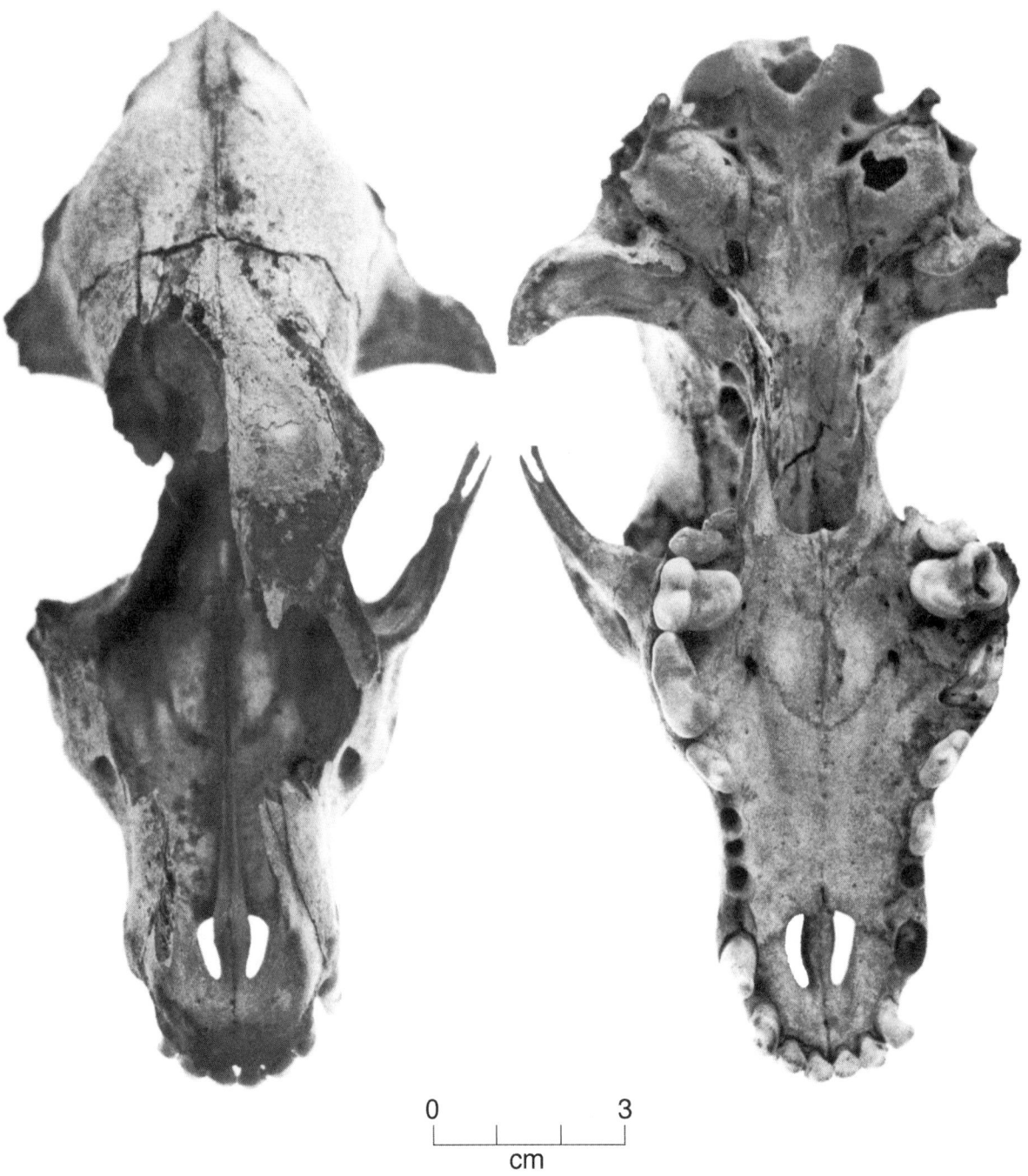

Figure 18.3. Two views of the cranium from the second of two dogs associated with looted burials in Zone A of Terrace 9, Quebrada 5a.

may have weighed no more than a cocker spaniel. We have at least three crania from the larger type, which was 75–80 percent the size of a golden retriever. At least one of the larger dogs we recovered was brown-and-tan.

There is no evidence for the eating of dogs. Their remains are rare to absent in Structures D and 9 and scarce even in the quebrada middens, where any dead dog would likely have been discarded.

Our best dog specimens came from looted burial cists dating to the latter part of the Late Intermediate. These dogs, all belonging to the larger type, may have served as guardians for buried nobles in the afterlife.

19

The Raising of Guinea Pigs

Christopher P. Glew and Kent V. Flannery

One of the domestic animals raised by the Late Intermediate occupants of Cerro Azul was the guinea pig, *Cavia porcellus*. Our information on this domestic rodent comes from four lines of evidence. First, Marcus found two small rooms in Structure D that had clearly been set aside for the raising of guinea pigs. Second, complete specimens of *Cavia* had been left as food offerings with a number of Late Intermediate burials. Third, complete specimens of sacrificed guinea pigs were discovered in several rooms of Structure D. Fourth, there were *Cavia* bones in the site's domestic refuse.

Rooms 9 and 10 of Structure D

As Marcus' crew carried out its excavation of the northwest quadrant of Structure D, they came upon two small rooms that had been created within the building's Northwest Canchón (Fig. 19.1). It appears that early in the occupation of the building, the architects had laid out a tapia-walled room measuring 4.18 m by 2.70 m. Its floor was simply the pre-existing floor of the Northwest Canchón. At a later date, this room was divided into two smaller units by the addition of a very informal wall of tapia blocks. Marcus designated the western unit Room 9; the eastern unit became Room 10.

Room 10 had upon its floor a layer of maize stalks, leaves, and loose kernels; unidentified grass; and a variety of unidentified small, leafy herbs. Room 9 had upon its floor several hundred pellets of guinea pig dung (Fig. 19.2). Marcus concluded that this pair of rooms had served as a *cuyero* or guinea pig nursery. Room 10, whose doorway provided access from the Northwest Canchón, was used to store food and bedding for the rodents. Room 9, which had no doorway, became the living quarters for the guinea pigs.

Guinea Pig Remains from Archaeological Contexts

In the Andes guinea pigs were not only eaten, they were considered appropriate animals for rituals of sacrifice, dedication, and divination (Sandweiss and Wing 1997). Nearly 300 guinea pig bones were recovered from Cerro Azul. Significantly, at least a dozen guinea pigs had either been used in dedicatory rituals or left with burials as food for the afterlife.

The Raising of Guinea Pigs

Figure 19.1. Room 9 (foreground) and Room 10 (background), seen from the southwest. Guinea pigs were kept in Room 9, while Room 10 was used as a place to store their bedding and food.

Figure 19.2. Pellets of guinea pig dung from the floor of Room 9, Structure D.

Structure D

Our largest sample of *Cavia* bones from Structure D came from **Feature 6**, a midden that Marcus' crew screened through both 1.5-mm and 0.6-mm mesh. This sample consisted of 181 fragments, representing a minimum of 11 individual guinea pigs. Virtually every part of the skeleton was represented and well preserved, making it unlikely that these animals had been cooked and eaten.

Marcus has interpreted Feature 6 as the final pile of debris swept up from the floor of the Southwest Canchón before Structure D was abandoned. If that is the case, there must have been guinea pig bones scattered all over the floor of the canchón prior to its final sweeping. The full list is given in Table 19.1.

We are not sure why there would have been so many guinea pig bones scattered over the floor of the Southwest Canchón; this was a large work space, open to the outside world, where (among other things) llama caravans were loaded and unloaded.

A number of the guinea pig bones had unfused epiphyses, indicating that they came from animals less than fully grown. None of the bones showed signs of having been cooked, and only one bone out of 181—a right femur—showed signs of having been burned. We suspect, therefore, that we may be dealing with 10–12 guinea pigs that gave their lives in rituals of healing or divination, after which their remains were discarded in the Southwest Canchón. This canchón was, after all, a place where refuse was allowed to accumulate before being carried off to the quebrada middens.

While cleaning the floor of the Southwest Canchón, Marcus' crew recovered the right femur of yet another guinea pig, perhaps a bone missed when the canchón was swept.

Table 19.1. Guinea Pig Bones from Feature 6 of Structure D.

Skull and mandible fragments
Skull caps	4
Auditory bulla	1
Petrous bones	16
Other skull fragments	16
Complete palate with both maxillae	1
Right maxillae	4
Left maxillae	4
Right premaxillae	5
Left premaxillae	3
Right mandibles	9
Left mandibles	11 (MNI = 11)
Loose upper incisors	2
Loose lower incisors	2
Loose cheek teeth	18
Fragments of teeth	3

Elements of the forelimb
Right scapulae	3
Left scapulae	8
Right humeri	5 (1 immature)
Left humeri	5
Right ulnae	5
Left ulnae	6
Right radii	3
Left radii	2
Right forepaw	1 (articulated carpals and metacarpals)

Elements of the hind limb
Left innominates	2
Right ilia	3 (epiphyses unfused)
Left ilia	2 (epiphyses unfused)
Right ischium	1 (epiphysis unfused)
Left ischia	3 (2 with epiphyses unfused)
Right femora	2
Left femora	3 (epiphyses unfused)
Left tibiae	2 (shafts only)
Left astragalus	1

Vertebrae
Atlases	3
Cervical vertebrae	3
Thoracic vertebrae	2
Lumbar vertebrae	6
Sacrum	1
Caudal vertebrae	8
Costal cartilage	1

Totals:
 NISP = 181
 MNI = 11

The **North Platform** of the Southwest Canchón supported a *llamkana pata*, or raised work platform. One left mandible of *Cavia* was found lying directly upon this *pata*, where we believe the canchón's overseer performed his duties. At a later date, the platform was covered with Late Intermediate fill. This fill produced one right mandible, one left mandible, and one left auditory bulla from a guinea pig.

The Northeast Canchón

The Northeast Canchón of Structure D was an unroofed work area where (among other things) weaving was carried out. A number of hollows in the floor in the northwest corner of this canchón held ashy deposits with refuse. Included in the refuse was one left innominate from an adult guinea pig.

Room 1

Room 1 had originally been an elite residential unit, but was converted to fish storage after it suffered earthquake damage. In its very latest fill (deposited after its period of storage had ended) Marcus' crew recovered three possible thoracic vertebrae from a guinea pig.

Room 3

Like Room 1, Room 3 was an elite residential unit that had been converted to storage after suffering earthquake damage. Its very latest fill produced 11 fragments of *Cavia* (possibly including the remains of two pregnant females). The list of bones is as follows:

Adult or juvenile remains
  Right mandible: 1
  Left mandible: 1
  Right premaxilla: 1
  Loose teeth: 2
  Right femur: 1 (epiphyses unfused)
  Left femur: 1 (epiphyses unfused)
Remains of late-term foetuses
  Left humerus: 1
  Left femora: 2 (MNI = 2)
  Right tibia: 1

Totals:
NISP = 11
MNI = 3–4

Room 4

Room 4 had a complex stratigraphic history. The room's original floor lay at a depth of 3.73 m below the datum point established for Structure D. At a later date, the lower part of the room was filled in and a second floor established at a depth of 2.23 m. Just above the room's original floor, Marcus' crew recovered one left mandible of *Cavia*.

Just below the room's second floor, the crew came upon what appeared to be the remains of a sacrificed guinea pig. This rodent's complete skull, right and left mandibles, and one cervical vertebra were found together, as if the animal had been decapitated. Not far away, the crew found the right and left scapulae; the right and left humeri; the right and left radii; the right and left ulnae; the complete pelvis; the right and left femora; the right tibia; 26 vertebrae (including the sacrum and one caudal vertebra); 20 ribs; and 4 other fragments, all clearly from the same individual. The age of this guinea pig was such that the epiphyses of all but two of its limb bones (one ulna and one femur) were completely fused.

This guinea pig may have been sacrificed as part of a ritual that accompanied the laying down of the second floor and the repurposing of Room 4. A few of its bones had been rendered virtually unrecognizable by the weight of the tapia overburden.

Room 5

Room 5 also had a complex stratigraphic history. Originally created as a storage room of some kind, it had been converted to a residential unit later in the building's history. *Cavia* remains were associated only with the room's later (residential) period of use.

One of the first steps in the conversion of Room 5 to a residential unit was the creation of a large tapia platform that filled the eastern half of the room. Embedded in this platform Marcus' crew found a virtually complete guinea pig, this one possibly sacrificed at the time the room was repurposed. Present were the cranium and both mandibles; the right scapula; the right and left humeri; the right and left radii; the right and left ulnae; the complete pelvis; the right and left femora; the left tibia and fibula; 2 metapodials; 23 vertebrae; 13 ribs; and 9 other fragments. Once again, the few missing bones can be attributed to the crushing weight of the later tapia overburden. This guinea pig was young enough to have unfused epiphyses on most of its limb bones.

After the creation of the tapia platform, a series of later "sleeping benches" were built in Room 5. The uppermost fill on these later benches produced one right mandible of *Cavia*.

Room 7

Room 7 was a small unit that had once been connected to the Southwest Canchón by a narrow passageway. After that passageway was deliberately blocked, Room 7 became a convenient place to dump refuse. Included in that refuse was one left femur of *Cavia*.

Room 8

Room 8 was a sand-filled fish storage room. After the abandonment of Structure D, wind erosion caused a layer of debris from the Southwest Canchón to drift down into Room 8. Found in this debris was the left mandible of a guinea pig.

A Late Horizon Squatters' House

Feature 4 of Structure D was a layer of ashy midden, associated with a Late Horizon squatters' house built in the "Room 2" patio after Structure D had ceased to function as an elite compound. This midden yielded one left mandible of a guinea pig.

Structure 9

Our largest sample of *Cavia* bones from Structure 9 came from **Feature 20**, a midden resting against the east side of the building. It is presumed that this midden (which Marcus' crew screened through both 1.5-mm and 0.6-mm mesh) represents domestic refuse from the kincha house of the building's commoner-class overseer.

Feature 20 produced some 44 bones, the remains of a minimum of four individual guinea pigs. The counts are given in Table 19.2.

None of the guinea pig bones from Feature 20 were burned, but many were damaged and incomplete. In contrast to the remains from Feature 6 of Structure D, virtually no vertebrae or ribs were found. It seems possible that most, if not all, of the guinea pig bones from Feature 20 were the remains of meals in which the animals were used in stews. Such stewing would have softened many bones, and might well account for the missing vertebrae and ribs. Two guinea pig limb bones had been gnawed by a smaller rodent (presumably the rice rat, *Oryzomys xanthaeolus*), indicating that they had been exposed to the elements for some time.

Room 4

Room 4 was a small storage unit in the northeast corner of Structure 9. It produced one lumbar vertebra of *Cavia*.

Room 5

Room 5 was a fish storage room on the west side of Structure 9. After its main period of use was over, this room became a convenient place to dump trash. Marcus suspected that the trash (which appeared to be domestic refuse) had come from the nearby kincha house of the building's administrator. Included in this domestic refuse were the following *Cavia* bones:

Left mandible: 1
Distal left humerus: 1 (adult)
Left innominate: 1 (adult)
Right femur: 1 (epiphyses unfused)
Left tibia: 1 (juvenile)

Totals:
 NISP = 5
 MNI = 2

Room 6

Room 6 was a storage unit immediately adjacent to Room 5. Trash thrown into the room late in its history included one petrous bone, one right premaxilla, and one unfused left scapula of guinea pig.

Table 19.2. Guinea Pig Bones from Feature 20 of Structure 9.

Skull and mandible fragments	
Skull cap	1
Right occipital condyle	1
Fragment of right zygomatic arch	1
Fragment of left zygomatic arch	1
Left auditory bulla	1
Petrous bones	4 (all from different individuals)
Right maxilla	1
Left maxilla	1
Right premaxillae	2
Left premaxillae	2
Left mandibles	3
Loose cheek teeth	7
Elements of the forelimb	
Right humeri	3 (2 with epiphyses unfused)
Left humerus	1 (shaft only)
Left radius	1
Right ulnae	2
Elements of the hind limb	
Left ilia	3
Right femora	2 (1 rodent-gnawed; 1 with epiphyses unfused)
Left femora	2 (1 rodent-gnawed; 1 with epiphyses unfused)
Right tibia	1 (shaft only)
Left tibiae	4 (2 with epiphyses unfused)
Vertebrae	
Atlases	2

Totals:
 NISP = 44
 MNI = 4

Guinea Pig Remains from Midden Deposits in the Quebradas of Cerro Camacho

Several terraces in the quebradas of Cerro Camacho had midden deposits with domestic refuse, almost certainly carried to the quebrada from one or more of the residential compounds. Most of the *Cavia* remains from these middens were poorly preserved and likely represented the refuse from meals.

Quebrada 5-south

Two of the most important discoveries made in Quebrada 5-south were Structure 11, a storage facility, and Structure 12, a looted burial cist. Between these two structures, however, was a midden deposit, some of whose ceramics appeared to antedate the occupation of Structure D. This midden produced the skull of a guinea pig.

Quebrada 5a, Terrace 9

Terrace 9 of Quebrada 5a had a complex stratigraphic history. It began as a venue for domestic refuse, and later became a convenient place to create subterranean burial cists. Of all the stratigraphic levels, the one producing the oldest guinea pig remains was Zone B—a midden from the early part of the Late Intermediate period—featuring abundant coquina clams and gray ash.

One sample of the Zone B midden, screened through both 1.5-mm and 0.6-mm mesh, produced one left mandible and one left humerus from an adult guinea pig. A second sample—consisting of Zone B matrix that had been reused as mortar in Structure 5, a burial cist—yielded one left humerus and one right ulna from an immature guinea pig.

Stratigraphically above Zone B lay Zone A of Terrace 9, an ashy midden dating to a later stage of the Late Intermediate. This zone had been penetrated by so many looters' pits that it proved difficult to get a reliable sample from it. One undisturbed area between two pits produced a complete skull of *Cavia*, two additional left maxillae, one right maxilla, and one right mandible, representing a minimum of three guinea pigs. The complete *Cavia* skull produced laughter on the part of Marcus' crew, since decomposition had given its hair what appeared to be a "punk-rock" hairdo. "*Este cuy salió muy 'punk'*," her workmen told her.

Quebrada 6, Terrace 11

On Terrace 11 of Quebrada 6, Marcus' crew discovered a richly organic midden, full of domestic refuse from one of the nearby residential compounds. At a depth of 30–50 cm in this midden, screening produced the right maxilla and mandible of a guinea pig.

Guinea Pigs Included in Burials

Another context in which the University of Michigan project recovered guinea pigs was as food for the afterlife. Whole guinea pigs were found in the gourd bowls that sometimes accompanied undisturbed Late Intermediate burials; often they shared the bowl with fish, shellfish, and plant foods. The disturbed remains of additional guinea pigs were found in burial cists that had been looted.

Burial 4, Individual 1a

Burial 4 was discovered in Stratigraphic Zone A of Terrace 9, Quebrada 5a. It was a multi-person interment, including at least three adults in mummy bundles (see Fig. 11 of Marcus 2015). A gourd vessel rested against the mummy bundle of Individual 1a, an adult woman.

The gourd vessel had originally contained two complete specimens of *Cavia*, now partially decomposed and incomplete. Guinea Pig 1 was represented by its cranium; its right and left mandibles; its right scapula; its right and left humeri; and its right ulna (NISP = 7, MNI = 1). Guinea Pig 2 was represented by its cranium; its right mandible; its right and left scapulae; its right and left humeri; its left radius; its left ulna; its sacrum; its right innominate; and patches of its brownish yellow hair (NISP = 11, MNI = 1). In addition, the gourd contained 21 loose bones or teeth that could be from either Guinea Pig 1 or Guinea Pig 2. Included were 3 loose incisors; 9 vertebrae; 7 ribs; and 2 complete axes (NISP = 21). Both of these guinea pigs were adults.

Burial 6, Individual 1

Burial 6 was found in Stratigraphic Zone B of Terrace 9, Quebrada 5a. It consisted of two individuals buried side by side: a man older than 50 years of age (Individual 1) and a child of 6–7 years (Individual 2) (Guillén n.d.). Individual 1 had a gourd bowl near his feet.

The gourd bowl associated with Individual 1 featured what Marcus' workmen called a *cuy planchado*—that is, a complete guinea pig flattened by the weight of the overburden.

Burial 7, Individual 2

Burial 7 was found in Stratigraphic Zone B of Terrace 9, immediately to the north of Burial 6. Like Burial 7, it consisted of an adult (Individual 2) and a child (Individual 1) buried side by side. According to Guillén (n.d.), Individual 1 was a 14-month-old child, while Individual 2 was a man approximately 45 years of age.

Resting on the feet of the adult male was a gourd bowl containing (among other things) a complete guinea pig flattened by the weight of the overburden (Fig. 19.3).

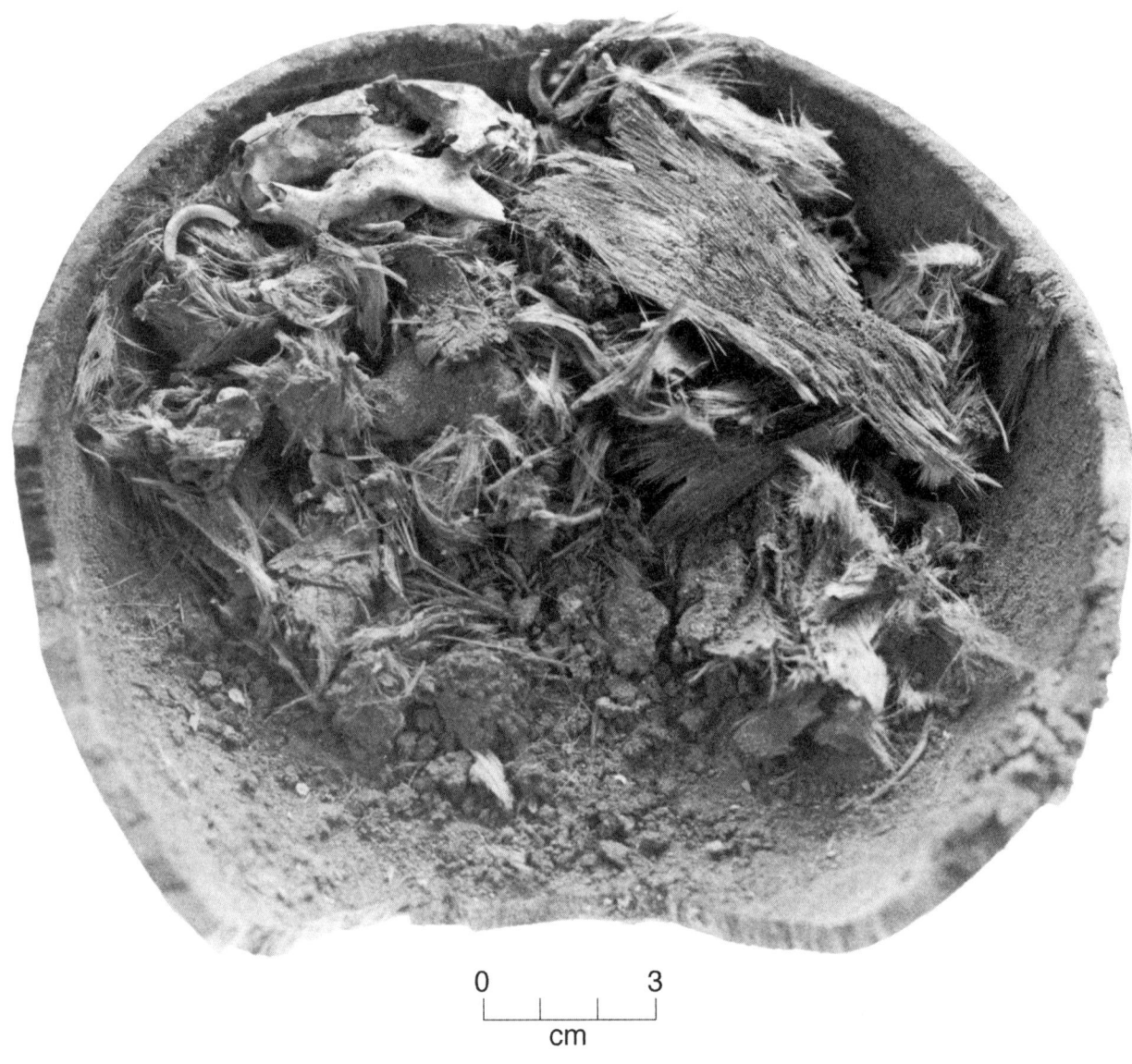

Figure 19.3. This gourd vessel, found with Individual 2 of Burial 7, contained the remains of a guinea pig (Terrace 9, Quebrada 5a).

Burial 8

Burial 8 was a woman approximately 50 years of age (Guillén n.d.), wrapped in a multi-layered mummy bundle. Near her chest was a gourd bowl containing (among other things) a complete guinea pig flattened by the weight of the overburden. Like Burials 6 and 7, this burial was found in Stratigraphic Zone B of Terrace 9, Quebrada 5a.

Burial 9

Burial 9 was discovered inside a small cobble-lined grave (Structure 8), near the bottom of Stratigraphic Zone A of Terrace 9. Looters had reached this grave but did not clean it out completely. As Marcus (personal communication) interprets the situation, Burial 9 had originally consisted of at least one adult (Individual 3) and two children or infants (Individuals 1 and 2). Individual 3 was a woman 35–39 years of age (Guillén n.d.). The children or infants were too damaged by looting to be aged accurately, but one had a deciduous cheek tooth that was just erupting.

Associated with Individual 3 was a gourd bowl containing two guinea pigs. Guinea Pig 1 was represented by its left mandible; its left scapula; its right humerus; its right radius; four cervical vertebrae; and one thoracic vertebra (NISP = 9, MNI = 1). Guinea Pig 2 was evidently a pregnant female, since it was accompanied by evidence of a late-term foetus. This guinea pig was represented by its cranium; its left mandible; its right and left innominates; its left femur; one lumbar vertebra; and bits of yellowish-brown fur (NISP = 6, MNI = 1). All the surviving bones of this adult guinea pig were fully fused, and included among the bones from its pelvic region was the left mandible of a late-term foetus.

Additional Guinea Pigs from Disturbed Burials

Structure 4 was one of the burial cists on Terrace 9 of Quebrada 5a. This cist had been badly looted, but the looters had left behind anything they did not consider commercially valuable. Included among the objects they left was the right mandible of a guinea pig that may once have been part of a burial offering.

Structure 5 was a looted burial cist on the same terrace, not far from Structure 4. Immediately below its floor, Marcus' crew found the remains of what appears to have been a pregnant female guinea pig. The list of surviving bones is as follows:

Cranium
Right and left mandibles
Left scapula
Right humerus
Left ulna
Left ilium

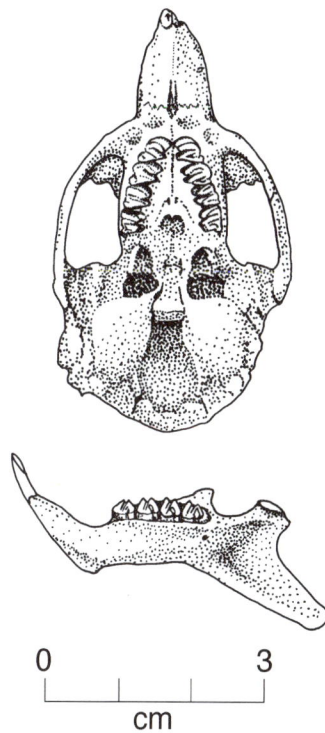

Figure 19.4. Cranium and mandible of guinea pig from a looted burial in Zone A of Terrace 9, Quebrada 5a. This individual's coat color ranged from 10 YR 6/6 to 10 YR 7/6 in the Munsell system.

Right and left femora
Right and left tibiae
Most of a foetal *Cavia*

Near the west profile of Marcus' excavation on Terrace 9 of Quebrada 5a, her crew encountered the badly looted remains of a burial too incomplete to be given a number. From the area of this looted burial came the *Cavia* specimens shown in Figure 19.4. In addition to the cranium and both mandibles, this guinea pig was represented by a patch of fur that would be classified as "brownish yellow" (10YR 6/6 to 10YR 7/6) in the Munsell Soil Color System (Munsell Color Company 1954).

Cavia Remains from Structure 12

One of the most significant discoveries made in Quebrada 5-south was a burial cist designated Structure 12. Although this cist had been looted, the looters had left two gourd bowls resting on the floor. One of these vessels, Gourd 2, contained the complete guinea pig shown in Figure 19.5. Its coat color would have been 10YR 6/4 in the Munsell system ("brownish yellow").

In addition to the remains found in Gourd 2, a few other *Cavia* specimens were recovered from the looters' debris in

Figure 19.5. This vessel, Gourd 2 from the floor of Structure 12, contained the remains of a guinea pig (Quebrada 5-south).

Structure 12. Included were the right mandibles of two additional guinea pigs; one left mandible; miscellaneous ribs and vertebrae; and most of the desiccated skin of a *Cavia* that had been multicolored (brown and white).

A Colonial-Era Guinea Pig from Structure 1

Ritual use of guinea pigs continued at Cerro Azul well into the Spanish Colonial era. Our clearest example came in the form of a guinea pig left as an offering in the ruins of Structure 1 (Fig. 13.44).

Structure 1 was a 12-room adobe building in Inca style, built on the summit of Cerro del Fraile after the Inca conquest of the Kingdom of Huarco (Marcus 1987a; Marcus et al. 1985). Shallowly buried in the ruins of this Late Horizon building was the guinea pig shown in Figure 19.6. Whoever buried this rodent had filled its mouth with coca leaves and wrapped the animal in a Colonial-era red-and-white polka dot bandana. Evidently the ruins of the Inca buildings on the sea cliffs at Cerro Azul were still considered sacred places, even after the Spanish Conquest.

Conclusions

The residents of Cerro Azul raised guinea pigs to eat, to use in rituals, and to leave with adult burials. Guinea pig bones left over from meals were more poorly preserved than those from burials or sacrifices, presumably because those animals had been cut up and added to stews. Slow boiling or simmering likely softened some of the smaller guinea pig bones to the point where they were not preserved at all.

Most of the guinea pig remains from Structure D, an elite residential compound, appeared to be from animals buried *in situ* after sacrifice, or discarded in the Southwest Canchón after they had given their lives in rituals of healing or divination. Those from Structure 9, a storage facility with a commoner-class overseer, had the appearance of food remains. Middens in the quebradas of Cerro Camacho also included *Cavia* bones that appeared to have resulted from meals.

On the other hand, gourd bowls from Late Intermediate burials in the quebradas often contained well-preserved guinea pigs, left as food for the afterlife. In the case of adults and children buried together, it was only the adults who received guinea pigs. Some of the specimens placed with burials were pregnant females; we do not know whether these females were chosen deliberately or accidentally.

Finally, it appears that even after the Spanish Conquest, guinea pigs were considered appropriate animals for sacrifice. At least one was buried in the ruins of an Inca building at Cerro Azul, wrapped in a bandana and given a mouthful of coca.

Figure 19.6. This guinea pig, wrapped in a red-and-white polka dot bandana, was left as an offering in the ruins of Structure 1. Its mouth was full of coca leaves.

20

Macrofossil and Palynological Analysis of the Coprolites from Cerro Azul

John G. Jones

In the course of excavating Cerro Azul between 1982 and 1986, the University of Michigan project recovered six coprolites. Three of these coprolites were of human origin, while the other three were probable dog coprolites.

One of the human coprolites came from within an undisturbed burial (Burial 19), while the other two were found in looted (or partially looted) burial cists (Structures 5 and 12). Marcus suspects that all three coprolites were originally inside corpses, and may reflect meals eaten not long before death.

As for the dog coprolites, two were from a partially looted burial cist (Structure 12), while the third was from a midden deposit on Terrace 11 of Quebrada 6. Dogs were sometimes included with Late Intermediate burials (see Chapter 18), and some of the dead dogs may have had coprolites within them. These coprolites have the potential to inform us about the diet of dogs at Cerro Azul.

Introduction

Scientists have long recognized the importance of coprolite analysis in archaeology. Fry (1976) demonstrated the potential of fecal studies in the interpretation of prehistoric diet and subsistence. Paleoenvironments, in terms of utilized plant taxa and pollen spectra, have also been studied from coprolite contents (Bryant 1974).

Coprolite studies, for the most part, are limited by the conditions necessary for coprolite preservation. Xeric conditions are generally necessary to allow for feces recovery, since coprolites are perishable. Caves and dry rockshelters, for example, often provide remarkably well-preserved coprolites. Open-air sites, on the other hand, rarely provide the conditions necessary for feces and other perishable item preservation. The Peruvian coastal desert, however, is an exception, with unusually well-preserved coprolites being recovered from many sites (Jones 1986a, 1986b, 1988).

Cerro Azul was a Late Intermediate community on the coast of central Peru; fishing was a specialized occupation. During the course of excavation, six coprolites were recovered and shipped to Texas A&M University for analysis (proveniences provided in Table 20.1).

Methodology: Macrofossil Analysis

The Cerro Azul coprolites were analyzed utilizing a technique outlined by Fry (1976) involving a wet-screen

Table 20.1. Proveniences of the Cerro Azul Coprolites.

Sample	Provenience
CA-1 human	Burial 19 (Quebrada 5a, Terrace 9, Zone A)
CA-2 human	Structure 5 (Quebrada 5a, Terrace 9)
CA-3 probably dog	Midden, Quebrada 6, Terrace 11, depth 50 cm
CA-4 probably dog (not analyzed)	Structure 12 (Quebrada 5-south)
CA-5 probably dog	Structure 12 (Quebrada 5-south)
CA-6 human	Structure 12, child burial (Quebrada 5-south)

reconstitution process in trisodium phosphate (Na_3PO_4). Samples were initially cleaned of all extraneous dirt and material, sketched, photographed, and measured. A quantified sample was then removed and placed in a 0.5 percent weight-to-volume solution of trisodium phosphate for a minimum of 72 hours. This process reconstitutes and disaggregates the coprolite, yet does not damage the macro- or micro-fossil components.

The samples were allowed to settle undisturbed during the last twenty-four hours to provide for standardized Munsell color readings. After we had recorded fluid color the jars were opened, and odor (if present) was noted. The samples were then wet-screened through a series of graded geological sieves, in mesh openings of 1.0 and 0.5 mm. Fluid and material smaller than 0.5 mm was retained in a catch pan and saved for later pollen analysis. Materials retained by the screens were transferred to filter paper for drying, and upon desiccation were transferred to Petri dishes for macrofossil component analysis.

Pollen Analysis

Material smaller than 0.5 mm retained in the catch pan was processed for fossil pollen. Palynologists generally employ a four-step process for pollen extraction, including the removal of carbonates with hydrochloric acid (HCl), the removal of silicates with hydrogen fluoride (HF), heavy density separation, and the removal of cellulose through acetylation. Concentrated HCl was added to the pollen samples for the removal of carbonates. With the abundance of shellfish in the diet and in the soil matrix, numerous very small chips were present in the samples; thus, this step was necessary. The samples were next rinsed and placed in concentrated HF for the removal of silicates. This was done in a vented fume hood, with the acid being allowed to work overnight. After the removal of the silicates, the samples were rinsed, sonicated, and rinsed again until a neutral pH was achieved.

The samples were next transferred to 50 ml tubes and placed in a solution of zinc chloride ($ZnCl_2$) (Sp. G. 1.90–1.95) and spun on a clinical centrifuge for 30 minutes at 2000 RPMs. This step separates the lighter organic materials from the heavier inorganic sediments. After separation, the organic fraction was removed, rinsed, and dehydrated in glacial acetic acid in preparation for the final treatment of acetylation. Here, cellulose and lipids were removed from the sediment through the addition of an acetolysis mixture consisting of nine parts acetic anhydride to one part concentrated sulphuric acid. The samples were placed in a heating block for approximately 15 minutes at 200° F. or until most of the plant material had been removed. The samples were next rinsed in water, rinsed in absolute ethanol, rinsed in tert-butyl alcohol, and finally transferred to 1 dram vials for storage in glycerin.

A single drop of material was placed on a microslide and examined under a compound stereo microscope at 400x. Identifications were made based on Texas A&M's extensive modern pollen reference collection and were noted to the finest taxonomic level possible.

Determination of Feces Origin

Coprolite analysts generally employ a series of four tests to determine feces origin. The first is a simple observation of the actual dry coprolite. Human feces can generally be distinguished from carnivores' or herbivores' coprolites on the basis of their shape and visible contents. Bryant (1974) states that large carnivore scats can often be identified by their coating of intestinal lubricant. Herbivore feces can often be identified by their form as well as their high plant content.

The second and third tests take place while the samples are in the trisodium phosphate solution, and consist of noting fluid color and odor. Fry (1976) reports that in solution, human feces generally turn the fluid an opaque dark brown to black color. Coprolites from herbivores and carnivores, on the other hand, often turn the solution a transparent light brown or yellowish hue. Odor can also be a factor, and may be a diagnostic human fecal characteristic. Bryant (1974) points out that coprolites of a nonhuman origin tend to produce a musty smell, whereas human feces often emit an intense fecal odor. However, it is my experience that coastal Peruvian feces from open-air sites rarely possess any odor. It is possible that some form of leaching occurs, possibly as a result of the coastal salt fog.

The final test takes place during the analysis process and consists of the actual macrofossil examination of the feces. Human feces may exhibit food items that would not be expected to occur in a nonhuman diet, such as milled or charred foods. However, this test is not infallible either, as the consumption of human coprolites by dogs and their subsequent redeposition in the archaeological record has been noted by Reinhard and Jones (1987).

Based on the above four tests, it was determined that three coprolites in this study were of probable human origin (Samples CA-1, CA-2, and CA-6) and three were probably from dogs (CA-3, CA-4, and CA-5). Samples CA-3 and CA-5, though of nonhuman origin, were analyzed. Owing to its small size, Sample CA-4 was not analyzed.

Macrofossil Results

Results of the Cerro Azul coprolite macrofossil analysis are presented in Table 20.2. Because of the small sample, quantification of macrofossil components was unwarranted. Rather, a simple presence/absence listing of components was chosen as the clearest means of data presentation.

Owing to a lack of comparative skeletal material, the fish bone from the Cerro Azul coprolites could not be identified at Texas A&M. Jeffrey Sommer, one of the University of Michigan

Table 20.2. Macrofossil Component Analysis of the Coprolites from Cerro Azul by Presence/Absence Listing.

Component	CA-1 human	CA-2 human	CA-3 dog	CA-5 dog	CA-6 human
Zea mays kernel frags.	X	X			X
Capsicum seeds		X			
Poaceae caryopsis		X			
Monocot fibers				X	X
Dicot fibers			X		X
Indeterminate fiber	X				
Indeterminate seed	X				X
Mammal bone		X			
Fish bone	X	X		X	X
Rodent bone				X	
Indeterminate bone			X		
Crayfish shell frags.	X	X		X	X
Insect chitin				X	X
Shell fragments				X	X
Gastropod radulae				X	
Hair	X	X	X	X	
Charcoal	X		X	X	X
Stone	X		X	X	X
Golden spider beetle		X			
Undiff. fecal debris	X	X			

zooarchaeologists, undertook this task. His results are given in Table 20.3.

Table 20.3. Fish Bone from the Cerro Azul Coprolites.

Sample	Identification
CA-1 human	9 unidentified small fish bones
CA-2 human	small sample (unidentified)
CA-5 dog	6 pharyngeal teeth, cf. *Anisotremus scapularis*
	3 unidentified bones
CA-6 human	10 vertebrae, *Sardinops sagax*
	2 unidentified haemal spines
	39 frags. very small fish

The results of Sommer's analysis are worthy of comment. Several of the Cerro Azul burials were accompanied by gourd bowls containing sardines (*Sardinops sagax*) for the afterlife. Coprolite CA-6 reinforces the elite preference for sardines over anchovetas (see Chapter 9). For its part, Coprolite CA-5 reveals that the Cerro Azul dogs sometimes helped themselves to discarded fish heads, in this case the grunt (*Anisotremus scapularis*).

Cultigens were recovered in all human coprolites. *Zea mays* (maize) kernel fragments were found in all three human feces. In addition, chile pepper seeds (*Capsicum* sp.) were recovered in Sample CA-2. Chile pepper seeds have been noted in other late period coastal sites (Jones 1986b), and botanist C. Earle Smith, Jr. found *Capsicum baccatum* to be common at Cerro Azul (Chapter 13).

Probable wild foods were also represented in the diet at Cerro Azul. A single grass (Poaceae) caryopsis was found in Sample CA-2. Though grass seeds were a major component in coprolites from many nearby preceramic sites (Jones 1988), the single seed here may represent an accidental ingestion. Plant fiber represented only a small fraction of the Cerro Azul diet, at least in the case of the last days of these three individuals. Traces of both monocot and dicot fiber were found in Sample CA-6, with an indeterminate fiber type being recovered in sample CA-1. Poorly preserved indeterminate seeds were found in Samples CA-1 and CA-6. Meat was clearly present in the diet, as marked by the presence of bone in sample CA-2 and hair in Samples CA-1 and CA-2. Faunal analysis reveals that both camelids and guinea pigs were eaten at Cerro Azul.

Despite the presence of fish in all three human coprolites, no fish scales were recovered in the Cerro Azul feces, suggesting that fish were scaled more frequently in the Late Intermediate than was typical in the Preceramic.

Crayfish fragments were also noted in all three human coprolites. Crayfish played an important role in the diet at El Paraíso, a Late Preceramic ceremonial center north of Lima (Jones 1986a, Quilter et al. 1991), and the large crayfish *Cryphiops caementarius* was collected at Cerro Azul (Chapter 5). Mollusc shell fragments were also found in Sample CA-6, though their ingestion may have been an accidental byproduct of mollusc consumption.

Other components probably represent accidental ingestions, and include insect chitin (present only as a trace amount), charcoal, and minor amounts of stone and grit (possibly from grinding stones). Undifferentiated fecal debris, consisting primarily of fine materials not disaggregated in the trisodium phosphate solution, was noted in Samples CA-1 and CA-2.

Post-depositional disturbance of the feces is indicated by the presence of a coprophageous golden spider beetle of the family Ptinnidae, characteristically found in dry feces.

The wide variety of components in the dog feces indicates a general foraging pattern. Very small amounts of fiber were found in both analyzed samples, though the bulk of the diet probably consisted of miscellaneous animal foods. Bones and hair were found in both analyzed dog coprolites, and there were fish and rodent bones in Sample CA-5. In the latter sample, a complete articulated rodent tail was recovered. One possibility is the rice rat *Oryzomys xanthaeolus*, which was common at the site.

Sample CA-6 also contained crayfish, shell fragments, insect chitin, and traces of shell. It is noteworthy that gastropod radulae were found in this sample; marine gastropods were an important part of the Cerro Azul diet.

A sharp distinction between human and nonhuman coprolites can be seen in examining the role of cultigens in the diet. All three human feces contained an abundance of maize, yet this food was conspicuously absent in the dog coprolites. The small amount of plant material present in the dog feces, as well as the variety of meat and miscellaneous items, probably reflects the scrounging of scraps from waste areas.

Though conclusions are difficult based on only three samples, the human diet reveals, as suspected, a reliance on cultigens and locally abundant marine resources.

Pollen Analysis Results

Results of the Cerro Azul coprolite pollen analysis are presented in Table 20.4.

Coprolites generally contain an abundance of pollen, owing to the human practice of ingesting flowers and fruit. A large quantity of pollen is similarly ingested by drinking water and undergoing normal respiration. Coastal Peruvian coprolites, however, tend to exhibit surprisingly low concentrations of pollen. The most likely reason for this is the relative lack of floral types that produce large amounts of anemophilous (windborne) pollen. This effect has been observed both in the Peruvian highlands and on the coast (Bryant and Weir 1984).

The Cerro Azul coprolites are no exception to this rule. Extraordinarily low concentration values were encountered in four of the five Cerro Azul feces examined. It has been found through experience that soils with concentration values lower

Table 20.4. Results of the Cerro Azul Coprolite Sample Pollen Analysis (numbers given are the number of grains observed in a one slide count).

Component ID	CA-1 human	CA-2 human	CA-3 dog	CA-5 dog	CA-6 human
Nolanaceae	1	1			
Anacardiaceae					1
Brassicaceae		1			
Solanaceae	2	3	1		
Solanum		1			
Poaceae	1	4	1	3	3
Zea mays		8			
Cheno-Am	20	16	2	5	2
Low-spine Asteraceae		1			
High-spine Asteraceae		1		2	1
Salix			1		
Typha/Sparganium				1	26
Heliotropium	1				
Tillandsia	6	5			
Waltheria		1			
Caesalpinia		2			
Isoetes-like	1				
Unknown A	2	10			
Indeterminate	1			1	3
Total	35	54	5	12	36
Lycopodium	42	803	30	126	200
Sample weight (grams)	2.4	5.0	3.4	2.6	3.0
Pollen concentration	7743.0	299.9	1093.1	816.8	1338.0

than 1000 grains per ml of sediment tend to produce unreliable counts due to the effects of differential preservation.

Low values must be examined with caution. As a result, pollen counts from the Cerro Azul samples are not statistically reliable, and, like the macrofossil data, are useful primarily in terms of species presence/absence.

A minimum of 17 taxa were identified in palynological examination of the Cerro Azul coprolites. Cultigens were represented by *Zea mays* pollen—eight grains were found in Sample CA-2 alone. Solanaceae pollen (found in Samples CA-1, CA-2, and CA-3) and pollen of the genus *Solanum* from CA-2 may also represent economic plants—for example, the potato (*Solanum tuberosum*), which was recovered at Cerro Azul.

Cheno-Am pollen, a group of morphologically similar grains belonging to members of the Chenopodiaceae and Amaranthaceae families, were recovered in all samples. Cheno-Am pollen often signals disturbed soils (such as around settlements and fields) and a generally drier climate; in addition, there is evidence that guinea pigs at Cerro Azul were fed Cheno-Am herbs. A single Brassicaceae (mustard family) grain in Sample CA-2 may also represent an economic plant.

Wind-borne pollen is present in the Cerro Azul samples, though in relatively low frequencies. Poaceae (grass family) pollen was found in all samples and probably reflects its ingestion through normal activities such as drinking water. The single low-spine Asteraceae and *Salix* (willow) grain, as well, are probably contaminants (willow was widely used as an industrial plant at Cerro Azul). The relative abundance of *Typha/Sparganium* (cattail or burr weed) in Sample CA-6 probably represents an economic usage. Cattail was present in the refuse at Cerro Azul, and its rhizomes may have been used as food. Furthermore, Yacovleff and Herrera (1934) report that parts of this plant were used in the embalming of mummies at Paracas.

Zoophilous (insect-pollinated) plant types are also represented but, owing to their low frequencies in the coprolites, probably represent accidental ingestions. These include grains from the Nolanceae and Anacardiaceae (cashew family) families, high-spine Asteraceae types, a grain from *Heliotropium* in the Boraginaceae family, and a single grain comparing favorably to *Waltheria* in the Sterculiaceae family. All these taxa are commonly found in the Pacific coastal region.

Tillandsia sp. is a common plant in the piedmont inland from Cerro Azul, and was used at the site for decorative tassels on slings and as stuffing for the false heads of mummy bundles. Several grains were recovered in Samples CA-1 and CA-2. This plant is not known to provide any dietary benefit; thus, these grains probably reflect simple contamination by people handling *Tillandsia*.

Two *Caesalpinia* grains were found in Sample CA-2 and are likewise probably contaminants.

A single *Isoetes*-like spore and several unknown and indeterminate grains were also found.

Conclusion

Based on the coprolite macrofossil and pollen analyses, a reliance on maize (which was both eaten and made into chicha), chile peppers, and maritime resources is indicated, with wild foods also being employed, possibly as a supplement. Palynological analysis reveals the prehistoric flora of Cerro Azul to be similar to that in the area today. Inferences on health or the medicinal use of plants, however, cannot be made. Studies of additional coprolites, if encountered, will augment our knowledge of this important time period.

21

The Economy of the Kingdom of Huarco

Joyce Marcus

Cerro Azul lay at the interface of two ecosystems. To the west was a marine system with five trophic levels, all of which contributed resources to the Kingdom of Huarco. We have considered Cerro Azul's access to seaweed, shellfish, crabs, fish, sea birds, sea lions, and dolphins, as well as its strategy for drying sardines and anchovetas for export.

To the east of Cerro Azul lay a terrestrial system, enriched by irrigation canals. That system produced edible resources such as maize, beans, squash, potatoes, and chile peppers, and industrial plants like cotton, bulrush, canes, and soapberry. The citizens of Huarco enhanced the connections between the terrestrial and marine ecosystems. They collected the guano of sea birds so that maize fields could be fertilized. The maize from those fields was then turned into beer, rewarding fishermen for their harvests of anchovetas. Millions of dried fish left Cerro Azul on their way to communities of farmers.

All these acts, and their ecological consequences, were the result of conscious economic decisions. So let us now consider the economy of the Kingdom of Huarco and the strategies that drove the movement of resources.

I will begin with the question that first directed my attention to the Cañete Valley: How economically specialized were the Late Intermediate occupants of Peru's south-central coast? This question—originally posed by my co-director, María Rostworowski de Diez Canseco—grew out of her study of ethnohistoric documents. In the Spanish Colonial sources she found an idealized model in which fishermen only fished and farmers only farmed.

Lizárraga (1946, Chapter XLVII, page 90), for example, states that fishermen in the Chincha Valley never tilled the land, because they could get all the agricultural products they wanted through exchanges with farmers. One possible exception to this rule was the cultivation of bulrush, which the fishermen needed to construct their caballitos de totora (Archivo General de la Nación, año 1684, Derecho Indígena, Cuaderno 140, folios 20v, 171v).

The idealized model was supported by documents stating that fishermen lived in their own communities (Rostworowski 1977c, 1978–1980), had their own specialized dialect (Rabinowitz 1980, 1983; Rostworowski 1986:130), traveled their own roads (Archivo General de la Nación, año 1764, Juzgado de Agua 3.3.10.78, folio 19; see *Aviso* 1570), and were further specialized into *challwa camayoc* ("those who exchanged the fish") and *challwa hapic* ("those who caught the fish") (Biblioteca Nacional de Madrid, año 1571, ms. No. 3042, folio 225v). So critical to the economy were these specialists that even after the Inca had conquered the coast, fishermen were not required to provide the usual obligatory labor to their overlords. Instead, they provided the Inca with fish and crayfish (Rostworowski 1977b:173).

Our excavations at Cerro Azul also support a model of fishermen who did not need to till the land. As reported in Chapters 13 and 14, the botanical remains from Cerro Azul included few of the plant parts that we would expect to see if crops had been grown near the site. Equally significant was the complete lack of agricultural implements. Figure 21.1 shows a series of wooden farming tools found in the Chincha Valley by Max Uhle (see Kroeber and Strong 1924: Fig. 16). Not a single one of these tools showed up in our excavations at Cerro Azul.

The Social Hierarchy of Small Coastal Polities

In small polities such as the Kingdom of Huarco, economic decisions were made at several levels of the political hierarchy. At the apex of that hierarchy was the supreme ruler, described as living in a palace at Canchari (Harth-Terré 1933, Villar Córdova 1935, Marcus 2008:7–9). One account of the Inca conquest claims that in A.D. 1470 that supreme ruler was a woman, referred to as a *coya*, "queen," or "*señora del valle*" (Rostworowski 1978–1980:156).

How many levels of hereditary nobles would there have been in the hierarchy below the ruler of Huarco? At Cerro Azul we found archaeological evidence for both (1) elite burials with gold and silver sumptuary goods and (2) residential apartments built for nobles. Who were the nobles at Cerro Azul, and how were they related to their supreme ruler?

Structure D was only one of 10 elite residential compounds at Cerro Azul (Marcus 2008:19–25). Assuming that all these compounds featured elite residential apartments, how were all those elite families related? Were the nobles in Structure A—the largest compound—hierarchically above those of Structure D, or were all of Cerro Azul's elite families equal in rank? Which of the economic decisions were they responsible for?

Our excavations also suggested the presence of commoner-class overseers, sometimes occupying *llamkana patakuna* in compounds like Structure D and sometimes occupying kincha houses in storage buildings like Structure 9. How many levels of Late Intermediate overseers existed, and with what economic decisions were they entrusted?

Even at the level of the commoner-class fisherman, there would have been economic decisions to make. For example, each fisherman likely decided where to fish each day, and what type of net to take with him (Chapter 7).

Finally, our excavations also suggest that there were ritual specialists at Cerro Azul. We do not know whether these specialists were men or women, nobles or commoners, nor do we know their place in the sociopolitical hierarchy. Some ritual specialists may have been curanderos or healers, like the one who put together the "medicine bundle" found in Structure 12 (Chapter 13). Other specialists may have practiced divination, or diagnosed illness, with guinea pigs (Chapter 19).

Unfortunately, there is no single ethnohistoric document that answers all these questions for the Kingdom of Huarco. What I

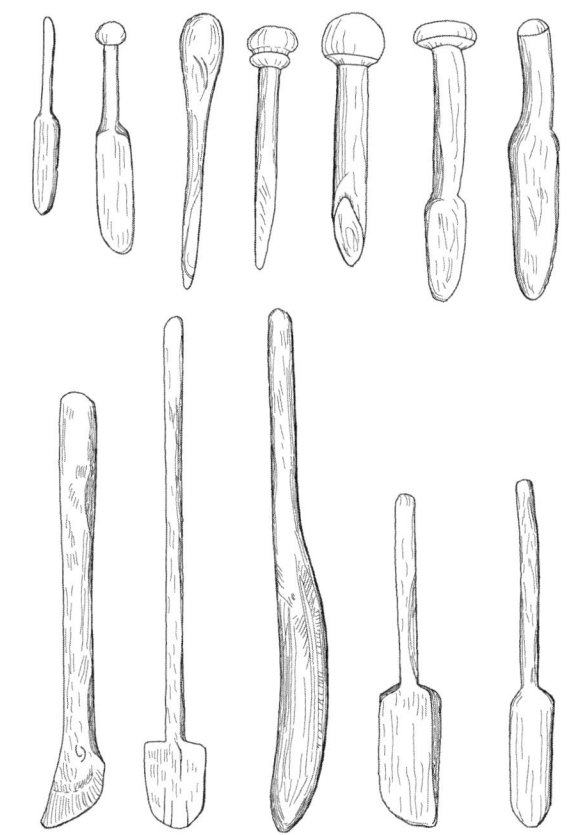

Figure 21.1. Wooden agricultural implements found in the Chincha Valley by Max Uhle (redrawn by Kay Clahassey from Fig. 16 in Kroeber and Strong 1924).

propose to do in this chapter, therefore, is to look at a sample of late coastal kingdoms, beginning on the north coast and ending in the south with the Valley of Chincha. Based on what I learn, I will then try to construct a plausible model for the socioeconomic organization of the Kingdom of Huarco.

The North Coast

Let us begin our survey on Peru's north coast, a region where there is clear ethnohistoric evidence for occupational specialization. Consider the following description of those who made chicha on that part of the coast:

> don Pedro Payampoyfel, lord and leader of the beer-making Indians of this district, says that we have no other job but making corn beer ... we do not have lands or fields to plant; rather, we subsist only by making corn beer and exchanging beer for maize, wool, and other things. (AGI, Justicia 458, folio 2090v; Rostworowski 1977a:241; 1977b:180)

An interesting point emerges from this document: the degree of specialization on the north coast may have been even greater than that seen in ancient Huarco. It appears that the occupants of Structure D at Cerro Azul engaged in both drying fish and brewing beer. On the north coast, on the other hand, it appears that chicha brewers were so specialized that they neither fished, nor farmed, nor produced anything but beer.

A second point made by the document is that chicha brewers had a "lord and leader." Each set of specialists, in other words, was under the command of a member of the nobility. It turns out that these lesser nobles were, in turn, hierarchically below a series of *curacas* who owed allegiance to a supreme ruler or *hatun curaca* (Rostworowski 1977c). Since the Spaniards did not fully understand Peru's small kingdoms, they sometimes referred to the supreme ruler as a *cacique principal*, using a Carib word they had picked up in the chiefdoms of the Antilles. In several cases they also mentioned a ruler's second-in-command, or segunda persona.

The 16th-century Spaniards applied a variety of terms to the north coast kingdoms ruled by a *hatun curaca*, calling them *curacazgos*, señoríos, or parcialidades (Duviols 1979; Espinoza Soriano 1969–1970; Masuda et al. 1985; Murra 1961, 1975, 1980; Netherly 1977, 1984, 1990; Rostworowski 1972, 1975, 1977a, 1977c, 1986, 1993; Shimada 1982). Parcialidad was a particularly imprecise term, since it could refer either to a polity or to a circumscribed ethnic group.

Netherly (1984) has argued that north coast political organization was built on two principles: hierarchy and duality. She concluded that north coast society had a dual corporate structure with ranked subdivisions at every level. Each polity was divided into two moieties (one more highly ranked than the other), with each moiety led by a noble whose position in the hierarchy was determined by the rank of the moiety he headed (Netherly 1977, 1984, 1990).

Each parcialidad combined social, economic, and religious functions and provided a social identity and (at the higher levels of integration) ethnic identity for its members. In some 16th-century documents, parcialidades were equated with the *ayllus* of the highlands (Archivo General de Indias, 1565–67, Justicia 458, folio 1942r). This may reflect the Spaniards' application of a highland Quechua term to a coastal institution; alternatively, it may result from emphasizing the kin dimension of the sociopolitical unit, since kinship still structured many relationships and exchanges (Archivo General de Indias, Justicia 458, folios 1803, 1804, 1834; Netherly 1977, 1984).

Specialized economic activities distinguished each parcialidad from the others (Netherly 1984:231; Rostworowski 1970, 1975, 1977a, 1977b). The parcialidades of farmers were the most numerous. The second most numerous were those of fishermen. In some north coast polities, fishermen were sufficiently differentiated from farmers to be considered ethnically and linguistically distinct (e.g., Netherly 2009; Rabinowitz 1980, 1983; Rostworowski 1986). Other full-time occupational specialists—fishermen, potters, goldsmiths, leatherworkers, and exchange specialists—comprised parcialidades under their own lords (Netherly 1984, Shimada 1982).

Parcialidades on the north coast were administered by a paramount lord or *cacique principal*. The paramount lord was simultaneously the head of the highest-ranking moiety and the head of the entire polity. Along with the *cacique principal* was a lord called segunda persona, a second-in-command who came from the complementary lower-ranking moiety. Next came a group of lesser lords, *principales*, apparently the lords of parcialidades.

Frequently two *principales* were associated with the paramount lord and his second-in-command. In that case, each moiety was subdivided into two divisions, creating four units. If the polity was large enough, these divisions might be further subdivided into eight parts (Netherly 1984).

Some polities were delimited by physical boundaries that could, potentially, be detected by archaeologists. Such boundaries were often irrigation canals or walls. North coast documents show that one way a lord could assert his claim to a parcialidad and its lands was to claim ownership of its main canal. For example, there is testimony by 16th-century informants from the Lambayeque and Pacasmayo valleys to the effect that both field size and standing crops were measured by the small feeder canals that watered them, with informants speaking of *cinco acequias de maíz* or "five canals' worth of corn."

The delimitation of a polity by the presence of a canal system can be seen both on the north coast (Netherly 1984) and in the lower Cañete Valley (Marcus 2008: Fig. 1.2). Several north coast canals that delimited polities retained their prehispanic names after the Conquest. In the mid-sixteenth century, for example, the Chicama Valley was still divided between two ancient polities—Chicama and Licapa—each with its own canals.

Netherly (1984) discusses "single polity canals" controlled by a single parcialidad; use and jurisdiction of such canals were not shared with another parcialidad. Canals of the Virú Valley (and some in the Chicama Valley) fell into this category of single polity canals; the canals of the Moche Valley (Farrington 1974, 1977, 1980, 1983; Farrington and Park 1978) belonged in this category as well. Not surprisingly, there was a close correlation between rights to water, rights to land use, and the boundaries of parcialidades. Groups claimed rights to a particular **canal** because that canal watered their lands; they also asserted claims to particular **lands** because those lands were watered by a particular canal.

The north coast also had some very large irrigation systems, and the way those systems were run may help us understand the way the Kingdom of Huarco managed the large canals coming off the Cañete River. A good example can be found in Netherly's (1977, 1984) analysis of documents on the Taymi Canal, which took off from the Chancay River and ran westward, eventually providing water for 12 smaller secondary canals.

In order to avoid conflicts among the users of smaller canals, a series of rules were established. Those dependent on the secondary canal farthest from the intake point were allowed to use the water first; those closest to the intake point went last. Users of the 12th canal were responsible for cleaning both their small canal and the main Taymi Canal, as far upstream as the takeoff point for the 11th canal; users of the 11th canal cleaned both that canal and the Taymi Canal, as far upstream as the takeoff point for the 10th canal; and so on. All this cleaning was supervised by hereditary nobles, who were carried to the place of work on litters.

While admittedly a practical task, the cleaning of canals was also an act with ritual significance, serving to reinforce the solidarity of the polity. References to the rituals associated with the cleaning and upkeep of canals are given both in the 16th-century documents and in the ethnographic literature (Cobo 1990:143, Guaman Poma de Ayala 1980, Hastorf 1993, Isbell 1978, Mitchell 1976, Sherbondy 1982; Treacy 1994). Sixteenth-century documents refer to north coast lords insisting that the canals could not be cleaned without their ritual supervision.

During months of abundant water, irrigators were free to use as much as they needed. As the water level in the rivers—and consequently the canals—diminished, a system of proportional distribution came into use.

The Central Coast

Ethnohistoric documents reveal the *curacazgos* of Peru's central coast to have been extremely fluid over time. Many saw their borders fluctuate in response to armed conflicts, negotiated truces, and alliances based on the mutual advantage of interpolity trade. This situation is epitomized by the Chillón Valley, where the Colli occupied a coastal setting similar to that of Huarco; the Quivi and Canta had midvalley settings similar to that of Lunahuaná; and the Yauyos lived still farther upriver (Rostworowski 1972, 1988).

At times the supreme leader of the Colli controlled not only most of the Chillón Valley but also part of the Rímac Valley (Justicia 413, see Rostworowski 1988). In his most powerful moments the *curaca* of the Colli made the Quivi ruler his vassal, and expected the Quivi to pay him in "cotton, coca, corn and other things" (Archivo General de Indios, Justicia 413; Probanza de Canta 1559, folio 198; see Rostworowski 1988).

The *curaca* of Colli protected his polity from hostile neighbors by constructing a fort and defensive walls—once again, a situation similar to that of Huarco. During sieges, when their access to the river was cut off, the Colli attempted to extend their control higher into the Andes, establishing new villages such as Yarutini and Huayquihusa. In response the Yauyos fought back and repulsed the Colli, taking their former overlords prisoner and incorporating them into the Yauyos polity.

The usual method of ending such conflict was for the Colli to send messengers to meet with the *curacas* of neighboring polities, asking for their help in arranging a truce. The documents indicate that Colli nobles and Colli commoners had different incentives for establishing peace. The commoners wanted to reinstate the exchange of agricultural goods between polities; the nobles wanted to reestablish their access to foreign sumptuary goods. Thus both of D'Altroy and Earle's (1985) idealized types of finance—wealth and staple—were in play at the same time. I suspect that a similar situation applied to the Kingdom of Huarco.

During Late Intermediate times, the Colli polity extended to Chuquicoto in the Chillón Valley; there a hill was used to mark the boundary between the lands of the Colli and the lands of the Canta. At times the Colli ascended into Canta territory; at times, the Canta descended into Colli lands (Archivo General de Indias, Probanza de Canta 1559, folio 254v).

These alternating cycles of expansion and contraction—during which highlanders expanded at the expense of coastal peoples and vice versa—were destined to change under the Inca. The Inca altered the cycles of expansion and contraction by inserting *mitmaqkuna* (colonies of laborers) in conquered territories. The Inca placed two such colonies on the highly desirable and contested lands of the Quivi, because conditions there were ideal for the growing of coca. The two colonies consisted of ethnic groups from Yauyos and Canta, rather than the Quivi or the Colli, their overlords at that time.

Another major change effected by the Inca was the creation of a new dynastic line. Following the death of the last Colli *curaca*, the Inca placed a lord of their own choosing in his place to start a new dynasty. This new lord or *yanacon yanayucu*—who was to serve the Inca—occupied a new hierarchical level, above that of the former Colli *curaca*.

It is worth noting that prior to the takeover by the Inca, relations between a local lord and the citizens of his polity were considered reciprocal (Espinoza Soriano 1963:66). The land within a *curacazgo* belonged to the *hatun curaca*, but it was there for his people to farm. While citizens produced crops for their leader, he was supposed to take care of their needs as well. One example of this was the way the *curaca* took care of the elderly in his polity, making sure that they were fed and clothed. In return for this care the elderly were asked only to perform light duties, such as collecting coca leaves on the lands of the *curaca* (Rostworowski 1977a:43). In the town of Guaravni in the Chillón Valley, for example, the *curaca* obtained 20 small baskets of coca as service from the elderly; in return, senior citizens received clothing, food, and drink (Espinoza Soriano 1963:66).

These acts of reciprocity characterized coastal polities before the arrival of the Inca (Espinoza Soriano 1963; Rostworowski 1977a). A similar system of reciprocity may also have characterized Huarco prior to A.D. 1470.

In addition to relatively modest polities, such as those belonging to the Colli, Canta, and Quivi, the central coast also featured the multiethnic polity headed by Ychsma (later called Pachacamac). This polity, described as a *macro-etnía* by Rostworowski, united the valleys of Rímac and Lurín and was large enough to feature two divisions, *hanan* (upper) and *lurín* (lower) (Rostworowski 1992, Eeckhout and Owens 2008, Ramos Giraldo 2011). The polity's political and religious capital was

Pachacamac, a huge coastal center to which multiple farming and fishing communities owed allegiance.

The ethnohistoric descriptions of Pachacamac provide insights not available from any other *curacazgo* in our sample (Makowski 2015). For example, the circulation of commodities in the Rímac and Lurín valleys does not always fit our standard notions of economic necessity and exchange. Much of the movement of products to Pachacamac reflects its role as a sacred pilgrimage center and the site of the most important oracle on the central coast (Eeckhout and Owens 2008:377; Ramos Giraldo 2011; Rostworowski 2003). To understand the importance of this "ritual economy," we have to consider cosmology as well as commerce.

To begin with, Pachacamac was named for the offspring of the Sun and Moon, a deity who "made the earth move" and was "the creator of the earth." Two of Pachacamac's divine wives were Pachamama (The Earth) and Urpay Huachac (The Creator of Fish). In the cosmology of various coastal communities we learn that in the beginning "there were no fish in the ocean"; Urpay Huachac is credited with raising the first fish in a small lagoon near one of the temples at Pachacamac. Eventually the rival deity Cuniraya became angry with Urpay Huachac, as a result of which he threw all her fish into the ocean. From that moment on Urpay Huachac became the goddess of all marine fish and birds, a goddess to be venerated by all who entered the sea to fish (Rostworowski 1992).

So widespread on the coast was the cult of Urpay Huachac that her name was even applied to an island off Chincha (Albornoz 1967[1580]:34, Duviols 1967, Rostworowski 1977c:265). The veneration of Urpay Huachac, Rostworowski suggests, began on the coast but was spread to the central highlands by those transporting dried fish to that region. Since the cult reached Cañete, Cerro Azul may well have been one of the communities involved in introducing Urpay Huachac to the highlands.

Among the traditional offerings made to Urpay Huachac—whose name means "she who gives birth to doves"—were the bones of the West Peruvian dove (*Zenaida meloda*) and the eared dove (*Zenaida auriculata*). The bones of both these doves have been identified at Cerro Azul (Chapter 11), and their role in the Urpay Huachac cult makes them more than just evidence for dove hunting.

One of the fishing villages near Pachacamac was named Quilcay. Colonial documents indicate that when there was a shortage of men to marry, the women of Quilcay did not turn to nearby farming communities for a mate. They looked, instead, for a husband from another fishing village, even if that village lay as far away as Santiago de Cao in the Moche Valley (Rostworowski 1986:130). This fact tells us that fishing people saw themselves as belonging to a separate segment of society and sought to marry within their group. Like so many fishing communities, Quilcay is described as having its own road for the transport of fish.

In addition to its prominent oracle, Pachacamac also had a series of *huacas* or sacred places. Some of these *huacas* were natural landmarks such as hills and rivers; others were man-made landmarks such as earthen mounds and pyramidal platforms (Bray 2015). Pilgrims from many ethnic groups traveled to Pachacamac's *huacas* (*wak'as*) to make sacrifices and leave offerings.

Both visitors and temple personnel, for example, made daily offerings of anchovetas and sardines to a group of condors and black vultures that were kept in the sanctuary of the main temple of Pachacamac (Pizarro 1944, 1978; Rostworowski 1992:47). This ritual is revealing, as it suggests that not all of the small fish dried on the coast were exported inland or to the highlands. Some of these fish may, in fact, have been sent to ritual centers in other coastal valleys.

Huacas like those of Pachacamac were seen to be related as brothers, sisters, sons, and daughters. This network of kin relationships linked *huacas* to each other and indirectly integrated the diverse ethnic groups that traveled the pilgrimage route. In any reconstruction of the Late Intermediate economy, therefore, we should take pains not to underestimate the volume and variety of commodities and sumptuary goods transported to ritual destinations.

The great oracle of Pachacamac in the Lurín Valley was considered the father of a lesser oracle in the Chincha Valley. Despite its subordinate status, this Chincha oracle was graced with a 40-meter-tall shrine (Morris and Santillana 2007).

An "extended family" of *huacas* thus ran south along the coast from the Lurín Valley. Many *huacas* in the valleys of Mala, Cañete, and Chincha were considered wives or offspring of Pachacamac. During the Late Horizon, Structure 3 at Cerro Azul —a stone masonry *ushnu* or temple platform, with a stairway leading down to the ocean—was considered such a *huaca* (Marcus 1987a: Fig. 69). Still other *huacas* lay on islands, such as the island of Concavilca (offshore from the Mala Valley) and the islands of Urpay Huachac, Quillairaca, and Churruyoc (off the coast of Chincha) (Rostworowski 1992:31).

To the South: The Chincha Valley

To the south of the Kingdom of Huarco lay the Kingdom of Chincha. According to ethnohistoric sources, this polity had 30 caciques administering 30,000 tribute-paying citizens. Of those, 12,000 were farmers, 6,000 were traders, and 10,000 were fishermen (Biblioteca del Palacio Real de Madrid, Miscelánea de Ayala, Tomo XXII, folio 271r-v; Rostworowski 1970:170–171; see also Rostworowski 1977a:99; Cieza de León 1932: Chapter LVIII). Lizárraga (1946 [1605]:90) deals with the missing 2,000 *tributarios* by dividing the 30,000 into three groups of identical size (i.e., 10,000 farmers, 10,000 fishermen, and 10,000 traders).

An A.D. 1558 document says that the citizens of Chincha lived in a state of continual conflict with their neighbors (Castro and Ortega Morejón 1936:236; 1974; see also Crespo 1978). The *curaca* of Chincha controlled *cien mil* balsas and conducted long-distance trade in *Spondylus* shell and copper. Some balsas were constructed from tree trunks, while additional seagoing vessels were made of bulrushes or inflated sea lion skins. During their journeys inland, the Chincha traders subsisted on some of their own dried fish. The remaining fish was delivered to the highlands, along with decorated gourds and coca. These data from Chincha are suggestive, since Cerro Azul also dried fish and decorated gourds (Figs. 21.2, 21.3).

The Chincha fishermen are described as living in a settlement that consisted of one long street, running parallel to the ocean. They are said to have had their own bay and inlet and—by unspoken agreement—to have fished without experiencing any competition. According to a 1570 Chincha *Aviso*, "cada día o los más de la semana entraban en la mar, cada uno con su balsa y redes y salían y entraban en sus puertos señalados y conocidos, sin tener competencia unos con los otros" (Rostworowski 1977a:214). "El hecho de poseer cada grupo sus playas y caletas particulares era una costumbre general en los llanos, de sur a norte" (Rostworowski 1977a:224–225). When not at work, the fishermen simply drank chicha and danced, because they had no obligation other than fishing (*Aviso* 1570; Rostworowski 1977a:214–215).

The street on which the Chincha fishermen lived is said to have begun "two leagues" before one arrived in Chincha and continued on until it reached "Lurín Chincha." This street has now been identified—and partially excavated—by Sandweiss (1992), who has given it the archaeological name of Lo Demás. Once he became convinced that Lo Demás was the long, narrow fishing community described in the 1570 *Aviso*, Sandweiss saw the opportunity to evaluate two models.

According to Sandweiss' Model One, the Chincha fishing settlement was expected to produce archaeological evidence for fishing, but no other kind of craft activity or production of materials to be exported. It was also expected to provide evidence for separate commoner and elite residential areas, since there would have been a *principal de los pescadores*. According to Model Two, the elite sector should have evidence for both auxiliary specialists and links to the Lord of Chincha.

Sandweiss' expectations were met when he found multiple residential sectors at Lo Demás. Based on his excavations, he concluded that Sector I was where the commoners lived. As expected, he did not find evidence of nonfishing activities; as at Cerro Azul, the artifact inventory included no farming tools. The fishing paraphernalia in Sector I included nets, a fishhook, and a mallero or net-making template (Sandweiss 1992:144). Actual fish remains included branchia, cranial elements, and whole fish heads.

Sandweiss considers it possible that Sector I occupants grew cotton for their nets; they appear to have harvested marsh plants for making mats and possibly watercraft as well. Sector I residents also cooked food, engaged in spinning and sewing, did some wood working, and raised guinea pigs. In a discovery similar to that made at Cerro Azul, Sandweiss found that Lo Demás may also have had curanderos.

In Sector IV of Lo Demás—a Late Horizon residential area —Sandweiss found evidence for elite residents with attached specialists. Sector IV, in other words, may have housed a local "lord of the fishermen" and his retainers. Sector IV had an abundance of cotton, and greater evidence of fiber processing than Sector I. It also seems to have been involved in the decoration of gourds, reinforcing information provided by the ethnohistoric sources.

Unlike Sector I, Sector IV produced camelid dung pellets. Their presence suggested to Sandweiss that the nobles of Lo Demás, like those of Cerro Azul, were shipping products by llama. The scarcity of wool textiles at Lo Demás, and the small number of camelid bones in the refuse, led him to conclude that llamas were present mainly or solely as beasts of burden. As at Cerro Azul, dried fish may have been the principal export.

The Indigenous View of a Coastal Kingdom

In the cosmology of the coast, as we have seen, the filling of the ocean with fish was the result of a divine conflict between Urpay Huachac and Cuniraya. This same cosmology reveals that an ideal coastal kingdom featured two attributes: duality and reciprocity. This idealized view, to be sure, contrasted with a reality in which asymmetrical structure and strong social stratification were present. Fishermen and farmers were not simply separate communities of specialists; each was under the authority of a *principal*, or noble, who in turn owed allegiance to the *curaca* of his kingdom. These *principales* gave orders that promoted or reinforced specialization. The hereditary lord of the fishermen represented his community before the polity's higher authorities, and in return received tributary support from the fishermen under his direction.

In 1540, a *principal* of Jayanca in the Lambayeque Valley was described as having 50 fishermen under his command; he himself took orders from the lord of Jayanca (Archivo General de las Indias, Justicia 418, folio 276r). At Camaná in southern Peru, a *principal* directed the work of 40 fishermen. A similar pattern characterized the Huaura Valley, where the fishing community was led by a noble named Mazo. In a 1549 *visita*, the señorío of Maranga in the Rímac Valley was said to have had a *curaca* named don Antonio Marca Tanta; subject to his authority were three *principales*, one of whom was the *jefe de los principales* (Rostworowski 1986).

We get hints in the documents of the kind of tribute that indigenous leaders expected. In the 1549 *visita* of Maranga in the Rímac Valley we learn that every four months, the *encomendero* received 100 *cargas* of dried fish from the fishing villages under his command. In the region of Paita, Alcedo (1967[1788]) states that there were 310 fishermen whose annual tribute payments to their leader amounted to 8,100 dried and salted smoothhound sharks, 172 birds, and a specified sum of Spanish Colonial coinage (Rostworowski 1977a:224).

In some cases, the ethnohistoric sources merely strengthen the conclusions we would have drawn from our archaeological evidence. In other cases, the documents give us a level of detail

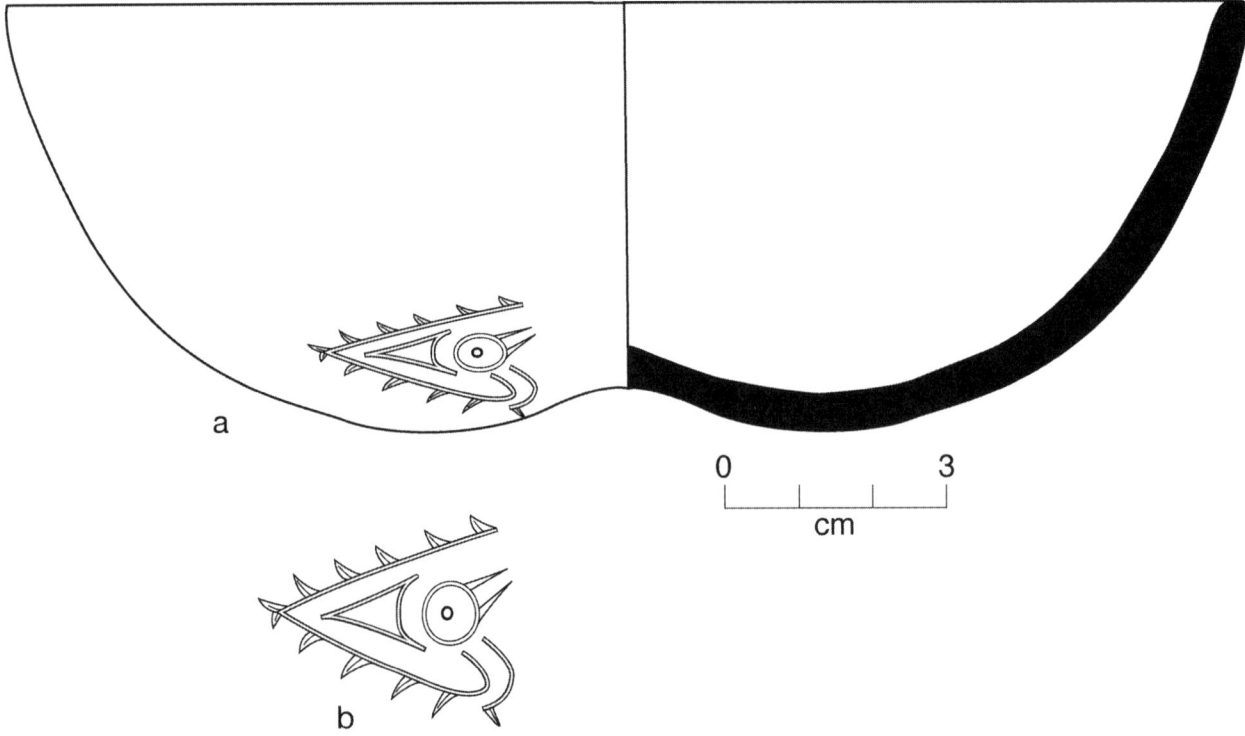

Figure 21.2. Pyroengraved gourd bowl, found with Individual 1 of Burial 7 (Terrace 9, Quebrada 5a). *a* shows the bowl in cross-section; *b* shows the pyroengraved motif.

The Economy of the Kingdom of Huarco

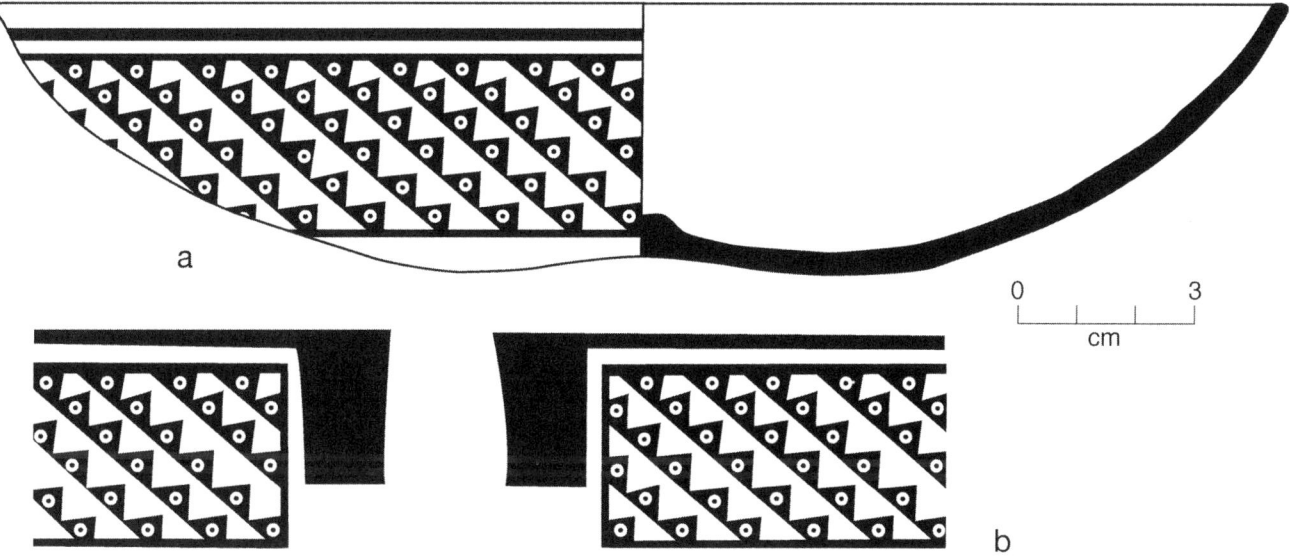

Figure 21.3. Pyroengraved gourd bowl, found in looters' debris just northeast of Structure 4, a burial cist on Terrace 9 of Quebrada 5a. *a* shows the bowl in cross-section; *b* is a rollout of the pyroengraved motif.

that no excavation could provide. Let us look at a few examples. Based on excavation data alone, we would have concluded (1) that Cerro Azul's fishermen were under the authority of nobles; (2) that the nobles relied on commoner overseers; (3) that commoner-class fishermen would not have concentrated heavily on sardines and anchovetas unless ordered to do so; and (4) that the "industrial" production of dried anchovetas was likely directed from the top down.

Without the documents, however, we would never have guessed (1) that there may have been one *principal* for every 40 or 50 fishermen, and (2) that the movement of dried fish may not have been limited to exchanges with agricultural communities.

Some percentage of the fish dried at Cerro Azul, for example, was almost certainly paid in tribute to noble overlords, or delivered to religious centers in the form of offerings. The documents may also answer our longstanding question about why the sand deposits from anchoveta storage rooms also produced a few bones of drums, grunts, and other medium-sized fish. Colonial documents force us to consider the possibility that some of these larger fish were part of the tribute paid to the *principales*.

A Model for the Late Intermediate Economy of Huarco

Having now looked at a sample of Peruvian coastal kingdoms, let us make a stab at constructing a model for the Kingdom of Huarco. Our socioeconomic model will be based on two sources of data: (1) relevant ethnohistoric documents and (2) empirical archaeological data from the Cañete Valley.

Let us begin with the sociopolitical hierarchy (Fig. 21.4). According to an A.D. 1558 document, Cañete had its own lord and separate government (Castro and Ortega Morejón 1936[1558], Crespo 1978). Based on other coastal documents, such an individual would likely have been referred to as a *hatun curaca*. Assuming that Cañete's government had the dual structure of other coastal kingdoms, the *hatun curaca* would have belonged to the most highly ranked moiety, and may have had a second-in-command drawn from the less highly ranked moiety.

Archaeological support for the presence of a *hatun curaca* is provided by the ruins of Canchari, a site whose location is indicated in Figures 1.1 and 1.2. Canchari was a defensible hilltop stronghold where the ruler, his staff, and his troops could all be protected (Marcus 2008:7–9). Two irrigation canals encircled the hilltop, serving as defensive moats as well as sources of water in case of siege. Harth-Terré (1933) reconstructs the palace of Canchari as having two grand plazas, two access ramps, and ample servants' quarters (Marcus 2008: Fig. 1.4). Because Canchari has never been excavated, however, we do not know how much of the site is Late Intermediate and how much is Late Horizon. Williams and Merino (1974) report finding sherds from both periods on the surface.

A question worthy of further investigation is the relationship between the *señor de Huarco* and the *señor de Lunahuaná*. It appears that during the Late Intermediate (A.D. 1000–1470), Huarco and Lunahuaná were separate polities, each with its own hereditary lord. The Inca conquered Lunahuaná first and built their own governmental center at Inkawasi (Fig. 1.1). Lunahuaná was then used as a staging area for the Inca invasion of Huarco. Huarco and Lunahuaná are said to have been on extremely friendly terms, facilitating the exchange of products between *yunga* and *chaupi yunga*.

Does this close relationship imply some special connection between the *curacas* of the two polities? Rostworowski (1978–1980) has suggested that there may have been times when Huarco and Lunahuaná were separate and independent, while Marcone Flores and Areche Espinola (2015) have argued that there were times when a single *hatun curaca* administered both. We do not know which situation was in effect during the period when Structures D and 9 were occupied.

Intermediate Levels of the Hierarchy

Below the level of the kingdom's second-in-command came a division into specialized farming and fishing communities. This division is shown in Figure 21.4, which contrasts Cerro Azul with a generic farming community. Not shown is a group of specialized merchants, like those who made up one-third of Chincha's citizenry. I have omitted such a group for Cañete because we as yet have no empirical archaeological data for specialized merchants.

Figure 21.4 assumes that each farming or fishing community had its own *curaca*, a hereditary leader who occupied the second tier of the sociopolitical hierarchy (or the third tier, if you consider the second-in-command to have occupied his own tier). To confirm the existence of a community-level *curaca* for Cerro Azul we will need further excavation. If such a leader existed, he might have lived in Structure A.

We suspect that, like Structure D, each of the 10 elite compounds at Cerro Azul had apartments for minor nobles. Confirming this will also require further excavation. We believe that many of the individuals buried in cists on Terrace 9 of Quebrada 5a may be just such minor nobles, and we regret that so many of their burials had been looted.

Archaeological evidence indicates that between the minor nobles and the commoners was an intermediate level of overseers. Such commoner-class overseers are barely mentioned in the ethnohistoric documents, but their presence is affirmed by our discovery of wattle-and-daub houses and modest work platforms in areas where fish were dried and stored.

Finally, at the base of the sociopolitical hierarchy were the commoners who performed the labor. Figure 21.4 divides Cerro Azul's laborers into fishermen, fish dryers, and brewers. Such a division is supported by some ethnohistoric documents, but, as discussed below, needs to be confirmed archaeologically.

In Stratigraphic Zones A and B, two middens on Terrace 9 of Quebrada 5a, we found refuse left by commoner fishermen

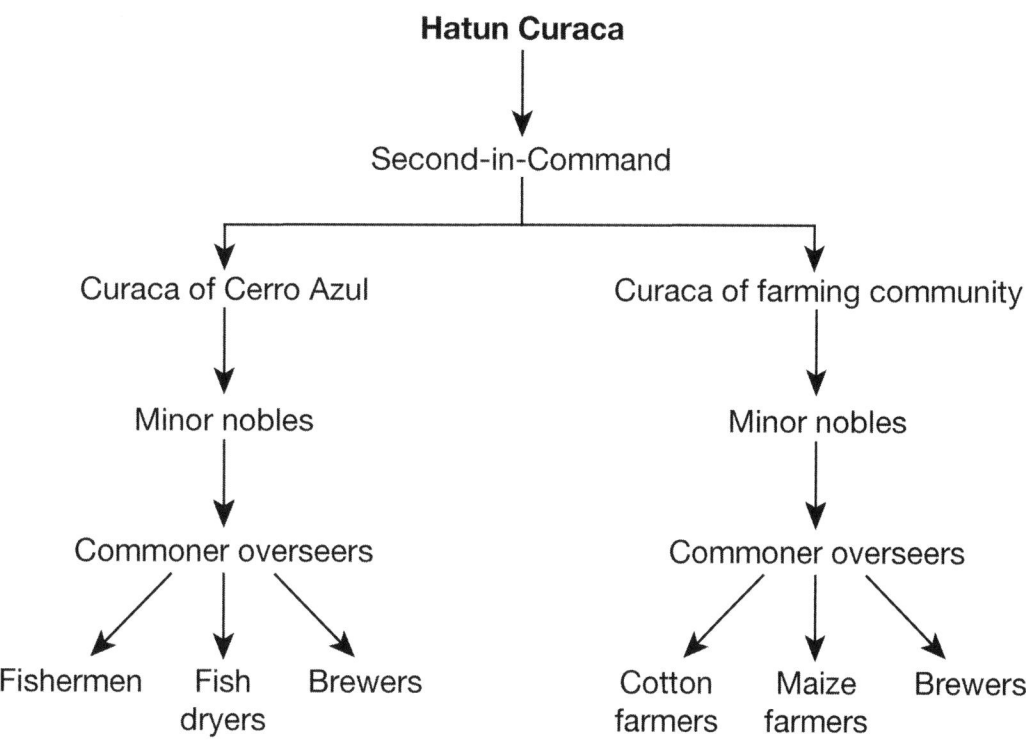

Figure 21.4. Working model of the sociopolitical hierarchy in the Kingdom of Huarco.

who paid little attention to anchovetas and sardines. Those commoners seem to have subsisted on medium-sized fish such as grunts, drums, mullets, and mackerels. In other middens, such as the one on Terrace 11 of Quebrada 6, we found refuse left by specialist fishermen who concentrated on anchovetas and sardines. Our discoveries in Structures D and 9 suggest that the drying and storing of these small fish for export was under the supervision of overseers who reported to the minor nobles. These discoveries support our division of fishermen into those who did subsistence fishing and those who harvested and dried anchovetas for export.

The data from Cerro Azul do not support the notion of brewing as a specialized craft; both brewing and fish drying were carried out in Structure D. In addition, whoever did the brewing in Structure D also left a number of spindles and spindle whorls on the floor of the brewery (Marcus 2016). What these artifacts suggest is that the brewing in Structure D was done by women; it may even have been directed by elite women. While we assume that men did the fish drying and women the beer making, we see no evidence that brewing was a specialist activity at Cerro Azul, in the way it was on the north coast.

A crucial question, still to be answered, is this: At what level was the decision made to concentrate on harvesting and drying millions of anchovetas? Did the orders come from the minor nobles, the *curaca* of Cerro Azul, or the *hatun curaca* of Huarco? And when the decision was made to convert more and more rooms to fish storage, what was the motivation? Was it in response to increasing demand from Cerro Azul's trading partners? Increasing demand for tribute among the nobles? Or the *curacas'* increasing need for a commodity that would bring them more sumptuary goods?

Exchange Spheres and the Circulation of Commodities

Once occupational specialization had been established, commodities would have circulated through four nested and overlapping spheres. Within the Kingdom of Huarco, communities like Cerro Azul would have sent fish, crabs, crayfish, shellfish, and possibly seaweed inland to farming communities. In return, they seem to have received maize, beans, squash, chile peppers, potatoes, peanuts, tropical fruits, cotton, and gourds from the farmers. It strikes me that this type of exchange might have been overseen by the *hatun curaca's* second-in-command, leaving the supreme leader free to engage in major diplomacy.

Exchanges between the Kingdom of Huarco and the Kingdom of Lunahuaná would have brought Cerro Azul camelids, camelid byproducts, and other resources of the *chaupi yunga*. Included among the latter were orchard crops, huarango and algarrobo wood, prickly pear spines, and possibly coca. The Kingdom of Lunahuaná had access to its own crayfish from the Río Cañete, but undoubtedly wanted dried anchovetas and sardines from the *yunga*.

Like other groups of fishermen mentioned in the ethnohistoric documents, the Cerro Azul fishermen probably had their own road, perhaps the one marked on the map published by Larrabure y Unanue (see Fig. 1.2). This would have allowed them to interact with Chincha to the south, Asia and Mala to the north, and perhaps even Pachacamac. Since fishermen up and down the coast shared in the cult of Urpay Huachac, the Cerro Azul fishermen had reasons to visit coastal oracles and *huacas* and exchange marine products, in spite of the potential overlap of their resources with those of their neighbors. Giant Chilean mussel shells from Nasca became pigment dishes in Huarco; dried fish from Cerro Azul may have been fed to the temple vultures at Pachacamac.

Finally, there would have been exchanges of products with the sierra and the puna. It is almost certainly from their highland trading partners that Huarco's nobles received prestige goods such as gold, silver, copper, and dyed alpaca wool. Both commoners and nobles at Cerro Azul may also have received *ch'arki* that had been dried in the highlands.

The Difference between Ecology and Economy

I opened this book by discussing the ecology of the Cañete coast. I am closing with a discussion of the economy of the Kingdom of Huarco. There is no doubt that the two subjects are linked. Human actors turned desert alluvium into irrigated gardens and made small wild fish the equivalent of a cash crop. In spite of these links, it would be a mistake to simply merge ecology and economy. For the Late Intermediate, at least, there were several differences between them.

To begin with, the workings of the ecosystem were strongly bottom-up. The primary producers of Trophic Level 1— phytoplankton—helped determine how many organisms of Level 2 there would be. Even when phytoplankton and zooplankton were abundant, the mix of species and their size ranges helped determine whether sardines or anchovetas would be the dominant small fish. The relative abundance of these Level 2 species, in turn, helped determine the success of cormorants, boobies, mackerels, bonitos, and sea lions. In 1983 and 1984, many Level 3–4 predators starved on the beach because there were too few Level 2 fish to eat.

In contrast, the Late Intermediate economy was directed from the top down. At the apex was a *hatun curaca* who was the ultimate decision maker. Below him were several tiers of lesser nobles who, with the aid of trusted overseers, made decisions on behalf of commoner-class specialists.

One reason for the top-down structure was the Andes' tradition of "command economies." Tons of products moved in ancient Peru, yet the Quechua language had no word for "market." It was in the interest of coastal rulers to maintain specialized communities, keeping the production of commodities so efficient that there was no incentive to backslide into self-sufficiency.

To be sure, many economic decisions had ecological consequences. The collection of guano so affected cormorant nesting sites that the Inca eventually declared certain islands "off limits" during the breeding season. The "industrial level" harvesting of anchovetas cut into the food supply of Level 3–4 predators, especially within 100 meters of the shore. Selective hunting of female sea lions may have made it difficult for some alpha males to create the harems they desired. Human harvesting of coquina clams could not help but affect the grey gulls' food supply.

None of these unintended consequences, however, would have changed the bottom-up structure of the ecosystem. Nor would human-induced shortages, or short-term perturbations in the natural environment, have altered the top-down structure of the Late Intermediate economy. The longer Structures D and 9 were occupied, the more rooms were devoted to drying fish. Coastal peoples turned sardines and anchovetas into tribute payments, ritual offerings to *huacas*, and a cash crop for long-distance trade.

The archaeological data from Cerro Azul indicate that left to their own devices, commoner-class fishermen would have subsisted on drums, grunts, mullets, mackerels, and other medium-sized fish. Only under elite supervision did they engage in the export of dried anchovetas. Once economic specialization had begun, it was in the interest of the nobles to maintain it. Its high level of productivity not only provided fishermen with plants they could not grow themselves; it also created so much tribute that nobles could acquire all the sumptuary goods they wanted, and justify their privileged status by supporting the sacred *huacas*.

Like any kingdom, Huarco featured economic inequality. In an attempt to mask the extent of that inequality, and to encourage loyalty and social solidarity, the nobles lavished beer on their fishermen and used some of their wealth to support the elderly.

In addition to their contrasting bottom-up and top-down structures, there is an additional difference between ecology and economy. Several generations of substantivist economists—most notably the late Karl Polanyi (1968, 1977)—have argued that economy is embedded in society, and cannot simply be seen as a response to supply and demand. Nowhere would Polanyi's view be more appropriate than in the Andes, where markets like those of the Aztec did not exist.

One cannot understand the economy of the Kingdom of Huarco, or any Andean polity, without taking into account the deeply embedded principles of reciprocity and hierarchy.

It is not enough to know where certain desirable products were produced; one's relationship to his partner in the dual structure, and his obligation to reciprocate with equivalent gifts, strongly influenced what he traded for and with whom. A fisherman's need to visit all the kin-related coastal *huacas*—not to mention the shrines to Urpay Huachac—meant that he would often be carrying offerings of fish to communities that already had their own fish. This made sense in terms of the fisherman's ritual life, but not in terms of supply and demand.

Further complicating our construction of a model for the Late Intermediate economy were the continually shifting alliances, expansions, contractions, and conquests of that period. These changes affected who one's duality partner was, and to whom one owed reciprocity. For us to simulate the dynamic features of coastal economies, it will surely be necessary to divide the Late Intermediate into shorter phases. And that will require collaboration by all of us who work on pre-Inca societies.

Appendix A

Artisanal Fishing at Cerro Azul, 1984–1986

Khalid Kattan, Robert G. Reynolds, and Joyce Marcus

During the decade of the 1980s, the waters off Cerro Azul were the scene of three different types of fishing. Individuals cast their atarrayas from the sea cliffs, or fished from cobble beaches using the espinel (see Chapter 7). Small teams of men—known as artisanal fishermen—went out at night in boats with oars or outboard motors, usually returning the next day. And all the while, one could see in the distance commercial trawlers that passed Cerro Azul without ever docking.

In March of 1984 the Capitanía del Puerto began to record the catches of artisanal fishermen returning to Cerro Azul. Aware that Marcus was conducting a study of ancient fishing at Cerro Azul, the officials of the Capitanía kindly shared some of these data with her.

From March 1984 until July 1986, thousands of trips made by artisanal watercraft were logged into the Capitanía. That period included the aftermath of the 1982–83 El Niño, two mild La Niña years, and a return to "normal" ocean temperatures.

The Capitanía's log supplied quantitative support for much of the anecdotal data Marcus had collected from her fishermen informants. It confirmed the fact that warm-water species like shrimp and pompano had visited Cerro Azul during the 1982–83 ENSO and its aftermath, gradually disappearing as the ocean cooled. It documented both the disappearance of cool-water fish like the silversides, and that species' return in 1986. The data from the Capitanía's log also showed some of the ways that Cerro Azul's artisanal fishermen responded to the 1982–83 ENSO. For example, faced with the loss of their preferred bony fish, some fishermen traveled as far north as Asia to catch sharks, rays, and chimaeras.

Kattan (n.d.) is now engaged in a statistical analysis of the 6013 most fully documented artisanal trips recorded by the Capitanía, using approaches such as cluster analysis and multidimensional scaling. In this Appendix, however, we make use only of his descriptive statistics, and emphasize only those data relevant to Late Intermediate fishing at Cerro Azul.

As useful as the Capitanía's log may be, let us make clear what its limitations are for interpreting ancient fishing. First, the artisanal fishermen of the 1980s used nylon nets and outboard motors; their boats could travel farther than a caballito de totora and return with a heavier load. Second, the log did not record the catches of individuals who fished on foot from the beach or the sea cliffs, since they never reported to the Capitanía. Third, artisanal fishermen showed virtually no interest in sardines and anchovetas; fourth, they harvested dolphins on a scale not seen in the Late Intermediate. Bearing these limitations in mind, however, let us look at some potential insights from the Capitanía's log.

The Rebound from El Niño

Figure A.1 compares monthly catches of shrimp (*Xiphopenaeus* sp.) and silversides (*Odontesthes regia*) from 1984–86. Shrimp had extended their range to Cerro Azul during the 1982–83 El Niño and were still available by the thousands in April of 1984. They effectively disappeared under La Niña conditions, except for a brief blip in April of 1985.

Silversides are one of the most sensitive fish to warming waters. They swam south to Chile during El Niño, slowly returning north in late 1985. Individual fishermen began to catch *Odontesthes* in small numbers even before the artisanal boatmen did. Silversides showed a modest spike in March 1986, then began their full recovery in June of that year.

Other fish reacted to El Niño in their own way. The Chilean jack mackerel (*Trachurus symmetricus*) spends a lot of time in the open sea, and anchovetas are one of its preferred foods. During ENSO conditions, however, the jack mackerel is known to follow anchovetas to cool-water refuge areas near the shore (Chapter 1). This approach to the shore places the jack mackerel within the range of artisanal fishing boats and increases the number taken.

Figure A.2 shows that artisanal catches of jack mackerel peaked in May 1984, during the aftermath of the 1982–83 El Niño. There was a lesser peak in October and November of 1984, after which anchoveta populations began to recover and the jack mackerel returned to deeper water. There it is taken by commercial trawlers.

The bonito (*Sarda sarda*) is cyclically available to Cerro Azul fishermen throughout the year, and does well under warm-water conditions. Figure A.3 shows half a dozen peaks in bonito capture over the length of our study. The strongest peak came in April 1984, while the residual warmth of the 1982–83 ENSO still lingered. During the two La Niña years of 1984–85 and 1985–86 bonito peaks became lower, and during the return to "normal" conditions the artisanal fishermen caught an average of fewer than 100 per month. Bonitos, in other words, had a more nuanced response to the El Niño-La Niña cycle than simply presence/absence.

Synchronicity between Species

Many Cerro Azul fishermen told Marcus about pairs of fish species whose presence/absence seemed to be correlated. The Capitanía's log confirmed several of these anecdotal correlations.

Figure A.4 compares the monthly catches of lorna (*Sciaena deliciosa*) and mismis (*Menticirrhus ophicephalus*) at Cerro Azul. The similarities in their peaks of abundance are evident.

Both the lorna and the mismis are members of the drum family. The lorna Marcus collected at Cerro Azul averaged 32 cm in length, while the mismis averaged 24 cm (Chapter 6). These two fish have similar diets (Vargas et al. 1999), and should have peaked in abundance at Cerro Azul when polychaete worms, pelecypod molluscs, mole crabs, and other benthic crustaceans were readily available. The fact that the mismis is three-quarters the size of the lorna raises the possibility that these two species reduce competition by taking prey of different sizes.

Figure A.5 compares the monthly catches of grunt (*Anisotremus scapularis*) and morwong (*Cheilodactylus variegatus*). Once again, the similarities in these two species' peaks of abundance are striking.

Both of these fish are attracted to sea cliffs and rocky offshore islands. Grunts come to the cliffs to eat kelp; morwongs hide from predators in kelp beds. Both species eat copepods, amphipods, small crabs, and polychaete worms (Berrios and Vargas 2004). We should not be surprised, in other words, to find that these two fish might be attracted to the same habitat and peak in abundance at the same time.

As for the possible means by which grunts and morwongs reduce competition, we can once again look to their relative sizes. The grunts Marcus collected at Cerro Azul averaged 38.6 cm in length, while the morwongs averaged 27.5 cm. These two species may therefore be taking prey of different sizes.

Conclusions

Although the artisanal fishermen of the 1980s had technologically superior nets and watercraft, competed with commercial trawlers and the fish meal industry, and showed many behaviors not seen in the Late Intermediate period, the log from the Capitanía del Puerto nevertheless provides data relevant to ancient Cerro Azul. It confirms anecdotal accounts of the warm-water species that visit Cerro Azul in ENSO years; reinforces anecdotal accounts of the cool-water species that emigrate; and shows that some species' response to changing water temperatures is more subtle than simply presence/absence. The log also reinforces the Cerro Azul fishermen's observations of species pairs whose peaks of abundance and scarcity seem to be correlated.

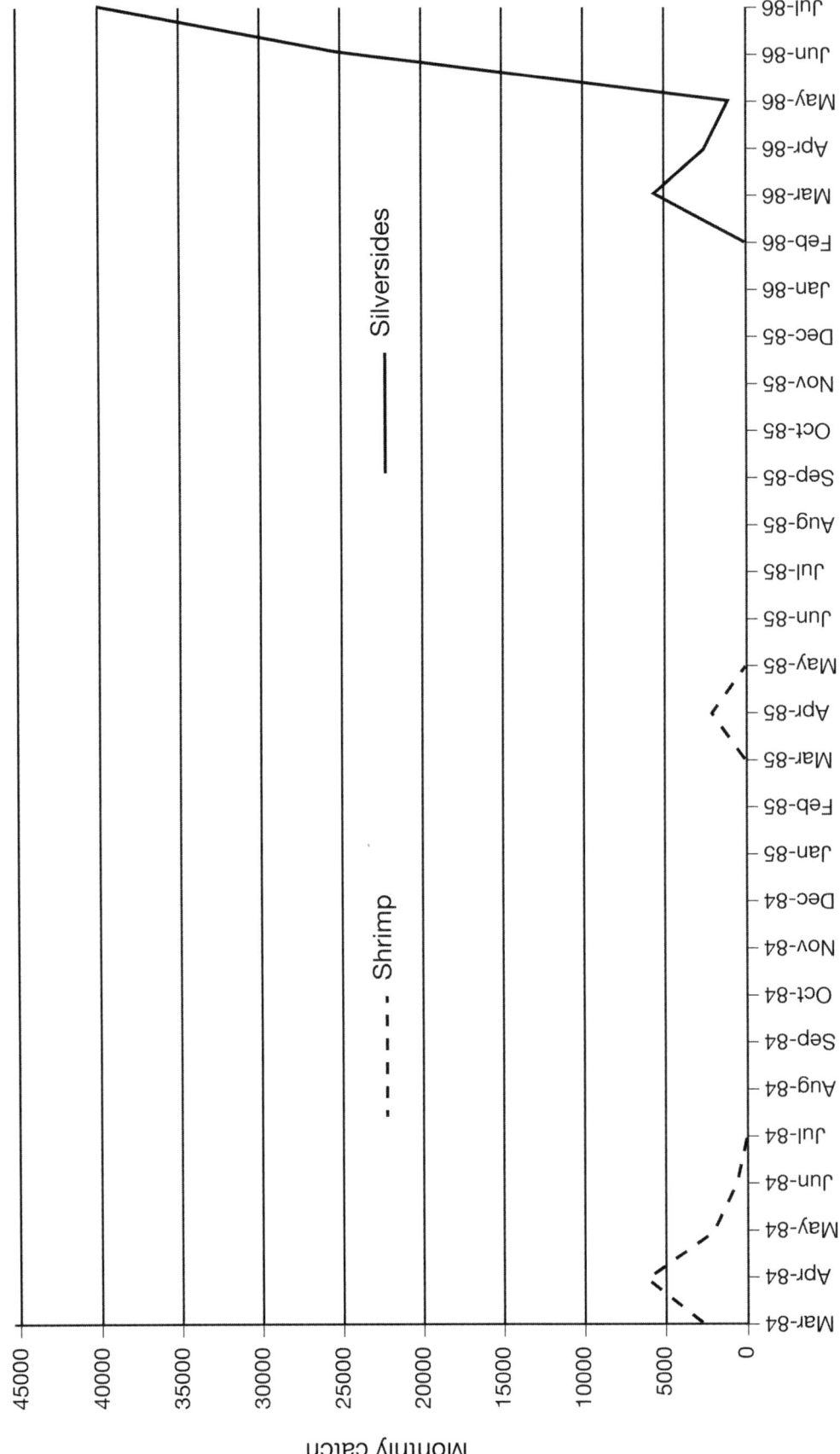

Figure A.1. A comparison of the monthly catches of shrimp (*Xiphopenaeus* sp.) and silversides (*Odontesthes regia*) at Cerro Azul, 1984–86. Shrimp extended its range to Cerro Azul during the 1982–83 El Niño and began to retreat in 1984. Pejerrey disappeared during El Niño, but returned in force in 1986. (The numbers given for shrimp are in dozens.)

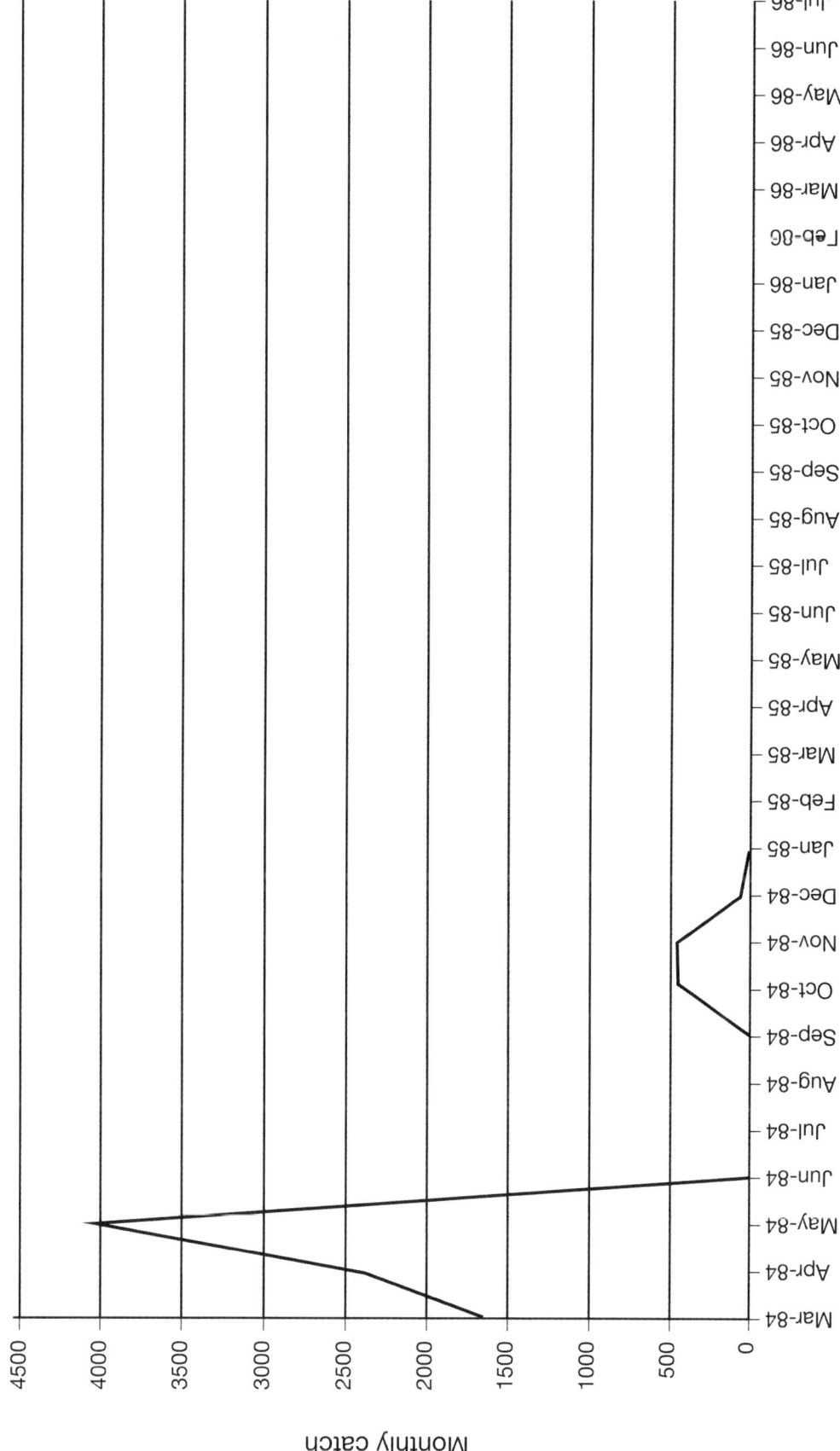

Figure A.2. Monthly catches of Chilean jack mackerel (*Trachurus symmetricus*) at Cerro Azul, 1984–86. During El Niño years, this fish follows the anchoveta to refuge areas near the shore, making it vulnerable to artisanal fishermen. As conditions return to normal it swims back to the open sea, where it falls prey to commercial trawlers.

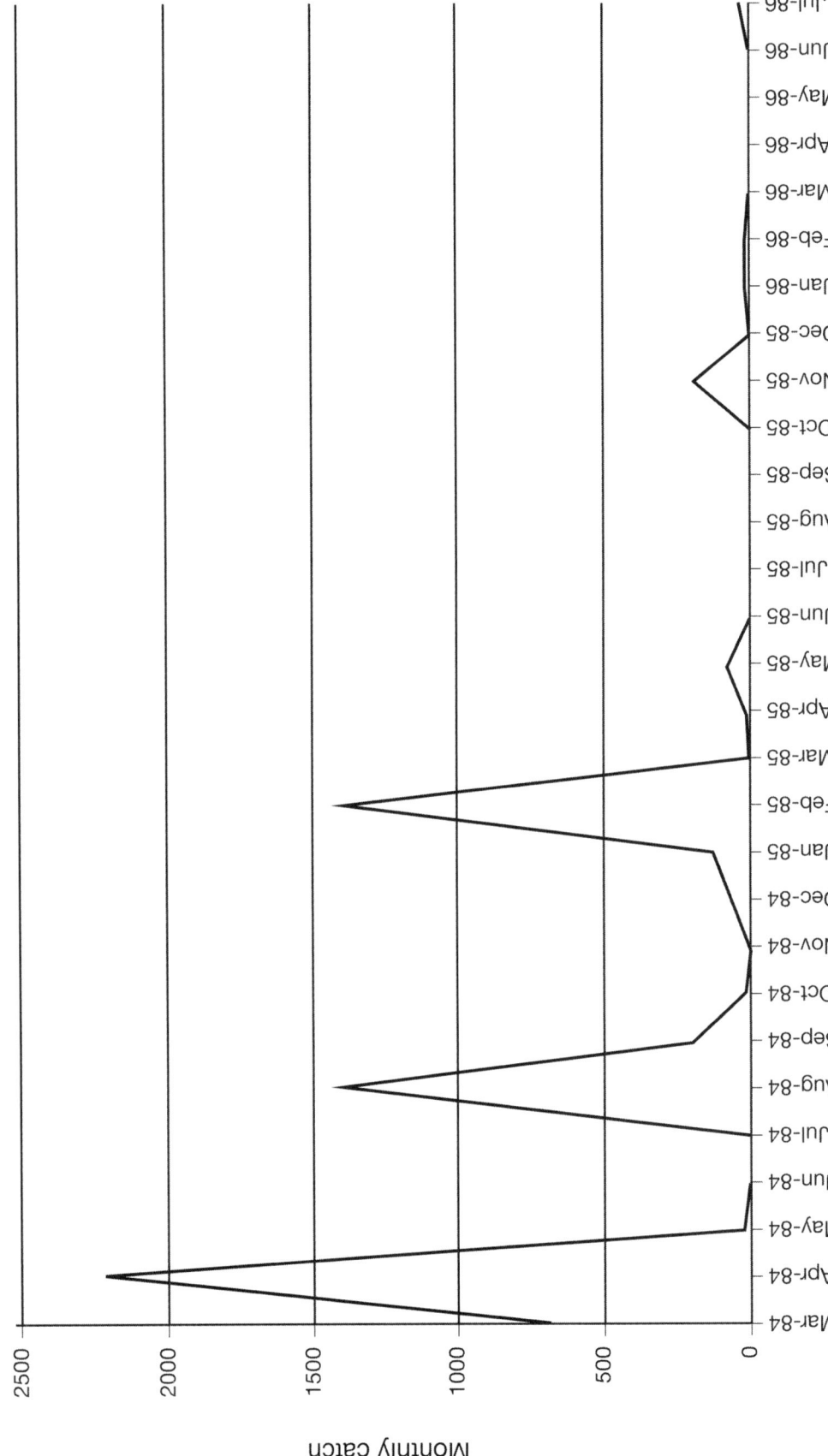

Figure A.3. Monthly catches of bonito (*Sarda sarda*) at Cerro Azul, 1984–86. The bonito underwent cycles of increase and decrease in availability. It was more available during the aftermath of the 1982–83 El Niño, when it came closer to the shore; it gradually became less available as normal conditions returned.

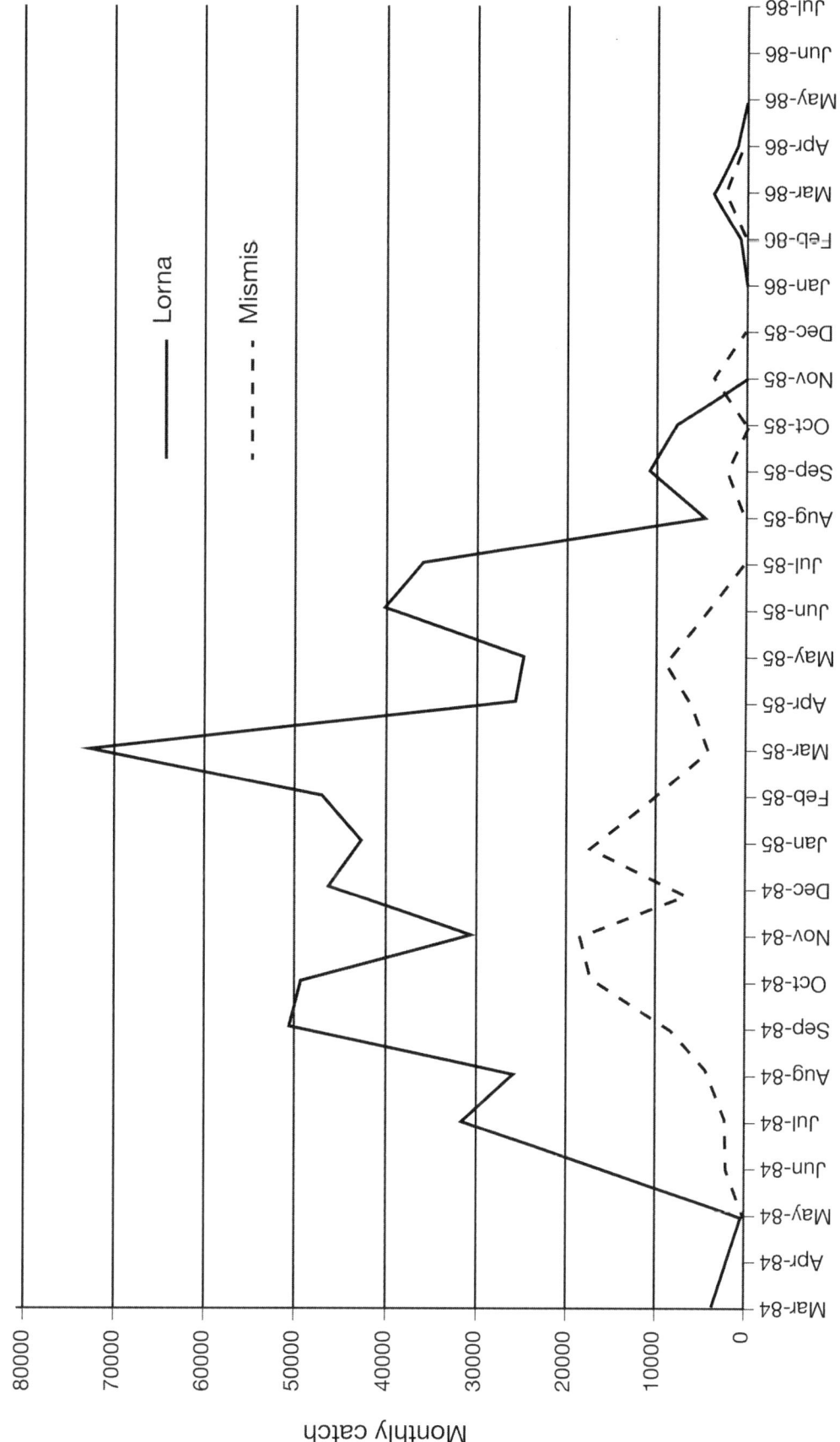

Figure A.4. A comparison of the monthly catches of lorna (*Sciaena deliciosa*) and mismis (*Menticirrhus ophicephalus*) at Cerro Azul, 1984–86.

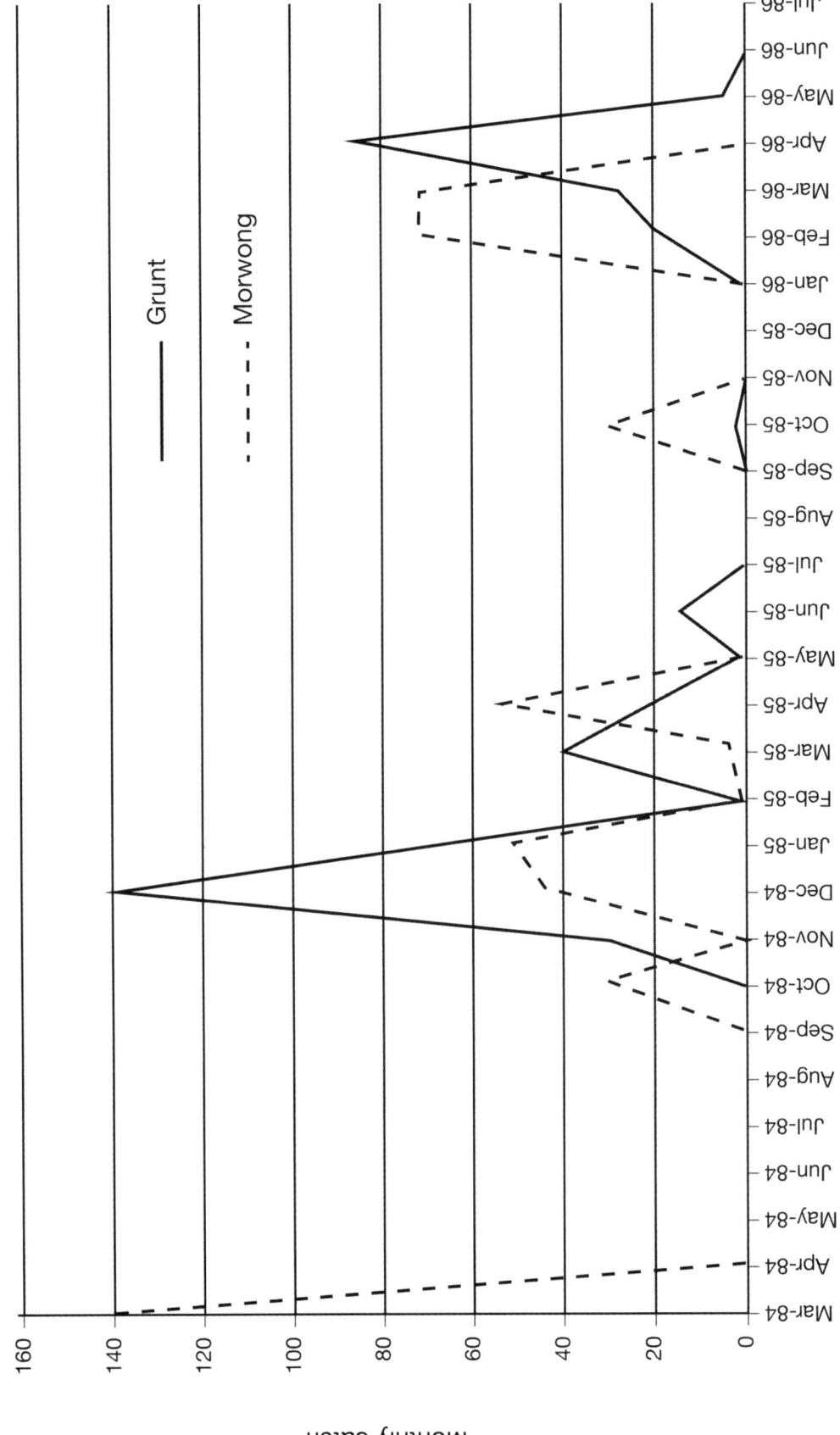

Figure A.5. A comparison of the monthly catches of grunt (*Anisotremus scapularis*) and morwong (*Cheilodactylus variegatus*) at Cerro Azul, 1984–86.

References Cited

Cited Documents

Archivo General de Indias (AGI), Sevilla, Spain
1558–67 Justicia 413. (See Rostworowski de Diez Canseco 1988:83–291)

Archivo General de Indias (AGI), Sevilla, Spain
1559 Probanza de Canta

Archivo General de Indias (AGI), Sevilla, Spain
1565–67 Justicia 458. Documentos procedentes de las visitas hechas por el Gregorio González de Cuenca a los corregimientos de Chuma, Saña, Cajamarca y Chachapoyas.

Archivo General de Indias (AGI), Sevilla, Spain
1573 Justicia 418. Autos entre Alonso Carrasco and Alonso Pizarro de la Rua, ambos vecinos de Trujillo del Perú, sobre que se dividen por mitad la encomienda de indios del valle de Jayanca.

Archivo General de Indias (AGI), Sevilla, Spain
1575 Escribanía de Cámara 498-B.

Archivo General de la Nación (AGN), Lima, Peru
1684 Derecho Indígena, Cuaderno 140

Archivo General de la Nación (AGN), Lima, Peru
1764 Juzgado de Agua 3.3.10.78

Aviso
1570 Aviso de el modo que havia en el govierno de los indios en tiempo del Inga y como se repartian las tierras y tributos. Tomo XXII, Miscelánea de Ayala, folios 261 al 273v. Biblioteca del Palacio Real de Madrid, reproduced in *Apéndice I* of Rostworowski de Diez Canseco [1970].

Biblioteca del Palacio Real de Madrid; see *Aviso*

Biblioteca Nacional de Madrid
1571 manuscript no. 3042

A

Acosta, fray José de
1940[1590] *Historia natural y moral de las Indias*. Fondo de Cultura Económica, México.
1954[1590] *Historia natural y moral de las Indias*. Biblioteca de Autores Españoles, Volume 73. Ediciones Atlas, Madrid.

Alamo Vásquez, Victor Raúl
1973 Datos ecológicos y pesquerías de los moluscos de importancia comercial en el Perú. Bachelor's thesis in Biological Sciences, Universidad Nacional Mayor de San Marcos, Lima.

Alamo Vásquez, Victor Raúl and Violeta Valdivieso
1987 Lista sistemática de moluscos marinos del Perú. *Boletín del Instituto del Mar del Perú*. Callao.
1997 Lista sistemática de moluscos marinos del Perú. Segunda edición. Publicación especial. *Boletín del Instituto del Mar del Perú*. Callao.

Albornoz, Cristóbal de
1967[1580] La instrucción para descubrir todas las guacas del Pirú y sus camayos y haziendas (fines del siglo xvi). ...In Duviols (1967) *Journal de la Société des Américanistes* LVI(1):7–39. Musée de l'Homme, Paris.

Albrecht, Elena, Dapeng Zhang, Robert A. Saftner, and John R. Stommel
2012 Genetic diversity and population structure of *Capsicum baccatum* genetic resources. *Genetic Resources in Crop Evolution* 59:517–538.

Alcedo, Antonio de
1967[1788] Diccionario geográfico de las indias occidentales o América (edited by Ciriaco Pérez-Bustamante). 4 volumes. *Biblioteca de autores españoles desde la formación del lenguaje hasta nuestros días*, tomos 205–208. Ediciones Atlas, Madrid.

Alheit, Jürgen and Miguel Ñiquen
2004 Regime shifts in the Humboldt Current ecosystem. *Progress in Oceanography* 60:201–222.

Alves-Araújo, Anderson, Ulf Swenson, and Marccus Alves
2014 A taxonomic survey of *Pouteria* (Sapotaceae) from the northern portion of the Atlantic rainforest of Brazil. *Systematic Botany* 39(3):915–938.

Arkush, Elizabeth and Charles Stanish
2005 Interpreting conflict in the ancient Andes. *Current Anthropology* 46(1):3–28.

Aronés, Katia, Patricia Ayón, Hans-Jürgen Hirche, and Ralf Schwamborn
2009 Hydrographic structure and zooplankton abundance and diversity off Paita, northern Peru (1994 to 2004)—ENSO effects, trends and changes. *Journal of Marine Systems* 78:582–598.

Austin, Daniel F.
1988 The taxonomy, evolution, and genetic diversity of sweet potatoes and related wild species. In *Exploration, Maintenance, and Utilization of Sweet Potato Genetic Resources: Report of the First Sweet Potato Planning Conference, 1987*, pp. 27–60. International Potato Center, Lima, Peru.

Austin, Oliver L., Jr.
1961 *Birds of the World*. Golden Press, New York.

Avendaño, Miguel and Marcela Cantillánez
2011 Reestablecimiento de *Choromytilus chorus* (Molina, 1782) (Bivalvia: Mytilidae) en el norte de Chile. *Latin American Journal of Aquatic Resources* 39:390–396.

Aviso
1570 Aviso de el modo que havia en el govierno de los indios en tiempo del Inga y cómo se repartian las tierras y tributos, Biblioteca del Palacio Real de Madrid, Miscelánea de Ayala, Tomo XXII, folios 261 al 273v. Reproduced in *Apéndice I* of Rostworowski de Diez Canseco [1970].

Ayala, Liliana, Luis Paz-Soldán, and Paola Gárate
2013 A mass mortality event of sooty shearwaters (*Puffinus griseus*) on the central coast of Peru. *Notornis* 60:258–261.

Ayón, Patricia and Gordon Swartzman
2008 Changes in the long-term distribution of zooplankton in the Humboldt Current ecosystem off Peru, 1961–2005, and its relationship to regime shifts and environmental factors. *Fisheries Oceanography* 17:421–431.

Ayón, Patricia, María I. Criales-Hernández, Ralf Schwamborn, and Hans-Jürgen Hirche
2008a Zooplankton research off Peru: a review. *Progress in Oceanography* 79:238–255.

Ayón, Patricia, Gordon Swartzman, Arnaud Bertrand, Mariano Gutiérrez, and Sophie Bertrand
2008b Zooplankton and forage food species off Peru: large-scale bottom-up forcing and local-scale depletion. *Progress in Oceanography* 79:208–214.

Ayón, Patricia, Gordon Swartzman, Pepe Espinoza, and Arnaud Bertrand
2011 Long-term changes in zooplankton size distribution in the Peruvian Humboldt Current system: conditions favoring sardine or anchovy. *Marine Ecology Progress Series* 422:211–222.

B

Banack, Sandra Anne, Xanic J. Rondón, and Wilfredo Diaz-Huamanchumo
2004 Indigenous cultivation and conservation of totora (*Schoenoplectus californicus*, Cyperaceae) in Peru. *Economic Botany* 58(1):11–20.

Bernardi, Giacomo, Yvette R. Alva-Campbell, João L. Gasparini, and Sergio R. Floeter
2008 Molecular ecology, speciation, and evolution of the reef fish genus *Anisotremus*. *Molecular Phylogenetics and Evolution* 48:929–935.

Berrios C., Viviana and Mauricio Vargas F.
2004 Estructura trófica de la asociación de peces intermareales de la costa rocosa del norte de Chile. *Revista de Biología Tropical* 52:201–212.

Biffi, Daniella and José Iannacone
2010 Variabilidad trófica de *Lontra felina* (Molina 1782) (Carnivora: Mustelidae) en dos poblaciones de Tacna (Perú) entre agosto y diciembre de 2006. *Mastozoología Neotropical* 1:11–17.

Bird, Junius B. (edited by John Hyslop and in collaboration with Milica D. Skinner)
1985 The Preceramic Excavations at the Huaca Prieta, Chicama Valley, Peru. *Anthropological Papers of the American Museum of Natural History*, Vol. 62, Part 1. New York.

Bird, Junius and Louisa Bellinger
1954 *Paracas Fabrics and Nazca Needlework*. National Publishing Co., Washington, D.C.

Bjerknes, Jacob
1969 Atmospheric teleconnections from the Equatorial Pacific. *Monthly Weather Review* 97:163–172.

Bohm, Bruce A., Fred R. Ganders, and Timothy Plowman
1982 Biosystematics and evolution of cultivated coca (Erythroxylaceae). *Systematic Botany* 7:121–133.

Bonavia, Duccio, Carlos M. Ochoa, Oscar Tovar S., and Rodolfo Cerrón Palomino
2004 Archaeological evidence of cherimoya (*Annona cherimolia* Mill.) and guanabana (*Annona muricata* L.) in ancient Peru. *Economic Botany* 58:509–522.

Bonner, W. Nigel
1981 Southern fur seals, *Arctocephalus* (Geoffroy Saint-Hilaire and Cuvier, 1826). In *Handbook of Marine Mammals, Vol. 1: The*

Walrus, Sea Lions, Fur Seals and Sea Otter, edited by Sam H. Ridgway and Richard J. Harrison, pp. 161–208. Academic Press, New York.

Bray, Tamara L. (editor)
2015 *The Archaeology of Wak'as: Explorations of the Sacred in the Pre-Columbian Andes*. University Press of Colorado, Boulder.

Brooke, Michael and John Cox
2004 *Albatrosses and Petrels across the World*. Oxford University Press, Oxford.

Brown Vega, Margaret and Nathan Craig
2009 New experimental data on the distance of sling projectiles. *Journal of Archaeological Science* 36:1264–1268.

Brusca, Richard C.
1980 *Common Intertidal Invertebrates of the Gulf of California*. 2nd edition. University of Arizona Press, Tucson.

Bryant, Vaughn M., Jr.
1974 The role of coprolite analysis in archaeology. *Bulletin of the Texas Archaeological Society* 45:1–28.

Bryant, Vaughn M., Jr. and Glendon H. Weir
1984 Human diets in the Ayacucho region of Peru. Unpublished manuscript.

Buse de la Guerra, Hermann, Ernesto Sarmiento, Franklin Pease, María Rostworowski, Yoshitaro Amano, and José Antonio del Busto
1977 *La pesca en el Perú prehispánico*. Editoriales Unidas S.A., Lima, Peru.

C

Cahlander, Adele
1980 *Sling Braiding of the Andes*. Colorado Fiber Center, Boulder.

Cancino, Juan M. and Juan Carlos Castilla
1988 Emersion behaviour and foraging ecology of the common Chilean clingfish *Sicyases sanguineus* (Pisces: Gobiesocidae). *Journal of Natural History* 22:249–261.

Cantillánez, Marcela, Miguel Avendaño, Manuel Rojo, and Alberto Olivares
2011 Parámetros reproductivos y poblacionales de *Thais chocolata* (Duclos, 1832) (Gastropoda, Thaididae) en la Reserva Marina La Rinconada, Antofagasta, Chile. *Latin American Journal of Aquatic Resources* 39:499–511.

Caro, Andrés U. and Juan Carlos Castilla
2004 Predator-inducible defenses and local intra-population variability of the intertidal mussel *Semimytilus algosus* in central Chile. *Marine Ecology Progress Series* 276:115–123.

Carstensen, Daniel W.
2010 Environmentally induced responses of *Donax obesulus* and *Mesodesma donacium* (Bivalvia) inhabiting the Humboldt Current system. PhD thesis, University of Bremen, Germany.

Castilla, Juan Carlos and L. R. Durán
1985 Human exclusion from the rocky intertidal zone of central Chile: the effects on *Concholepas concholepas* (Gastropoda). *Oikos* 45:391–399.

Castro, fray Cristóbal de and Diego de Ortega Morejón
1936[1558] Relación y declaración del modo que en este valle de Chincha... . In *Quellen zur Kulturgeschichte des präkolumbischen Amerika*, III, edited by Hermann Trimborn, pp. 236–246. Strecker und Schröder, Verlag, Stuttgart, Germany.
1974[1558] Relación y declaración del modo que en este valle de Chincha y sus comarcanos se gobernaron antes que hobiese ingas y después que los hobo hasta que los cristianos entraron en esta tierra (1558). Edited by Juan Carlos Crespo. *Historia y Cultura* 8:91–104. Museo Nacional de Historia, Lima.

Caviedes, César N.
1984 El Niño 1982–83. *Geographical Review* 74:267–290.

Chaudhary, Bhupendra, Ran Hovav, Ryan Rapp, Neetu Verma, Joshua A. Udall, and Jonathan F. Wendel
2008 Global analysis of gene expression in cotton fibers from wild and domesticated *Gossypium barbadense*. *Evolution and Development* 10:567–582.

Chiou, Katherine L. and Christine A. Hastorf
2014 A systematic approach to species-level identification of chile pepper (*Capsicum* spp.) seeds: establishing the groundwork for tracking the domestication and movement of chile peppers through the Americas and beyond. *Economic Botany* 68:316–336.

Chiou, Katherine L., Christine A. Hastorf, Duccio Bonavia, and Tom D. Dillehay
2014 Documenting cultural selection pressure changes on chile pepper (*Capsicum baccatum* L.) seed size through time in coastal Peru (7,600 B.P.-present). *Economic Botany* 68:190–202.

Chong, Javier, N. Cortés, and R. Bustos
2000 Hábitos alimenticios de la corvina *Cilus gilberti* (Abbott, 1889) (Pisces-Sciaenidae) frente a la costa de Talcahuano. *Biología Pesquera* 28:29–35.

Chu, Alejandro
2015 La plaza y el *ushnu* mayor de Incahuasi, Cañete. *Cuadernos de Qhapaq Ñan* 3(3):92–110.

Cieza de León, Pedro
1932[1550] *La crónica del Perú*. Espasa-Calpe, S. A., Madrid.

Cisneros, Fausto H., Roberto Zevillanos, and Luis Cisneros-Zevallos
2009 Characterization of starch from two ecotypes of Andean achira roots (*Canna edulis*). *Journal of Agricultural Food Chemistry* 57:7363–7368.

Clark, Daniel, Maribel Tupa, Andrea Bazán, Lily Chang, and Wilfredo L. Gonzáles
2012 Chemical composition of *Apodanthera biflora*, a cucurbit of the dry forest in northwestern Peru. *Revista Peruana de Biología* 19(2):199–203.

Cobo, Bernabé
1990 *Inca Religion and Customs* (translated by Roland Hamilton). University of Texas Press, Austin.

Crespo, Juan Carlos
1978 Chincha y el mundo andino en la *Relación* de 1558. *Histórica* 2(2):185–212.

Cuadernos del Qhapaq Ñan volume 3 (see Alejandro Chu 2015, José Luis Díaz Carranza 2015, Giancarlo Marcone Flores and Rodrigo Areche Espinola 2015)

Cushman, Gregory T.
2005 "The most valuable birds in the world": international conservation science and the revival of Peru's guano industry, 1909–1965. *Environmental History* 10:477–509.

D

Dall, William Healey
1909 Report on a collection of shells from Peru, with a summary of the littoral marine molluscs of the Peruvian zoological province. *Proceedings of the U.S. National Museum* 37(1704):147–294.

D'Altroy, Terence N. and Timothy K. Earle
1985 Staple finance, wealth finance, and storage in the Inka political economy. *Current Anthropology* 26(2):187–206

Davies, Norman de Garis
1943 *The Tomb of Rekh-mi-re' at Thebes*, Volume II. Metropolitan Museum of Art, New York.

Delgado-Salinas, Alfonso, Tom Turley, Adam Richman, and Matt Lavin
1999 Phylogenetic analysis of the cultivated and wild species of *Phaseolus* (Fabaceae). *Systematic Botany* 24:438–460.

Díaz, Pablo E. and Gabriela Muñoz
2010 Diet and parasites of the insular fish *Scartichthys variolatus* (Blenniidae) from Robinson Crusoe Island, Chile: how different is this from two continental congeneric species? *Revista de Biología Marina y Oceanografía* 45:293–301.

Díaz Carranza, José Luis
2015 Hallazgos de coca en colcas del valle medio del río Cañete correspondientes al Horizonte Tardío. *Cuadernos del Qhapaq Ñan* 3:128–147.

Donnan, Christopher B.
1995 Moche funerary practice. In *Tombs for the Living: Andean Mortuary Practices*, edited by Tom D. Dillehay, pp. 111–159. Dumbarton Oaks Research Library and Collection, Washington, D.C.

Donnan, Christopher B. and Donna McClelland
1999 *Moche Fineline Painting: Its Evolution and Its Artists*. UCLA Fowler Museum of Cultural History, Los Angeles.

Duffy, David Cameron
1983 The foraging ecology of Peruvian seabirds. *The Auk* 100:800–810.

Duviols, Pierre
1967 Un inédit de Cristóbal de Albornoz: la instrucción para descubrir todas las guacas del Pirú y sus camayos y haziendas. *Journal de la Société des Américanistes* 56(1):7–39. Paris.
1973 Huari y llacuaz, agricultores y pastores: un dualismo prehispánico de oposición y complementaridad. *Revista del Museo Nacional* 39:153–192. Lima.
1979 La dinastia de los incas: ¿Monarquía o diarquía? Argumentos heurísticos a favor de una tesis estructuralista. *Journal de la Société des Américanistes* 66:67–83. Paris.

E

Eeckhout, Peter and Lawrence Stewart Owens
2008 Human sacrifice at Pachacamac. *Latin American Antiquity* 19(4):375–398.

ElMahi, Ali Tigani
2000 Traditional fish preservation in Oman: the seasonality of a subsistence strategy. *Proceedings of the Seminar for Arabian Studies* 30:99–113.

Emery, Irene
1966 *The Primary Structures of Textiles: An Illustrated Classification*. The Textile Museum, Washington D.C.

Engel, Frédéric
1963 A Preceramic Settlement on the Central Coast of Peru: Asia, Unit I. *Transactions of the American Philosophical Society* 53(3):3–139. Philadelphia.

Espinoza, Pepe and Arnaud Bertrand
2008 Revisiting Peruvian anchovy (*Engraulis ringens*) trophodynamics provides a new vision of the Humboldt Current system. *Progress in Oceanography* 79:215–227.
2014 Ontogenetic and spatiotemporal variability in anchoveta *Engraulis ringens* diet off Peru. *Journal of Fish Biology* 84:422–435.

Espinoza, Pepe, Arnaud Bertrand, Carl B. van der Lingen, Susana Garrido, and Blanca Rojas de Mendiola
2009 Diet of sardine (*Sardinops sagax*) in the northern Humboldt Current system and comparison with the diets of Clupeoids in this and other eastern boundary upwelling systems. *Progress in Oceanography* 83:242–250.

Espinoza Soriano, Waldemar
1963 La guaranga y la reducción de Huancayo. Tres documentos inéditos de 1571 para la etnohistoria del Perú. *Revista del Museo Nacional* 32:8–80. Lima.
1969–70 Los mitmas yungas de Collique en Cajamarca, siglos XV, XVI y XVII. *Revista del Museo Nacional* 36:9–57. Lima.

Etcheverry, Ángela V. and Carlos E. Trucco Alemán
2005 Reproductive biology of *Erythrina falcata* (Fabaceae: Papilionideae). *Biotropica* 37:54–63.

F

Farias, Nahuel E., Tomás A Luppi, and Eduardo D. Spivak
2014 Habitat use, relative growth and size at maturity of the purple stone crab *Platyxanthus crenulatus* (Decapoda: Brachyura), calculated under different models. *Scientia Marina* 78:568–578.

Farrington, Ian S.
1974 Irrigation and settlement pattern: preliminary research from the north coast of Peru. In *Irrigation's Impact on Society*, edited by Theodore Downing and McGuire Gibson, pp. 83–94. University of Arizona Press, Tucson.
1977 Land use, irrigation, and society on the north coast of Peru in the prehispanic era. *Zeitschrift für Bewässerungswirtschaft* 12:151–186.
1980 The archaeology of irrigation canals, with special reference to Peru. *World Archaeology* 11:287–305.
1983 The design and function of the intervalley canal: comments on

a paper by Ortloff, Moseley, and Feldman. *American Antiquity* 48:360–375.

Farrington, Ian S. and C. C. Park
1978   Hydraulic engineering and irrigation agriculture in the Moche Valley, Peru: ca. A.D. 1250–1532. *Journal of Archaeological Science* 5(3):255–268.

Fernández, Camila and Ciro Oyarzún
2001   Trophic variations of the Chilean croaker *Cilus gilberti* during the summer period 1997–98 (Perciformes, Sciaenidae). *Journal of Applied Ichthyology* 17:227–233.

Fischer, Sönke and Matthias Wolff
2006   Fisheries assessment of *Callinectes arcuatus* (Brachyura, Portunidae) in the Gulf of Nicoya, Costa Rica. *Fisheries Research* 77:301–311.

Fitzhugh, William W. and Susan A. Kaplan
1982   *Inua: Spirit World of the Bering Sea Eskimo*. Smithsonian Institution Press, Washington, D.C.

Flannery, Kent V., Joyce Marcus, and Robert G. Reynolds
1989   *The Flocks of the Wamani: A Study of Llama Herders on the Punas of Ayacucho, Peru*. Academic Press, San Diego.

Fry, Gary F.
1976   Analysis of prehistoric coprolites from Utah. *University of Utah Anthropological Papers* No. 97. Salt Lake City.

G

Gade, Daniel W.
1966   Achira, the edible canna, its cultivation and use in the Peruvian Andes. *Economic Botany* 20:407–415

Garcilaso de la Vega, "El Inca"
1966[1609]   *Royal Commentaries of the Incas and General History of Peru* (translated by Harold V. Livermore). University of Texas Press, Austin.

Garrido, M. V., O. R. Chaparro, R. J. Thompson, Orlando Garrido, and J. M. Navarro
2012   Particle sorting and formation and elimination of pseudofaeces in the bivalves *Mulinia edulis* (siphonate) and *Mytilus chilensis* (asiphonate). *Marine Biology* 159:987–1000.

Glynn, Peter W.
1988   El Niño-Southern Oscillation 1982–1983: nearshore population, community, and ecosystem responses. *Annual Review of Ecology and Systematics* 19:309–345.

Griffiths, W. B. and P. Carrick
1994   Reconstructing Roman slings. *Arbeia Journal* 3:1–11.

Guaman Poma de Ayala, Felipe
1936[1613]   *Nueva corónica y buen gobierno*. Institut d'Ethnologie, Paris.
1980[1613]   *El primer nueva corónica y buen gobierno*, edited by John V. Murra and Rolena Adorno. Translated by Jorge L. Urioste. 3 volumes. Siglo Veintiuno Editores, Mexico City.

Guillén, Sonia E.
n.d.   Cerro Azul—informe e inventario del material oseo humano. Manuscript in possession of the volume editor.

Gutiérrez, D., A. Siffedine, D.B. Field, L. Ortlieb and others
2009   Rapid reorganization in ocean biogeochemistry off Peru towards the end of the Little Ice Age. *Biogeosciences* 6:835–848.

H

Haig, Janet
1980   Anthropoda: Crustacea: superfamily Hippoidea: families Hippidae and Albuneidae (mole and sand crabs). In *Common Intertidal Invertebrates of the Gulf of California* (second edition), by Richard C. Brusca, pp. 286–291. University of Arizona Press, Tucson.

Harth-Terré, Emilio
1933   Incahuasi. Ruinas inkaicas del valle de Lunahuaná. *Revista del Museo Nacional* II(2):101–125. Lima.

Hastorf, Christine A.
1987   Archaeological evidence of coca (*Erythroxylum coca*, Erythroxylaceae) in the upper Mantaro Valley, Peru. *Economic Botany* 41:292–301.
1993   *Agriculture and the Onset of Political Inequality before the Inka*. Cambridge University Press, Cambridge, England.

Heiser, Charles B.
1974   Totoras, taxonomy, and thor. *Plant Science Bulletin* 20:22–26.
1978   The totora (*Scirpus californicus*) in Ecuador and Peru. *Economic Botany* 32:222–236.

Heizer, Robert F. and Irmgard W. Johnson
1952   A prehistoric sling from Lovelock Cave, Nevada. *American Antiquity* 18(2):139–147.

Hutchinson, Thomas J.
1874   Explorations amongst ancient burial grounds (chiefly on the sea-coast valleys) of Peru, Part I. *The Journal of the Anthropological Institute of Great Britain and Ireland* 3:311–326.

Huyer, Adriana, Robert L. Smith, and Theresa Paluszkiewicz
1987   Coastal upwelling of Peru during normal and El Niño times, 1981–1984. *Journal of Geophysical Research* 92:14297–14307.

Hyslop, John
1985   *Inkawasi, the New Cuzco: Cañete, Lunahuaná, Peru*. British Archaeological Reports, International Series, volume 234. Oxford, England.

I

Iannacone, José
2014   Metazoos parásitos de la mojarrilla *Stellifer minor* (Tschudi) (Ostcichthyes, Sciaenidae) capturados por pesquería artesanal en Chorrillos, Lima, Perú. *Revista Brasileira de Zoologia* 21(4):815–820.

Ibáñez, Ana Laura and A. Colín
2014   Reproductive biology of *Mugil curema* and *Mugil cephalus* from western Gulf of Mexico waters. *Bulletin of Marine Science* 90:941–952.

Ibarcena Fernández, Walter, Luis Muñante Angulo, Luis Muñante Melgar, and Juana Vásquez Flores

2005 La explotación de la macha (*Mesodesma donacium* Lamarck 1818) en el litoral de Tacna. *Ciencia y Desarrollo* 8:12–22.

Isbell, Billie Jean
1978 *To Defend Ourselves: Ecology and Ritual in an Andean Village*. Institute of Latin American Studies, University of Texas, Austin.

J

Jones, John G.
1986a Preliminary analysis of the El Paraíso coprolites. Manuscript on file, Department of Anthropology, Texas A&M University.
1986b Preliminary analysis of the Lo Demás coprolites. Manuscript on file, Department of Anthropology, Texas A&M University.
1988 Middle to Late Preceramic (6000–3000 BP) Subsistence Patterns on the Central Coast of Peru: The Coprolite Evidence. MA thesis, Department of Anthropology, Texas A&M University.

Jordán, Rómulo
1963 Un análisis del número de vertebras de la anchoveta peruana (*Engraulis ringens J.*). *Boletín del Instituto de Investigación de los Recursos Marinos* 1:25–43. Callao.

K

Kaplan, Lawrence, Thomas F. Lynch, and C. Earle Smith, Jr.
1973 Early cultivated beans (*Phaseolus vulgaris*) from an intermontane Peruvian valley. *Science* 179:76–77.

Kattan, Khalid
n.d. Learning the Impact of an El Niño on Local Fishing Behavior in Cerro Azul, Peru Using Cultural Algorithms. PhD dissertation, Department of Computer Sciences, Wayne State University, Detroit, Michigan.

Keen, Angeline Myra
1958 *Sea Shells of Tropical West America*. Stanford University Press, Stanford, California.

King, Judith E.
1983 *Seals of the World*. 2nd edition. British Museum (Natural History), London.

Kochert, Gary, H. Thomas Stalker, Marcos Giménez, Leticia Galgaro, Catalina Romero Lopes, and Kim Moore
1996 RFLP and cytogenetic evidence on the origin and evolution of allotetraploid domesticated peanut, *Arachis hypogaea* (Leguminosae). *American Journal of Botany* 83:1282–1291.

Koepcke, María
1964 *Las aves del departamento de Lima*. Gráfico Morsom, S.A., Lima.

Koford, Carl B.
1957 The vicuña and the puna. *Ecological Monographs* 27:153–219.

Korfmann, Manfred
1973 The sling as a weapon. *Scientific American* 229(4):34–42.

Kroeber, Alfred Louis
1937 *Archaeological Explorations in Peru. Part IV: Cañete Valley*. First Marshall Field Archaeological Expedition to Peru. Anthropology Memoirs Vol. II, No. 4, pp. 220–273. Field Museum of Natural History, Chicago, Illinois.

Kroeber, Alfred Louis and William Duncan Strong
1924 The Uhle collections from Chincha. *University of California Publications in American Archaeology and Ethnology* 21(1):1–54. Berkeley.

Kroeber, Alfred Louis and Dwight Wallace
1954 Proto-Lima: a middle period culture of Peru. *Fieldiana Anthropology* 44, no. 1. Chicago Natural History Museum, Chicago.

Kubler, George
1948 Towards absolute time: guano archaeology. In "A Reappraisal of Peruvian Archaeology," assembled by Wendell C. Bennett. *Memoirs of the Society for American Archaeology* 4:29–50.

L

Lack, David
1945 The ecology of closely related species with special reference to cormorant (*Phalacrocorax carbo*) and shag (*P. aristotelis*). *Journal of Animal Ecology* 14(1):12–16.

Larrabure y Unanue, Eugenio
1935[1893] *Historia y arqueología—valle de Cañete. Manuscritos y publicaciones*, Tomo II. Imprenta Americana, Lima.

Lim, Teck-Kah
2012 *Edible Medicinal and Non-Medicinal Plants: Vol. 2, Fruits*. Springer Netherlands, Dordrecht.

Lizárraga, fray Reginaldo de
1946[1605] *Descripción de las indias*. Los Pequeños Grandes Libros de América. Editorial Loayza, Lima.

Llompart, F. M., D. C. Colautti, T. Maiztegui, A. M. Cruz-Jiménez, and C. R. M. Baigún
2013 Biological traits and growth patterns of pejerrey *Odontesthes argentinensis*. *Journal of Fish Biology* 82:458–474.

Löfgren, Lars
1984 *Ocean Birds*. Crescent Books, New York.

Loud, Llewellyn Lemont and Mark Raymond Harrington
1929 Lovelock Cave. *University of California Publications in American Archaeology and Ethnology* Vol. 25, No. 1. Berkeley.

M

Macía, Manuel J. and Henrik Balslev
2000 Use and management of totora (*Schoenoplectus californicus*, Cyperaceae) in Ecuador. *Economic Botany* 54(1):82–89.

Makowski, Krzysztof
2015 Pachacamac—old wak'a or Inka syncretic deity? Imperial transformation of the sacred landscape in the lower Ychsma (Lurín) Valley. In *The Archaeology of Wak'as: Explorations of the Sacred in the Pre-Columbian Andes*, pp. 127–166. University Press of Colorado, Boulder.

Marcone Flores, Giancarlo and Rodrigo Areche Espinola
2015 El valle de Cañete durante los períodos prehispánicos tardíos:

perspectivas desde El Huarco–Cerro Azul. *Cuadernos del Qhapaq Ñan* 3:48–68.

Marcus, Joyce
1987a *Late Intermediate Occupation at Cerro Azul, Perú: A Preliminary Report.* Technical Report 20, University of Michigan Museum of Anthropology, Ann Arbor.
1987b Prehistoric fishermen in the Kingdom of Huarco. *American Scientist* 75(4):393–401.
2008 *Excavations at Cerro Azul, Peru: The Architecture and Pottery.* Monograph 62, Cotsen Institute of Archaeology, University of California at Los Angeles.
2009 A world tour of breweries. In *Andean Civilization: A Tribute to Michael E. Moseley,* edited by Joyce Marcus and Patrick Ryan Williams, pp. 303–324. Cotsen Institute of Archaeology, University of California at Los Angeles.
2015 Studying the individual in prehistory: a tale of three women from Cerro Azul, Peru. *Ñawpa Pacha: Journal of Andean Archaeology* 35(1):1–22.
2016 Barcoding spindles and decorating whorls: how weavers marked their property at Cerro Azul, Peru. *Ñawpa Pacha: Journal of Andean Archaeology* 36(1):1–21.

Marcus, Joyce and Kent V. Flannery
2010 En búsqueda de la mentalidad andina: aventuras en el Perú con Ramiro Matos. *Arqueología y Vida* 3:9–22. Lima.

Marcus, Joyce, Kent V. Flannery, Jeffrey D. Sommer, and Robert G. Reynolds
In press Maritime adaptations at Cerro Azul, Peru: a comparison of Late Intermediate and 20th-century fishing. In *Andean Maritime Communities,* edited by Gabriel Prieto and Daniel H. Sandweiss, University Press of Florida, Gainesville.

Marcus, Joyce, Ramiro Matos Mendieta, and María Rostworowski de Diez Canseco
1985 Arquitectura inca de Cerro Azul, valle de Cañete. *Revista del Museo Nacional* 47:125–138. Lima.

Marcus, Joyce and Jorge Silva
1988 The Chillón Valley "Coca Lands": archaeological background and ecological context. In *Conflicts Over Coca Fields in XVIth-Century Perú,* edited by María Rostworowski de Diez Canseco. Memoir 21 of the Museum of Anthropology University of Michigan. Studies in Latin American Ethnohistory & Archaeology IV:1–52. Ann Arbor.

Marcus, Joyce, Jeffrey D. Sommer, and Christopher P. Glew
1999 Fish and mammals in the economy of an ancient Peruvian kingdom. *Proceedings of the National Academy of Sciences* 96:6564–6570.

Martínez de Compañón, Baltazar Jaime
1936 *Trujillo del Perú a fines del siglo XVIII.* Edited by and prólogo by Jesús Domínguez Bordona. Madrid, Spain.

Masuda, Shozo
1981 Cochayuyo, macha, camarón e higos charqueados. In *Estudios etnográficos del Perú meridional,* edited by Shozo Masuda, pp. 173–192. University of Tokyo Press, Tokyo, Japan.
1985 Algae collectors and *lomas.* In *Andean Ecology and Civilization: An Interdisciplinary Perspective on Andean Ecological Complementarity,* edited by Shozo Masuda, Izumi Shimada, and Craig Morris, pp. 233–250. University of Tokyo Press, Tokyo, Japan.

Masuda, Shozo, Izumi Shimada, and Craig Morris
1985 *Andean Ecology and Civilization: An Interdisciplinary Perspective on Andean Ecological Complementarity.* University of Tokyo Press, Tokyo, Japan.

McClelland, Donna, Donald McClelland, and Christopher B. Donnan
2007 *Moche Fineline Painting from San José de Moro.* Cotsen Institute of Archaeology, University of California at Los Angeles.

McConnaughey, Bayard H. and Robert Zottoli
1983 *Introduction to Marine Biology.* Fourth edition. The C.V. Mosby Co, St. Louis.

Means, Philip A.
1919 *Distribution and use of slings in pre-columbian America, with descriptive catalogue of ancient Peruvian slings in the United States National Museum.* In Proceedings of the United States National Museum, vol. 55:317–349. Smithsonian Institution, Government Printing Office, Washington, D.C.

Medina, Marianela and Hugo Arancibia
2002 Dinámica trófica del jurel (*Trachurus symmetricus murphyi*) en el norte de Chile. *Investigaciones Marinas* 30:45–55.

Medina, Marianela, Hugo Arancibia, and Sergio Neira
2007 Un modelo trófico preliminar del ecosistema pelágico del norte de Chile (18° 20' S–24°00' S). *Investigaciones Marinas* 35:25–38.

Medina Chauca, Wenceslao
1982 Ecoanálisis de los peces osteichthyes comunes de las aguas costeras del Perú según la forma de dentición, biotopo, y obtención de alimento. *Revista Peruana de Biología* 2:77–133.

Méndez G., Matilde
1981 Claves de identificación y distribución de los langostinos y camarones (Crustacea: Decapoda) del mar y ríos de la costa del Perú. *Boletín del Instituto del Mar del Perú,* vol. 5. Callao.

Miller, George R.
1979 An Introduction to the Ethnoarchaeology of the Andean Camelids. Ph.D. dissertation, Department of Anthropology, University of California at Berkeley. University Microfilms International, Ann Arbor, Michigan.

Mitchell, William P.
1976 Irrigation and community in the central highlands. *American Anthropologist* 78(1):25–44.

Moore, Jerry D.
1989 Pre-hispanic beer in coastal Peru: technology and social context of prehistoric production. *American Anthropologist* 91(3):682–695.

Moreira, Priscila Ambrósio, Juliana Lins, Gabriel Dequigiovanni, Elizabeth Ann Veasey, and Charles R. Clement
2015 The domestication of *annatto* (*Bixa orellana*) from *Bixa urucurana* in Amazonia. *Economic Botany* 69(2):127–135.

Morris, Craig
1979 Maize beer in the economics, politics and religion of the Inca Empire. In *Fermented Food Beverages in Nutrition,* edited by Clifford F. Gastineau, William J. Darby, and Thomas B. Turner, pp. 21–34. Academic Press, New York.

1982 The infrastructure of Inka control in the Peruvian central highlands. In *The Inca and Aztec States, 1400–1800: Anthropology and History*, edited by George A. Collier, Renato I. Rosaldo, and John D. Wirth, pp. 153–171. Academic Press, New York.

Morris, Craig, R. Alan Covey, and Pat Stein
2011 The Huánuco Pampa Archaeological Project. Volume I: The Plaza and Palace Complex. *American Museum of Natural History Anthropological Papers* No. 96. New York.

Morris, Craig and Julián Idilio Santillana
2007 The Inka transformation of the Chincha capital. In *Variations in the Expression of Inka Power: A Symposium at Dumbarton Oaks 18 and 19 October 1997*, edited by Richard L. Burger, Craig Morris, and Ramiro Matos Mendieta, pp. 135–163. Dumbarton Oaks, Washington, D.C.

Munsell Color Company
1954 *Munsell Soil Color Charts*. Munsell Color Company, Baltimore, Maryland.

Murphy, Robert Cushman
1920 The seacoast and islands of Peru I. *Brooklyn Museum Quarterly* 7(2):69–95.
1923 Fisheries resources in Peru. *The Scientific Monthly* 16(6):594–607.
1951 Review of *The Biogeochemistry of Vertebrate Excretion* by G. E. Hutchinson. *Ecology* 32:567–569.

Murra, John V.
1960 Rite and crop in the Inca state. In *Culture in History: Essays in Honor of Paul Radin*, edited by Stanley Diamond, pp. 393–407. Columbia University Press, New York.
1961 Social structural and economic themes in Andean ethnohistory. *Anthropological Quarterly* 34(2):47–59. George Washington University, Institute for Ethnographic Research.
1975 *Formaciones económicas y políticas en el mundo andino*. Instituto de Estudios Peruanos, Lima.
1980 *The Economic Organization of the Inka State*. JAI Press, Greenwich, Connecticut.

N

National Research Council
1989 *Lost Crops of the Incas: Little Known Plants of the Andes with Promise for World-Wide Cultivation*. National Academy Press, Washington, D.C.

Nee, Michael
1990 The domestication of *Cucurbita* (Cucurbitaceae). *Economic Botany* 44(3):56–68.

Nelson, Edward W.
1899 The Eskimo about Bering Strait. *Bureau of American Ethnology Annual Report* 1:1–518. Smithsonian Institution, Washington, D.C.

Netherly, Patricia J.
1977 Local Level Lords on the North Coast of Peru. PhD dissertation, Department of Anthropology, Cornell University, Ithaca, New York.
1984 The management of late Andean irrigation systems on the north coast of Peru. *American Antiquity* 49(2):227–254.
1990 Out of many, one: the organization of rule in the north coast polities. In *The Northern Dynasties: Kingship and Statecraft in Chimor*, edited by Michael E. Moseley and Alana Cordy-Collins, pp. 353–377. Dumbarton Oaks, Washington, D.C.
2009 Landscapes as metaphor: resources, language, and myths of dynastic origin on the Pacific coast from the Santa Valley (Peru) to Manabí (Ecuador). In *Landscapes of Origin in the Americas: Creation Narratives Linking Ancient Places and Present Communities*, edited by Jessica Joyce Christie, pp. 122–152. University of Alabama, Tuscaloosa.

O

Oliva, Doris
1988 *Otaria byronia* (de Blainville, 1920), the valid scientific name for the southern sea lion (Carnivora: Otariidae). *Journal of Natural History* 22:767–772.

Oliva, Marcelo E., Raúl E. Castro, and Rodrigo Burgos
1996 Parasites of the flatfish *Paralichthys adspersus* (Steindachner, 1867) (Pleuronectiformes) from northern Chile. *Memorias do Instituto Oswaldo Cruz* 91(3):301–306. Río de Janeiro.

O'Neale, Lila M.
1937 Appendix VI. Middle Cañete Textiles. In *Archaeological Explorations in Peru. Part IV: Cañete Valley*. First Marshall Field Archaeological Expedition to Peru. Anthropology Memoirs Vol. II, No. 4, pp. 268–273. Field Museum of Natural History, Chicago, Illinois.

ONERN (Oficina Nacional de Evaluación de Recursos Naturales)
1970 *Inventario, evaluación y uso racional de la costa: cuenca del Río Cañete (Junio)*, Vol. 1. Lima.

Orrego, Henry and Jaime Mendo
2012 Variación interanual de la dieta de la merluza *Merluccius gayi peruanus* (Guitchenot) en la costa peruana. *Ecología Aplicada* 11(2):103–116.

Ortiz Grisales, Sanín, Lucy Viviana Bastidas Burbano, Ginna Alejandra Ordóñez Narváez, Magda Piedad Valdés Restrepo, Diosdado Baena García, and Franco Alirio Vallejo Cabrera
2014 Inbreeding and gene action in butternut squash (*Cucurbita moschata*) seed starch content. *Revista de la Facultad Nacional Agropecuaria de Medellín* 67:7169–7175.

Ottsen, Hendrick
1617 Iovrnael oft daghelijcx-register van de voyagie na Rio de Plata, ghedaen met het schip ghenoemt de Silveren Werelt, het welcke onder 't admiraelschap van Laurens Bicker, ende het bevel van Cornelis van Heems-kerck als commis die custen van Guinea versocht hebbende, ende van den admirael daer na versteken zijnde, alleen voorts seylande na Rio de Plata, daer in de voorsz. riviere by de 60 mijlen opwaerts ghekomen wesende, tot Bonas Aeris ... Colijn, Amsterdam. Second edition.

P

Paredes, Carlos
1974a Moluscos del Perú. Ph.D. dissertation in Biological Sciences, Lima, Peru.
1974b El modelo de zonación en la orilla rocosa del Departamento de Lima. *Revista Peruana de Biología* 1(2):168–191.

Paredes, Rosanna, Carlos B. Zavalaga, Gabriella Battistini, Patricia

Majluf, and Patricia McGill
2003 Status of the Humboldt penguin in Peru, 1999–2000. *Waterbirds* 26(2):129–137.

Parsons, Mary Hrones
1970 Preceramic subsistence on the Peruvian coast. *American Antiquity* 35:292–304.

Paul, R. K. G.
1981 Natural diet, feeding and predatory activity of the crabs *Callinectes arcuatus* and *C. toxotes* (Decapoda, Brachyura, Portunidae). *Marine Ecology Progress Series* 6:91–99.

Peña G., G. Mario
1970 Zonas de distribución de los gastrópodos marinos del Perú. *Revista Anales Científicos*, Universidad Nacional Agraria "La Molina" 8(34):153–170. Lima.

Philander, S. George
1990 *El Niño, La Niña, and the Southern Oscillation*. Academic Press, New York.

Pickersgill, Barbara
1969 The archaeological record of chili peppers (*Capsicum* spp.) and the sequence of plant domestication in Peru. *American Antiquity* 34:54–61.

Pizarro, Pedro
1944[1571] *Relación del descubrimiento y conquista de los reynos del Perú*. Editorial Futuro, Buenos Aires.
1978[1571] *Relación del descubrimiento y conquista de los reinos del Perú*. Edición de Guillermo Lohmann Villena and Pierre Duviols. Pontificia Universidad Católica del Perú, Lima.

Polanyi, Karl
1968 *Primitive, Archaic, and Modern Economies: Essays of Karl Polanyi*. Edited by George Dalton. Doubleday, Garden City.
1977 *The Livelihood of Man*. Edited by Harry W. Pearson. Academic Press, New York.

Prado, Luis and Juan Carlos Castilla
2006 The bioengineer *Perumytilus purpuratus* (Mollusca: Bivalvia) in central Chile: biodiversity, habitat structural complexity and environmental heterogeneity. *Journal of the Marine Biological Association of the United Kingdom* 86(2):417–421.

Prieto B., O. Gabriel
2011 Chicha production during the Chimú period at San José de Moro, Jequetepeque Valley, north coast of Peru. In *From State to Empire in the Prehistoric Jequetepeque Valley, Peru*, edited by Colleen M. Zori and Ilana Johnson, pp. 105–128. British Archaeological Reports, International Series 2310. Archaeopress, Oxford, England.

Proulx, Donald A.
2006 *A Sourcebook of Nasca Ceramic Iconography: Reading a Culture through its Art*. University of Iowa Press, Iowa City.

Q

Qhapaq Ñan (see Alejandro Chu 2015; José Luis Díaz Carranza 2015; Giancarlo Marcone Flores and Rodrigo Areche Espinola 2015)

Quilter, Jeffrey, Bernardino Ojeda E., Deborah M. Pearsall, Daniel H. Sandweiss, John G. Jones, and Elizabeth S. Wing
1991 Subsistence economy of El Paraíso, an early Peruvian site. *Science* 251(4991):277–283.

Quinn, William H., Victor T. Neal, and Santiago Antúnez de Mayolo
1990 El Niño occurrences over the past four and a half centuries. *Journal of Geophysical Research* 92(C13):14449–14461.

R

Rabinowitz, Joel Bezalel
1980 Pescadora: The Argot of Chimu Fishermen. Master's thesis, Department of Anthropology, University of Texas, Austin.
1983 La lengua pescadora: the lost dialect of Chimu fishermen. In *Investigations of the Andean Past: Papers from the First Annual Northeast Conference on Andean Archaeology and Ethnohistory*, edited by Daniel H. Sandweiss, pp. 243–267. Cornell University, Ithaca, New York.

Ramos Giraldo, Jesús A.
2011 *Santuario de Pachacámac: cien años de arqueología en la costa central*. Editorial Cultura Andina, Lima.

Reinhard, Karl J. and John G. Jones
1987 Dietary and parasitological analysis of Turkey Pen Cave: a Basketmaker II village in the Grand Gulch, Utah. Manuscript in possession of John G. Jones.

Ricklefs, Robert E.
1979 *Ecology*. Second edition. Chiron Press, New York.

Rimachi, Luis Fernando, D. Andrade, Milusqui Verástegui, Jaime Mori, Victor Soto, and Rolando Estrada J.
2012 Variabilidad genética y distribución geográfica del maní, *Arachis hypogaea* L. en la región Ucayali, Perú. *Revista Peruana de Biología* 19:241–248.

Rojas de Mendiola, Blanca, Noemí Ochoa, Ruth Calienes, and Olga Gómez
1969 Contenido estomacal de anchoveta en cuatro áreas de la costa peruana. *Informe No. 27, Instituto del Mar del Perú*. Callao.

Rostworowski de Diez Canseco, María
1970 Mercaderes del valle de Chincha en la época prehispánica: un documento y unos comentarios. *Revista Española de Antropología Americana* 5:135–178. Madrid. (includes the ***Aviso***).
1972 Las etnías del valle del Chillón. *Revista del Museo Nacional* 38:250–314. Lima.
1975 Pescadores, artesanos y mercaderes costeños en el Perú prehispánico. *Revista del Museo Nacional* 41:311–349. Lima.
1977a *Etnía y sociedad: costa peruana prehispánica*. Instituto de Estudios Peruanos, Lima.
1977b Coastal fishermen, merchants, and artisans in prehispanic Peru. In *The Sea in the Pre-Columbian World: A Conference at Dumbarton Oaks, October 26th and 27th, 1974*, edited by Elizabeth P. Benson, pp. 167–188. Dumbarton Oaks, Washington, D.C.
1977c La estratificación social y el hatun curaca en el mundo andino. *Histórica* 1(2):249–286.
1978–80 Guarco y Lunaguaná: dos señoríos prehispánicos de la costa sur central del Perú. *Revista del Museo Nacional* 44:153–214. Lima.
1981 *Recursos naturales renovables y pesca. Siglos XVI y XVII*. Instituto de Estudios Peruanos, Lima.

1986 La región del Colesuyu. *Revista Chungara* 16–17:127–135.
1988 *Conflicts over Coca Fields in XVIth-Century Perú*. Memoir 21 of the University of Michigan Museum of Anthropology. Studies in Latin American Ethnohistory & Archaeology, Volume IV. Ann Arbor, Michigan.
1992 *Pachacamac y el señor de los milagros: una trayectoria milenaria*. Instituto de Estudios Peruanos, Lima.
1993 *Ensayos de historia andina: elites, etnías, recursos*. Instituto de Estudios Peruanos, Lima.
2003 Peregrinaciones y procesiones rituales en los Andes. *Journal de la Société des Américanistes* 89(2):97–123.

Rowe, John H.
1946 Inca culture at the time of the Spanish Conquest. In *Handbook of South American Indians*, edited by Julian H. Steward, Bureau of American Ethnology 143, Vol. 2, pp. 183–330. Smithsonian Institution, Washington, D.C.

Ruiz, Hipólito
1952 *Relación histórica del viage que hizo a los reynos del Perú y Chile*, edited by Jaime Jaramillo-Arango. Real Academia de Ciencias Exactas, Físicas y Naturales, Madrid, Spain. Two volumes.

S

Sánchez Romero, Jorge
1973 *Historia marítima del Perú*. Tomo 1, Volumen 2: *Aspectos Biológicos y Pesqueros del Mar Peruano*. Editorial Ausonia, Lima.

Sandweiss, Daniel H.
1992 *The Archaeology of Chincha Fishermen: Specialization and Status in Inka Peru*. Bulletin of the Carnegie Museum of Natural History, No. 29. Carnegie Museum of Natural History, Pittsburgh.

Sandweiss, Daniel H. and María del Carmen Rodríguez de Sandweiss
1991 Moluscos marinos en la prehistoria andina: breve ensayo. *Boletín de Lima* 75:55–63.

Sandweiss, Daniel H. and Elizabeth S. Wing
1997 Ritual rodents: the guinea pigs of Chincha, Peru. *Journal of Field Archaeology* 24:47–58.

Sauer, Jonathan and Lawrence Kaplan
1969 *Canavalia* beans in American prehistory. *American Antiquity* 34:417–424.

Schafleitner, Roland, Raymundo Gutiérrez, Ricardo Espino, Amelie Gaudin, José Pérez, Mariano Martínez, Alejandro Domínguez, Luz Tincopa, Carlos Alvarado, Giannina Numberto, and Merideth Bonierbale
2007 Field screening for variation of drought tolerance in *Solanum tuberosum* L. by agronomical, physiological and genetic analysis. *Potato Research* 50:71–85.

Schulenberg, Thomas S., Douglas F. Stotz, Daniel F. Lane, John P. O'Neill, and Theodore A. Parker III
2007 *Birds of Peru*. Princeton University Press, Princeton, New Jersey.

Serra, Gianluca, Guido Chelazzi, and Juan Carlos Castilla
1997 Effects of experience and risk of predation on the foraging behavior of the south-eastern Pacific Muricid *Concholepas concholepas* (Mollusca: Gastropoda). *Journal of Animal Ecology* 66:876–883.

Sherbondy, Jeanette Evelyn
1982 The Canal Systems of Hanan Cuzco. PhD dissertation, University of Illinois, Urbana-Champaign.

Shimada, Izumi
1982 Horizontal archipelago and coast-highland interaction in north Peru: archaeological models. In *El Hombre y su Ambiente en los Andes Centrales*, edited by Luis Millones and Hiroyasu Tomoeda, pp. 137–210. Senri Ethnological Studies no. 10. National Museum of Ethnology, Senri Expo Park, Osaka, Japan.

Silverman, Helaine
1993 *Cahuachi in the Ancient Nasca World*. University of Iowa Press, Iowa City.

Simpson, Michael G.
2010 *Plant Systematics* (2nd edition). Elsevier Inc., New York.

Siracusano, Gabriela
2011 *Pigments and Power in the Andes: From the Material to the Symbolic in Andean Cultural Practices 1500–1800*. Translated by Ian Barnett. Archetype Publications, London, England.

Smith, C. Earle, Jr.
2009 Preceramic plant remains from Guilá Naquitz. In *Guilá Naquitz: Archaic Foraging and Early Agriculture in Oaxaca, Mexico* (updated edition), edited by Kent V. Flannery, pp. 265–274. Left Coast Press, Walnut Creek, California.

Spooner, David M., Karen McLean, Gavin Ramsay, Robbie Waugh, and Glenn J. Bryan
2005 A single domestication for potato based on multilocus amplified fragment length polymorphism genotyping. *Proceedings of the National Academy of Sciences* 102:14694–14699.

Stephens, S. G.
1975 A reexamination of the cotton remains from Huaca Prieta, north coastal Peru. *American Antiquity* 40:406–419.

Stoltman, James B.
2008 Petrographic analyses of the Cerro Azul pottery types. In *Excavations at Cerro Azul, Peru: The Architecture and Pottery*, by Joyce Marcus, pp. 63–71. Monograph 62 of the Cotsen Institute of Archaeology, University of California at Los Angeles.

Sumich, James L.
1976 *An Introduction to the Biology of Marine Life*. Second edition. William C. Brown Publisher, Dubuque, Iowa.

Swartzman, Gordon, Arnaud Bertrand, Mariano Gutiérrez, Sophie Bertrand, and Luis Vásquez
2008 The relationship of anchovy and sardine to water masses in the Peruvian Humboldt Current system from 1983 to 2005. *Progress in Oceanography* 79:228–237.

T

Tam, Jorge, Marc H. Taylor, Verónica Blaskovic, Pepe Espinoza et al.
2008 Trophic modeling of the northern Humboldt Current

ecosystem, Part I: comparing trophic linkages under La Niña and El Niño conditions. *Progress in Oceanography* 79:352–365.

Tarazona, Juan, H. Salzwedel, and Wolf Arntz
1988   Positive effects of "El Niño" on macrozoobenthos inhabiting hypoxic areas of the Peruvian upwelling system. *Oecologia* 76:184–190.

Taylor, Marc H., Jorge Tam, Verónica Blaskovic, Pepe Espinoza, R. Michael Ballon, Claudia Wosnitza-Mendo, Juan Argüelles, Eric Díaz et al.
2008   Trophic modeling of the northern Humboldt Current ecosystem, Part II: elucidating ecosystem dynamics from 1995 to 2004 with a focus on the impact of ENSO. *Progress in Oceanography* 79:366–378.

Tohme, Joseph, Orlando Toro Ch., Jaime Vargas, and Daniel G. Debouck
1995   Variability in Andean *"nuña"* common beans (*Phaseolus vulgaris*, Fabaceae). *Economic Botany* 49(3):78–95.

Tosi, Joseph A., Jr.
1960   Zonas de vida natural en el Perú: memoria explicativa sobre el mapa ecológico del Perú. Instituto Interamericano de Ciencias Agrícolas de la OEA, Zona Andina, *Boletín Técnico 5*. 4 mapas. Lima.

Tovar Serpa, Humberto
1979   Informe del censo de lobos marinos efectuado en abril-mayo de 1979. *Informe* No. 64. Instituto del Mar del Perú. Callao.

Towle, Margaret A.
1961   *The Ethnobotany of Pre-Columbian Peru*. Viking Fund Publications in Anthropology 30. Aldine Publishing Company, Chicago.

Treacy, John M.
1994   *Las chacras de Coporaque: andenería y riego en el valle del Colca*. Instituto de Estudios Peruanos, Lima.

U

Ugent, Donald, Tom Dillehay, and Carlos Ramírez
1987   Potato remains from a Late Pleistocene settlement in southcentral Chile. *Economic Botany* 41:17–27.

Ugent, Donald, Shelia Pozorski, and Thomas Pozorski
1984   New evidence for ancient cultivation of *Canna edulis* in Peru. *Economic Botany* 38:417–432.

Uhle, Max (see Kroeber and Strong 1924)

Urton, Gary and Alejandro Chu
2015   Accounting in the king's storehouse: the Inkawasi khipu archive. *Latin American Antiquity* 26(4):512–529.

V

Valdivieso Milla, Violeta
1987   *Lista sistemática de moluscos marinos del Perú*. Boletín del Instituto del Mar del Perú, Callao.

Valdivieso Milla, Violeta and Hugo Alarcón
1983   Los moluscos en la pesquería peruana. *Documenta* 11(91):5–22.

van Zonneveld, Maarten, Xavier Scheldeman, Pilar Escribano, María A. Viruel, Patrick Van Damme, William García, Cesar Tapia, José Romero, Manuel Siguenas, and José I. Hormaza
2012   Mapping genetic diversity of cherimoya. *PLoS ONE* 7(1):e29845.

Vargas F., Mauricio, Sandra Cifuentes P., and Esteban Emparanza M.
1999   Espectro trófico de peces concurrentes al área de crianza Playa Chipana (21°19' S–70°04' W) del norte de Chile. *Revista de Biología Tropical* 47(3):597–600.

Vaz Ferreira, Raúl
1981   South American sea lion, *Otaria flavescens* (Shavo, 1800). In *Handbook of Marine Mammals, Vol. 1: The Walrus, Sea Lions, Fur Seals and Sea Otter*, edited by Sam H. Ridgway and Richard J. Harrison, pp. 39–65. Academic Press, New York.

Vegas Vélez, M.
1968   Revisión taxonómica y zoogeográfica de algunos gasterópodos y lamelibranquios marinos del Perú. *Anales Científicos*, Universidad Nacional Agraria "La Molina" 6(1/2):1–29. Lima.

Veldmeijer, André J. and Janine Bourriau
2009   The carrier nets from a burial at Qurna. *The Journal of Egyptian Archaeology* 95:209–222.

Vélez Diéguez, Juan
1980   Clave artificial para identificar los peces marinos comunes en la costa central del Perú. *Boletín de Lima* 9:1–16.

Viacava C., Moisés, Ricardo Aitkin S., and Jorge Llanos U.
1978   Estudio del camarón en el Perú, 1975–1976. *Boletín del Instituto del Mar del Perú* 3(5):161–232. Callao.

Villagómez, don Pedro de
1649   *Carta pastoral de exortación e instrucción contra las idolatrias de los indios del arçobispado de Lima*. Jorge López de Herrera, Impresor de Libros, en la calle de la cárcel de corte, Lima.

Villar Córdova, Pedro Eduardo
1935   *Las culturas pre-hispánicas del departamento de Lima*. First edition. Talleres gráficos de la escuela de la guardia civil y policía, Lima.

W

Walker, Sir Gilbert T.
1924   Correlation in seasonal variations of weather, IX: a further study of world weather. *Memoirs of the Indian Meteorological Department* 24:275–332.

Weberbauer, August
1945   *El mundo vegetal de los Andes peruanos*. Ministerio de Agricultura, Lima.

Weisburd, Stefi
1984   El Niño brought the blues, but bliss too. *Science News* 126:228.

Wendrich, Willemina
1989   Chapter 9: Preliminary report on the Amarna basketry and cordage. In *Amarna Reports V*, edited by Barry J. Kemp, pp. 169–201. Egypt Exploration Society, London, England.

Wheeler, Jane C.
1982 Aging llamas and alpacas by their teeth. *Llama World* 1:12–17.
1984a La domesticación de la alpaca (*Lama pacos*) y la llama (*Lama glama*) y el desarrollo temprano de la ganadería autóctona en los Andes centrales. *Boletín de Lima* 36:74–84.
1984b On the origin and early development of camelid pastoralism in the Andes. In *Animals and Archaeology*, Vol. 3: *Early Herders and their Flocks*, edited by Juliet Clutton-Brock and Caroline Grigson, pp. 395–410. *British Archaeological Reports, International Series* 202. Oxford, England.

Wheeler, Jane C., Eduardo Pires-Ferreira, and Peter Kaulicke
1976 Preceramic animal utilization in the central Peruvian Andes. *Science* 194:483–490.

Williams, Carlos and Manuel Merino
1974 *Inventario, catastro y delimitación del patrimonio arqueológico del valle de Cañete*. Tomos I y II. Instituto Nacional de Cultura, Centro de Investigación y Restauración de Bienes Monumentales. Lima.

Wing, Elizabeth S.
1972 Utilization of animal resources in the Peruvian Andes. In *Andes 4: Excavations at Kotosh, Peru, 1963 and 1966*, edited by Seiichi Izumi and Kazuo Terada, pp. 327–351. University of Tokyo Press, Tokyo.

Y

Yacovleff, E. and F. L. Herrera
1934 El mundo vegetal de los antiguos peruanos. *Revista del Museo Nacional* 3:241–322. Lima.

York, Robert and Gigi York
2011 *Slings and Slingstones: The Forgotten Weapons of Oceania and the Americas*. The Kent State University Press, Kent, Ohio.

Z

Zhang, Dapeng, Jim Cervantes, Zosimo Huamán, Edward Carey, and Marc Ghislain
2000 Assessing genetic diversity of sweet potato (*Ipomoea batatas* [L.] Lam.) cultivars from tropical America using AFLP. *Genetic Resources and Crop Evolution* 47:659–665.

Index

A

abalone, 37; see also false abalones
acacia, 7, 30, 262; see also huarango
acequia, 65, 342; see also irrigation, irrigation canals, Taymi Canal
Acequia de Chome, 5, 6, 9
Acequia de Cuyba, 5, 6
Acequia de Lloclla, 6
Acequia de Pachamama, 5, 6
Acequia de la Quebrada de Hualcará, 5, 6, 9, 65
Acequia de San Miguel, 6, 9
Acequia María Angola, 6, 9, 65
achiote (*Bixa orellana*), 201, 208, 209, 215, 263, 279, 282; achiote domestication, 208; achiote pod, 208
achira (Queensland arrowroot) (*Canna edulis*) 201, 202, 203, 215, 221, 231
agave family, 159, 168
agave fiber, 104, 158, 159, 162
agricultural communities, 214, 231, 348
agricultural implements, 341, 345
Alaska, 166
albatross (*Diomedea* sp.), 173, 182
algae, 11, 12, 14, 43, 58, 77, 95, 98, 99; blue-green, 11; brown, 11, 12, 36, 37, 41; red, 11, 41
algarrobo (*Prosopis*), 7, 266, 282, 350
algodón blanco (*Gossypium*), 221, 279, 282; see also cotton
algodón pardo, 221, 279, 283; see also cotton
almendrillo (huevo de coto), 263, 281
alpaca, 301, 303
alpaca fibers, 158
alpargatas (sandals), 226
Amaranthaceae, 339
Amazon Basin, 208, 210, 279
amphipods, 11, 14, 80, 85, 86, 89, 92, 353
amphora, 103, 104, 203, 204, 205, 224
Anacardiaceae, 338, 339
Ancash, 204, 209
anchoveta, 8, 15, 16, 18, 19, 24, 30, 32, 73, 77, 79, 80, 82, 85, 89, 94, 95, 110, 115, 116, 117, 121, 122, 124, 148, 153, 154, 157, 172, 174, 187, 189, 337, 344, 348, 349, 350, 352, 353; bones, 27, 32, 120; eggs, 82; heads, 146, 148, 150, 153, 156, 337; scales, 25, 27, 146; skin sticking to clay surfaces, 25, 27; anchoveta storage, 22, 25, 26, 116–119, 142, 146, 147, 148, 157, 348; see also dried fish, fish storage
anchoveta blanca (*Anchoa nasus*), 11, 73, 79, 95, 120, 121
anchoveta negra (*Engraulis ringens*), 11, 73, 77, 79, 95, 120, 121, 135, 136, 142, 143, 144, 146, 147, 149, 150, 151, 152, 153, 154, 155, 156
Ancón, 256, 257, 259, 261, 272
Ancón Bay, 19
Andean coot (*Fulica ardesiaca*), 177, 181, 182, 185
andesites, 6, 10
angelote, 76
Anisotremus scapularis, 135, 151, 337, 358; see grunt
Annona cherimolia, see cherimoya
Antilles, 342
Apodanthera, 201, 208, 209
Apodanthera biflora, 207
aquifer, 6
Arachis duranensis, 206
Arachis hypogaea, see peanut
Arachis ipaensis, 206
arched blue crab, 12, 55, 72, 157
Arctocephalus australis, 186, 190, 191; see fur seal
Argentina, 168, 204, 206, 207, 209, 264
Argentine pampas, 161

artisanal boats, 44, 89, 95, 97, 98, 120, 352
artisanal fishermen, 44, 55, 72, 73, 74, 76, 79, 86, 89, 95, 189, 352–358
Asia, the site, 44, 101, 159, 352; Valley of Asia, 10, 76, 77, 186, 350
Asteraceae, 338, 339
atarraya, 98, 99, 100, 103, 104, 110, 113, 115, 352; see also net, red de cortina, trammel net
atlatl (spearthrower), 158
atlatl point, 158
Aulacomya ater, 12, 40, 41, 45, 46, 48, 49, 50, 51; see also mussel
avifauna, 172–185
Aviso, 345
ayanque (*Cynoscion analis*), 73, 87, 88, 89, 121, 128–129, 131, 142, 144, 147, 149, 155
ayllu, 342
Aztec market, 350

B

bagre, 80; see sea catfish
Balanus, 37, 38, 50; see barnacles, chanques
balloon vine (*Cardiospermum* sp.), 263, 281
balsas, 101, 102, 104, 344,
band-tailed gull (*Larus belcheri*), 173, 179, 180, 181, 185
bandana, 333
barbudo, 80, 81; see Pacific threadfin, *Polynemus approximans*
barley (*Hordeum* sp.), 226, 228, 230, 231
barnacles, 12, 37, 38, 39, 41, 50, 95; see also *Balanus*
barquillo, see chiton, *Enoplochiton niger*
barrilete, 19
barrio (or ward), 102
basil (*Ocimum basilicum*), 226, 228, 229, 231
Basin of Mexico, 204
basket, 32, 79, 115, 120, 233, 262, 263, 264, 266
basketry, 212, 263, 265, 284
bast fibers, 158
bat (*Tadarida* sp.), 289
battles, mock, 159
beads, 281, 283, 284
beans, 3, 22, 30, 214, 220, 226, 254–261, 340
bean pods, 213, 215, 216; see *Canavalia, Phaseolus*
bedding, 324, 325
beer, 7, 232, 340, 341, 342, 349, 350; see also brewery, chicha, maize
Bering Straits, 158, 166
Bignoniaceae, 281
biodiversity, 43
birds, 3, 93, 116, 166, 172–185
Bixa orellana, 263; see achiote
blackruff (*Seriolella violacea*), 12, 73, 82, 84
black turban shell, 38, 54; see *Tegula atra*
blenny, 72, 73, 92, 132; see also trambollo, *Labrisomus philippii*
blenny, scaleless, 73, 92, 132, 135, 145, 147, 150; see also borracho or *Scartichthys gigas*
blue crab, see arched blue crab
blue-footed booby (*Sula nebouxi*), 177
bobbins, 282, 284
Boca del Río, 67, 71
bolas, 158, 166, 171, 185, 186; two-stone bolas, 166, 167, 168, 169; three-stone bolas, 166
bolas hunters, 166
Bolivia, 204, 206, 210
boll of cotton, 279, 281, 283
bonito (*Sarda sarda*), 12, 16, 19, 28, 94, 115, 121, 129, 132, 135, 138, 142, 145, 146, 147, 148, 149, 153, 155, 156, 157, 350, 353; see also Pacific bonito

bony fish, 76, 77, 122, 133, 142, 148, 149, 189, 352
booby, 12, 17, 19, 76, 172, 173, 174, 183, 350
Boraginaceae, 339
borracho, 11, 14, 92
bottle gourd (*Lagenaria siceraria*), 221, 223, 262, 263, 272, 275, 283
bottlenose dolphin, 189, 190, 192, 197
bottom feeders, 80
bracelet, 283
Brassicaceae, 338, 339
Brazil, 206
Brevoortia, 76, 120, 121; see also machete
brewery, 18, 23, 46, 61, 62, 214, 215, 220, 232, 246, 253, 289, 349
brewing beer, 233, 342, 348, 349; see also brewery, maize, maize varieties
brewing facilities, 217
bristle worms, 7, 11, 12, 14, 19
broad spectrum diet, 14
bromeliads, 158, 159, 168, 262, 264
bulrush (*Scirpus californicus*), 6, 30, 71, 80, 101, 177, 262, 263, 264, 266, 267, 281, 282, 283, 340, 345; bulrush marshes, 177, 263; see also Huaca Chola, totora
burial, 3, 20, 98, 102, 109, 153, 154, 185, 203, 224, 232, 233, 234, 251, 265, 282, 329, 331, 334; with gourd bowls, 275, 276, 277, 337; burials with nets, 102
burial cists, stone-lined and stone masonry, 28, 30, 49, 50, 51, 64, 65, 103, 104, 108, 114, 150, 152, 153, 154, 156, 159, 168, 184, 185, 197, 203, 204, 206, 209, 211, 221, 223, 224, 225, 226, 251, 252, 253, 259, 260, 265, 272, 273, 283, 300, 318, 319, 323, 329, 334, 347
burned bone, 26, 288, 298
burned camelid bone, 26, 288
burr grass, 263, 266, 281, 339
burro (*Sciaena fasciata*), 73, 87, 88, 89, 121, 128–129, 131, 149, 150
burrowing owl, 9, 179
burrows, 179
butchering, sea lion, 189, 195, 196, 197; of llamas, 290; see also earth oven or pachamanca
butchering marks on bones, 189, 196, 288, 303
butternut squash, 207, 208, 213, 214, 215, 220, 221, 223, 224, 226, 228, 231, 272; see *Cucurbita moschata*

C

caballa (*Scomber japonicus peruanus*), 14, 73, 93 ; see also mackerel, Pacific mackerel
caballitos de totora, 95, 97, 101, 102, 115, 120, 142, 166, 171, 172, 190, 264, 340, 352
cabinza (*Isacia conceptionis*), 14, 86, 120
cabletera, 100, 103; see also atarraya
cabrilla (Peruvian rock bass), 14, 82, 83
cache, 104
cacique, 342, 344
cacique principal, 342
cactus, see agave family
Caesalpinia, 338, 339
Cahuachi, 43
Calidris, 173, 179
Callao, 17, 45, 79, 81
Callejón de Huaylas, 204
Callianassa islagrande, 58, 59
Callinectes arcuatus, 55
Callinectes sapidus, 55
Calyptraea trochiformis 37, 49, 50, 51, 53; see also sea snail
Camacho Black jar, 298
Camacho Reddish Brown jar, 205, 215, 224

Camaná, 345
camelid, 158, 220, 287–317, 337, 350
camelid bladder, 314
camelid wool, 158–159, 166, 168, 341
canals, irrigation, 5, 6, 9, 45, 67, 68, 342; see also acequia
canasta, 67, 71; see crayfish trap
Canavalia, 9, 30, 201, 204, 206, 213, 214, 215, 216, 217, 218, 220, 221, 222, 223, 226, 231, 254
Canavalia ensiformis, 204
Canavalia plagiosperma, 204
Canchari, 3, 4, 5, 341, 348
canchones, open-air work areas, see Figures 2.2, 2.4, 2.5, 2.6
cane, 68, 69, 152, 265, 269, 283, 340; see also caña brava
cane impressions, 49, 265, 270–271
cane walls, 30, 270–271; see kincha
cangrejo, 54, 55; cangrejo moro, 54; see also crab
Canna edulis, see achira
Canta, 343
Cantagalllo, 5, 9
caña brava (*Gynerium sagittatum*), 7, 71, 263, 265, 266, 272, 273, 281, 282, 283
caña de Guayaquil (*Guadua angustifolia*), 263, 266, 283
caña hueca (*Phragmites australis*), 221, 263, 265, 268, 270, 281, 282, 283
cannibalism, 95
Capsicum baccatum, 201, 337; see also chile
Capsicum chinense, 210, 226, 229, 231; see also chile pepper
caracoles, 38
caravans, 22, 287, 294; see llama caravans
Carcharhinus, 121, 156; see also cartilaginous fish, requiem sharks
Carib, 342
carnivore, 37, 38, 95
carrier nets, 103, 104, 109, 110, 111, 114
carrizo (*Phragmites australis*), 6, 71, 221, 263, 265, 268, 269, 282
cartilaginous fish, 74, 120, 121, 122, 129, 130, 133, 135, 136, 143, 144, 146, 147, 150, 152, 153, 154, 155, 156; see chimaera, rays, sharks
cartilaginous vertebrae, 74
cast nets, 86, 98, 99, 103, 104, 114, 115, 120; see also atarraya
catfish, 55, 65, 73, see also sea catfish
cattails (*Typha* cf. *angustifolia*), 6, 80, 262, 263, 264, 283, 339
caves, 334
Centinela, La; see Cerro Centinela
ceramic assemblage, 26, 224
Cereus macrostibas (columnar cactus), 7
Cerro Azul Bay, 3, 8, 10, 12, 14, 30, 54, 57, 58, 64, 65, 74, 80, 98, 101, 115, 120, 178, 179, 180
Cerro Camacho, 10, 20, 21, 28, 32, 36, 45, 48, 63, 121, 148, 150, 154, 189, 190, 195, 214, 221, 282, 300, 329
Cerro Centinela, 10, 11, 12, 13, 17, 19, 20, 21, 98, 99, 101, 120, 185, 186
Cerro del Fraile, 10, 11, 20, 21, 174, 178, 181, 185, 186, 197, 226, 254, 259, 261, 283, 333
Cerro del Oro, 4, 5, 6, 9, 159, 162
ceviche, 37, 48, 94, 195
Chan Chan, 101
Chancay River, 342
chanque, 37; see also *Concholepas concholepas*
ch'arki, 10, 30, 74, 119, 197, 287, 290, 300, 303, 312, 313, 314, 317, 350; dolphin ch'arki, 197
ch'arki de raya, 74
ch'arki parcels, 287, 301, 312
chaskis, 102
chauchu, see crayfish trap, 68, 70
chaupi yunga, 3, 10, 65, 68, 203, 204, 348, 350

Cheilodactylus variegatus, 92; see morwong
chelate hands, pincers, 65, 66, 71; see also crayfish
chelipeds, see crab, crayfish, ghost shrimp
chenopod-amaranth, 282
Chenopodiaceae, 339
cherimoya (*Annona cherimolia*), 201, 202, 213, 214, 220, 226, 228, 231
Chesapeake Bay blue crab, 55
Chicama Valley, 342
chicha, 7, 9, 18, 22, 32, 104, 202, 214, 215, 217, 226, 232, 242, 246, 252, 290, 339, 341, 342, 345; chicha morada, 224, 232; see also beer, brewery, maize varieties
chicha production, 242, 251; see also brewery, maize
chicha storage, 27, 233, 290
chifa (Chinese restaurant), 95
Chilca, 17
Chile, 14, 16, 17, 19, 38, 45, 55, 77, 82, 89, 91, 92, 93, 94, 95, 172, 353
Chilean coast, 12
Chilean model, 12
chile, 212, 226
chile pepper (*uchu*) (*Capsicum baccatum*), 9, 210, 212, 213, 228, 231, 336, 339, 340, 349
chile pepper seeds, 220, 337
Chilean jack mackerel, 73, 79, 82, 83, 85, 95, 130, 142, 143, 146, 147, 148, 149, 150, 152, 153, 155, 156, 353, 355
Chilean mussel, 41, 46, 350; see *Choromytilus chorus*
Chillón Valley, 343, 345
chimaera, 74, 120, 121, 352
chimaera, plow-nosed, 74, 76; Fig. 6.5
Chimbote, 95, 121
Chimu, 101
Chincha, 17, 79, 82, 102, 173, 344, 350
Chincha Island, 17
Chincha Sur, 186
Chincha Valley, 79, 102, 114, 340, 341, 344
chinchi uchu, 210
chinchorro, 100, 104, 106, 115; see also net, trammel net
chita (*Anisotremus scapularis*), 14, 86, 98, 120, 121, 122, 125, 131, 135, 142, 143, 149, 150, 152, 153; see also grunt
chitin, 337; see insect chitin
chiton, 12, 35, 36, 95
chiwa (net bag), 60, 67, 70, 71
chocchikuna, 314, also see weaving
chonta palm (*Gulielma ciliata*), 114, 263
chonta wood, see mallero
Choromytilus chorus, 41, 43, 46, 49
Chorrillos, 89
chuitas, 174, 175
Chukchi, 166, 171
Chumbe, 6
chuño (chuñu), 119, see also potato
Chuquicoto, 343
Churruyoc, 344
Cilus gilberti, 11, 14, 58, 87, 89, see also corvina
circular cast net, see atarraya
ciruela del fraile (*Bunchosia armeniaca*), 9, 201, 204, 205, 213, 214, 215, 221, 226
cists, burial, 28, 30, 49, 50, 51, 63, 110, 154, 197, 221, 232, 235, 348
clams, 8, 11, 30, 35, 41, 43, 44, 45, 47, 80, 154; coquina, 35, 43, 49, 153, 184, 224, 329; wedge, 35; white, 35; see also *Donax obesulus*, *Mesodesma donacium*, *Mulinia edulis*
claw, see purple stone crab
cliffs, 37, 41, 94, 98, 101, 173, 179, 352; see *peña*
clingfish, 72, 94, 95, 97, 132, 139, 145, 149, 157; see pejesapo

cloth bundle, 203, 221
clubs, 171, 186
clupeids, 120, 121, 130
coastal upwelling, 10
cobs, see maize
coca (*Erythroxylum coca*), 201, 203, 205, 231, 233, 343, 345, 350
coca leaves, 203, 204, 205, 224, 333, 343
coca tea, 204, 224
cochayuyu, 11
coco (*Paralonchurus peruanus*), 73, 87, 88, 89, 121, 127, 131, 144, 149, 150, 153, 155, 156, 157
cojinova, see also blackruff (*Seriolella violacea*), 12, 82, 84, 85
collcas, 22, 46, 61, 65, 67, 68, 104, 109, 136, 168, 191, 208, 209, 215, 216, 231, 242, 246, 282; see storage
Colli, 343
Colombia, 204
Columbidae, 179; see also doves, pigeons
combs, 265
comienzo, 103, 104
commercial fishing, 17
common beans, 9, 201, 204, 214, 254–261
common gallinule (*Gallinula chloropus*), 173, 177, 181, 182, 184
common mullet, 81
commoner households, 30, 52, 220
commoner-class fishermen, 348
commoner-class midden, 26, 47, 52, 116, 214, 220
commoner-class overseer, 26, 27, 36, 46, 47, 52, 63, 117, 135, 143, 145, 146, 147, 148, 152, 157, 183, 195, 214, 220, 246, 317, 328, 341, 348, 349, 350
commoner-class squatters, 282
commoner-class supervisor, 117, 282, 288
commoner-class support staff, 22
comparative collection, 72
Concavilca, island of, 344
Concholepas concholepas, 35, 37, 45, 46, 47, 48, 49, 50, 51, 53, 95, 187; see also chanque, 38, 39; false abalones
condor, 344
contexts, 20, 23, 30; primary, 20, 23, 25, 27, 28, 30, 32; secondary, 20, 22, 23, 25, 26, 27, 28, 32; tertiary, 20, 23, 25, 26, 30, 32
Confite Puntiagudo, 224, 226, 228, 232, 234, 242, 246, 253; see also maize, maize varieties
conglomerates, 6
cooking sea lions, 195, 197
cooking vessels, 28, 195
coots, 173, 181
copepods, 11, 14, 15, 16, 76, 77, 79, 80, 82, 85, 86, 91, 92, 93, 95, 353
copper, 102, 344, 350
coprolites, 3, 30, 63, 64, 65, 71, 334–339; human coprolites, 64, 71; dog coprolites, 64, 335, 336, 337
coquina clams, 12, 30, 35, 43, 44, 45, 46, 47, 49, 50, 51, 52, 55, 76, 152, 153, 283, 329, 350; see *Donax obesulus*
coral reefs, 86
cordage, 32, 262, 264, 265
Cordia, 263, 281; see also almendrillo (huevo de coto)
cormorants, 12, 17, 19, 76, 171, 172, 174, 177, 182, 185, 350; red-legged, 183; see also guanay cormorant
corn, 342; see beer, brewery, chicha, maize, maize varieties
corpses, 334; see burial, burial cists
corvina (*Cilus gilberti*), 11, 14, 28, 58, 73, 86, 89, 90, 92, 101, 117, 119, 121, 129, 135, 136, 145, 148, 149, 157
Coryphaena hippurus, 73, 85; see also dolphinfish
costa (cobble beach), 6, 7, 10, 12, 50, 52, 72, 89, 97, 98, 101, 116, 352
cotton, 9, 104, 158, 159, 203, 221, 224, 262, 263, 279–280, 283, 340, 343, 345, 349; brown, 203 221, 281, 282, 283; white, 203, 221, 281, 282, 283
cotton bolls, 279, 282
cotton boom, 7, 67, 80, 177
cotton fiber, 103, 105, 283, 345
cotton seeds, 279, 280, 281, 282, 283
cotton yarns, 115, 283
cotton-producing haciendas, 45
coya (queen), 341
crab, 8, 12, 19, 22, 58, 61, 74, 80, 85, 86, 89, 95, 340, 349, 353; purple stone crabs, 12, 54, 61; variety of crabs, 55
crab claw, removed before cooking, 54, 61
crayfish, 6, 19, 54 58, 60, 61, 63, 64, 65, 66, 67, 70, 71, 89, 119, 336, 337, 340, 349, 350; see also *Cryphiops caementarius*
crayfish trap, 65, 67, 68, 69, 71, 262, 281; catching crayfish by hand, 71
crickets, 179
croakers, 86, 89; see coco
crookneck squash, 207
croplands, see irrigation, irrigation canals
crusher, 54, 56; see also purple stone crab
crusher plate, 74
crustaceans, 3, 7, 11, 12, 16, 54–64, 74, 77, 79, 80, 85, 87, 89, 95, 189, 353
Cryphiops caementarius, 58, 65, 66, 337
cuculí (*Zenaida meloda*), 9, 173, 179
cucurbit, 201, 207, 208, 213, 214
Cucurbita moschata, 201, 207, 208
cueros de lobomarino, 101; see sea lion skins, watercraft
Cuniraya, 344, 345
cupule, see corn, maize
curaca, 342, 343, 344, 345, 348, 349
curacazgos, 342, 343, 344; see also Huarco, Lunahuaná, parcialidades, petty kingdoms, señoríos
curandero, 341, 345; see divination, guinea pig, healing
curtain nets or red de cortina, see nets
custard apple, 202; see cherimoya
cut marks, on bone, 129, 179, 182, 183, 184, 185, 189, 288, 289, 296, 298, 299, 303, 309, 310, 311, 312, 313
cutter, 54; see purple stone crab, 63
cuy, see guinea pig
cuyero (guinea pig nursery), 25
cycles of sardine and anchoveta dominance, 77, 79, 148
cynipid wasp, 210, 214
Cynoscion analis, see ayanque
Cyperus spp. (sedges), 263, 264, 268, 281

D

decorated gourds, see pyroengraved gourd bowls
deer bone, not represented at Cerro Azul, 158
deity, 344; see Urpay Huachac
demersal fish, 95
dent corn, 232, 234, 253,
detritus feeders, 14, 54, 55
detritus food chain, 10, 12, 14
diatoms, 11, 14, 16, 36, 76, 77, 79, 80
diet, nobles vs commoner, 51, 79, 80, 97, 148–150, 157, 204, 287, 303, 314, 317, 341
dinoflagellates, 11, 15, 16, 77, 79
Diomedea, see albatross
Distichlis spicata, 7, 102
dive-bombing boobies, 172, 177, 185, 350
divination, 324, 326, 341; see also guinea pig
dogs (*Canis familiaris*), 318–323

dogs in burials, 318
dog coprolites, see coprolites
dogfish shark, 8, 73, 74, 101, 155, 157
dolphins, 93, 186, 189, 190, 192, 194, 197, 340, 352
dolphinfish, 19, 72, 73, 85, 157
domestic chicken (*Gallus gallus domesticus*), 173, 181
Donax obesulus, 12, 43, 44, 45, 46, 47, 48, 49, 50, 51, 52, 53, 55, 152, 184
dorado, 19, 85; see also dolphinfish (*Coryphaena hippurus*)
dove, 182, 185, 344
doves, eared (*Zenaida auriculata*), 9, 173, 179, 344
doves, West Peruvian (*Zenaida meloda*), 9, 173, 179, 181, 182, 344
dried camelid meat, 10, 30, 119, 287, 303, 317; see *ch'arki*
dried fish, 10, 22, 23, 24, 26, 27, 32, 78, 79, 102, 117, 118, 119, 122, 124, 142, 143, 146, 147, 154, 157, 272, 287, 340, 344, 345, 348
dried mollusc meat, 45
dried smoothhound sharks, 345
drum (Sciaenidae), 12, 30, 55, 72, 73, 97, 101, 115, 116, 121, 143, 150, 348, 349, 350, 353; see ayanque, burro, coco, corvina, lorna, mismis, mojarrilla, róbalo, zorro
drum otoliths, 132, 135
drying fish, 24, 30, 62, 94, 116–119, 148, 340, 342
drying fish for export, 116, 148, 340, 344, 345, 349, 350
drying maize, 46
dry weight, of shellfish meat, 45, 51
ducks, 6, 166, 173, 177, 182, 184, 185
dung, guinea pig, 282, 326
dung pellets, 226, 289; guinea pig, 325, 326; owl, 179; rodent, 282; llama, 22, 226, 281, 287, 288, 289, 299, 345
dyed alpaca wool, 350

E

eagle rays (*Myliobatis peruvianus*), 44, 74, 75, 76, 157
eared dove (*Zenaida auriculata*), 9, 181, 344
earth oven, 28, 30, 51, 156, 189, 195, 202, 226, 318, 319
ecological communities, 43, 350
ecological niche construction, 43
economic specialization, 102, 116, 340–351
Ecuador, 19, 77, 85, 89, 93, 204, 209
edible gum, 266
edible plants, 3, 201, 207, 212, 213, 231, 262, 300; see also beans, cherimoya, maize, potato, squash
edible rhizomes, 202, 262, 263, 339
eggs, 11, 14, 80, 82
Egypt, 104, see carrier nets
El Fraile, 36, 37, 45, 82, 98; see also Cerro del Fraile
elite families, 22, 23, 206, 207, 284, 341, 345
elite residences, 22, 27, 32, 50, 61, 116, 148, 154, 157, 182, 183, 185, 197, 214, 215, 217, 220, 226, 232, 233, 251, 253, 265, 282, 290, 298, 327, 328, 333, 341, 345
elite vs. commoners, 26–27, 44, 46, 49, 50, 52, 148–150, 337; see also diet
El Niño, 16, 17, 19, 43, 44, 45, 55, 72, 76, 77, 79, 80, 82, 84, 85, 89, 93, 94, 95, 157, 172, 173, 175, 186, 187, 189, 192, 352, 353, 355, 356; see also ENSO
El Paraíso, 337
embalming, 339
embroidery, 23
Emerita analoga, 54, 58, 179; see also mole crabs, *muy muy*
energy-transfer levels, 10
Engraulis eggs, 77
Engraulis ringens, 11, 77, 78, 79, 120, 121, 130, 143, 146, 147, 148, 154; see anchoveta

Enoplochiton niger, 35, 95
ENSO (El Niño-Southern Oscillation), 14, 16, 19, 44, 55, 72, 79, 80, 81, 82, 85, 157, 186, 187, 352
Erythrina, 201, 204, 206, 213, 224, 254, 259, 261
Erythrina edulis, 261
Eskimos, 166
espinel, 7, 89, 98, 100, 101, 115, 352
estuaries, 79, 80, 89
euphausiids, 11, 15, 77, 79, 82, 93, 95
Euphorbia, 201, 202, 203, 221
exostosis, 122, 127, 133, 136, 142, 152
export of fish, 116–119, 148, 154, 157, 340, 350; see also dried fish, drying of fish, fish storage

F

false abalones, 12, 28, 35, 43, 45, 46, 47, 48, 51, 52, 154, 195, 197
farmers, 102, 340, 341, 344, 348, 349
feast, 189, 197, 294, 317
feasting, 48
feathers, 89, 181, 185
feces, 3, 334–339; see also coprolites, dung, pellets
fertilizer, 17, 19, 340; see guano
Festuca, 10
fiber, textile, 7, 279; see also camelid wool, cotton
figurines, 208, 281, 283
filter feeders, 11, 14, 15, 41, 55, 76, 80
fingernail clippings, 22–23
fish, 3, 11, 22, 72–97, 329; see ayanque, bonito, caballa
fish bait, 58
fish bone, 10, 30, 337; see anchoveta, sardine
fish eggs, 14, 16, 79
fish heads, 121, 148, 150, 153, 156, 345
fishhooks, 89, 98, 100, 101, 115, 345; bone fishhook, 101; shell fishhook, 101
fish larvae, 11, 80, 82, 86, 93, 95; crustacean larvae, 14, 16, 54, 65, 82
fish oil, 154, 185, 226
fish scales, 142, 337
fish skin, 142
fish storage, 23, 24, 26, 27, 28, 47, 48, 61, 63, 116, 117, 142, 143, 146, 147, 148, 150, 157, 182, 183, 184, 191, 220, 246, 290, 296, 299, 327, 327, 348, 349
fish-storage facility, 36, 47, 63, 195, 221, 246, 328
fisherman's kit, 104
fishermen, 72, 98–102, 187, 342, 344, 345; see *peña*, *playa*, dried fish, fish storage, nets, watercraft
fishermen's dialect, 340, 342; see also Pescadora
fishing nets, 98–104, 341, 353; see also net
Fissurella crassa, 35, 36, 45, 46, 47, 48, 51, 95; competition with *F. limbata*, 35–36; see also limpet, 12, 45
Fissurella limbata, 35, 36, 38, 46, 47, 48, 51; competition with *F. crassa*, 35–37; see also limpet
flax fiber (Egypt), 104
flounder, left-eye (*Paralichthys adspersus*), 12, 14, 72, 73, 95, 97, 101, 121, 129, 132, 135, 139, 145, 147, 149, 155, 157
flying fish, 85
fodder, 10, 25, 47, 282, 287
fog, salt, 25, 150, 154, 221, 335
food chain, marine, 10, 11
food stress, 187, 189; see also El Niño, ENSO
forager, central-place, 37
foraging habits, of crab, 55; of birds, 172
foraminifera, 11, 14, 77, 80, 82
Fortress of Ungará, 3, 65, 179; see also Ungará

freshwater crayfish, see crayfish
fruits, 22; tropical, 30, 349
Fucoid, 36
fur seals, 186, 187, 190, 191, 196, 197; see also *Arctocephalus*
Furcraea, 158

G

Galápagos Islands, 173
Galeichthys peruvianus, 73, 80, 121; see sea catfish
gall, wasp, 214
garuma, 44, 58, 179, 180; see also gull, grey gull, *Larus modestus*
gastropod, 36, 337; see limpet
geese, 166
genetic abnormality, see polydactyly
geology of the Cañete Valley, 6
ghost shrimp (*Callianassa*), 8, 55, 58, 59, 60, 74, 85, 89
gill net, 74, 92; see net
glume, see maize
gold, 341, 350
goldsmiths, 342
Gossypium barbadense, 263, 272, 279, 280; see also cotton
gourd, 9, 102, 217, 349
gourd, decorated, 345
gourd bowls, 30, 49, 154, 221, 272, 275, 276, 277, 283, 329, 331
gourd floats, 104, 106
gourd lid, 272, 275, 283
gourd vessel, 154, 329, 330, 332
grape-eye sea bass (*Hemilutjanus macrophthalmos*), 73, 82
Grapsus grapsus, see Sally Lightfoot crab, 55
grass, 158, 263, 282, 283, 324, 337; quids of grass, 266, 281, 287
grass fiber, 159, 168, 337
grasshoppers, 179
grass stems, 168, 283
grazing food chain, 10, 12, 14, 58
Greater Southwest U.S., 254
Greece, 158
gremio de pescadores, 102
grey gull (*Larus modestus*), 12, 44, 58, 64, 173, 179, 185, 350; see garuma
grunt, 11, 14, 30, 72, 73, 86, 98, 99, 103, 115, 116, 120, 121, 126, 135, 136, 142, 144, 147, 157, 337, 348, 349, 350, 353; as yuyu-eaters, 11; see also chita
guanay cormorant (*Phalacrocorax bougainvillii*), 12, 17, 95, 173, 174, 175, 176, 181, 182, 183, 184, 185; see also cormorant
guanaco, 166, 186, 301
guanaco hunting, 166
guano (*wanu*), 17, 18, 19, 172, 340, 350
guano deposits, 17, 18
guano island, 17, 172, 174, 179
Guaravni, 343
guava (*Psidium guajava*), 9, 201, 208, 213, 214, 231, 262, 263, 281; guava branches, 281
Guayas Basin (Ecuador), 272
guinea pig (*Cavia porcellus*); 22, 25, 30, 32, 47, 62, 143, 154, 183, 272, 282, 290, 294, 324–333, 337, 339, 341; bedding, 47, 282; sacrifice of, 23, 24, 32, 290, 294, 324, 327, 333; see also fodder, 287
guinea pig nursery (cuyero), 22, 25, 298, 324
guitarfish (*Rhinobatos planiceps*), 76, 119
Gulf of Guayaquil, 85
gull, 36, 43, 44, 172, 179, 181, 184, 185; see also grey gull
Gynerium sagittatum, 7, 265, 269; see also caña brava

H

habanero pepper, 210, 226, 228
hake, 12, 72, 73, 95, 120, 157, 189
hammerhead shark, 74
hanan, 343; see also *lurín*
hatun curaca, 342, 343, 348, 349, 350
hatun maccma, 233
Hawaii, 85
headbands, 159
healing, 326, 341; see curanderos; see also *cuy* or guinea pig
hearth, 28, 32, 48, 166, 195, 300, 318
Heliotropium, 338, 339
Herbay, 4
herbivore, 14, 36, 37, 92, 95
herbs, 229, 282, 324, 339
herders, 158; see also camelids, llamas
hermaphroditic *Calyptraea*, 38
herring (Clupeidae), 73, 93
heterotrophic, 11, 12
heterotrophy, 12
hides, 166, 189, 345; see sea lion
Hippoidea, 58
hooks, 92, see fishhooks
huaca (*wak'a*), 344, 350, 351
Huaca Chivato, 5, 6
Huaca Chola, 6, 7, 80, 102, 177
Huaca Prieta, 30, 154, 272
Hualcará, 9
Huancavelica, 209
Huanchaco (Moche Valley), 101
Huaura Valley, 345
huarango (*Acacia* spp.), 7, 263, 266, 282, 283, 350
huarango posts, 282, 283
Huarco, 3, 7, 14, 17, 283, 287, 341, 342, 343, 349; see also Kingdom of Huarco
Huarmey Valley, 202
Huayquihusa, 343
human coprolites, 3; see coprolites
Humboldt Current, 10, 14, 15, 16, 19, 77, 79, 80, 82, 173, 179
Humboldt penguins (*Spheniscus humboldti*), 17, 172, 173, 174, 181
hunting birds, 158, 166, 168, 171; see also bolas, cormorants
hunting mammals, 11, 158, 166, 186, 187
husks, 213, 221, 224, 226, 228, 232, 233, 235, 237, 251; see maize
hygroscopic qualities of sand, 117

I

Ica, 102, 159, 166, 209
igneous rock, 92
Ihuanco, 4, 7, 9
Ilo, 121
inbreeding of llamas, 300, 301, 317; see polydactyly, three-toed llama, four-toed llama
imbricated kernels, 213, 232, 234, 242, 253
Inca, 11, 94, 121, 166, 185, 197, 226, 259, 283, 284, 333, 343, 348, 350
Inca conquest, 3, 22, 23, 197, 214, 226, 341
Inca tern (*Larosterna inca*), 173, 179, 180, 181, 182, 185; *zarcillo*, 179, 180
Indian Ocean, 16
industrial level, processing of sardines and anchovetas at the, 157, 348; see also dried fish for export
industrial plants, 7, 9, 80, 212, 221, 262–284; see also cotton
Inkawasi, 287, 348; location of, 4

insect chitin, 336
interments, see burial, burial cists
intertidal zone, 11, 12, 14, 36, 38, 43, 44, 95
Inuit, 166
irrigation, 7, 9, 179, 203, 204, 206, 208, 209, 210, 233, 342, 350
irrigation canals, 9, 10, 58, 65, 233, 340, 342, 343, 348; see also acequia
Isacia conceptionis, see cabinza
Isla Corriente, 86
island, rocky, 17, 45, 72, 74, 97, 166, 186, 187
Isoetes, 338, 339

J

jack mackerel, 12, 16, 143, 157; see also Chilean jack mackerel
jack-pompano, 72, 82
jaiba, 55
Jalisco, 204
Jayanca, 345
jellyfish, 11
jora (sprouted maize kernels), 215, 232; see also brewery, chicha, maize beer
junco (sedge), 263, 265, 268, 283
Juncus (rushes), 263, 264
jurel or Chilean jack mackerel (*Trachurus symmetricus murphyi*), 79, 82, 83, 85, 121, 355

K

kayaks, 166
kelp, 86, 353
kelp beds, 11, 14, 92, 93, 353
kelp gull (*Larus dominicanus*), 44
kellu uchu, 210
kernels, 22, 25, 32, 221, 224, 242, 246, 251, 252, 253; see also beer, chicha, maize
keyhole limpets, 35, 36; as central-place foragers, 37; see also limpet
kidney cotton, 263, 279, 280, 283, 284; see also cotton
kincha, see also carrizo or *Phragmites australis*
kincha house, 27, 28, 30, 49, 117, 143, 145, 147, 148, 150, 152, 195, 220, 221, 223, 265, 266, 270, 282, 283, 299, 300, 328; see also squatters' house, wattle-and-daub
Kingdom of Huarco, 3, 4, 5, 6, 9, 10, 17, 18, 19, 54, 65, 102, 121, 185, 197, 204, 266, 284, 301, 333, 340–351
Kingdom of Lunahuaná, 3, 4, 7, 10, 17, 233, 266, 284, 287, 317, 348, 350
kitchen, 23, 61, 62, 215, 289; see also brewery, chicha
knife, tumi, 104
knots, 103; see also mallero, net
krill, 11, 15, 16, 77

L

Labrisomus philippii, see also scaled blenny, trambollo
La Centinela, 12, 20, 21, 36, 45, 82, 86, 98; see also Cerro Centinela
Lagenaria siceraria, 223; see gourd bowl
La Niña, 16, 17, 19, 72, 352, 353, see also ENSO
Lagenorhynchus, 193, 194
Lagenorhynchus obliquidens, see dolphin
lagoon, 80, 344; see lisa or mullet, 80, 81
Lake Ticllacocha, 6
Lake Titicaca, 303
Lambayeque, 207, 342, 345
lapa, see keyhole limpets
Larus modestus, 44, 180

larvae, 80, 82, 89, 91
lavas, 6
leatherworkers, 342
left-eye flounder, see flounder, lenguado
legume, 212, 220, 226
lenguado or left-eye flounder (*Paralichthys adspersus*), 95, 96, 142, 151
lesser yellowlegs (*Tringa flavipes*), 173, 177, 184
Licapa, 342
lightfish (bristlemouth), 82; see also *Vinciguerria*, 82
lighthouse, 186, 187
lima beans, 9, 201, 214, 254, 257, 258, 261
limpet, 12, 35–38, 45, 47, 48, 52, 95
lisa, 14, 98, 114; see mullet
litter, 265, 283, 284, 343
litter pole, 153, 265, 273
"Little Ice Age," 16, 148, 157
liwi, 166; see also bolas
llama, 298, 300, 301; four-toed llama, 300, 301, 302; three-toed, 301, 302
llama bones, 22, 298, 301; see also camelid, llama
llama caravans, 22, 117, 143, 226, 266, 287, 289, 326
llama dung pellets, 3, 10, 22, 30, 143, 288, 289; see also dung, pellets
llama herding, 10, 287
llama hide, 300
llamkana pata, 22, 45, 327, 341
llama pack trains, 10, 117, 287; see also llama caravans
lobomarino, cueros de, 166; see also balsas, caballitos de totora, sea lions, watercraft
Lo Demás (site in the Chincha Valley), 114, 345
lomas, 7, 10
loom parts, 23, 242
lorna (*Sciaena deliciosa*), 11, 14, 19, 87, 88, 89, 121, 122, 128–129, 131, 136, 142, 143, 144, 146, 149, 150, 151, 152, 153, 155, 157, 353, 357
Los Balsares de Huanchaco, 101
Los Gavilanes, 202
Lovelock Cave, Nevada, 158; see also slings
lúcuma (*Pouteria lucuma*), 9, 154, 201, 208, 210, 213, 214, 215, 217, 219, 220, 221, 223, 224, 226, 228, 231
lugs on chicha-storage vessel, 246
Lunaguaná, see Lunahuaná, Kingdom of Lunahuaná
Lunahuaná, 3, 6, 7, 17, 68, 214, 231, 317, 343, 348
Lurín Valley, 102, 344
Lurinchincha, 102, 345
Lutra felina, 187; see sea otters, 65, 92

M

macha, 43, 44
machete, 76, 120, 121, 130, 149
mackerel, 12, 30, 72, 79, 93, 97, 116, 157, 189, 349, 350
macroalgae, 38
mahi mahi, 85
maize (*Zea mays*), 9, 22, 201, 202, 213, 215, 232 253, 336, 337, 340, 343, 349; maize cobs, 26, 30, 136, 214, 215, 217, 220, 221, 224, 226, 228, 231, 232, 240, 241, 242, 243–250, 289, 298; maize kernels, 22, 25, 215, 221, 224, 226, 236, 324, 336; maize leaves, 25, 203, 214, 221, 226, 232, 251; maize tassels, 224, 226, 232, 233, 237, 250, 251; maize stalks, 25, 214, 226, 232, 233, 237, 251, 324; see also beer, brewery
maize beer (chicha), 9, 214
maize cob, 22, 214, 217, 221, 236, 238, 239, 240, 241–248; whole ears of maize, 30, 221, 232, 234, 235, 249
maize-cob lugs, 246

maize consumption, see diet
maize fields, 18, 340, 341; see canals, irrigation
maize varieties, 215, 226, 233, 242, 251, 253
Mala Valley, 17, 344, 350
mallero, 114–115, 263, 345
Mamacona, 102
maní, see peanut
Manihot esculenta, see manioc
manioc (*Manihot esculenta*), 231
mantis shrimp, 95
Mapuche, 166
Maranga, 345
María Angola, 6, 9; see also acequia
marine habitats, 10, 72
marine worms, 80, 89
marisqueros, 52, 53; use of marisqueros, 53
marlin, 19
maruchas, 12, 58, 59; see also ghost shrimp
mate (gourd), 272
matting, 262, 263, 264, 265, 283, 345
medicinal plants, 201, 202, 203, 204, 212, 213, 224, 231, 262; see also medicine bundle
medicine bundle, 203, 221, 341
menhaden, 76; see also machete
Menticirrhus ophicephalus, 121, 353; see mismis
Menticirrhus rostratus, see also zorro
Merluccius gayi peruanus, 73, 95, 120
merluza or Peruvian hake (*Merluccius gayi peruanus*), 73, 95, 96
mesh bag, 104; see also carrier net
mesh size, 110–114; see also mallero
Mesodesma donacium, 43, 44, 45, 46, 47, 48, 49, 50, 51, 53, 54
Mesopotamia, 158
Mexico, 208, 209
mismis (*Menticirrhus ophicephalus*), 14, 19, 73, 87, 89, 91, 104, 121, 131, 143, 144, 147, 149, 150, 157, 353, 357
mitmaqkuna, 343
MNI, minimum number of individuals, 121
Moche, 17, 101
Moche Valley, 342, 344
Mochica pottery, 101, 257, 261
moieties, 342, 348
mojarrilla (*Stellifer minor*), 73, 87, 91, 95, 104, 121, 129, 135, 145, 146, 149, 150, 155, 157
mole crabs (*Emerita analoga*), 8, 12, 14, 44, 58, 62, 64, 82, 87, 89, 179, 180, 353
molluscs, 11, 12, 14, 35–53, 54, 74, 76, 80, 85, 87, 89, 92, 99, 224, 283, 337, 353
morwong or pintadilla (*Cheilodactylus variegatus*), 14, 73, 91, 92, 103, 121, 149, 157, 353, 358
Mugil cephalus, 80, 81, 121
Mugil curema, 80, 81, 121
Mulinia edulis, 12, 35, 44, 45, 46, 47, 48, 49, 50, 51, 52
mullet (*Mugil cephalus, Mugil curema*), 6, 14, 19, 30, 55, 73, 80, 98, 102, 114, 115, 116, 121, 130, 149, 150, 152, 153, 157, 348, 350; reducing competition between *M. cephalus* and *M. curema*, 80
mullu (*Spondylus*), 102
mummies, 159, 224, 339
mummy bundle, 224, 265, 274, 282, 283, 329, 331, 339
mummified burials, 30; see burial cists
mummified tern, 179, 180
mussel, 12, 14, 30, 35, 41, 43, 45, 47, 50, 51, 92, 95, 98, 154; mussel colonies, 43; mussel shells, 300; see also *Aulacomya ater, Choromytilus chorus, Perumytilus purpuratus, Semimytilus algosus*

Mustelus mento, 74; see also smoothhound
muy muy, 12, 44, 58, 62, 95, 179; see also mole crab
Myliobatis peruvianus (eagle ray), 73, 119
mysids (opossum shrimp), 11, 91

N

nail clippings, 23; see fingernail and toenail clippings, Fig. 2.3
Nasca, 43, 82, 101, 202, 204, 281, 350
Nasca Valley, 43
nauplius, 11, 14, 16, 54, 55, 77, 79
needles, 7, 159, 262, 272, 278, 283
nematode worms, 80
neritic zone, 11
net, 102, 157, 158, 166, 345; see also carrier nets, red de cortina, trammel nets
net bag, 60, 104; see also carrier net, chiwa
net-making, see net
netting, 32
net weights, 103, 104, 105, 108, 115
New Zealand, 82, 91, 173
NISP, number of identified specimens, see 121; see crayfish
noble families, 116, 279, 281, 284, 317, 341, 345, 348
noble patrons, 252
Nolanceae, 338, 339
Nuevo Imperial, 4, 9
nuña, 204; see also beans

O

ocean temperature, 17; see El Niño, ENSO
octopus, 12, 74, 76, 82, 89, 187, 189
Odontesthes regia regia, 73, 82, 353; see also pejerrey, silversides
ojo de uva (*Hemilutjanus macrophthalmos*), 82, 83; see also sea bass
Oman, 119
omnivore, 14, 77, 80, 86
opossum shrimp, 14, 82, 89, 91, 93
oracles, 344, 350
Orinoco River, 209
Opuntia, 7, 282
ossified vertebral centrum, 74, 75, 122, 127, 129, 130, 135, 136, 142, 143, 144, 146, 147, 150, 152, 153, 154, 155, 156
ostracod, 11, 14, 77, 80, 86
Otaria byronia, 12, 186, 187, 188, 189, 190, 191, 195
otoliths, 129, 131, 132, 144, 145, 151, 155, 156
Otuma, 159
owl (*Speotyto cunicularia*), 9, 179

P

Pacarán, 231
Pacasmayo, 41, 342
Pacatnamú, 114
pacay (*Inga feuillei*), 9, 201, 206, 208, 213, 215, 217, 224, 225, 226, 228, 231, 274, 283
pacay leaves, 206, 208, 224, 225, 226, 228, 283
Pachacamac, 173, 202, 209, 343, 344, 350
Pachamama, 344
pachamanca, 32, 48, 189, 195, 197; see also sea lion cookout, 32, 189, 195, 300
Pacific bonito (*Sarda sarda chiliensis*), 12, 28, 73, 93, 94, 136, 142, 143, 146, 147, 154, 356
Pacific hake (*Merluccius gayi peruanus*), 96, see also merluza
Pacific mackerel (*Scomber japonicus peruanus*), 79, 93, 95, 121, 143, 150, 153; see also caballa, mackerel

Pacific sardine (*Sardinops sagax*), 11, 16, 17, 30, 73, 77, 78, 85, 89, 121, 122, 123, 124, 130, 135, 136, 142, 143, 144, 146, 147, 148, 149, 150, 152, 153, 154, 155, 156; sardine heads, 17, 152
Pacific sea lion, 197; see sea lion
Pacific threadfin, 73, 80, 82
Pacific white-sided dolphin (*Lagenorhynchus*), 189, 190, 192, 196
Paita, 95, 345
palm phytoliths, 262, 282
palms, 262, 262
paloma del cabo (*Daption capensis*), 174
paloma pompano (*Trachinotus paitensis*), 19, 84, 85
Panama, 85, 89
Panama Star and Herald, 179, 180
Paracas, 159, 257, 261, 263, 265, 339
Paracas Peninsula, 186
Paraguay, 206
Paralabrax humeralis, 73, 82, 83, 121; see also cabrilla, Peruvian rock bass
Paralichthys adspersus, 96, left-eye flounder; see also flounder, lenguado
Paralonchurus peruanus, see coco
parcialidad, 342; see also curacazgos, señoríos
parrot, 173, 181, 184, 185
pata, 135, 327; see *llamkana pata* or raised work platform, 22, 45, 46, 348
Patagonia, 166
peanut, 9, 201, 204, 213, 215, 220, 231, 349
peanut hulls, 207, 215, 220, 226, 231
pejegallo (*Callorhinchus callorynchus*), 77; see also chimaera or plow-nosed chimaera
pejerrey (*Odontesthes regia regia*), 14, 19, 82, 354; see also silversides
pejesapo (*Sicyases sanguineus*), 14, 73, 94, 95, 121, 129, 139; see also clingfish
pelecypod, 41, 45, 52, 76, 89, 353
pelicans (*Pelecanus thagus*), 12, 17, 76, 171, 172, 177, 178, 179, 181, 182, 183, 184, 185
pellets, of llama dung, 10, of guinea pig, 3, 25
peña, 10, 13, 55, 98, 99, 102, 115, 119, 186
penguin, 17, 172, 173, 181, 184, 187; penguin skin, 173, 174; see Humboldt penguin
pepper tree (*Schinus molle*), 7; used to flavor chicha, 7
periwinkles, 12
Perumytilus purpuratus, 35, 42, 43, 45, 46, 47, 48, 49, 50, 51, 52, 53, 95; as bioengineer, 43; competition with *Semimytilus,* 43
Peruvian banded croaker, see coco
Peruvian booby (*Sula variegata*), 17, 93, 173, 174, 178, 182, 183, 184, 185; see also booby, piquero
Peruvian morwong, see pintadilla
Peruvian pelican (*Pelecanus thagus*), 17, 173; see pelican
Peruvian rock bass, 121, 130, 143, 149, 150, 157; see also cabrilla, rock bass
Peruvian weakfish, 89; see also ayanque
Pescadora, language spoken by north coast fishermen, 58
petrels, 174
petty kingdoms, 3
pez dama, 19
Phragmites australis, see caña hueca
pharyngeal teeth, 86, 122, 125, 135, 143, 144, 152, 155
Phaseolus beans, 30, 204, 213, 215, 216, 217, 221, 224, 231, 254, 259
Phaseolus lunatus, 201, 204, 206, 254, 255, 256, 258, 259, 261
Phaseolus pods, 30, 205, 217, 218, 220, 224, 258, 259; valves, 205, 217, 220
Phaseolus vulgaris, 201, 204, 218, 220, 221, 223, 226, 228, 229, 254, 255, 256, 257, 259, 260, 261
phytoliths, 282
phytoplankton, 11, 12, 14, 15, 16, 19, 76, 77, 79, 350
pichirrata, 74; see also sharks, thresher shark
pigeons (*Columba* spp.), 9, 173, 179, 181, 182, 185
pigment, dry, 208, 314; red pigment, 208, 314
pigment dishes, 40, 41, 43, 350
pigment pouch, 314
pilgrimage center, 344
pilidium, 11
pillows, 224, 264, 265, 266, 274, 283, 284; see willow branches
Pinnipedia, 186
pinnipeds, 186, 191, 195, 197
pintadilla (*Cheilodactylus variegatus*), 14, 91, 129, 132, 136, 150
piquero, 174, 177, 178
Pisco Valley, 82, 204
plankton, 16, 19, 38, 41, 54, 77
plants, 10, 20; see beans, cotton, maize, squash
planula, 11
Platyxanthus orbignyi, 54, 55; see also purple stone crab
playa (sandy beach), 6, 8, 10, 12, 30, 43, 44, 45, 50, 58, 59, 60, 72, 89, 93, 97, 101, 102, 179, 186
Playa El Hueso, 4, 44, 45, 58, 101
Pleistocene foragers, 166
plow-nosed chimaera, 73, 76, 77; see chimaera, pejegallo
plumbeous rail (*Pardirallus sanguinolentus*), 173, 177, 182, 184
pluteus, 11
Poa, 10
Poaceae, 263, 266, 336, 337, 338, 339
Polanyi, Karl, 350
pollen, 335, 337, 338, 339
pollen spectra, 334
polychaete worms, 11, 12, 54, 74, 80, 86, 87, 89, 92, 95, 353
polydactyly, in camelids, 287, 300, 303, 317; see also llamas
Polynemus approximans, 73, 80, 81; see barbudo
pompano (*Paloma pompano*) 72, 73, 85, 157, 352
porcelain crabs, 14, 55, 82, 92
Porcellanidae, 55
potato, 9, 119, 209, 226, 228, 340, 349; white potato (*Solanum tuberosum*), 201, 209, 211, 224, 229, 339; potato fields, 19
potsherd weights, 104, 108, 115; see nets
potters, 342
pottery vessels, 62, 63, 147, 148, 152, 159, 195, 203, 215, 282, 283, 290, 294
Pouteria, see lúcuma
prickly pear cactus (*Opuntia* spp.), 7, 263, 272, 282
prickly pear spines, 7, 262, 272, 278, 281, 282, 283, 350
principales, see nobles
projectile shafts, 265
protozoans, 11, 16, 77, 79
Prosopis juliflora (algarrobo), 7, 262, 263
Proyecto Qhapaq Ñan, 20, 231
pseudofeces, 44
pseudopods, 77
Puebla, 204
pukios, 101
puna, 301, 350
Puno, 166
Punta los Mártires, 186
purple corn, 234, 240, 249, 253; see maize, maize varieties
purple stone crab, 12, 54, 55, 56, 60, 61, 62, 63, 64; see also *Platyxanthus orbignyi*
puya, 158, 159, 264
pyroengraved gourd bowls, 346, 347

Q

quebradas, 45, 48–51, 64, 65, 121, 122, 148, 150, 154, 157, 179, 185, 189, 190, 195, 206, 214, 221, 226, 233, 249, 251–252, 282, 300, 319, 323, 326, 329; location on map, see Fig. 2.1
Quebrada de Hualcará, see Hualcará
Quebrada de Pocotó, 7
quid, see grass
Quilcay, 102, 344
Quillairaca, 344
Quilmaná, 4, 10
Quivi, 343; see coca
Qurna, Egypt, 104, 109

R

rabiblanca, 179; see also eared dove
radiolaria, 11
radula, 36, 41, 337
rail family, 173, 177, 182, 183
raising guinea pigs, 62, 143, 298, 345; see guinea pig, llama
ramp, 23, 228, 282
raya águila, see eagle ray, *Myliobatis peruvianus*
rays, 72, 74, 101, 120, 121, 155, 189, 352
red de cortina, 95, 98, 100, 101, 103, 115; see also net, trammel nets
red-legged cormorant (*Phalacrocorax gaimardi*), 173, 174, 175, 177, 181, 182, 183
red paint, 208, 279
reed, 6, 68, 69, 265, 269, 283
residential quarters, 32, 142, 150, 184
requiem sharks, 74, 142, 156, 157; see sharks
rhizomes, 283
rhyolites, 6
rice rat (*Oryzomys xanthaeolus*), 9, 179, 221, 224, 233, 236, 252, 328, 337
Rímac Valley, 343, 344, 345
Río de la Plata, Argentina, 168
Río Seco, 159
ritual, 159, 201, 314, 343, 344, 351
ritual economy, 344
ritual plant, 203, 208, 212, 213, 231, 262; see coca
ritual sacrifice, 303, 317, 327, 344; see guinea pig sacrifice, sacrifice
ritual sling, 159, 165
ritual specialists, 341
roads, 102, 344, 350
roasting sea lion, 28, 48, 166, 189, 195, 197, 300, 318; see also earth oven, hearth, *pachamanca*
róbalo (*Sciaena starksi*), 7, 12, 73, 86, 89, 90, 92, 114, 115, 121, 129, 133, 134, 136, 142, 148, 149, 150, 155, 156, 157
rock bass (*Paralabrax humeralis*), see also cabrilla
rock snail, 12, 38, 41, 45, 51, 52
rockshelter, 334
rocky island, 14, 45, 86, 91, 353
rocky ledges, 174, 175
rocky substrate, 38, 41, 53, 74, 82, 86
Rome, 158
root crop, 209, 211; see also potato, sweet potato
ropes, 262, 263, 264, 266, 282
rotifers, 11, 93
rue (*Ruta graveolens*), 226, 228, 229, 231
rush (*Juncus* spp.), 283

S

sacrifice; sacrifice of a guinea pig, 290, 324; of a llama, 301
sacrificed guinea pig, 290, 324; see guinea pig
sacuara, 263; see also cattail
Salicornia fruticosa, 7
salitre, 48, 150, 156, 319
Salix, 263, 338, 339; see willow
Salix humboldtiana, 7
Sally Lightfoot crab, 12, 55
salt, 116
salt fog, 7, 30, 117, 185, 226
saltgrass, 7, 102
salty crust, 48
San Isidro, 9
San Luis, 4, 9
San Vicente Cañete, 51, 81, 82
sand in storage rooms, 24, 26, 27, 28, 116, 117, 142, 143, 146, 147, 148, 150, 157, 183, 184; see also drying fish, fish storage, storage facility
sand dunes, 7
sandpiper family, 173, 177, 181, 182, 183
Santiago de Cao, 344
saponin, 281
Sapindus saponaria, see soapberry
Sarda sarda chiliensis, 93, 150; see bonito, Pacific bonito
sardine-anchoveta cycles, 15, 19, 77, 97; sardine/anchoveta dominance, 15, 79, 89, 157
sardines, 12, 17, 19, 77, 85, 86, 89, 93, 94, 95, 97, 110, 115, 118, 119, 121, 135, 148, 153, 154, 157, 187, 337, 340, 344, 348, 349, 350, 352
Sardinops sagax, 77, 79, 120, 122, 337; see Pacific sardine, sardines
scaled blenny (*Labrisomus philippii*), 103, 121, 129, 137, 145, 149, 157
scaleless blenny (*Scartichthys gigas*), 11, 14, 43, 73, 121, 129, 132, 145, 149, 150, 157
scallop (*Argopecten purpuratus*), 16, 45
Scartichthys gigas, 11, 43; see borracho and scaleless blenny
scavengers, 12, 54, 55, 58, 80
Sciaena deliciosa, 73, 121, 357; see lorna
Sciaena fasciata, see burro
Sciaena starksi, 87; see róbalo
Sciaenidae (drum family), 14, 58, 86, 121, 132, 135, 136, 142, 145, 147, 148, 149, 153, 155
Scirpus, 269; see bulrush, totora
Scomber japonicus peruanus, see Pacific mackerel
Scomberomorus maculatus sierra, 73; see sierra mackerel
Scombridae, see mackerel, tuna
sea basses, 72, 73, 82
sea birds, 12, 18, 58, 172–185, 340
sea catfish (*Galeichthys peruvianus*), 6, 12, 14, 72, 80, 144, 149, 150, 157
sea caves, 172, 186
sea cliffs, 14, 17, 18, 20, 45, 53, 55, 72, 82, 86, 91, 92, 95, 97, 98, 129, 175, 178, 179, 180, 185, 197, 226, 283, 352, 353
sea cucumbers, 74
seagulls, 76
sea lions, 12, 28, 32, 93, 166, 168, 171, 173, 186, 187, 188, 189, 190, 195, 196, 340, 350; flippers, 32, 102, 189, 191, 195, 197; sea lion roast, 37, 48, 189, 190, 318
sea lion cookout, 28, 37, 189, 300
sea lion feasts, see roasting sea lion
sea lion skins, 32, 102, 115, 166, 195, 197, 345; inflatable watercraft, 32, 101, 102, 166, 189, 195, 197
sea lion pups, see sea lions
seals, see fur seals
sea otters, 36, 41, 65, 173, 187, 197
sea snails, 35, 37, 38, 40, 45, 47, 52, 92, 95

sea urchins, 38, 74, 86
seaweed, 11, 12, 14, 91, 92, 95, 340, 349
sedges, 283; see *Cyperus*
Semele, 51; see also white clam
Semimytilus algosus, 35, 42, 43, 45, 46, 47, 48, 49, 50, 51, 52, 53, 54, 92, 95; competition with *Perumytilus*, 43; see also mussel
señorío, 3, 342; see also Kingdom of Huarco, Kingdom of Lunahuaná, parcialidad, petty kingdom
Seriolella violacea, see blackruff
sewing, 7, 272, 283, 284, 345; see also needles
sexual dimorphism in sea lions, 186–187, 188
sharks, 12, 72, 74, 82, 93, 119, 120, 121, 122, 142, 156, 345, 352; tooth, 74
shearwater (*Puffinus* sp.), 173, 174, 182
shellfish, 10, 22, 35, 45, 48, 50, 74, 153, 189, 329, 340, 349; shellfish collectors, 41; meat contributed by, 51–53; shellfish meat, 43, 335
shipment, 242, 251, 345; see llama caravans, see maize varieties
shrimp, 55, 72, 74, 80, 89, 157, 189, 352, 353
Siberia, 166
Sicyases sanguineus, 94, 95; see clingfish or pejesapo
sierra mackerel, 73, 93, 94; see also *Scomberomorus maculatus sierra*
silver, 341, 350
silversides, 16, 19, 72, 82, 95, 157, 352, 353, 354; the return of the cool-water silversides, 72; see pejerrey
skipjack (*Katsewonus pelamis*), 19
sleeping bench, 22, 23, 24, 27, 142, 296, 327
slings, 7, 158, 159, 168, 185, 186, 262, 264 ; sling tassels, 7, 159, 160, 161, 162, 164, 262, 264, 339; sling's owner, 159
sling stone, see slings
slipper shell, 38, 49; see *Calyptraea trochiformis*, 40, 49
smoothhound (*Mustelus mento*), 73, 74, 75, 345
snail, see rock snail, sea snail
soapberries (*Sapindus saponaria*), 262, 263, 280, 281, 283, 340; soapberry seeds, 282
Solanaceae, 228, 339
Solanum tuberosum, 338
soups, see stew
Southern Oscillation, 16; see also ENSO, El Niño, La Niña
Spanish Conquest, 283
Spanish olive jars, 226
spears, 171, 186
spearthrower, see atlatl
specialization, see also fishermen, goldsmiths, potters
spider crab, 55
spindles, 7, 262, 265, 272, 349
spindle whorls, 349
spines, algarrobo and huarango, 7, 262, 266, 274, 282
spinning, 23, 215, 272, 282, 284, 345
Spondylus, 344
sponges, 55
squash, 9, 207, 213, 214, 217, 223, 231, 340, 349
Squatina, 76
squatters, 181, 183, 185, 226, 228, 229, 259, 282, 283, 298
squatters' house, 23, 47, 143, 217, 218, 219, 271, 282
squid, 12, 74, 82, 85, 89, 93, 95, 187, 189
staple finance, 343
starfish, 11, 37, 74
Stellifer minor, see mojarrilla
stew(s), 328, 333
stone crab, 54, 57, 60, 64, 82; see also *Platyxanthus orbignyi*
stone masonry burial cists, 153, 154, 159, 224, 252; see burial, cists
storage, 24, 30, 48, 156, 208, 220, 221, 249
storage, fish, 24–26, 62, 221; see also drying fish, fish storage
storage bins, 23, 46, 65, 102, 103, 104, 136, 168, 191, 222, 282, 289; see also *collcas*

storage cist, 30, 32, 51, 156, 208, 226, 319
storage facility, 48, 150, 195, 223, 231, 280, 283
storage room, 24, 25, 26, 142, 182, 184, 195, 215, 217, 296
storage vessel, 195, 233, 282, 290, 299
storing fish, 26; see fish storage
subsistence economy, 20, 334, 341
substantivist economist, 350
suirucu, see soapberry (*Sapindus saponaria*)
Sultanate of Oman, see Oman
suspension feeder, 41, 58
sweet potato (*Ipomoea batatas*) or camote, 9, 201, 209, 211, 215; see also potato
swimming crab, 55
swordfish, 19, 82, 93
sympatric competition, 43

T

Tacna, 44
Tambo de Mora (Chincha Valley), 16, 79
tannin, 208
tapeworms, 207
Taymi Canal, 342–343
Tegula atra, 37, 38, 40, 45, 46, 48, 49, 50, 51, 52, 54, 95; see also sea snail
Tehuelche, 166
teleost, 77
temblador (*Discopyge*), 76
terns, 179, 181, 184, 185
terraces, artificial, 28, 32, 36, 48, 63, 64, 150, 152, 154, 185, 221, 224, 226, 234, 282, 283; see also quebradas
Thais chocolata, 35, 37, 40, 41, 45, 46, 47, 48, 49, 50, 51, 52, 53; see also sea snail
thatching, 266
Thebes, 104
thresher shark, see sharks
thrush family, 173, 182
tiburón azul, 74
tiburón martillo, 74
Tierra del Fuego, 41
Tillandsia, 7, 158, 159, 263, 264, 281, 338, 339; see also bromeliads
tollo or smoothhound shark (*Mustelus* sp.), 74, 75
tomasa, 209; see potato
totora, 102, 263, 264, 265, 281, 282; see caballito de totora
totora rhizomes, 101
totorales, 101
Trachinotus paitensis, 19, 73, 82, 84; see paloma pompano
Trachurus symmetricus murphyi, 82
trambollo or scaled blenny (*Labrisomus philippii*), 14, 73, 92
trammel nets, 95, 98, 100, 103, 104, 105, 107, 108, 110, 114, 115, 120
trapezoidal niches, 185, 226, 228
traps, 58, 67; see crayfish trap
trash fish, 177
tribute, 18, 344, 345, 348, 349, 350
trochorphore, 11
trophic food chain, 10, 11, 89; short trophic chain, 14
trophic levels, 10, 11, 12, 17, 19, 186, 340, 350
tubers, see potato, sweet potato
Tumbes, 207
tumi, see knife
tuna, 16, 19, 72, 82, 93, 119
tunicate, 16, 79, 93
Tuquillo, 202
turban shell, 12, 54

Tursiops, 189, 190, 192, 194
twilled mat, 264, 267
Typha, 338, 339; see cattails

U

Ucayali, 206
uchu, 214; see *Capsicum,* chile pepper
Ungará, Fortress of, 3, 4, 5, 9
upwelling system, 14, 16; see coastal upwelling
Urpay Huachac, 344, 345, 350, 351; see also deity, islands, pilgrimage, ritual
ushnu, 344

V

Valley of Chincha, 79, 102, 273; see also Chincha
vegetal fiber, 164, 262, 272
veliger, 11
Venezuela, 209
vessels, cooking, 299; storage, 299
vetch (*Vicia* sp.), 226, 228, 231
vicuña, 301; see also alpaca, camelid, llama
Virú Valley, 342
volcanic ash, 6
volcanics, 6
vulture, 344, 350

W

wachaques balsares, 101
wall, defensive, 6, 7, 9, 20, 343
warak'a, see sling, 158
wasp gall, 210, 212, 214, 215
watercraft, 11, 98, 101, 115, 157, 345, 352, 353; inflatable, 189
waterfowl, 80, 166, 168; see bolas, hunting birds
wattle-and-daub house, 63, 150, 152, 195, 217, 265, 348; kincha house, 23, 26, 30, 32, 47, 49, 143, 150, 152, 183, 298, 299, 341
wealth finance, 343
weapon, see bolas, slings
weavers, 104, 212, 242
weaving equipment, 46, 115, 215
weaving, 23, 46, 136, 215, 262, 266, 282, 284, 327, 345
weaving textiles, 46, 345

wedge clams, 35, 43, 44, 54
weights, see net weights
whale, 12, 197
whale shark, 19
whelks, 95
white clams, 44, 45, 46, 47, 49, 51, 52; no longer found at Cerro Azul, 44, 45; see also *Mulinia edulis,* 45
white mullet, see mullet
white potato, 209, 211, 224
wild bottle gourd, 272
willow, 7, 263, 266
willow branches, 224, 266
willow twigs, 7, 266, 274, 283
wooden floats, 104, 115
wooden post, 49
workbasket, 272, 283; in burial of woman, 272, 278; with cactus spines, 278

X

Xiphopenaeus, 19, 353, 354; see shrimp

Y

yanacon yanayucu, 343
Yarutini, 343
Yauyos, 6, 343
Ychsma, 343; see also Pachacamac
yuca de monte, 201, 207, 209, 215
Yucatán Peninsula, 209
Yukon, 166
yunga, 3, 10, 203, 348, 350; see also *chaupi yunga*

Z

zapallo (*Cucurbita maxima*), 231
Zárate Island, 186
Zea mays, 154, 269, 338, 339; see beer, brewery, corn, maize, maize varieties
Zenaida auriculata, 9, 173, 179; see also dove
Zenaida meloda, 9, 173, 179, see also dove
zooplankton, 11, 12, 14, 15, 16, 19, 55, 76, 77, 79, 93, 95, 350
zorro (*Menticirrhus rostratus*), 73, 87
Zúñiga, 68, 231